AF900347

Aging Mechanisms

Nozomu Mori • Inhee Mook-Jung
Editors

Aging Mechanisms

Longevity, Metabolism, and Brain Aging

Springer

Editors
Nozomu Mori
Department of Anatomy
and Neurobiology
Nagasaki University School of Medicine
Nagasaki
Japan

Inhee Mook-Jung
Department of Biochemistry and Biomedical
Sciences, College of Medicine
Seoul National University
Seoul
South Korea

ISBN 978-4-431-55762-3 ISBN 978-4-431-55763-0 (eBook)
DOI 10.1007/978-4-431-55763-0

Library of Congress Control Number: 2015955138

Springer Tokyo Heidelberg New York Dordrecht London
© Springer Japan 2015
This work is subject to copyright. All rights are reserved by the Publisher, whether the whole or part of the material is concerned, specifically the rights of translation, reprinting, reuse of illustrations, recitation, broadcasting, reproduction on microfilms or in any other physical way, and transmission or information storage and retrieval, electronic adaptation, computer software, or by similar or dissimilar methodology now known or hereafter developed.
The use of general descriptive names, registered names, trademarks, service marks, etc. in this publication does not imply, even in the absence of a specific statement, that such names are exempt from the relevant protective laws and regulations and therefore free for general use.
The publisher, the authors and the editors are safe to assume that the advice and information in this book are believed to be true and accurate at the date of publication. Neither the publisher nor the authors or the editors give a warranty, express or implied, with respect to the material contained herein or for any errors or omissions that may have been made.

Printed on acid-free paper

Springer Japan KK is part of Springer Science+Business Media (www.springer.com)

Preface

According to a predictive report on the increase in the elderly population in various countries, Japan and Korea are poised to become the world's top two nations, with constantly high elderly populations approaching 35–40 % in the latter half of this century. The rapidly aging populations in Japan and Korea promote movement toward a "senescent" society. Amazingly, however, if we examine the numbers of people living to over 100 years old, more than 60,000 centenarians currently reside in Japan. In Korea, the centenarian population is approximately 3,300, but it is estimated to grow to almost 40,000 by the year 2050. Japan and Korea are thus growing old so rapidly that we are indeed moving "into the unknown," as described in a special report in *The Economist* several years ago (November 18, 2010). In such nations, research on aging is important not only sociologically but also biologically to understand the background of aging in our population.

During the last several decades, our knowledge concerning aging research has been growing rapidly. However, we are still far from achieving a complete understanding of the whole mechanism of aging: how we, as animals, age and how our lifespan is determined. Many researchers are working in this field in both Japan and Korea, and our scientific societies in relation to biomedical aging in the two countries have been interacting with each other. When one of the editors (N.M.) moved from the National Institute for Longevity Science (Research Institute, National Center for Geriatrics and Gerontology, Nagoya) to Nagasaki in 2006, we met on the small island of Ioujima in Nagasaki to discuss and exchange ideas on the basic science of aging. We called this discussion forum the "Asian Aging Core for Longevity" (AACL), and we have continued this series of AACL meetings for the last 10 years, alternating between Japan and Korea. In the last 5 years, it was supported by the Asian CORE program of the Japan Society for the Promotion of Science (JSPS). The studies described in this book were supported, at least in part, by the fund of the JSPS-Asian CORE program, through an intimate discussion at the AACL meetings.

This book was initially planned to summarize the aging research among all of the members of the AACL. However, owing to space and time limitations, we could not include the work of all the AACL members. The book is composed of a series of topics in aging research which were explored through the daily dedicated laboratory work of our AACL members and their colleagues. Based on the topics of each manuscript, we, the editors, ordered each chapter such that the stories would facilitate the understanding of the scientific outcomes of our latest research on aging from the basic biology of longevity, metabolism, and brain aging with the goal of developing potential therapeutics for age-related diseases, such as Alzheimer's.

For convenience, we subdivided the chapters into seven parts. Part I provides an overview and re-investigation of various hypotheses of aging. Part II introduces studies of centenarians and cells of progeria patients. Part III discusses the usefulness of induced pluripotent stem (iPS) cells and primary cultured neurons for the study of aging and further explores the potentials of using lower organisms, such as budding yeast and nematodes, for the study of longevity and as a model for neurodegenerative disorders. Part IV introduces several key regulatory factors affecting tissue aging, such as α-Klotho, FOXO transcription factors, neuropeptide Y, uremic toxins, and vitamin K. Part V describes the studies of the physiological nature of brain aging. It includes how aging affects potentials of adult neurogenesis in the hippocampus, signaling molecules for cognitive decline, neuroplasticity in the hippocampus and the cerebellum, and volume analysis of aging brains using a large brain database. Part VI addresses the pathophysiology of the aged brain. Alzheimer's disease (AD) is the most debilitating disease in the elderly. The processing mechanisms of amyloid precursor protein (APP), the accumulation and aggregation of amyloid-β, and its degradation mechanisms are discussed here. A novel approach exploring new biomarkers of AD is also discussed. Finally, Part VII explores issues of antiaging, particularly focusing on therapeutic approaches for neurodegenerative diseases, including AD and polyglutamine diseases.

How do we age, and how is longevity determined? These questions are fundamental issues in biology. Solving the mechanisms of biological aging and the determination of longevity is not easy and may never be completely achieved. However, the studies described in this book are on the cutting edge of the science of biological mechanisms for each aspect of the biology of aging. We hope this book will be useful for researchers and graduate students in the field worldwide.

At this point, we would like to express our sincere thanks to all the authors for their valuable contributions to this volume. We are also grateful to Dr. Sang Chul Park, Dr. Eun Seong Hwang, Dr. Isao Shimokawa, and Dr. Yong-Sun Kim for their cooperation and encouragement during the course of our past meetings of the JSPS-AACL. Finally, we would like to thank Ms. Kanako Ishimaru and Ms. Noriko Tominari for their excellent secretarial assistance throughout the course of all the AACL meetings and also for putting this book together for final publication.

Nagasaki, Japan Nozomu Mori
Seoul, South Korea Inhee Mook-Jung

Contents

Contributors

Yasumichi Arai Center for Supercentenarian Research, Keio University School of Medicine, Shinjuku, Tokyo 160-8582, Japan
yasumich@keio.jp

Shogo Endo Memory Neuroscience, Aging Neuroscience Research Team, Tokyo Metropolitan Institute of Gerontology, Sakae-cho 35-2, Itabashi, Tokyo 173-0015, Japan
sendo@tmig.or.jp

Sataro Goto Institute of Health and Sports Science & Medicine, Juntendo University Graduate School, Inzaishi, Chiba 270-1695, Japan
sgotou@juntendo.ac.jp

Reiko Inagi Division of Chronic Kidney Disease Pathophysiology, The University of Tokyo, Graduate School of Medicine, Hongo, Bunkyo-ku, Tokyo 113-8655, Japan
inagi-npr@umin.ac.jp

Satoshi Inoue Department of Geriatric Medicine, Graduate School of Medicine, The University of Tokyo, Hongo, Bunkyo-ku, Tokyo 113-8655, Japan
INOUE-GER@h.u-tokyo.ac.jp

Naoaki Ishii Department of Molecular Life Science, Tokai University School of Medicine, Isehara, Kanagawa 259-1193, Japan
nishii@is.icc.u-tokai.ac.jp

Ken-ichi Isobe Department of Immunology, Nagoya University Graduate School of Medicine, Tsurumai, Showa-ku, Nagoya 466-8550, Japan
kisobe@med.nagoya-u.ac.jp

Atsushi Iwata Department of Neurology, Graduate School of Medicine, The University of Tokyo, Hongo, Bunkyo-ku, Tokyo 113-8655, Japan
iwata-tky@umin.ac.jp

Jung-Hee Jang Department of Pharmacology, School of Medicine, Keimyung University, Daegu 704-701, South Korea
pamy202@kmu.ac.kr

Shozo Jinno Department of Anatomy and Neuroscience, Graduate School of Medical Sciences, Kyushu University, Maidashi, Higashi-ku, Fukuoka 812-8582, Japan
sjnno@med.kyushu-u.ac.jp

Sho Kakizawa Department of Biological Chemistry, Graduate School of Pharmaceutical Sciences, Kyoto University, Kyoto 606-8501, Japan
sho-kaki@pharm.kyoto-u.ac.jp

YoungSoo Kim Center for Neuro-Medicine, Brain Science Institute, Korea Institute of Science and Technology, Hwarangno 14-gil 5, Seongbuk-gu, Seoul, Republic of Korea

Cheol-Koo Lee Division of Biotechnology, College of Life Sciences and Biotechnology, Korea University, Seoul 136-701, Republic of Korea
cklee2005@korea.ac.kr

Seung-Jae Lee Neuroscience Research Institute and Department of Medicine, Seoul National University College of Medicine, Seoul 110-799, Korea
Sjlee66@snu.ac.kr

Seung-Jae V. Lee Department of Life Sciences, Pohang University of Science and Technology, Pohang, Gyeongbuk, South Korea

Baohua Liu Health Science Center, Shenzhen University, Shenzhen, China
ppliew@szu.edu.cn

Inhee Mook-Jung Department of Biochemistry and Biomedical Sciences, College of Medicine, Seoul National University, Seoul 110-799, South Korea
inhee@snu.ac.kr

Nozomu Mori Department of Anatomy and Neurobiology, Graduate School of Biomedical Sciences, Nagasaki University, 1-12-4 Sakamoto, Nagasaki 852-8523, Japan
morinosm@nagasaki-u.ac.jp

Yo-ichi Nabeshima Laboratory of Molecular Life Science Foundation for Biomedical Research and Innovation, Kobe, Japan
nabemr@lmls-kobe.org

Takaomi C. Saido Laboratory for Proteolytic Neuroscience, RIKEN Brain Science Institute, 2-1 Hirosawa, Wako, Saitama 351-0198, Japan
saido@brain.riken.jp

Isao Shimokawa Department of Pathology, Graduate School of Biomedical Sciences, Nagasaki University, 1-12-4 Sakamoto, Nagasaki 852-8523, Japan
shimo@nagasaki-u.ac.jp

Toshiharu Suzuki Laboratory of Neuroscience, Graduate School of Pharmaceutical Sciences, Hokkaido University, Sapporo 060-0812, Japan
tsuzuki@pharm.hokudai.ac.jp

Yasuyuki Taki Department of Nuclear Medicine & Radiology, Institute of Development, Aging and Cancer, Tohoku University, 4-1 Seiryo-cho, Aoba-ku, Sendai 980-8575, Japan
ytaki@idac.tohoku.ac.jp

Zhongjun Zhou Department of Biochemistry, LKS Faculty of Medicine, The University of Hong Kong, Hong Kong
zhongjun@hku.hk

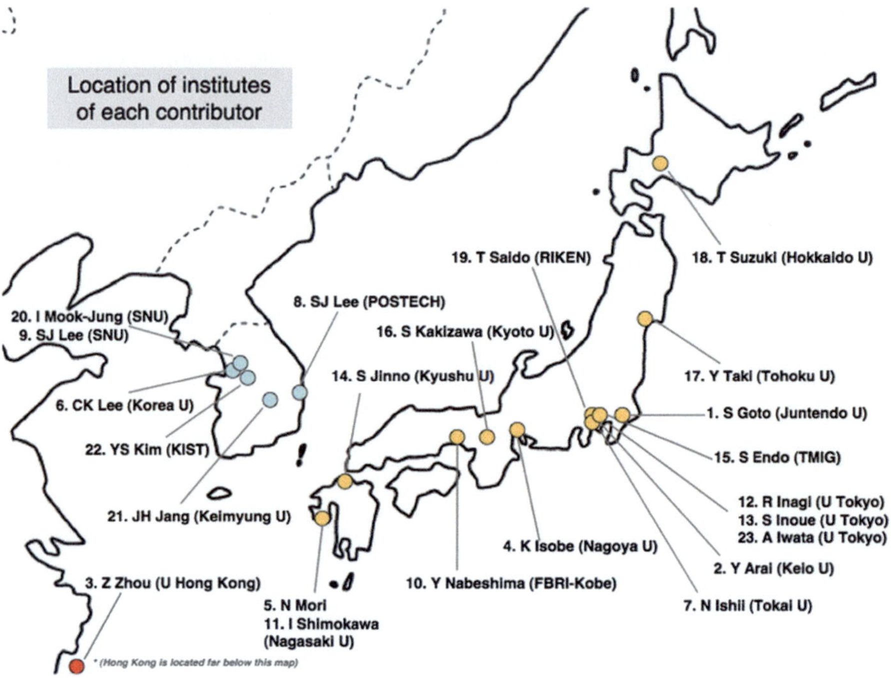
Location of institutes
of each contributor
19. T Saido (RIKEN)
18. T Suzuki (Hokkaido U)
8. SJ Lee (POSTECH)
20. I Mook-Jung (SNU)
9. SJ Lee (SNU)
16. S Kakizawa (Kyoto U)
17. Y Taki (Tohoku U)
14. S Jinno (Kyushu U)
6. CK Lee (Korea U)
1. S Goto (Juntendo U)
22. YS Kim (KIST)
15. S Endo (TMIG)
12. R Inagi (U Tokyo)
13. S Inoue (U Tokyo)
23. A Iwata (U Tokyo)
21. JH Jang (Keimyung U)
4. K Isobe (Nagoya U)
2. Y Arai (Keio U)
3. Z Zhou (U Hong Kong)
10. Y Nabeshima (FBRI-Kobe)
7. N Ishii (Tokai U)
5. N Mori
11. I Shimokawa
(Nagasaki U)
* (Hong Kong is located far below this map)

Part I
From Hypotheses to Mechanisms

Chapter 1
The Biological Mechanisms of Aging: A Historical and Critical Overview

Sataro Goto

Abstract This overview describes the historical background and modern versions of selected theories of the mechanisms of biological aging, including the mutation or genome instability theory, the free radical or oxidative stress theory, the mitochondrial theory, the error catastrophe theory, the altered protein or protein homeostasis dysregulation theory, the dysdifferentiation or epigenetic theory and the hyperfunction theory. The author has been involved in the development of some of these theories, which are therefore described in more detail. A discussion on the definition of aging and general comments on aging theory are included. The most popular theory of aging, the free radical or oxidative theory, was proposed more than 50 years ago but has recently faced severe criticism. To date, no single theory has been able to successfully explain the mechanisms of aging. We are thus awaiting the appearance of a new paradigm or an integration of the existing theories for a better understanding of the mechanisms of aging.

Keywords Mechanisms of aging • Mutation theory of aging • Free radical theory of aging • Mitochondria theory of aging • Error catastrophe theory of aging • Altered protein theory of aging • Dysdifferentiation theory of aging • Hyperfunction theory of aging

---

This article is dedicated to the memory of my friend Denham Harman, the Father of the Free Radical Theory of Aging and the Founder of the International Association of Biomedical Gerontology. He made great contributions to the progress of research on the mechanisms of aging and passed away on November 25, 2014, at the age of 98. He was an Honorary Member of the Japan Society for Biomedical Gerontology.

S. Goto, Ph.D. (✉)
Institute of Health and Sports Science & Medicine, Juntendo University Graduate School, Inzaishi, Chiba 270-1695, Japan
e-mail: sgotou@juntendo.ac.jp

© Springer Japan 2015

N. Mori, I. Mook-Jung (eds.), *Aging Mechanisms*,
DOI 10.1007/978-4-431-55763-0_1

1.1 Introduction

The average human lifespan in developed countries has increased by more than 20 years in the past several decades. Our major concern has shifted from a pure increase in the lifespan to an extension of the health-span, by reducing the risks of so-called lifestyle diseases, such as cancer, cardiovascular disease, stroke, and type 2 diabetes mellitus. Currently, most people are more interested in maintaining a high quality of life by delaying frailty that results from conditions such as sarcopenia and osteoporosis, even when such conditions are not fatal. However, it is often stated that the major risk factor for developing these diseases is aging itself. This means that the biological mechanisms of aging are likely to underlie the etiology and development of age-related diseases, although aging itself is not technically a disease.

Since Peter Medawar stated in 1952 that aging is an unsolved problem in biology (Medawar 1952), the mechanisms of aging have been the subject of intensive research interest, and a large number of papers have been published on the mechanisms of aging. Half a century after Medawar's statement, leading scientists of biogerontology claimed that aging is no longer an unsolved problem in biology (Hayflick 2007; Holliday 2006). Robin Holliday writes that recently published major books on aging agree that the biological reasons for aging in mammals are now well understood and that the process of aging is therefore no longer an unsolved problem in biology. It is true that there appears to be similar, apparently common or conserved, senescent phenotypes in different species of animals in which longevity differs by several hundred-fold; however, the very basic problem of the mechanisms behind such species differences in longevity are not clear nor have they been studied deeply. In this chapter, I provide an overview of selected theories of the mechanisms of biological aging. The overview includes theories of historical interest that are no longer popularly believed and theories that have since been transformed into newer versions of themselves. The latter group is presented under the same sections as the original theories from which they were developed.

1.2 The Definition of Aging

There are two words with somewhat similar meanings that are commonly used in gerontology and that are often confused: aging and senescence. Caeb Finch writes in his influential book that the term aging is mainly used to describe any changes that occur during the passage of physical time, during which there need be not common mechanisms, such as the aging of collagen, the aging of diploid cells in culture or of erythrocytes in circulation, the aging of populations or societies, or the aging of genes and species during evolution. In contrast, the term senescence is used to describe age-related changes in an organism that adversely affect its vitality and functions and, most importantly, increase its mortality rate as a function of time

(Finch 1990). Robert Arking states that "the terms aging and senescence seem to overlap considerably, and the difference between them may be one of emphasis rather than fundamentals" (Arking 1998). Because the term aging is often used to convey what he describes as senescence in most current gerontology writing, I use the term aging to discuss the mechanisms of aging (senescence) in this chapter.

To cite a few examples of the definition of aging (senescence) by leading scientists in biomedical gerontology, Medawar wrote, as cited by Bernard Strehler in his book, "Senescence may be defined as that change of the bodily faculties and sensibilities and energies which accompanies aging, and which renders the individual progressively more likely to die from accidental causes of random incidence." Strehler himself defines it as "the changes which occur (1) generally in the postreproductive period and (2) which result in a decreased survival capacity of the part of the individual organism." He further notes that "different evolutionary lines might very well decline in their survival capacities for entirely different immediate reasons. It may also be, however, that there are one or more dominant mechanisms of aging, common to all higher forms of life" (Strehler 1977). Alex Comfort defines senescence (aging) as a decrease in viability (leading to an increasing probability of death) with increasing chronological age and an increase in vulnerability (Comfort 1964). The term vulnerability may be rephrased as frailty, a term more commonly used in geriatric medicine.

After surveying the definition of aging in the literature, aging can be defined as a progressive functional decline that occurs in every individual within a population of a species, beginning around the time of reproductive maturity and leading to an increased probability of death over time. Theories of the mechanisms of aging that fit with this definition will be examined in this chapter.

1.3 Aging Theories

In 1990, Zhores Medvedev wrote that more than 300 theories about the biological mechanisms of aging could be found in the literature (Medvedev 1990). Among the theories cited in his review, some are still popular and some have disappeared or have been transformed, while other new theories have emerged and are currently being tested for validity. Theories of aging are mixed in that there are different levels of aging phenomena at the molecular, cellular, tissue, organ or systemic levels. George Martin has proposed a classification of the mechanisms of aging into two categories: public and private mechanisms (Martin et al. 1996a). The public mechanisms are those that could potentially be applied to the aging of different animals and tissues or cells, while the private mechanisms are those that are only true in specific species, cells, tissues or organs. For example, the immunological theory can only be true in animals with the appropriate system, such as mammals, but may not be true in nematodes or insects. When thinking about the aging that occurs in any somatic cells of different species of animals, it is more appropriate to focus on "public mechanisms" rather than "private mechanisms" for the purposes

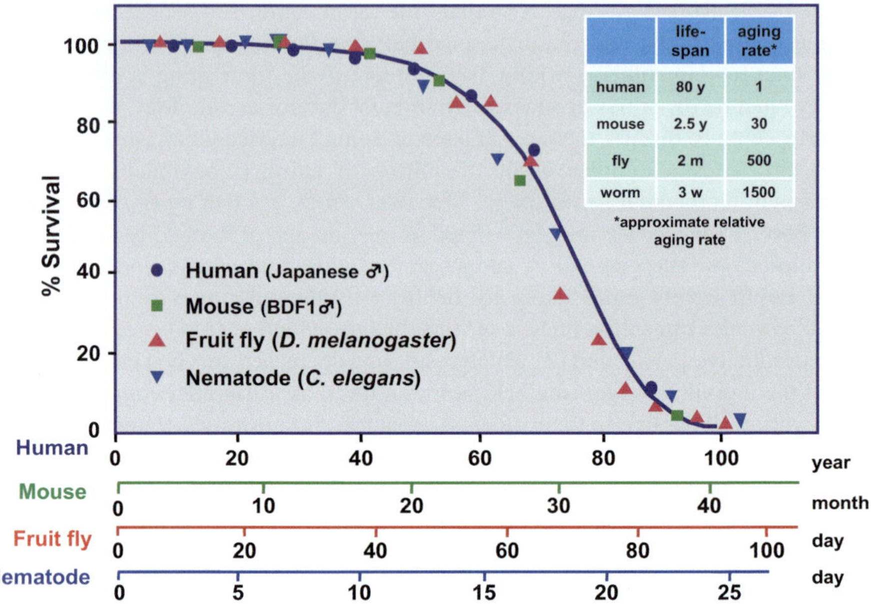

Adapted and modified from Goto S (2002) Saibo kogaku 21: 704-708 (in Japanese)

Fig 1.1 Survival curves of human, mouse, fruit fly and nematode

of our discussion. In this chapter, I therefore discuss the mechanisms of aging that can mainly, although not exclusively, be seen as public. The private mechanisms of aging, however, are by no means unimportant. Indeed, they are useful by themselves to explain particular etiologies or the progress of individual age-related diseases. It should be noted that private mechanisms often involve public mechanisms. For instance, endocrinological decline with age is caused by public mechanisms, such as oxidative stress or protein alteration. It should be noted that each theory is naturally not mutually exclusive or incompatible, but can instead be regarded as a part of other theories. Figure 1.1 illustrates age-related changes in the mortality rate of different animal species, with lifespan differences of more than 1000-fold (between humans and nematodes, for example). The apparent similarity of the survival curves may suggest that the underlying mechanisms of aging are common among the shown animal species. In fact, many aging phenotypes are conserved in model animals, as shown in Table 1.1 (Vijg and Campici 2008). It should be noted, however, that no overall correlation of age regulation was found in the gene expression database, at least between mice and humans, and therefore, aging processes in mice and humans may be fundamentally different, despite certain commonalities in the observed transcriptional profiles in the genes of electron transport chain for aging mice, humans, flies and nematodes (Zahn et al. 2007). In the following sections, I examine selected public mechanisms of aging.

Table 1.1 Common and uncommon aging phenotypes in human and model animals

phenotype	human	mouse	fly	nematode
Decreased cardiac function	Yes	Yes	Yes	NA
Apoptosis, cellular senescence	Yes	Yes	Yes	?
Cancer, hyperplasia	Yes	Yes	No	No
Genome instability	Yes	Yes	Yes	Yes
Macromolecular aggregates	Yes	Yes	Yes	Yes
Reduced memory & learning	Yes	Yes	Yes	NA
Decline in GH, DHEA, testosterone, IGF	Yes	Yes	?	?
Increase in gonadotropins, insulin	Yes	Yes	?	?
Decreased thyroid function	Yes	Yes	NA	NA
Decrease in innate immunity	Yes	Yes	Yes	Yes
Increase in inflammation	Yes	Yes	No	No
Skin morphology changes	Yes	Yes	?	Yes
Decreased mitochondrial function	Yes	Yes	Yes	Yes
Sarcopenia	Yes	Yes	Yes	Yes
Osteoporosis	Yes	Yes	NA	NA
Abnormal sleep	Yes	Yes	Yes	?
Decrease in vision	Yes	Yes	?	NA
Demyelination	Yes	Yes	?	No
Decreased fitness	Yes	Yes	Yes	Yes
Arteriosclerosis	Yes	No	NA	NA
Changes in fat	Yes	Yes	?	?

Adapted from Vijg and Campisi, 2008

Highlight by yellow are common aging phenotypes in listed animals
NA not applicable

1.4 Mutation Theory of Aging • Genome Instability Theory of Aging

This theory predicts that mutations accumulating in the genome are responsible for aging. One of the early proponents of this theory was Leo Szilard. As a nuclear physicist, he proposed that somatic cell mutations induced by ionizing radiation would accelerate aging (Szilard 1959). Ionizing radiation shortened the life-span of mice and rats, shifting the survival curves to the left, with similar shapes as un-irradiated controls, apparently suggesting an acceleration of normal aging (Lindop and Rotblat 1961). It was later shown that the major cause of the observed lifespan-shortening was an increased rate of carcinogenesis rather than an acceleration of physiological aging. Irradiated rodents have therefore not been used as models of accelerated aging. In the meantime, it has been reported that the DNA repair activity of skin fibroblasts in cultures irradiated with ultraviolet light depend on an animal's maximum lifespan (Hart and Setlow 1974). The activity of cells from long-lived animals, such as humans, elephants and cows, is nearly five times greater than that in short-lived rats and mice. Although the repair capacity and lifespan were not proportional, it was thought that long-lived species may have a more active repair system that could therefore play a role in the mechanisms of aging. More recently, it was reported that base excision repair activity declines with age in mice in the brain, liver, spleen and testes (Cabelof et al. 2002). To study the mutation frequency in vivo, selectable markers, such as hypoxanthine phosphoribosyltransferase (HPRT), of purine metabolism have been used for the detection of 6-thioguanine resistant cells that are defective in the HPRT gene. Using this method, it was reported that the mutation frequency increased with age (from 2 to 94 years of age) in cultured human kidney tubular epithelial cells (Martin et al. 1996b). To overcome the limitation that the cells to be assayed must proliferate in vitro in the assay, transgenic mice with reporter genes, such as the bacterial LacZ gene, have been developed. The DNA recovered from the transgenic mouse tissues, including the brain and heart, consisting of mainly post-mitotic cells, were screened for mutations in the integrated shuttle vector in a bacterial host (Dollé et al. 2000). Significant age-related increases in mutations accumulated in the young animals and were found in the small intestine ($10–25 \times 10^{-5}$) and heart ($5–10 \times 10^{-5}$) in mice from 3 to 33 months of age. However, no changes were observed in the brain (5×10^{-5}). It is noted that the increase is linear from young to old ages, with no larger changes at older ages.

Because functional decline with age is apparently more significant in the brain and heart than in the intestine and because the frequency of mutation is not high enough to account for this decline, it appears to be difficult to ascribe a cause of aging to the age-related accumulation of mutations. In fact, the serious proponents of this theory recognize one important question about this theory, stating that "it is not known whether the frequency of the random changes is sufficient to cause the phenotypic effects generally associated with aging" (Vijg and Suh 2013).

1.5 Free Radical Theory of Aging • Oxidative Stress Theory

The free radical theory of aging is one of the most well-known and popular theories of aging proposed thus far. The theory has now been transformed into the "oxidative stress theory of aging" because it most frequently involves reactive oxygen species (ROSs) and because the causative agents of stress are not only free radicals but also include non-radical ROS (Martin et al. 1996a). The principle of the theory was originally proposed by Denham Harman more than half a century ago (Harman 1956). The history of the theory and the inside story of how the idea came to him are found in a recent interview (Harman and Harman 2003). He was a chemist specializing in free-radicals who later became interested in aging and became a medical scientist. In the beginning, the theory apparently did not attract as much interest from scientists working on aging as other theories, such as the mutation theory and the protein cross-linking theory. This is likely because radicals were not very familiar to biologically oriented scientists, and the theory appeared to be too simple and straight forward to explain the complex aging phenomena. However, after superoxide dismutase (SOD), which catalyzes dismutation of superoxide radicals, was found to be widely distributed in mammalian tissues (McCord and Fridovich 1969), more researchers became interested in the capacity of free radicals to damage a variety of cellular constituents, potentially leading to aging. The major targets of free radical damage were believed to be membrane lipids, which contain many unsaturated fatty acids that are easily attacked by radicals to produce lipid peroxides. Lipid peroxides were thought to be components of the lipofuscin age-pigment, a well-known morphological marker of aged cells that consume substantial amounts of oxygen, such as neurons and kidney cells. Lipid peroxidation has readily been measured as thiobarbituric acid reactive substances (TBARS), although the method may be problematic in terms of its specificity and, recently, such substances have been evaluated as isoprostanes. DNA was another molecule of interest for oxygen radical attack. It can form 8-hydroxy-2′-deoxyguanosine (8-oxodG), which is relevant to cancers that increase with age (Fraga et al. 1990). Proteins have attracted the least interest mainly because of limitations in the methods used to detect them despite the fact that the catalytic activities of enzymes have long been known to decrease with age (Stadtman 1988). Stadtman and his collaborators established a relatively easy method for the detection of oxidatively modified proteins that have reactive carbonyl derivatives as oxidation products in amino acid residues, such as lysine, arginine and proline, which are easily measured by spectrophotometric or immunological methods by the reaction of the proteins with 2,4-dinitrophenylhydrazine. All cellular components (e.g., membrane phospholipids, DNA and proteins) have been reported to have oxidative damage with age, which could potentially cause the physiological decline of the organisms (Cutler and Rodriguez 2003). The free radical theory of aging has prompted researchers to study radical scavengers and anti-oxidants to see if such chemicals can extend the lifespan of animals. Harman himself showed in his early studies that the synthetic antioxidants 2-mercaptoethylamine and butylated hydroxytoluene can extend the lifespan of mice (Harman 1968). Numerous studies have been conducted since then to try to extend the lifespan of experimental animals or to ameliorate age-related diseases in humans that are likely caused by ROS, mostly using antioxidant vitamins, such as vitamin C and E, or natural products, such as polyphenols and carotenes. The results have been rather disappointing in human clinical trials attempting to reduce the risks of age-related diseases, although antioxidant supplements are reported to be promising in experimental animals (Sadowska-Bartosz and Bartosz 2014). In human studies, it has been reported in a systematic review and meta-analysis of randomized trials with total of 232,606 participants that antioxidant supplements (β-carotene, vitamin A and E) can significantly increase all-cause mortality (Bjelakovic et al. 2007). In animal studies, for example, the popular "anti-aging" polyphenol resveratrol, which is not necessarily supposed to act as an antioxidant, has been shown to not extend the lifespan of genetically heterogeneous mouse strains in multiple laboratories (Strong et al. 2013).

The free radical theory of aging appeared to explain the rate-of-living theory of aging, which was first proposed many years ago, that suggests that there is an inverse relationship between the metabolic rate and longevity in different animal species. However, it turned out that this does not apply to mammals. The opposite is even true intraspecifically in mice, in that the higher the energy expenditure (indicating a larger consumption of oxygen), the longer the life span, contrary to what is expected from the free radical theory of aging (Speakman et al. 2004). Based on studies of genetically modified mice showing under- or overexpression of genes of antioxidant enzymes (e.g., cytoplasmic or mitochondrial superoxide dismutase, catalase, glutathione peroxidase), it was concluded that all of the anti-oxidant enzymes studied separately or in combination do not significantly influence the lifespan in mice (Pérez et al. 2009). On the other hand, it is true that oxidative damage in lipids, DNA and proteins increases with age, as described above, suggesting an involvement of free radicals in aging. Additionally, many mutant animals with longer life spans show increased resistance to oxidative damage. Thus, potential roles of ROS in driving aging should not be underestimated, although they may not play a crucial role in life span determination.

It has often been stated that the major source of ROS generation is mitochondria, as discussed later in Mitochondrial Theory of Aging. However, apart from ROS generated in the mitochondria as byproducts, oxidants can be generated as normal products in multiple enzyme reactions catalyzed by oxidases, such as NADPH oxidase, xanthine oxidase and monoamine oxidase, contributing to overall cellular oxidative stress. Such oxidants can be important in damaging cellular molecules and also signaling factors (Finkel 2011). Although the involvement of ROSs in signal transduction have attracted more interest in recent years than their potential detrimental role in aging, I do not discuss the details of this topic as it is beyond the scope of this overview.

I instead discuss the hormetic roles of ROSs that are relevant to aging. Hormesis is a dose-response relationship that exhibits stimulation at low doses and inhibition at higher doses, although whether a response is beneficial or harmful can be complex and is often not immediately obvious (Calabrese and Mattson 2011). Exposure to a variety of stressors, such as toxins, heat, ROS, and radiation can induce an adaptive response if they are not too strong, making an organism more resistant to subsequent stronger challenges (Gems and Partridge 2008). Nematodes pretreated with hyperbaric oxygen became more resistant to semilethal oxygen exposure (Cypser and Johnson 2002). Interestingly, an oxidative stressor (juglone) could induce substantial resistance to a lethal challenge. The lifespan of the pretreated worms was increased compared to naive counterparts. We have shown that regular moderate exercise in old rats can reduce oxidative stress, as measured by protein and DNA oxidation, by upregulating anti-oxidation systems, including the glutathione level, proteasome and DNA repair enzymes (Goto and Radák 2009; Radák et al. 2001). Others have also demonstrated that exercise induces anti-oxidant enzymes (Gomez-Cabrera et al. 2008) and that anti-oxidant vitamins C and E can ameliorate the beneficial effects of exercise (Ristow et al. 2009). Exercise hormesis is well recognized, as the ROS induced by moderate exercise constitute a significant mechanism of beneficial effects of the regimen (Gomez-Cabrera et al. 2008; Radak et al. 2005). See also the discussion on mitohormesis in the Mitochondrial theory of aging section.

Thus, ROS have two sides, making this Theory somewhat complex. On one hand, ROS are believed to have detrimental effects, as proposed in the original theory. On the other hand, they are also thought to have beneficial effects as signaling factors and factors that can protect an organism against stresses that could be encountered in life.

1.6 The Mitochondrial Theory of Aging

Mitochondria have long been known to be the power station of eukaryotic cells, generating the majority of ATP and therefore being vital to life. After the proposal of the Free Radical Theory of Aging, these organelles attracted increased interest, as they use most of oxygen taken up by cells that could potentially converted to damaging reactive oxygen species (ROSs) in the respiratory chain. Harman was the first to suggest that mitochondria can be a major source of free radicals and also a major target of the damage that drives aging as an obvious extension of Free Radical Theory of Aging (Harman 1972). In fact, mitochondrial DNA (mtDNA) and proteins are more vulnerable to oxidation than cytoplasmic or nuclear proteins and DNA, likely due to their proximity to the electron transport chain, their lack of histones to protect the DNA, and their low repair activities. Miquel proposed the Mitochondrial theory of aging (Miquel et al. 1980). A number of papers in support of this theory have been published. It has often been cited that ROS (such as hydrogen peroxide) generated in the mitochondria account for 1–2 % of the total oxygen uptake (Chance et al. 1979). Even higher values of 4–5 % have also been reported (Luft and Landau 1995). However, later studies have criticized these reports, and the current estimation for these values is as low as 0.15 % (St-Pierre et al. 2002).

Point mutations that may occur due to oxygen radicals accumulate in mtDNA with aging, possibly due to mtDNA polymerase errors, suggesting that this process may cause the age-related functional decline of cells and tissues (Michikawa et al. 1999). For this reason, mice with defective mtDNA polymerase have been constructed as a model of premature aging to prove or disprove this theory (Trifunovic et al. 2004). Studies of these mice demonstrated that mice with a homozygous mutation (mtDNA mutator mice) expressing proof-reading deficient mtDNA polymerase γ show reduced lifespans. They also show phenotypes of accelerated aging at 6–9 months of age, such as hair loss and graying, sarcopenia, osteoporosis, heart enlargement, and reduced subcutaneous fat, all of which are features that are typical of human aging (Trifunovic et al. 2004). Despite these premature aging phenotypes and the accumulation of mtDNA mutations, no increases in hydrogen peroxide production and oxidative stress markers (protein carbonyl, 8-OHdG and F2-isoprostane) have been observed in isolated mitochondria and tissues of the mice. Thus, these findings did not support the idea that mtDNA mutations cause increased ROS production that could drive aging. One criticism of this research is that these mice may not represent natural human aging because the levels of mtDNA mutations in human tissues are an order of magnitude lower than in mutator mice (Khrapko et al. 2006). It should, however, be noted that a recent report on the mtDNA mutator mice showed that the hydrogen peroxide levels in the aged animals were increased relative to the young mutator or wild type mice, suggesting that prolonged exposure to higher concentrations of ROSs could contribute to accelerated aging (Logan et al. 2014). Thus, the possible contribution of ROSs to aging in the mtDNA mutator mice remains controversial. Interestingly, however, 5 months of endurance exercise can rescue premature mortality in the mutator mice by inducing mitochondrial biogenesis, thereby mitigating the emergence of sarcopenia, brain atrophy, cardiac hypertrophy and pathologies (Safdar et al. 2011). Endurance exercise rescued mtDNA depletion in multiple tissues and reduced the frequency of point mutations in the mutant mice. These data support the view that lifestyle can improve the systemic deterioration of mitochondrial function that could increase morbidity and mortality with aging. Supporting evidence for the mitochondrial theory of aging has been obtained in transgenic mice overexpressing human catalase in the mitochondria, which exhibit increased lifespans with reduced cardiac pathologies and cataract severity (Schriner et al. 2005). These mice exhibited higher aconitase activity in the heart and lower 8-OHdG in the DNA of the skeletal muscle, suggesting that oxidative stress can be ameliorated by the overexpression of catalase targeted to mitochondria.

In view of the controversy regarding the contribution of mitochondrial ROS in aging, it is worth referring to the concept of mitochondrial hormesis (or mitohormesis) (Ristow 2014; Schulz et al. 2007). It was found that nematodes treated with 2-deoxy-D-glucose (2DG), an inhibitor of glycolysis, exhibited a prolongation of their lifespan with a compensatory increase in mitochondrial respiration, which is associated with increases in the level of ROS, followed by increased expression of catalase, which scavenges hydrogen peroxide (Schulz et al. 2007). When the worms were pretreated with VC, VE or other antioxidants, the elevation of catalase was abolished and the extension of lifespan of the worms treated with 2DG was blocked. It thus appears that mitochondrial oxidants induced an increased defense against oxidative stress as a hormetic response because excess oxidants are obviously detrimental.

The mitochondrial theory of aging has thus developed into a theory evaluating the roles of ROS generated from organelles as signals for cellular homeostasis rather than simply as damage markers, as originally suggested.

1.7 Error Catastrophe Theory of Aging

This theory was most prominently advanced by Orgel (1963) after the occurrence of remarkable developments in the field of molecular biology, such as the elucidation of the mechanisms of replication, transcription and translation. This theory predicted that nucleic acids and proteins can contain errors when they are synthesized because the information transfer in each step of gene expression and maintenance is not perfectly accurate and the synthesizing machineries consisting of error-containing molecules would make further errors, thus producing a vicious cycle of error propagation that could result in the gradual loss of cellular function (catastrophe). Although this theory is usually regarded as being advocated by Orgel, it should be noted that Medvediev presented a similar idea independently (Medvediev 1962). This theory has attracted particular attention from scientists interested in the mechanisms of aging because it suggests a hypothesis that is actually testable by means of emerging theoretical and methodological developments.

Possible detrimental consequences of the propagation of errors are likely more serious in non-dividing cells than in dividing cells because error-containing dividing cells can be eliminated and replaced by new cells or can be diluted by cell division, while error-containing molecules may be repaired or replaced by metabolic turnover.

Of the types of errors in information transfer, translational errors have been most extensively studied. These errors can occur in two independent steps of translation: (1) in the charging of individual tRNAs by cognate amino acids and (2) in the decoding of codon of mRNA. The former step is catalyzed by amino acyl tRNA synthetases that can mischarge amino acids to tRNAs by imperfect enzymes. The latter step occurs on ribosomes by matching codons with anti-codons of charged tRNA. A number of studies on mistranslation (error frequency) in aging have been conducted mainly using young and senescent cells in culture. For example, the error frequency of actin synthesis was studied in human fibroblasts at different replicative ages (Harley et al. 1980). Histidinol, an analogue of histidine, was added to the culture medium and thereby blocked the charging of tRNAs for histidine. The decrease in the histidine-charged tRNA concentration induces an incorporation of glutamine into actin in the place of histidine because the codons for glutamine (CAA or CAG) are similar to those for histidine (CAU or CAC) so that errors of translation can occur due to codon-anticodon mispairing at the third position. Late-passage cells from fetal, young and old donors cultured in vitro showed similar or lower error frequencies than the corresponding early passage cells, suggesting that error propagation does not occur and thus fails to support the error catastrophe theory of aging. In another study, age-related changes in the charging error were examined in vivo by the incorporation of ^{14}C-methionine and ^{3}H-ethionine, an analogue of methionine into proteins of young and old mouse livers (Ogrodnik et al. 1975). It was expected that ethionine could be mischarged to tRNA for methionine by methionyl tRNA synthase if the fidelity of the enzyme decreased with age. The misincorporation of ethionine in the place of methionine was 10–50 % higher in the ribosomal proteins of old animals, indicating that the charging fidelity indeed declined in older animals, although it was not clear if these errors propagate with age. As for the recognition of natural amino acids in young and old animals, age-related changes in the fidelity of aminoacylation by tyrosyl-tRNA synthetase in rats have been reported (Takahashi and Goto 1988). The enzymes were purified from the livers of young (4–7 month-old) and old (27–29 month-old) rats, such that no detectable phenylalanyl-tRNA synthetase was contaminated to study misrecognition phenylalanine vs. tyrosine by the enzyme. The error frequency of the tyrosyl-tRNA synthetase (on the order of 10^{-8}) from the older animals was slightly lower than that from the younger animals, but this difference was not significant. Thus, the fidelity of aminoacyl tRNA synthetase did not appear to change in old age, again suggesting that errors in translation would not increase with aging at the stage of tRNA charging with the amino acid. The fidelity of decoding on ribosomes from young and old animals has been mostly studied by assessing the misincorporation of non-cognate amino acids using synthetic mRNA of homopolymers, such as poly(U), which codes for phenylalanine polymers. The misincorporation of leucine into the poly(U)-dependent synthesis of polypeptides using ribosomes of tissues did not differ between young and old mice. We have studied codon recognition fidelity using a unique group of natural mRNAs that code for limited species of amino acids. The protamines are highly basic nuclear proteins from fish sperm consisting of 33 amino acid residues. They contain only seven different amino acids, of which approximately two-thirds are arginine. It was therefore possible to study the incorporation of radioactive amino acids that were not coded in the mRNAs in vitro. The fidelity of the decoding of the mRNAs on ribosomes from the livers of mice between 2 and 29 months of age was found to not change significantly (Mori et al. 1983). Thus, these findings are not consistent with the error catastrophe theory of aging in terms of the predicted age-related changes in translational fidelity. This is probably because the proofreading mechanisms of translation are maintained, keeping the fidelity high enough that propagation errors do not occur. The high fidelity of translation has been discussed from evolutionary perspectives as important for survival as it allows for an avoidance of protein misfolding (Drummond and Wilke 2009) (see also: Sect. 1.8). Another possibility is that error-containing proteins do not increase with age, such proteins are efficiently degraded and replaced by intact molecules by metabolic turnover as discussed in the next session (Sect. 1.8).

Other steps of information transfer in which error catastrophe can occur are DNA replication and transcription. No age-dependent differences have been found between the fidelity of nuclear DNA polymerase-α and -β partially that were purified from the regenerating livers of young (6-month-old) and older (28-month-old) mice when the enzymes were used to copy φX174 DNA (Silber et al. 1985). The same group of investigators showed that the fidelity of highly error-prone DNA polymerase -β in non-dividing neurons from young and old mice was not significantly different when copying the same bacteriophage DNA (Subba Rao et al. 1985). Thus, although available reports on the possibility of age-related changes in the fidelity of DNA polymerases are limited, it appears that the error catastrophe theory of aging is not supported by the current information on nuclear DNA replication. Although Orgel implied that transcription errors can lead to catastrophe (Orgel 1963), I am not aware of a published paper on age-related changes in the fidelity of nuclear gene expression or of RNA polymerases in the nucleus (Imashimizu et al. 2013). The integrity of RNA coded in mitochondrial DNA has been studied in the brains of young (1-month-old) and older (18-month-old) mice (Wang et al. 2014). The transcriptional error of the mitochondrial RNA polymerase was site-specific and varied greatly among different genes. The error levels in two age groups were not significantly different, suggesting that error propagation does not occur during aging. It is noted that transcriptional errors were independent of the DNA mutation frequency and were up to 200-fold more frequent than replication errors.

Thus, the error catastrophe theory of aging, which was once a popular hypothesis, is not supported by the current experimental evidence. This theory thus seems to have been largely forgotten, but it should be noted that pathologist George Martin has argued that "it may have been given a premature death certificate" because drifts in gene expression may be responsible for the "quasi-stochastic" distribution of lesions in geriatric pathologies, such as Alzheimer's disease and atherosclerosis and that errors in information transfer could feasibly contribute to this process (Martin 2012).

Although it is unlikely that error catastrophe occurs in genetic information transfer, it should be noted that errors in protein synthesis can occur as the misfolding of higher structures during translation. In fact, the rate of folding errors can be as high as 30 % of newly synthesized proteins, even though misfolding can be mostly prevented by chaperons (Schubert et al. 2000) (see also: Sect. 1.8).

1.8 Altered Protein Theory of Aging • Protein Homeostasis or Proteostasis Theory

The origin of this theory may be traced back to Friz Verzár, who reported an age-related increase in collagen cross-linking in rat tail tendons. A large number of studies have confirmed that changes in collagen occur with age in various tissues and animals (Robert 2006). However, because collagen is an extracellular protein and its relevance to cellular metabolisms is limited, researchers interested in aging and inspired by the findings became more concerned about the age-related changes of enzymes and other proteins involved in intracellular functions. In the meantime, studies on the error catastrophe theory of aging failed to support the predicted propagation of errors in translation, as described above, and instead suggested the presence of altered forms of enzymes in aged cells and tissues. Thus, altered enzymes were interpreted to be formed not by translational errors but instead by post-translational modifications. Altered forms of enzymes existing in old cells and animals have been detected by various means. They have been shown to have lower specific activity (by between 30 and 70 %) per unit weight of purified enzyme (Rothstein 1981). One problem with finding altered forms of an enzyme through purification is that altered enzymes are often lost during the purification process, as the purification protocol depends on enzymatic activity. Altered enzymes have been detected in crude extracts without purification, as they can cross-react with enzymes with no or reduced activity that remain as immunologically reactive as the native enzymes (Gershon and Gershon 1970). Another frequently used method was to examine the heat-stability of an enzyme in cell or tissue extracts. An enzyme is likely to become heat-labile if it is altered such that the mixture of native and altered enzymes has a biphasic or quasi-biphasic heat-inactivation kinetic curves for the activity so that the percentage of the altered form of an enzyme could be evaluated (Houben et al. 1984). Thus, many altered proteins, mainly enzymes, have been reported to increase in cells and tissues with aging, suggesting that they may be responsible for the age-related decline of physiological functions.

The causes of these alterations have been suggested to be post-translational modifications, including oxidation or nitrosylation by ROSs or RNSs (reactive nitrogen species) and glycation by glucose, rather than the synthetic errors predicted by the Error Catastrophe Theory of Aging. In some cases, reactive aldehydes derived from lipid peroxidation are responsible for the modifications. We and other investigators have shown that the heat-labile enzymes described above are generated by a reaction with ROSs in vitro (Takahashi and Goto 1990). The chemistry of modifications has been studied extensively, proving that the side-chains of specific amino acid residues, such as lysine and arginine, are modified (Stadtman 1993). Notably, carbonyl moieties generated by oxidation have most frequently been used to evaluate oxidative stress on proteins by biochemical or immunochemical methods (Levine et al. 1990; Nakamura and Goto 1996), although this method is not without problems (Fedorova et al. 2014; Goto and Nakamura 1997). In addition to a correlative relationship between the oxidative modification of proteins and aging, a causal relationship between age-related increases in oxidative stress and functional decline has been suggested (Martin et al. 1996a; Martin and Grotewiel 2006). However, despite numerous reports on the possible involvement of protein oxidation in aging, it is hard to predict its contribution, as multiple effects of oxidative stress on other molecules, such as DNA and membrane phospholipids, occur in parallel.

The glycation caused by non-enzymatic chemical reactions of proteins with glucose is another well-recognized post-translational modification that increases in long-lived proteins, such as collagens and elastin as well as lens crystallins. The glycation of proteins ends up generating a variety of products collectively called AGEs (Advanced Glycation End products). Because proteins exposed to a high concentration of glucose for a long period of time are susceptible to this modification, it accumulates frequently in extracellular matrix proteins and proteins with very low turnover rates. Glycation appears to be less involved in the age-related functional decline of cells as a general cause than other post-translational modifications that occur more frequently inside cells. Nevertheless, there is no question that glycation is involved in age-related diseases of endothelial cells, such as atherosclerosis, cardiovascular pathologies and renal disorders, in which tissue microvessel dysfunction is involved.

More recently, apart from the post-translational modifications described above, specific altered proteins with abnormal conformational structures in age-related neurodegenerative diseases, such as Alzheimer's disease (hyperphosphorylated tau), Parkinson's disease (mutant α-synulein), Huntington's disease (mutant huntingtin) and amyotrophic lateral sclerosis (misfolded SOD1), have been studied extensively (Stefani 2004). More generally, amyloid diseases that impair the functions of different organs are also protein conformation diseases that increase with age. There are many other examples of protein misfolding and aggregation causing age-related diseases (Stefani 2004). While numerous cases, especially in neurodegenerative diseases, have been reported in which protein alterations produce age-related pathologies, it is not clear whether such changes also contribute to the functional decline of cells and tissues in physiological aging. It is possible that minor alterations of individual proteins cause undetected changes, yet result in significant physiological deterioration during aging.

The accumulation of altered proteins with age can be driven by either increases in the formation, the decline of degradation, or both processes. While the mechanisms involved in the formation of such proteins have been extensively studied, the decrease in degradation or elimination has attracted less interest. Schoenheimer described the dynamic state of body constituents, such as lipids and proteins, as early as the late 1930s, when the stable isotope technique became available, thereby highlighting the importance of metabolic turnover as a homeostatic life maintenance mechanism for the first time. Due to the difficulty of the access to the famous book "The dynamic State of Body Constituents" (Harvard University Press, Cambridge, MA, 1949) by him, I cite here instead the excellent overview on this topic (Kennedy 2001). This view, however, was challenged by Monod (a Noble laureate famous for the operon theory) and collaborators, who studied the turnover of β-galactosidase in growing E. coli and concluded that most proteins in the cells are static rather than in a dynamic state (Hogness et al. 1955). They further suggested that the proteins in mammalian tissues would also be stable because the apparent dynamic state in these cells may be interpreted as some proteins being secreted or lost by cell death. However, it was shown that proteins in rabbit macrophages, non-dividing cells, and culture conditions do actually turnover, thus not supporting this hypothesis (Harris and Watts 1958). Even so, protein degradation has not attracted the same intense research interest as other more positive biological processes.

The degradation of intracellular proteins was thought to be mainly dependent on lysosomes, which were found to contain proteolytic enzymes (cathepsins) with different specificities at acidic pH values (de Duve 1983). While lysosomal proteolysis is thought to be nonspecific with regard to the protein substrates it degrades, the half-life of different proteins was reported to vary considerably. This fact facilitated studies on non-lysosomal protein degradation that were first performed in rabbit reticulocytes that do not have lysosomes. The extensive research on non-lysosomal protein degradation has established the mechanisms of the ubiquitin-proteasome system of proteolysis, showing that substrate proteins are marked with ubiquitin for degradation and digested by proteasomes (in the case of 26S proteasome, see below) (Ciechanover 2005). The proteasome is a multi-catalytic protease complex that exists in two forms, 26S and 20S, that differ in subunit composition but share a common catalytic specificity. The 26S proteasome degrades proteins tagged with ubiquitin chains and ATP dependently, while the 20S proteasome degrades non-ubiquitinated proteins without ATP.

On the other hand, the lysosomal pathway of proteolysis has developed into the elucidation of autophagy-lysosome systems, in which protein aggregates and damaged organelles are specifically recognized and destroyed, contrary to what was originally believed to be non-specific (Koga et al. 2011). Both systems of protein degradation have profound impacts on aging and age-related diseases, particularly in neurodegenerative diseases (Rubinsztein et al. 2011; Saez and Vilchez 2014).

The Altered Protein Theory of Aging prompted studies on protein turnover in aging (Goto et al. 2001; Van Remmen et al. 1995). For example, it was demonstrated that the half-lives of enolase in nematodes and aldolase in mice are extended in old animals compared with their younger counterparts, as determined by pulse-chase experiments. We found that the half-life of the various proteins introduced into mouse hepatocytes in primary culture were extended by 40–60 % in the cells from old animals (Goto et al. 2001; Ishigami and Goto 1990). It was also shown that prematurely terminated puromycinyl peptides, as a model of altered proteins, are much more slowly degraded in the livers of old mice than in those of younger animals (Lavie et al. 1982). Thus, the degradation of normal and abnormal proteins was shown to be impaired in old animals, and these findings were comparable with the age-related accumulation of altered proteins in different tissues. In the meantime, it was firmly established that the ubiquitin-proteasome system and the autophagy-lysosome system are responsible for intracellular protein degradation as described above. Many studies have demonstrated that proteasome activity declines with age (Saez and Vilchez 2014; Shibatani et al. 1996). We have shown that the activities of both the 20S and 26S forms of the liver proteasome decline similarly with aging in three age groups of rats of from 8–10 to 25–28 months of age (Hayashi and Goto 1998). Despite the decline in the enzyme activities, the amount of catalytic subunits measured by immunoblot did not change with age, suggesting that posttranslational modifications or subunit replacement are responsible for the decreased activities. In fact, other investigators have reported that the subunit composition of the proteasome is altered in aged tissues. Furthermore, a subunit of the proteasome is sensitive to oxidative modification (Ishii et al. 2005), suggesting that oxidative stress can accelerate the accumulation of oxidized proteins in aging by reducing the efficiency of damaged proteins. It is interesting to note that the 20S proteasome degrades oxidatively modified proteins selectively (Davies 2001) and that the 26S proteasome can be reversibly dissociated to produce the 20S proteasome by removing 19S regulators upon oxidative challenge, thereby facilitating adaptation to stress (Grune et al. 2011). It should be mentioned that the Lon protease plays an important role in the degradation of oxidized mitochondrial proteins, the activity of which decline with age and contribute to the accumulation of damaged proteins in the organelles (Ngo et al. 2013).

When the damage to proteins is extensive, forming insoluble cross-linked aggregates that are not degraded by proteasomes, the autophagy-lysosome system degrades them in addition to removing the damaged organelles (Wong and Cuervo 2010). The autophagy-lysosome system is considered to act via microautophagy, macroautophagy and chaperon-mediated autophagy, and the latter two systems are the predominant mechanisms of autophagy in animals. Macroautophagy refers to the digestion of contents of cytoplasmic regions engulfed in membrane vesicles, which then fuse with lysosomes for degradation. Chaperon-mediated autophagy is the digestion of substrates bound to the chaperon heat-shock cognate protein (hsc70), which is recognized by lysosomes via an interaction with the receptor protein on the surface. Substrates translocated across the lysosomal membrane are then digested. The substrates are often protein aggregates that are not digestible by proteasomes. The activities of these autophagic processes decline with aging (Rubinsztein et al. 2011). The age-associated decline in the chaperon-mediated autophagy can be caused by decreased content of the substrate receptor (lysosome-associated membrane protein type 2a) (Cuervo and Dice 2000) and the age-associated impairment of lysosomal function (Kurz et al. 2008).

A number of studies have established the extensive involvement of altered protein conformation in age-associated neurodegenerative diseases. These are mainly due to the impaired functions of ubiquitin-proteasomes and/or autophagy-lysosome systems and the chaperon dysfunctions described in many excellent reviews (Takalo et al. 2013). However, I do not go into the details of these studies as this subject is beyond the scope of this overview, although it is conceivable that these mechanism are also involved in the general age-related functional decline of house-keeping proteins.

Thus, the original idea that an accumulation of altered proteins produces the aged phenotype has expanded to include a variety of life processes. This has now become one of the most widely accepted theories used to explain the mechanisms of aging.

1.9 Dysdifferentiation Theory of Aging • Epigenetic Theory of Aging

Richard Cutler suggested that differentiated cells can undergo changes in transcription during aging, such that the strict pattern of gene expression is gradually relaxed, leading to the deterioration of the functions of cells and tissues (Cutler 1991). This idea, called the dysdifferentiation theory of aging, was based on the finding that the expression of globin or its related mRNA and murine leukemia virus RNA is increased in the brains and livers of aged mice compared to their younger counterparts (Ono and Cutler 1978). More recently, it has been shown that gene expression is more heterogeneous in the tissues of individuals with advanced aging, including the cerebral cortex and rat hippocampus. These findings are compatible with the dysdifferentiation theory of aging (Somel et al. 2006).

This theory has never been popular, but has been recently revived as the epigenetic theory of aging. Epigenetics is a phenomenon in which a fixed pattern of gene expression in a cell or an organism is inherited from one generation to the next without any changes in the genomic nucleotide sequence. This definition has been broadened to include the long-term stable control of gene expression in differentiated cells without changes in the nucleotide sequence, as manifested in various physiological and pathological situations, including aging and age-related diseases. The epigenetic regulation of long-term cell-specific gene expression is determined by a variety of mechanisms, including DNA methylation, histone modifications and microRNA expression. These epigenetic mechanisms of gene modulation are influenced throughout life by both internal and external stimuli, such as energy metabolism, nutrition and exercise, and can therefore impact aging and the risk of age-related diseases.

It has been shown that there are far more differences in the patterns of DNA methylation and histone acetylation in the circulating lymphocytes of older (50 years of age) monozygotic twins compared with younger (3 years of age) twins (Fraga et al. 2005). Interestingly and consistently with these differences, the differences in the gene expression between the older pairs were much greater than those in the younger pairs. These findings suggest that an identical genome in early life could undergo different epigenetic modifications throughout life, potentially resulting in differences in the aging rates and/or in their vulnerability to diseases. This type of variable epigenetic modifications may partly explain the relatively low contribution (approximately 30 %) that genes are believed to have on longevity compared with environmental factors (Ljungquist et al. 1998). Frailty is a common manifestation of physiological aging. It has been reported that a worsening frailty status, as measured by the loss of body weight, the development of sarcopenia and muscle weakness, and the reduction in physical activity, is associated with decreased global DNA methylation in the peripheral blood cells of individuals aged 65–105 over a 7-year-follow-up period (Bellizzi et al. 2012). Aging is often associated with reduced levels of global DNA methylation, mostly in CpG sequences, but its physiological implications remain unclear. However, it is known that the age-related hypermethylation of the promoter regions of tumor suppressor genes increase the risk of carcinogenesis. Changes in the post-translational modification of histones occur with age, which can lead to reduced gene expression, as decreased acetylation allows the chromatin to more tightly condense by increasing its interactions with DNA. In fact, reduced acetylation of lysine 9 in histone H3 in aged rat livers has been documented (Kawakami et al. 2009). Memory impairment is a common feature of old animals and a serious problem for elderly individuals. It has been reported that the acetylation of specific lysine residues in histone H3 and H4 are transiently increased in the hippocampus of young (3-month-old) mice subjected to contextual fear conditioning but not in their older (16-month-old) counterparts (Peleg et al. 2010). These findings suggest that memory impairment in old animals is correlated with defects in learning-induced histone acetylation. Intriguingly, the administration of histone deacetylase inhibitors, such as sodium butylate, to old mice prior to the conditioning increased the acetylation significantly in the coding regions of learning-regulated genes. These findings suggest that the dysregulation of histone acetylation is causally related to age-associated memory impairment, raising a possible mechanism for the treatment of this disorder. Reports on the involvement of microRNAs in aging are currently emerging, but are still limited. The involvement of epigenetic factors, including miRNAs, in the brains of patients with neurodegenerative diseases, such as Alzheimer's disease and Parkinson's disease, are increasing of interest to researchers in the field (Grasso et al. 2014).

As research on this theory is still in its immature stages, further research developments will be required to evaluate it fully.

1.10 Hyperfunction Theory of Aging

This recently proposed and unique theory of aging deserves mention, as it seriously opposes the influential Free Radical Theory of Aging and may open up a new door to explain the mechanisms of aging. In most of the aging theories described above, aging is believed to be due to an accumulation of molecular damage that is induced by ROSs and other chemicals or by errors in critical life maintenance processes. Mikhail Blagosklonny proposed that aging is instead caused by the hyperfunction of growth, such as hypertrophy and hyperplasia, rather than an increase in the damage that continues later in life, leading to age-related pathologies (Blagosklonny 2008). His claim is based on reports that contradict the ideas that aging is caused by an accumulation of molecular damage. According to such ideas, this molecular damage is mainly due to ROS. The reduced translation activity due to the deletion of ribosomal S6 protein kinase 1, a component of the Target of Rapamycin (TOR) pathway, is then believed to lead to an increased life span and resistance to age-related pathologies (Selman et al. 2009). TOR is an evolutionarily conserved protein kinase that regulates growth and metabolism and is involved in the modulation of aging (Kapahi et al. 2010). Blagosklonny admits that damage accumulation can cause the deterioration of cellular functions over time but also predicts that an organism could not live long enough to accumulate a lethal level of damage (Blagosklonny 2008). It is possible, however, that damage accumulation would increase the probability of death when exposed to internal and external stress, thus constituting a mechanism of aging. He stresses the role of the TOR pathway by placing it in the center of the hyperfunction theory of aging because most factors that appear to reduce the activity of TOR retard aging and extend the lifespan of model organisms (Blagosklonny 2012). Gems and Partridge support the idea of hyperfunction as a mechanism of aging, but state that it remains unclear how the pathway controls the rate of aging and lifespan (Gems and Partridge 2013). This theory predicts a form of antagonistic pleiotropy in which hyperfunction increases fitness early in life but is harmful in old age. The identity of the intrinsic or extrinsic factors that maintain hyperfunction in the face of declining metabolic activity with age remain unknown. It should be noted that a recent report describes that rapamycin extends the life span of mice but ameliorates few aging phenotypes, such that its effects are not due to a modulation of aging but are instead related to aging-independent drug effects (Neff et al. 2013).

1.11 Summary and Perspectives

Despite extensive efforts to solve the unsolved problems of biology for more than half a century, no single theory has yet successfully explained the mechanisms of aging. As all animals are considered to be the products of evolution, it is assumed that there are conserved aging mechanisms even between species with remarkably different lifespans, such as humans, mice, fruit flies and nematodes (see Fig. 1.1, Table 1.1). Although there appear to be conserved pathways that potentially drive aging (Kenyon 2010), it is not known how these very basic molecular mechanisms result in such great lifespan variation. This has remained an unsolved problem in gerontology. The leading theories that have so far been proposed are apparently acceptable, but not without objections, and different theories interrelate with each other by one theory being a part of the others, suggesting that each one contributes partly to be integrated into explaining the whole process of aging. In addition, it has been proposed that chance or stochasticity in addition to genes and environments can play a role in aging regardless of the mechanisms in both humans and model organisms (Kirkwood and Finch 2002; Vaupel et al. 1998). Nevertheless, no one would think that a lucky mouse can live for 100 years and an unlucky human would die of aging in 3 or 4 years. A major target of future studies of aging will be how to integrate the different theories to understand the mechanisms of varied aging rates in different animal species and individual differences of aging rates within a species.

We are perhaps in the stage of awaiting a new paradigm or an integration of the existing theories to provide us with an improved understanding of the mechanisms of aging.

References

Arking R (1998) Biology of aging. Observations ad principles, 2nd edn. Sinauer Associates, Sunderland

Bellizzi D, D'Aquila P, Montesanto A, Corsonello A, Mari V, Mazzei B et al (2012) Global DNA methylation in old subjects is correlated with frailty. Age (Dordr) 34:169–179

Bjelakovic G, Nikolova D, Gluud LL, Simonetti RG, Gluud C (2007) Mortality in randomized trials of antioxidant supplements for primary and secondary prevention: systematic review and meta-analysis. JAMA 297:842–857

Blagosklonny MV (2008) Aging: ROS or TOR. Cell Cycle 7:3344–3354
Blagosklonny MV (2012) Once again on rapamycin-induced insulin resistance and longevity: despite of or owing to. Aging (Albany NY) 4:350–358
Cabelof DC, Raffoul JJ, Yanamadala S, Ganir C, Guo Z, Heydari AR (2002) Attenuation of DNA polymerase beta-dependent base excision repair and increased DMS-induced mutagenicity in aged mice. Mutat Res 500:135–145
Calabrese EJ, Mattson MP (2011) Hormesis provides a generalized quantitative estimate of biological plasticity. J Cell Commun Signal 5:25–38
Chance B, Sies H, Boveris A (1979) Hydroperoxide metabolism in mammalian organs. Physiol Rev 59:527–605
Ciechanover A (2005) Proteolysis: from the lysosome to ubiquitin and the proteasome. Nat Rev Mol Cell Biol 6:79–87
Comfort A (1964) Ageing. The biology of senescence. Routledge & Kegan Paul Ltd, London
Cuervo AM, Dice JF (2000) Age-related decline in chaperone-mediated autophagy. J Biol Chem 275:31505–31513
Cutler RG (1991) Recent progress in testing the longevity determinant and dysdifferentiation hypotheses of aging. Arch Gerontol Geriatr 12:75–98
Cutler RG, Rodriguez H (2003) Critical review of oxidative stress and aging. Advances in basic science, diagnostics and intervention, vol I & II. World Scientific Pub Co LTD, Singapore
Cypser JR, Johnson TE (2002) Multiple stressors in Caenorhabditis elegans induce stress hormesis and extended longevity. J Gerontol A Biol Sci Med Sci 57:B109–B114
Davies KJ (2001) Degradation of oxidized proteins by the 20S proteasome. Biochimie 83:301–310
de Duve C (1983) Lysosomes revisited. Eur J Biochem 137:391–397
Dollé ME, Snyder WK, Gossen JA, Lohman PH, Vijg J (2000) Distinct spectra of somatic mutations accumulated with age in mouse heart and small intestine. Proc Natl Acad Sci U S A 97: 8403–8408
Drummond DA, Wilke CO (2009) The evolutionary consequences of erroneous protein synthesis. Nat Rev Genet 10:715–724
Fedorova M, Bollineni RC, Hoffmann R (2014) Protein carbonylation as a major hallmark of oxidative damage: update of analytical strategies. Mass Spectrom Rev 33:79–97
Finch CE (1990) Longevity, senescence, and the genome. The University of Chicago Press, Chicago
Finkel T (2011) Signal transduction by reactive oxygen species. J Cell Biol 194:7–15
Fraga CG, Shigenaga MK, Park JW, Degan P, Ames BN (1990) Oxidative damage to DNA during aging: 8-hydroxy-2′-deoxyguanosine in rat organ DNA and urine. Proc Natl Acad Sci U S A 87:4533–4537
Fraga MF, Ballestar E, Paz MF, Ropero S, Setien F, Ballestar ML et al (2005) Epigenetic differences arise during the lifetime of monozygotic twins. Proc Natl Acad Sci U S A 102: 10604–10609
Gems D, Partridge L (2008) Stress-response hormesis and aging: "that which does not kill us makes us stronger". Cell Metab 7:200–203
Gems D, Partridge L (2013) Genetics of longevity in model organisms: debates and paradigm shifts. Annu Rev Physiol 75:621–644
Gershon H, Gershon D (1970) Detection of inactive enzyme molecules in ageing organisms. Nature 227:1214–1217
Gomez-Cabrera MC, Domenech E, Viña J (2008) Moderate exercise is an antioxidant: upregulation of antioxidant genes by training. Free Radic Biol Med 44:126–131
Goto S, Nakamura A (1997) Age-associated, oxidatively modified proteins: a critical evaluation. Age (Omaha) 20:81–89
Goto S, Radák Z (2009) Hormetic effects of reactive oxygen species by exercise: a view from animal studies for successful aging in human. Dose Response 8:68–72
Goto S, Takahashi R, Kumiyama A, Radák Z, Hayashi T, Takenouchi M et al (2001) Implications of protein degradation in aging. Ann N Y Acad Sci 928:54–64

Grasso M, Piscopo P, Confaloni A, Denti MA (2014) Circulating miRNAs as biomarkers for neurodegenerative disorders. Molecules 19:6891–6910

Grune T, Catalgol B, Licht A, Ermak G, Pickering AM, Ngo JK et al (2011) HSP70 mediates dissociation and reassociation of the 26S proteasome during adaptation to oxidative stress. Free Radic Biol Med 51:1355–1364

Harley CB, Pollard JW, Chamberlain JW, Stanners CP, Goldstein S (1980) Protein synthetic errors do not increase during aging of cultured human fibroblasts. Proc Natl Acad Sci U S A 77: 1885–1889

Harman D (1956) Aging: a theory based on free radical and radiation chemistry. J Gerontol 11: 298–300

Harman D (1968) Free radical theory of aging: effect of free radical reaction inhibitors on the mortality rate of male LAF mice. J Gerontol 23:476–482

Harman D (1972) The biologic clock: the mitochondria? J Am Geriatr Soc 20:145–147

Harman D, Harman H (2003) "I thought, thought, thought for four months in vain and suddenly the idea came" – an interview with Denham and Helen Harman. Interview by K. Kitani and G.O. Ivy. Biogerontology 4:401–412

Harris H, Watts JW (1958) Turnover of protein in a non-multiplying animal cell. Nature 181: 1582–1584

Hart RW, Setlow RB (1974) Correlation between deoxyribonucleic acid excision-repair and life-span in a number of mammalian species. Proc Natl Acad Sci U S A 71:2169–2173

Hayashi T, Goto S (1998) Age-related changes in the 20S and 26S proteasome activities in the liver of male F344 rats. Mech Ageing Dev 102:55–66

Hayflick L (2007) Biological aging is no longer an unsolved problem. Ann N Y Acad Sci 1100: 1–13

Hogness DS, Cohn M, Monod J (1955) Studies on the induced synthesis of beta-galactosidase in Escherichia coli: the kinetics and mechanism of sulfur incorporation. Biochim Biophys Acta 16:99–116

Holliday R (2006) Aging is no longer an unsolved problem in biology. Ann N Y Acad Sci 1067: 1–9

Houben A, Raes M, Houbion A, Remacle J (1984) Alteration of enzymes in ageing human fibroblasts in culture. I. Conditions for the appearance of an alteration in glucose 6-phosphate dehydrogenase. Mech Ageing Dev 25:23–34

Imashimizu M, Oshima T, Lubkowska L, Kashlev M (2013) Direct assessment of transcription fidelity by high-resolution RNA sequencing. Nucleic Acids Res 41:9090–9104

Ishigami A, Goto S (1990) Age-related change in the degradation rate of ovalbumin microinjected into mouse liver parenchymal cells. Arch Biochem Biophys 277:189–195

Ishii T, Sakurai T, Usami H, Uchida K (2005) Oxidative modification of proteasome: identification of an oxidation-sensitive subunit in 26 S proteasome. Biochemistry 44:13893–13901

Kapahi P, Chen D, Rogers AN, Katewa SD, Li PW, Thomas EL et al (2010) With TOR, less is more: a key role for the conserved nutrient-sensing TOR pathway in aging. Cell Metab 11: 453–465

Kawakami K, Nakamura A, Ishigami A, Goto S, Takahashi R (2009) Age-related difference of site-specific histone modifications in rat liver. Biogerontology 10:415–421

Kennedy EP (2001) Hitler's gift and the era of biosynthesis. J Biol Chem 276:42619–42631

Kenyon CJ (2010) The genetics of ageing. Nature 464:504–512

Khrapko K, Kraytsberg Y, de Grey AD, Vijg J, Schon EA (2006) Does premature aging of the mtDNA mutator mouse prove that mtDNA mutations are involved in natural aging? Aging Cell 5:279–282

Kirkwood TB, Finch CE (2002) Ageing: the old worm turns more slowly. Nature 419:794–795

Koga H, Kaushik S, Cuervo AM (2011) Protein homeostasis and aging: the importance of exquisite quality control. Ageing Res Rev 10:205–215

Kurz T, Terman A, Gustafsson B, Brunk UT (2008) Lysosomes in iron metabolism, ageing and apoptosis. Histochem Cell Biol 129:389–406

Lavie L, Reznick AZ, Gershon D (1982) Decreased protein and puromycinyl-peptide degradation in livers of senescent mice. Biochem J 202:47–51
Levine RL, Garland D, Oliver CN, Amici A, Climent I, Lenz AG et al (1990) Determination of carbonyl content in oxidatively modified proteins. Methods Enzymol 186:464–478
Lindop PJ, Rotblat J (1961) Shortening of life and causes of death in mice exposed to a single whole-body dose of radiation. Nature 189:645–648
Ljungquist B, Berg S, Lanke J, McClearn GE, Pedersen NL (1998) The effect of genetic factors for longevity: a comparison of identical and fraternal twins in the Swedish Twin Registry. J Gerontol A Biol Sci Med Sci 53:M441–M446
Logan A, Shabalina IG, Prime TA, Rogatti S, Kalinovich AV, Hartley RC et al (2014) In vivo levels of mitochondrial hydrogen peroxide increase with age in mtDNA mutator mice. Aging Cell 13:765–768
Luft R, Landau BR (1995) Mitochondrial medicine. J Intern Med 238:405–421
Martin GM (2012) Stochastic modulations of the pace and patterns of ageing: impacts on quasi-stochastic distributions of multiple geriatric pathologies. Mech Ageing Dev 133:107–111
Martin I, Grotewiel MS (2006) Oxidative damage and age-related functional declines. Mech Ageing Dev 127:411–423
Martin GM, Austad SN, Johnson TE (1996a) Genetic analysis of ageing: role of oxidative damage and environmental stresses. Nat Genet 13:25–34
Martin GM, Ogburn CE, Colgin LM, Gown AM, Edland SD, Monnat RJ (1996b) Somatic mutations are frequent and increase with age in human kidney epithelial cells. Hum Mol Genet 5:215–221
McCord JM, Fridovich I (1969) Superoxide dismutase. An enzymic function for erythrocuprein (hemocuprein). J Biol Chem 244:6049–6055
Medawar PB (1952) An unsolved problem of biology. Lewis, London
Medvedev ZA (1990) An attempt at a rational classification of theories of ageing. Biol Rev Camb Philos Soc 65:375–398
Medvediev ZA (1962) Ageing at the molecular level and some speculation concerning maintaining the functioning of systems for replication of specific macromolecules. Colombia University Press, New York
Michikawa Y, Mazzucchelli F, Bresolin N, Scarlato G, Attardi G (1999) Aging-dependent large accumulation of point mutations in the human mtDNA control region for replication. Science 286:774–779
Miquel J, Economos AC, Fleming J, Johnson JE (1980) Mitochondrial role in cell aging. Exp Gerontol 15:575–591
Mori N, Hiruta K, Funatsu Y, Goto S (1983) Codon recognition fidelity of ribosomes at the first and second positions does not decrease during aging. Mech Ageing Dev 22:1–10
Nakamura A, Goto S (1996) Analysis of protein carbonyls with 2,4-dinitrophenyl hydrazine and its antibodies by immunoblot in two-dimensional gel electrophoresis. J Biochem 119:768–774
Neff F, Flores-Dominguez D, Ryan DP, Horsch M, Schröder S, Adler T et al (2013) Rapamycin extends murine lifespan but has limited effects on aging. J Clin Invest 123:3272–3291
Ngo JK, Pomatto LC, Davies KJ (2013) Upregulation of the mitochondrial Lon protease allows adaptation to acute oxidative stress but dysregulation is associated with chronic stress, disease, and aging. Redox Biol 1:258–264
Ogrodnik JP, Wulf JH, Cutler RG (1975) Altered protein hypothesis of mammalian ageing processes-II. Discrimination ratio of methionine vs ethionine in the synthesis of ribosomal protein and RNA of C57BL/6J mouse liver. Exp Gerontol 10:119–136
Ono T, Cutler RG (1978) Age-dependent relaxation of gene repression: increase of endogenous murine leukemia virus-related and globin-related RNA in brain and liver of mice. Proc Natl Acad Sci U S A 75:4431–4435
Orgel LE (1963) The maintenance of the accuracy of protein synthesis and its relevance to ageing. Proc Natl Acad Sci U S A 49:517–521

Peleg S, Sananbenesi F, Zovoilis A, Burkhardt S, Bahari-Javan S, Agis-Balboa RC et al (2010) Altered histone acetylation is associated with age-dependent memory impairment in mice. Science 328:753–756
Pérez VI, Bokov A, Van Remmen H, Mele J, Ran Q, Ikeno Y et al (2009) Is the oxidative stress theory of aging dead? Biochim Biophys Acta 1790:1005–1014
Radák Z, Kaneko T, Tahara S, Nakamoto H, Pucsok J, Sasvári M et al (2001) Regular exercise improves cognitive function and decreases oxidative damage in rat brain. Neurochem Int 38: 17–23
Radak Z, Chung HY, Goto S (2005) Exercise and hormesis: oxidative stress-related adaptation for successful aging. Biogerontology 6:71–75
Ristow M (2014) Unraveling the truth about antioxidants: mitohormesis explains ROS-induced health benefits. Nat Med 20:709–711
Ristow M, Zarse K, Oberbach A, Klöting N, Birringer M, Kiehntopf M et al (2009) Antioxidants prevent health-promoting effects of physical exercise in humans. Proc Natl Acad Sci U S A 106:8665–8670
Robert L (2006) Fritz Verzár was born 120 years ago: his contribution to experimental gerontology through the collagen research as assessed after half a century. Arch Gerontol Geriatr 43:13–43
Rothstein M (1981) Posttranslational alteration o proteins. CRC Press, Boca Raton
Rubinsztein DC, Mariño G, Kroemer G (2011) Autophagy and aging. Cell 146:682–695
Sadowska-Bartosz I, Bartosz G (2014) Effect of antioxidants supplementation on aging and longevity. Biomed Res Int 2014:404680
Saez I, Vilchez D (2014) The mechanistic links between proteasome activity, aging and age-related diseases. Curr Genom 15:38–51
Safdar A, Bourgeois JM, Ogborn DI, Little JP, Hettinga BP, Akhtar M et al (2011) Endurance exercise rescues progeroid aging and induces systemic mitochondrial rejuvenation in mtDNA mutator mice. Proc Natl Acad Sci U S A 108:4135–4140
Schriner SE, Linford NJ, Martin GM, Treuting P, Ogburn CE, Emond M et al (2005) Extension of murine life span by overexpression of catalase targeted to mitochondria. Science 308: 1909–1911
Schubert U, Antón LC, Gibbs J, Norbury CC, Yewdell JW, Bennink JR (2000) Rapid degradation of a large fraction of newly synthesized proteins by proteasomes. Nature 404:770–774
Schulz TJ, Zarse K, Voigt A, Urban N, Birringer M, Ristow M (2007) Glucose restriction extends Caenorhabditis elegans life span by inducing mitochondrial respiration and increasing oxidative stress. Cell Metab 6:280–293
Selman C, Tullet JM, Wieser D, Irvine E, Lingard SJ, Choudhury AI et al (2009) Ribosomal protein S6 kinase 1 signaling regulates mammalian life span. Science 326:140–144
Shibatani T, Nazir M, Ward WF (1996) Alteration of rat liver 20S proteasome activities by age and food restriction. J Gerontol A Biol Sci Med Sci 51:B316–B322
Silber JR, Fry M, Martin GM, Loeb LA (1985) Fidelity of DNA polymerases isolated from regenerating liver chromatin of aging Mus musculus. J Biol Chem 260:1304–1310
Somel M, Khaitovich P, Bahn S, Pääbo S, Lachmann M (2006) Gene expression becomes heterogeneous with age. Curr Biol 16:R359–R360
Speakman JR, Talbot DA, Selman C, Snart S, McLaren JS, Redman P et al (2004) Uncoupled and surviving: individual mice with high metabolism have greater mitochondrial uncoupling and live longer. Aging Cell 3:87–95
Stadtman ER (1988) Protein modification in aging. J Gerontol 43:B112–B120
Stadtman ER (1993) Oxidation of free amino acids and amino acid residues in proteins by radiolysis and by metal-catalyzed reactions. Annu Rev Biochem 62:797–821
Stefani M (2004) Protein misfolding and aggregation: new examples in medicine and biology of the dark side of the protein world. Biochim Biophys Acta 1739:5–25
St-Pierre J, Buckingham JA, Roebuck SJ, Brand MD (2002) Topology of superoxide production from different sites in the mitochondrial electron transport chain. J Biol Chem 277: 44784–44790

Strehler BL (1977) Time, cells and aging. Academic, New York
Strong R, Miller RA, Astle CM, Baur JA, de Cabo R, Fernandez E et al (2013) Evaluation of resveratrol, green tea extract, curcumin, oxaloacetic acid, and medium-chain triglyceride oil on life span of genetically heterogeneous mice. J Gerontol A Biol Sci Med Sci 68:6–16
Subba Rao K, Martin GM, Loeb LA (1985) Fidelity of DNA polymerase-beta in neurons from young and very aged mice. J Neurochem 45:1273–1278
Szilard L (1959) On the nature of the aging process. Proc Natl Acad Sci U S A 45:30–45
Takahashi R, Goto S (1988) Fidelity of aminoacylation by rat-liver tyrosyl-tRNA synthetase. Effect of age. Eur J Biochem 178:381–386
Takahashi R, Goto S (1990) Alteration of aminoacyl-tRNA synthetase with age: heat-labilization of the enzyme by oxidative damage. Arch Biochem Biophys 277:228–233
Takalo M, Salminen A, Soininen H, Hiltunen M, Haapasalo A (2013) Protein aggregation and degradation mechanisms in neurodegenerative diseases. Am J Neurodegener Dis 2:1–14
Trifunovic A, Wredenberg A, Falkenberg M, Spelbrink JN, Rovio AT, Bruder CE et al (2004) Premature ageing in mice expressing defective mitochondrial DNA polymerase. Nature 429: 417–423
Van Remmen H, Ward WF, Sabia RV, Richardson A (1995) Gene expression and protein degradation. Oxford University Press, New York
Vaupel JW, Carey JR, Christensen K, Johnson TE, Yashin AI, Holm NV et al (1998) Biodemographic trajectories of longevity. Science 280:855–860
Vijg, Campisi (2008) Puzzles, promises and a cure for ageing. Nature 454:1065–1071
Vijg J, Suh Y (2013) Genome instability and aging. Annu Rev Physiol 75:645–668
Wang W, Scheffler K, Esbensen Y, Strand JM, Stewart JB, Bjørås M et al (2014) Addressing RNA integrity to determine the impact of mitochondrial DNA mutations on brain mitochondrial function with age. PLoS One 9:e96940
Wong E, Cuervo AM (2010) Integration of clearance mechanisms: the proteasome and autophagy. Cold Spring Harb Perspect Biol 2:a006734
Zahn JM, Poosala S, Owen AB, Ingram DK, Lustig A, Carter A et al (2007) AGEMAP: a gene expression database for aging in mice. PLoS Genet 3:e201

Part II
Human Longevity: Centenarianism and Progeria

Chapter 2
Centenarian Studies: An Interdisciplinary Research on Healthy Longevity

Yasumichi Arai, Michiyo Takayama, Hiroki Inagaki, Yasuyuki Gondo, Yukie Masui, and Nobuyoshi Hirose

Abstract The scope and purpose of this chapter is to summarize aims, methods, and findings of centenarian studies, mainly from our own as an interdisciplinary research. Although most of centenarians remain independent in daily living until over 90s, about 97 % of them contracted chronic diseases including hypertension (63.6 %) and bone fracture (46.4 %). The prevalence of diabetes mellitus (DM) and carotid atherosclerotic plaque were peculiarities of centenarians, which could be associated with high adiponectin levels. While conducting the Tokyo centenarian study (TCS), we found that only 20 % of them enjoyed physical and cognitive independence at the age of 100 years, this elite subpopulation were highly likely to become semisupercentenarians (over 105 years) or even supercentenarians (beyond 110 years). Therefore, we began to think that 100 years of age is not a model of longevity, but over 105 years is. We describe the preliminary results of the Japan Semisupercentenarian Study which led to our conviction that semi-supercentenarians are a more appropriate model for the study of human longevity.

Keywords Centenarian • Supercentenarian • Healthy longevity • Physical function • Inflammation • Adipose tissue • Personality

Y. Arai, M.D., Ph.D. (✉) • N. Hirose
Center for Supercentenarian Research, Keio University School of Medicine, Shinjuku, Tokyo 160-8582, Japan
e-mail: yasumich@keio.jp

M. Takayama
Center for Preventive Medicine, Keio University School of Medicine, Tokyo, Japan

H. Inagaki • Y. Masui
Tokyo Metropolitan Institute of Gerontology, Tokyo, Japan

Y. Gondo
Department of Clinical Thanatology and Geriatric Behavioral Science, Osaka University, Suita, Tokyo, Japan

© Springer Japan 2015

N. Mori, I. Mook-Jung (eds.), *Aging Mechanisms*,
DOI 10.1007/978-4-431-55763-0_2

2.1 Introduction

One of the most striking features of modern society is the steady increase in life expectancy, accompanied by the rapid growth of the "oldest old" population, such as centenarians, those who reach the age of 100 years (Oeppen and Vaupel 2002; Vaupel 2010). Centenarians are virtually delayed or even escaped from life threatening diseases such as cardiovascular disease or cancers until the very late life, thus provide models for healthy aging (Perls 1995). During the recent two decades, the number of centenarian studies has appreciably increased, expanding our knowledge of biomedical and genetic underpinnings as well as psychosocial correlates of healthy longevity. In the first part of this review, we summarize functional and biomedical characteristics of centenarians mainly based on the results from the Tokyo Centenarian Study (TCS), an interdisciplinary research on the oldest old. We propose several hypotheses on human aging based on these findings. Centenarians are generally characterized by delayed onset of age-related diseases or disabilities into their 90s; however, upon reaching the age of 100 years, substantial evidence has demonstrated that frailty (Gondo et al. 2006; Motta et al. 2005), multimorbidity (Takayama et al. 2007; Andersen-Ranberg et al. 2001), and a high rate of hospitalization (Mandawat et al. 2012) are commonplace. These observations raise an essential question regarding whether one can reach the limit of life span with good health and function.

Therefore, in the second part of this review, we will briefly describe supercentenarians, an emerging phenotype of longevity elite, with our interim results of the Japanese Semisupercentenarian Study (JSS).

2.2 Background and Goals of the Tokyo Centenarian Study

The size of the aged population is rapidly increasing around the world. Because the very elderly are vulnerable to age-related multiple chronic conditions and disability, and are at a high risk for losing independence, considerable concern has arisen about the health care spending consumed by the very old. It is obvious that explication of multidimensional structure of health and its life-course trajectory is a prerequisite for planning preventive interventions and health promotion. In this regard, centenarians provide good opportunities to investigate mechanisms of healthy aging. Our goals of the centenarian study have been: (1) to characterize the ultimate aging phenotype in humans, (2) to identify the factors associated with longevity and healthy aging, and (3) to describe adaptation to functional decline by centenarians and their families. When a human approaches the limits of life span, what happens physically, psychologically, and cognitively to them? We also want to know what factors are associated with longevity. Is genetic make-up, life style, attitude, avoiding diseases and relationship with other people important factors for longevity and healthy aging? Are these factors related to each other? We think that human aging should be considered from two aspects: aging as an organism (biologic aging), and aging in a society. As humans age, functional decline is inevitable, though the period, rate and impact of decline varies. Older people must adapt to their functional decline and the resulting inconveniences, and the family or the people close to them must also understand and adapt to aging-related declines. This aspect is important in human aging. We are trying to find answers to these questions.

Between July 2000 and May 2002, a total of 304 Japanese centenarians (66 men, 238 women) living in the 23 wards of metropolitan Tokyo participated in a survey in which they were visited by Tokyo Centenarian Study staff members (Gondo et al. 2006; Takayama et al. 2007). We randomly chose centenarians from the basic registry of residence and sent a letter inviting participation to 1,194 centenarians, accounting for 66.9 % of an estimated 1,785 centenarians living in this area during the study period. A total of 513 (43.0 %) individuals agreed to participate in mailed survey. Among them, 304 individuals (25.5 % of the recruiting letter recipients) participated in the visit survey for medical examination. The remainder of the participants (209 or 11.7 % of the letter recipients) participated in a mail survey without a visit by study staff. Women outnumbered men in our sample 3.5–1, which was not significantly different from the ratio for the total centenarian population in this area (3.9–1). Of the participants ($N = 513$), most of the centenarians (317: 61.8 %) lived with their families, 184 (35.9 %) were institutionalized, and 12 (2.3 %) lived alone.

A geriatrician, a psychologist, and a nurse performed the medical, cognitive and psychological evaluations and phlebotomy. After completing a medical examination and providing a blood sample, the psychologist conducted a cognitive assessment and psychological test battery. The activities of daily living (ADL) were evaluated using the Barthel Index. Psychologists assessed cognitive function using the Mini-Mental State Examination (MMSE) at the time of the visit. Anxiety related personality traits were evaluated using the NEO-FFI. An informant version of the NEO was also administered to an informant who knew the subject well. The Barthel Index and MMSE are commonly used in centenarian studies to evaluate physical function and cognitive status, respectively.

2.3 Function of Centenarians

2.3.1 Physical and Cognitive Function

Maintaining physical and cognitive function is indispensable for independent life at old age. The ADL level was categorized according to the total score using the criteria of Silver et al. (2001) and the results appear in Fig. 2.1.

Out of 105 independent participants, only 33 (6.4 %) were fully independent (Barthel index = 100). Nearly 80 % of centenarians need care in daily living, and

Fig. 2.1 Activities of daily living of centenarians living in Tokyo. ADL was assessed with Barthel index

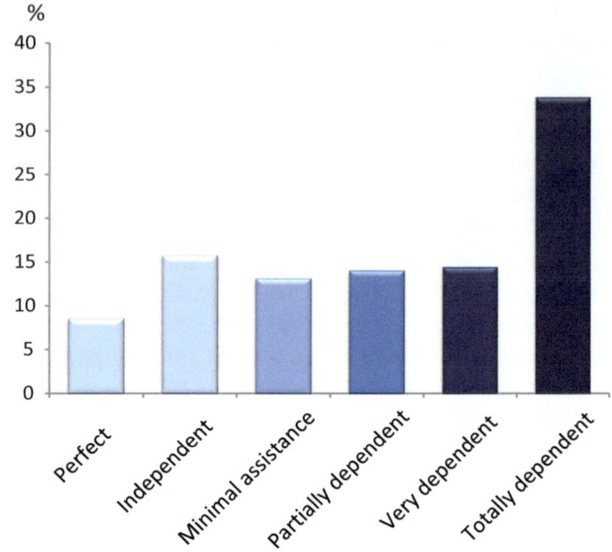

40 % were totally dependent. The mean Barthel score for men was higher than for women (54.3 ± 35.2 vs. 34.4 ± 32.7, respectively; $p < 0.01$).

Sensory dysfunction is deemed to contribute physical disability in the oldest old. In TCS, 172 (33.5 %) and 113 (22.0 %) participants, respectively, had "no problem" with vision and hearing. The others had moderate to severe problems with these senses, but only 1.9 % (n = 10) were blind and only 2.1 % were deaf (n = 11).

2.3.2 *Cognitive Function*

The 304 visit survey participants underwent a battery of cognitive tests in order to construct a clinical dementia rating (CDR) score and MMSE. The results of this testing is summarized in Fig. 2.2. The 304 visit survey participants were grouped into four categories according to their functional status and the results of this categorization are also shown in Fig. 2.2.

Women were more likely than men to have dementia ($p < 0.01$). The MMSE score of the men (16.1 ± 8.9) was generally higher than that of the women (11.5 ± 8.3).

2.4 Medical Histories of Centenarians

The present and past diseases of centenarians are summarized in Fig. 2.3.

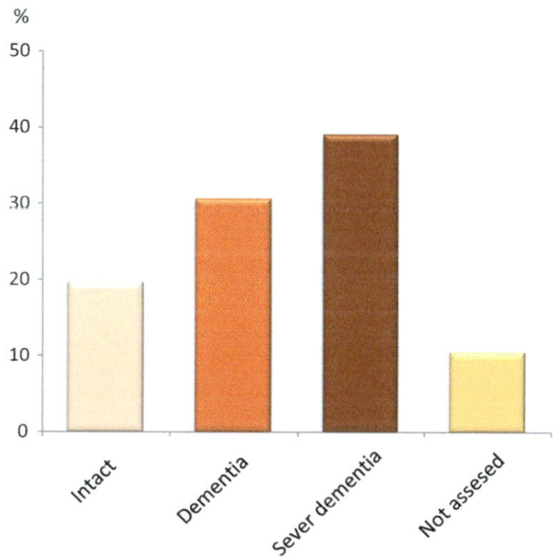

Fig. 2.2 Cognitive function of centenarians living in Tokyo. Cognitive function was assessed with clinical dementia rating and mini-mental state examination

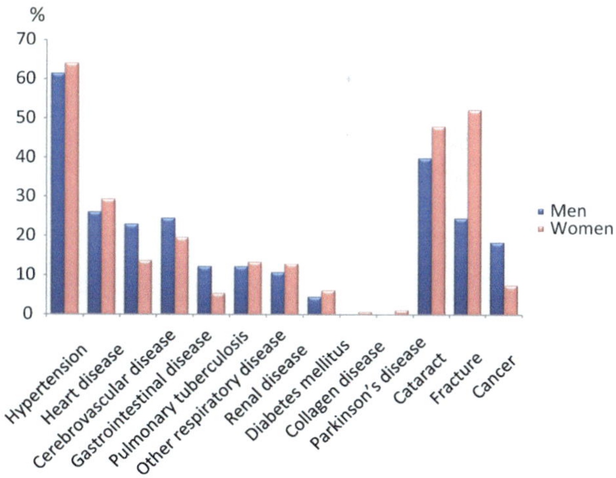

Fig. 2.3 Medical history of centenarians living in Tokyo

About 97 % of the centenarians contracted chronic diseases (Takayama et al. 2007). The most frequent illness was hypertension (63.6 % in all individuals) followed by cataract and bone fracture (46.4 % in all individuals). The prevalence of heart disease (heart failure, angina, MI) and cerebrovascular disease were 28.8 % and 15.9 %, respectively. The prevalence of diabetes mellitus (DM) was 6.0 %. The Ministry of Health, Labor and Welfare conducted a nation-wide survey of the prevalence of circulatory disease in Japanese adults (over 30 years old) in 2000. According to the survey, the prevalence of DM was 15.3 % people in their 60s, and 14.7 % for those age 70 years and older. These data show a substantially lower prevalence of DM in Japanese centenarians. The prevalence of cancer was 9.9 %. Since the mortality rate due to cancer was 22.1 % in male octogenarians and 14.5 %

Table 2.1 Morbidity profile of centenarians

Morbidity profile	Total	Men	Women	Barthel index	p	CDR	p
Survivors	16.2 %	15.4 %	16.5 %	48.6 ± 38.5	0.176	1.84 ± 1.72	0.632
Delayers	48.7 %	43.1 %	50.2 %	39.1 ± 32.5		1.79 ± 1.61	
Escapers	18.2 %	23.1 %	16.9 %	48.4 ± 37.6		1.66 ± 1.78	
Unclassified	16.9 %	18.5 %	16.5 %	NA		NA	

CDR clinical dementia rating, *NA* not applicable

Table 2.2 Correlations between major six diseases and physical and/or cognitive function of centenarians

	Correlation coefficients (r)	
Major 6 diseases	Barthel index	CDR
Stroke	−0.167**	0.157**
Heart disease	0.048	−0.086
Hypertension	0.203**	−0.194**
Diabetes mellitus	−0.098	0.068
Cancer	0.107	−0.092
Fracture	−0.232**	0.135*

Spearman correlation test. *$p < 0.05$, **$p < 0.01$

in female octogenarians, the prevalence of at least clinically evident cancer in centenarians is also relatively low.

Everts et al. (2003) showed that centenarians can be divided into three groups as survivor, delayer, and escaper according to when in their lives they incurred age-associated illnesses. In a similar way, We divided Japanese centenarians into three morbidity profiles; survivors, delayers, and escapers, according to the age of onset for six major age-related lethal illnesses (stroke, heart disease, hypertension, DM, cancer, and fracture). The definition of survivors, delayers and escapers are those with the age of onset <80 years, those with the age of onset 80–99 years, and those with no history of any of the diseases prior to age 100, respectively. Table 2.1 shows the prevalence of three categories and the level of ADL (as indicated by Barthel Index score) and cognitive function in each category. The prevalences of survivors, delayers and escapers were 16.2 %, 48.7 %, and 18.2 %, respectively. The prevalence of escapers in men was higher than in women. The level of ADL and cognitive function was not significantly different between the three categories (for ADL, $p = 0.176$. for MMSE, $p = 0.632$).

The effect of six major illnesses on ADL and cognitive function was analyzed by correlation coefficients (Table 2.2).

The significant negative effect of stroke and fracture on both Barthel index and CDR was apparent. Surprisingly, the undesirable effect of fracture did not depend on location. The level of ADL and cognitive function was not different between hip/femur, vertebral, and other location (data not shown), indicating that fracture has detrimental effects on ADL. This result clearly shows that the prevention of stroke and bone fracture in any location is important in maintaining the autonomy in the oldest old.

2.5 Molecular Basis of Healthy Longevity

2.5.1 Low Susceptibility of Atherosclerosis in Centenarians

There is a famous saying by Prof. William Osler "A man is as old as his arteries". The prevalence of carotid artery atherosclerosis was assessed by B mode ultrasonography (Homma et al. 2001). Two hundred and eighty-nine controls (age range: 21–98 years) and 30 centenarians were included in this study. The thickness of the intima-media complex (IMT) and the prevalence of plaque in the carotid artery were assessed. We found that the IMT increases linearly with age. The equation of correlation between IMT and age was IMT (mm) $= 0.009 \times \times \text{age} + 0.116$ ($r = 0.83$). However, the prevalence of plaque showed a different trend. Plaque begins to appear after 50 years of age and then increases in prevalence linearly with the highest prevalence observed among nonagenarians. However, the prevalence in centenarians was significantly lower than that in nonagenarians and similar to that of octogenarians (60 % in centenarians, 83.3 % in nonagenarians, and 57.6 % in octogenarians). The regression of plaque does not usually take place. We hypothesize that the nonagenarians with severe atherosclerosis may die before 100 years of age and that a low prevalence of clinically demonstrable atherosclerosis is critical to attain 100 years of age. The reason why centenarians have a low prevalence of atherosclerosis is not clear. Their genetic make-up and/or their lifestyle contribute to a low prevalence of atherosclerosis. The genome of centenarians is being analyzed in several laboratories. Within several years, we will have more information about how the genetic background and life style contributes to the low prevalence of atherosclerosis.

2.5.2 Aging and Inflammation: Two Hypotheses

The biomedical findings in centenarians are shown in Table 2.3. The characteristics were surprising and included (1) a low concentration of albumin, (2) low total cholesterol and HDL-cholesterol concentrations, (3) elevated CRP and IL-6 levels (4) elevated homocysteine levels, (5) a high thrombin-antithrombin complex (TAT) concentration, and (6) anemia. These results show that centenarians are malnourished, and in proinflammatory and prothrombotic state. We wanted to determine the relationship between these parameters and certain covariates. First, the effect of nutrition was analyzed. Centenarians were divided into two groups, according to the albumin concentration (Table 2.4). Well-nourished centenarians (high albumin group) had high concentrations of cholesterol, HDL-C, high hemoglobin and red blood cell counts. The concentrations of CRP and IL-6 were lower in well-nourished centenarians. The Barthel index and MMSE scores were significantly higher in well-nourished centenarians. These results suggest that nutritional status plays an important role in maintaining high ADL and cognitive function. The other

Table 2.3 Biochemical characteristics of centenarians

	Unit	n	Centenarian n = 273 (f = 199, m = 72)	Control n = 3698 (f = 1316, m = 2382)
Age		273	100.9 ± 1.5	55.4 ± 10.8
Nutritional parameter				
BMI		73	19.2 ± 3.3	23.0 ± 2.9
Albumin	g/dl	264	3.6 ± 0.4	4.7 ± 0.7
Serum lipid				
Total cholesterol	mg/dl	273	164.0 ± 32.8	212.1 ± 22.7
HDL cholesterol	mg/dl	273	51.1 ± 14.0	57.5 ± 15.2
Blood sugar				
BS	mg/dl	37	116.1 ± 27.3	<110
HbA1C	g/dl	183	5.4 ± 0.7	<5.8
Homocysteine				
Homocysteine	nm/ml	93	15.0 ± 5.7	3–14
Coagulation factors				
TAT	ng/dl	67	9.0 ± 9.0	<3
Endothelial function				
von Willbrandt factor	%	70	188.4 ± 71.7	60–170
Thrombomodulin	U/ml	53	5.0 ± 1.3	1.8–3.9
Inflammation				
CRP	mg/dl	185	0.639 ± 1.54	<0.3
IL-6	pg/ml	56	9.1 ± 9.6	<2.7
Peripheral blood				
RBC	$\times 10^6$/ml	267	356.7 ± 51.8	466.7 ± 41.4
Hb	g/dl	267	11.1 ± 1.6	14.4 ± 1.5
WBC	/μl	266	5416 ± 1508	5570 ± 1510
MMSE & ADL				
Cognitive function (MMSE)		104	15.0 ± 6.9	
ADL (Barthel)		116	42.9 ± 34.4	

interesting relationship is that between nutrition and inflammation. A high nutritional status was associated with low levels of the inflammatory markers. It is well known that proinflammatory cytokines induce malnutrition (cytokine-induced malnutrition). We speculate that the cause of malnutrition observed in centenarians is due partly to inflammation.

Next, the effect of homocysteine on the coagulation system was analyzed. The concentration of homocysteine is high compared to the reference range of Keio University Hospital. The centenarians were divided into two groups according to the homocysteine concentration: high and low homocysteine groups (data not shown). The concentrations of TAT, thrombomodulin and E-selectin in the high homocysteine group were significantly higher than that in the low homocysteine

Table 2.4 Nutritional status and biomarkers of centenarians

	unit	n (f/m)	Well nourished	Malnourished	p value
Nutritional parameters					
BMI		29/38	20.3 ± 10.0	18.7 ± 10.3	0.047
Albumin	g/dl	144/103	3.91 ± 0.23	3.18 ± 0.27	<0.001
Lipid parameter					
Total cholesterol	mg/dl	144/103	174.3 ± 31.6	152.4 ± 30.0	<0.001
HDL cholesterol	mg/dl	144/103	56.4 ± 14.5	45.2 ± 10.9	<0.001
Inflammatory parameter					
CRP	mg/dl	109/59	0.29 ± 0.47	0.89 ± 1.46	<0.001
IL-6	ng/ml	15/34	4.4 ± 2.8	11.8 ± 11.5	0.018
Peripheral blood					
RBC	$\times 10^6$/ml	144/99	369 ± 48	339 ± 51	<0.001
Hb	g/dl	144/99	11.5 ± 1.6	10.5 ± 1.6	<0.001
ADL (Barthel index)		70/30	54.6 ± 33.4	31.8 ± 28.5	0.020
Cognitive function (MMSE)		74/30	16.1 ± 7.0	12.5 ± 6.0	0.002

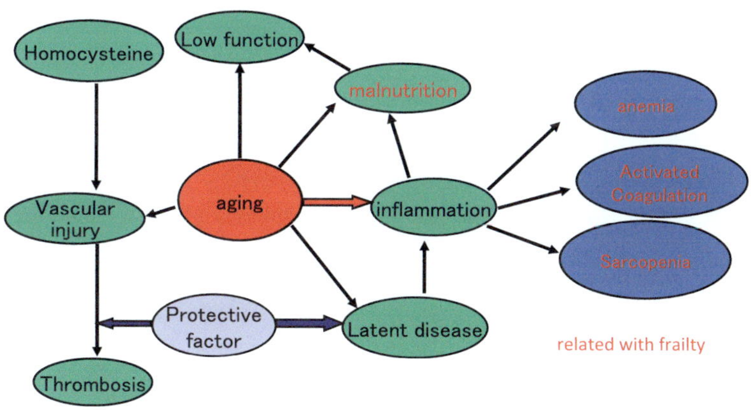

Fig. 2.4 Aging-inflammation hypothesis. Centenarians could have potent protective factors to counteract aging-associated inflammation and lessen disability

group, suggesting that the coagulation system is activated and endothelial cells are damaged in the high homocysteine group. It is known that the increased homocysteine concentrations are associated with prothrombotic conditions in centenarians.

From these results, we proposed two hypotheses: the aging-inflammation hypothesis and the presence of potent protective factors hypothesis (Fig. 2.4) (Hirose et al. 2004). Aging is associated with a proinflammatory state, which in turn is associated with impaired nutritional status. Also, nutritional status has a profound effect on ADL and cognitive function. The mechanism by which inflammation could lead to poor nutritional status is partly explained by proinflammatory cytokines, such as IL-6 and TNFα (Scoub 1994). However, why aging leads to a proinflammatory state is not clear. One possibility is the association between visceral and intramuscular fat accumulation with age. This fat, in turn, is a potent source of atherogenic cytokines and other inflammatory substances. This observation is expanded upon in the following section. Centenarians could have clinically non-evident diseases such as cancer and chronic infection. These diseases have been shown to cause an inflammatory reaction. Several investigators have reported that aging itself contributes to a proinflammatory state. Claudio Francheschi, who studies immune function in centenarians, proposed the "inflamm-aging" hypothesis (Franchschi et al. 2000). The aging-inflammation hypothesis is based on the analysis of the nutritional status of centenarians. If this hypothesis is correct, aging-associated phenotypes such as decline of ADL, cognitive function, and nutritional status can be modified by the suppression of excessive inflammatory reactions. Regulation of inflammatory reactions in as people age will be a fascinating target for future research.

Even though centenarians have many pathologic biomedical findings, they maintain their functional independence for the vast majority of their lives (Hitt et al. 1999). It is possible that potent protective factors are necessary for individuals to become centenarians in order to overcome many pathologic conditions observed in centenarians.

2.5.3 *Adipose Tissue Function and Longevity*

Advancing age is frequently associated with impaired glucose tolerance (IGT), insulin resistance and the development of diabetes, predominantly type 2 Diabetes. Current advances in obesity research has revealed that adipose tissue synthesizes a number of bioactive molecules termed as adipokines, and dysregulation of adipokines such as tumor-necrosis factor-α (TNF-α) and plasminogen activator inhibitor type 1 (PAI-1) may contribute to the development of insulin resistance and the metabolic syndrome (MetS). Epidemiological evidence also suggested crucial roles of adipokines on the progression of MetS not only in the middle age, but in the very old. Intriguingly, there is increasing evidence that preservation of insulin sensitivity and a low prevalence of MetS and DM is one of the striking physiological characteristics of centenarians (Barbieri et al. 2001; Bik et al. 2006). As far as insulin sensitivity and adipocytokines are concerned, a great deal of interest has been generated by the discovery of adiponectin, which is shown to exert anti-diabetic, anti-atherogenic, and anti-inflammatory effects in rodents and humans (Hotta et al. 2000; Trujillo and Scherer 2006). In cross-sectional studies, plasma adiponectin concentrations were significantly lower in obese individuals and in those having diabetes, MetS, and CVD, being inversely associated with body adiposity and insulin resistance (Stefan et al. 2002). Recently, a possible association between adiponectin and extended lifespan has been indicated in several animal models (Wang et al. 2006; Kloting and Bluher 2005). Based on these findings, we focused on adiponectin as a protective factor in centenarians against age-related metabolic alterations and vascular injury. We measured the adiponectin concentration in the plasma of 66 female centenarians and a younger control group (Arai et al. 2006). Since the adiponectin concentration is regulated by adiposity, the BMI of the young control group was matched to that of the centenarian group. The mean concentration of plasma adiponectin in female centenarians was almost twice as high as that in the BMI-matched female control group (20.3 ± 7.4 vs.10.8 ± 3.9, respectively, $p < 0.001$). In contrast, the serum leptin concentration in female centenarians was significantly lower compared to the control group. In centenarians, the adiponectin concentration was negatively correlated with BMI ($r = -0.245$, $p < 0.05$) as well as with the leptin concentration ($r = -0.346$, $p < 0.01$). Furthermore, it showed strong negative correlations with HbA1c ($r = -0.311$, $p < 0.01$) and CRP concentrations ($r = -0.316$, $p < 0.01$) and a positive correlation with HDL-C concentration ($r = 0.270$, $p < 0.05$). Adiponectin concentration was also negatively correlated with E-selectin concentration ($r = -0.261$, $p < 0.05$). These results raised a possibility that adiponectin might be a protective factor against age-related insulin resistance and atherosclerosis.

The classical perception of adipose tissue as an inert energy reservoir has been replaced over the last years by the notion that adipose tissue has a central role in producing a large number of bioactive substances called adipokines (Fig. 2.5) (Arai et al. 2011). During the development of obesity, adipose endocrine function is markedly changed. Increased levels of free fatty acids (FFA) and TNF-α were

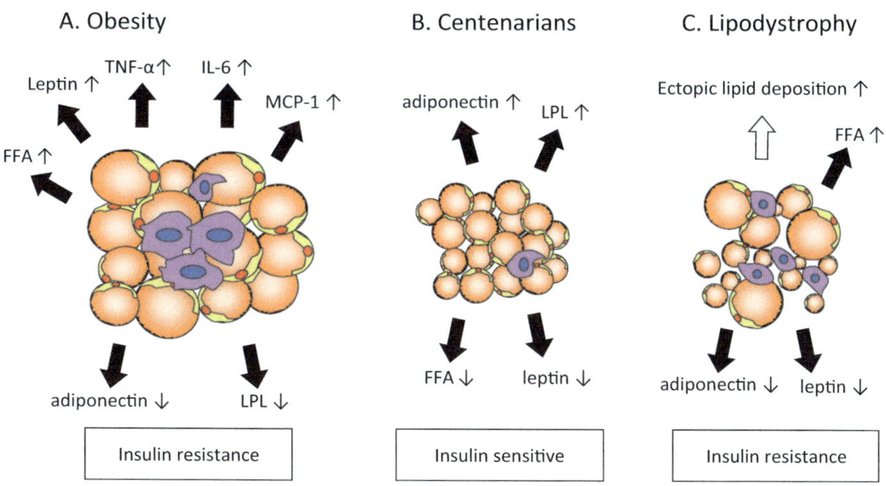

Modified from Arai et al. Adipokines and aging. J Atheroscler Thromb 2011; 18: 545-550.

Fig. 2.5 Hypothetical link between adipokines, insulin sensitivity, and longevity

released from adipocytes/macrophages with reciprocal decline in adiponectin, leading to insulin resistance. In contrast, adipokines secreted by small size adipocytes such as adiponectin, ameliorate insulin resistance. Lipodystrophy is a disease caused by atrophy of adipose tissue, in which profound reduction in leptin and adiponectin production is associated with severe insulin resistance and unfavorable to longevity. Taken together, well-functioning adipose tissue may have favorable effects on longevity. To examine this hypothesis, we aimed to describe adipose tissue function by profiling a series of adipokines including adiponectin, leptin, and TNF-α, and investigated the association between adipose tissue function and mortality beyond 100 years of age (Arai et al. 2008). We followed a cohort of 252 centenarians for a period of 6.2 years and demonstrated that cumulative dysregulation of multiple adipokines constitutes a strong marker of poor prognosis among centenarians, independent of conventional risk factors such as low serum albumin, interleukin-6, and HDL-C concentrations. Moreover, a graded relationship has been shown to exist between the extent of impairment of adipose endocrine function and the decline in the many key pathways responsible for health maintenance, including those for physical and cognitive function, IGF-1 axis, HDL metabolism, and nutrient synthesis and hepatic function (Arai et al. 2008). These findings suggest that adipose tissue function could be linked with multiple physiological functions, which are indispensable for survival at extremely old age. Therefore, understanding the interaction of these metabolic pathways is likely to be important to promote healthy aging and longevity.

2.6 Personality and Longevity

Epidemiological association between personality and longevity has been demonstrated so far, however, data on centenarians are quite limited. Therefore, we assessed personalities of centenarian by the NEO-FFI. The NEO-FFI assesses five dimensions of personality: neuroticism, extraversion, openness, agreeableness, and conscientiousness (Costa and McCrae 1985). We found significant traits difference in centenarians. Male centenarian showed higher openness in comparison to excepted score which consider age related change from sexagenarian to octogenarian. In female centenarians, a high score of openness, extraversion and conscientiousness were observed (Masui et al. 2006). Recent epidemiological researches reported personality trait is a significant predictor of longevity, therefore, we are currently surveying pathway which link personality longevity relationship in younger age cohort.

We also reported interesting finding which might link between personality-longevity relationships. It has been reported that negative affects such as depression and anxiety, are associated with morbidity and mortality. The serotonin transporter (5HTT) gene-linked polymorphism region has been linked with negative affect and personality. The two most common alleles are a 44-bp insertion (l allele) or deletion (s allele) in 5HTT promoter region (Lesch et al. 1996). The s allele reduces the efficiency of transcription of the gene, resulting in decreased serotonin transporter expression and serotonin uptake. Possession of the s allele has been associated with higher levels of neuroticism, harm avoidance, and depression, including geriatric depressive symptoms. We hypothesize that the 5HTT gene polymorphism could have association with human longevity. We compared polymorphic variations of the 5HTT gene between 265 Japanese centenarians and younger control subjects (Gondo et al. 2005). In addition, we evaluated the relationships between the 5HTT genotype and the physical, cognitive, and biologic status of centenarians, as indicated by the Barthel index, the MMSE, and serum albumin concentration.

The distribution of genotypes for 5HTT differed significantly between centenarians and younger control subjects. Comparison between genotypes indicated that the l/l genotype was more prevalent than the l/s or s/s genotypes in centenarians and this difference was observed only for women. The effects of the 5HTT genotype on NEO-FFI, Barthel Index, MMSE, and serum albumin also were analyzed in centenarians (Gondo et al. 2005). A significant main effect of genotype was observed only for the serum albumin concentration ($p < 0.05$). Further multiple comparisons indicated that serum albumin concentrations were higher in association with the l/l genotype than with the s/s genotype (in men: l/l, 4.1 ± 0.2 g/dl, s/s, 3.6 ± 0.2 g/dl; in women: l/l, 3.8 ± 0.4 g/dl, s/s, 3.6 ± 0.4 g/dl). Thus, the l allele is associated with a longevity advantage, especially in women. Although we do not know why 5HTT polymorphism is related to longevity, there are several possibilities. There is much psychological stress in human life. Centenarians living in Tokyo have experienced catastrophic life events including a severe earthquake, two world wars, and personal life events such as bereavement resulting from the deaths of spouses and children. The negative impacts of these life events on the affective state are stronger in individuals with the s allele. Next possibility is the susceptibility to becoming depressed after experiencing negative life events. The trend to become depressed is higher in individuals with s alleles than in those with l alleles. Centenarians who survived twentieth century might need strong stress resilience based on genetic background. We believe that genes associated with personality and behavior will be a fascinating future research target (Caspi et al. 2003).

2.7 Supercentenarians: An Emerging Model of Healthy Longevity

Centenarian studies have been conducted for over three decades. While conducting TCS (the Tokyo Centenarian Study), we began to think that centenarians are no longer an ideal model of human longevity. One of the reasons is the rapid increase in the number of centenarians. In Japan, the number of centenarians in 2013 was about 54,397 – 9 times higher than that in 1990 – and this number is expected to increase in the future (Demographic trends in Japan and world 2005); this trend is observed globally. The next reason is the functional decline; a decline in the physical function of centenarians was reported to have occurred between 1976 and 1994 in Okinawa (Suzuki et al. 1995). As noted in Sect. 2.3, centenarians are a heterogeneous group, from bed-ridden to having complete autonomy, with and without a history of familial longevity and coming from diverse backgrounds. Centenarians cannot be considered as a single entity. Therefore many investigators are employing new strategies including classification of centenarians. Tom Perl's, the PI of the New England Centenarian Study is focusing on the families of long lived individuals (Perls et al. 2002). The research target of Nil Barzilai and colleagues' Ashkenazi Jewish Centenarian Study is the founder population, Ashkenazi Jews, a genetically and culturally homogeneous population (Barzilai et al. 2003). European Union researchers are conducting the Genetics of healthy aging study (GEHA) to identify genetic factors associated with healthy aging through an exhausting recruitment of long-lived sibling pairs (Abott 2004).

A few studies have targeted to supercentenarians, individuals who have reached their 110th birthday. The number of validated supercentenarians worldwide is extremely rare, and has been monitored by an international collaborative effort involving the International Database on Longevity: the database includes a total of 12 supercentenarians in Sweden, 2 in Denmark, 49 in France, 78 in Japan, and 341 in the United States (Maier et al. 2010). According to the cohort life table and census in Japan, we calculated the provability become supercentenarians (Fig. 2.6). As expected, the chance to survive to 110 years of age is quite different between men and women. In men, the provability to become supercentenarians was about 1 per 300,000–400,000 between 1995 and 2010, which suggests the chance to become supercentenarian may be stagnated in men. In contrast, the chance for women becomes higher and higher in recent years, but still only 1 in 20,000 people within the same birth cohort reaches 110. Because of the extraordinarily low probability of reaching that age even in current low-mortality countries,

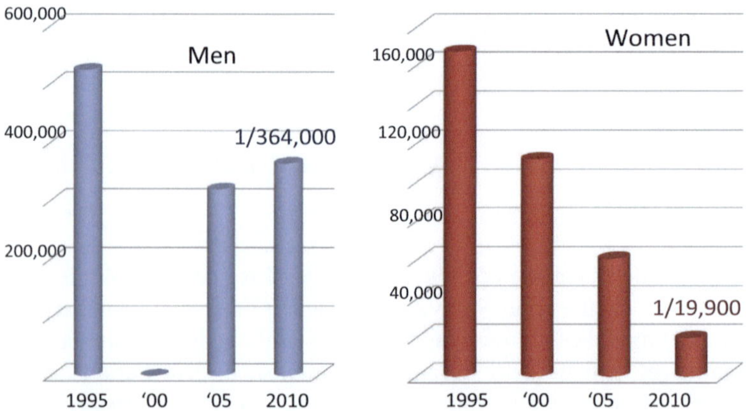

Fig. 2.6 Probability to become Supercentenarians (1995–2010)

epidemiological observations of supercentenarians are sparse and conflicting. The early phase of the New England Centenarians Study reported that of 32 supercentenarians enrolled, 41 % required minimal or no assistance in ADL (Schoenhofen et al. 2006). In contrast, the Okinawan Centenarian Study reported that the majority of 12 age-validated supercentenarians were physically independent at the age of 100 years; however, they rapidly lost functionality between 105 and 109 years, and consequently none were independent in basic ADL at the age of 110 years (Willcox et al. 2008). A precise description of what happens to individuals' health and independence at the tail end of the human survival curve has assumed an important role in accurately estimating future disability trends and health care demands, which has enormous social, economic, and medical implications in our aging society. To address this issue, we started Japanese Semi-supercentenarian study (JSS) (Arai et al. 2014) in which we conduct a nationwide survey of semisupercentenarians involving direct examination using a standardized procedure.

2.8 Preliminary Findings from Japan Semisupercentenarian Study

Given an extremely low probability of supercentenarians such as 1 per 1.5 million of general population, identification of them from basic registry database is extraordinary difficult, if not impossible. Therefore, we have targeted recruitment of semisupercentenarians, individuals aged 105 years or older, of whom provability is approximately 33 times as high as supercentenarians. Since 1963, a list of centenarians has been, and this list was linked with the resident registration system and the periodic census. According to the centenarian list compiled annually by the Ministry of Health, Welfare and Labour, 849 semi-supercentenarians were living in Japan in 2002, including 23 supercentenarians. They had been successively recorded on the annual centenarians list since 1997 or earlier. We identified 543 individuals (82 males and 461 females) among the 849 semi-supercentenarians and sent all of them an invitation letter for a home visit examination (Arai et al. 2014). As a result, 135 (115 females and 20 males) agreed to participate in our visiting survey. Because the list was discontinued in 2002, our subsequent recruitment strategy has relied on responses to local governments and nursing homes in the whole country, and direct inquires by our research team. Consequently, a total of 429 centenarians (90.2 % older than 105 years) were enrolled in the JSS by the end of November 2011.

Our first aim of JSS was to elucidate whether extended longevity beyond 100 years is accompanied with elongation of disabling process or remarkable compression of disability. We hypothesized that maintaining independence and delayed onset of disability is the key determinant of extended longevity beyond 100 years of age, because functional capability might highly correlate with mortality even at the extreme limit of lifespan. To test our hypothesis, we classified centenarians into three groups according to age at enrollment: Younger centenarians: individuals who died between 100 and 104 years of age.

Semi-supercentenarians: individuals who died between 105 and 109 years of age.

Supercentenarians: individuals who survived beyond 110 years old.

Then, we compared ADL levels among the three centenarian categories. Unsurprisingly, we found that supercentenarians are virtually characterized by marked postponement of the age-related debilitating process and maintenance of physical independence for an extraordinarily long period. We also followed up three centenarian categories for mortality and found that functional capability correlates significantly with mortality, and found that the most dependent fraction of individuals die first, and those with better fitness remain alive. The finding suggests that maintaining functional capability and delayed onset of disability is the key feature of supercentenarians. Explication of the biological and genetic architecture underpinning this phenotype will further our understanding of the ultimate aging process, which has assumed a central position in terms of refining preventive intervention and health promotion at advanced ages. Finally, our results highlight the potential of our longitudinal database of centenarians and supercentenarians to provide invaluable opportunities to test the central hypotheses of human aging.

2.9 Summary and Conclusion

We have summarized the results of our centenarian study to-date. We propose a new framework of phenotypic classification based on functional status. Several hypotheses also are proposed based on the biomedical characteristics of centenarians: the aging-inflammation hypothesis and adipose tissue as pro-longevity hypothesis. Despite many pathologic disadvantages, centenarians can live active lives or they have at least done so for the majority of their very long lives. Therefore, we hypothesize that potent protective factors are present in centenarians. We think that identification of protective factors is one of the purposes of centenarian studies. One protective factor is likely adiponectin. Adipose tissue, a main target of insulin, secretes adipokines, which modulate insulin sensitivity. Adipose tissue is therefore a promising and important target for further research. We also mentioned effect of personality on longevity and introduced a possibility that the relationship might be based on gene polymorphisms (5HTT gene). This issue should be investigated more thoroughly.

Our final and important hypothesis is that supercentenarians, not 100 years old, are perhaps a more fruitful model of human longevity than people who reach 100 years of age. To support this hypothesis, we demonstrated that supercentenarians are virtually characterized by marked postponement of the age-related debilitating process and maintenance of physical independence for an extraordinarily long period.

We do not describe the genetic factors associated with longevity. As mentioned previously, the research identifying the longevity genes is being conducted in several studies including our own, and these studies are in a rapid progress. Within several years, we will know whether longevity genes exist and what they are. These hypotheses and the results of centenarian studies will induce the fusion of human aging science with basic aging science, which will help us to understand how to achieve healthy aging.

Finally we want to stress that the centenarian studies are rewarding. Centenarians are incredibly interesting people and their life stories are fascinating and memorable. We hope that young investigators will be inspired by our centenarian studies and those of our international colleagues and go into this fascinating area of research.

Acknowledgement We sincerely appreciate centenarians, their family and care-givers. The data presented in this chapter owed their kind and willing corporation. We also express sincere gratitude to Ms. Yukiko Abe and Ms. Miho Shimura for their arrangement of field works.

References

Demographic Trends in Japan and World (2005) Tokyo: National Institute of Population and Social Security Research, Health and Welfare Statistics Association

Abott A (2004) Ageing: growing old gracefully. Nature 428:116–118

Andersen-Ranberg K, Schroll M, Jeune B (2001) Healthy centenarians do not exist, but autonomous centenarians do: a population-based study of morbidity among Danish centenarians. J Am Geriatr Soc 49:900–908

Arai Y, Nakazawa S, Kojima T et al (2006) High adiponectin concentration and its role for longevity in Japanese female centenarians. Geriatr Gerontorol Int 6:32–39

Arai Y, Takayama M, Gondo Y et al (2008) Adipose endocrine function, insulin-like growth factor-1 axis, and exceptional survival beyond 100 years of age. J Gerontol A Biol Sci Med Sci 63A:1209–1218

Arai Y, Takayama M, Abe Y, Hirose N (2011) Adipokines and aging. J Atheroscl Thromb 18:545–550

Arai Y, Inagaki H, Takayama M et al (2014) Physical independence and mortality at the extreme limit of Lifespan: supercentenarians study in Japan. J Gerontol A Biol Sci Med Sci 69:486–494

Barbieri M, Rizzo MR, Manzela D, Paolisso G (2001) Age-related insulin resistance: is it an obligatory finding? The lesson from healthy centenarians. Diabetes Metab Res Rev 17:19–26

Barzilai N, Atzmon G, Schechter C, Schaefer EJ, Cupples AL, Lipton R, Cheng S, Shuldiner AR (2003) Unique lipoprotein phenotype and genotype associated with exceptional longevity. JAMA 290:2030–2040

Bik W, Baranowska-Bik A, Wolinska-Witort E et al (2006) The relationship between adiponectin levels and metabolic status in centenarian, early elderly, young and obese women. Neuro Endocrinol Lett 27:493–500

Caspi A, Sugden K, Moffitt TE et al (2003) Influence of life stress on depression: moderation by a polymorphism in the 5-HTT gene. Science 301:386–389

Costa PTJ, McCrae RR (1985) The NEO personality inventory manual. Psychological Assessment Resources, Lutz

Everts J, Lawler E, Bogan H, Pearls T (2003) Morbidity profiles of centenarians: survivors, delayers, and escapers. J Gerontol Biol Sci Med Sci 58A:232–237
Franchschi C, Bonafe M, Valensin S et al (2000) Inflamm-aging. An evolutionary perspective on immunosenescence. Molecular and cellular gerontology (Toussaint O, Osiewacz HD, Lithgow GJ, Brack C eds). Ann N Y Acad Sci 908: 244–254.
Gondo Y, Hirose N, Arai Y et al (2005) Contribution of affect-associated gene to human longevity: prevalence of the long allele genotype of the serotonin transporter-linked gene in Japanese centenarians. Mech Ageing Dev 126:1178–1184
Gondo Y, Hirose N, Arai Y et al (2006) Functional status of centenarians in Tokyo, Japan: developing better phenotypes of exceptional longevity. J Gerontol A Biol Sci Med Sci 61:305–310
Hirose N, Arai Y, Gondo Y et al (2004) Tokyo centenarian study: aging inflammation hypothesis. GGI 4:S182–S185
Hitt R, Young-Xu Y, Silver M, Perls T (1999) Centenarians: the older you get, the healthier you have been. Lancet 354:652–652
Homma S, Hirose N, Ishida H et al (2001) Carotid plaque and intima-media thickness assessed by B-mode ultrasonography in subjects ranging from young adults to centenarians. Stroke 32:830–835
Hotta K, Funahashi T, Arita Y et al (2000) Plasma concentrations of a novel, adipose-specific protein, adiponectin, in type 2 diabetic patients. Arterioscler Thromb Vasc Biol 20:1595–1599
Kloting N, Bluher M (2005) Extended longevity and insulin signaling in adipose tissue. Exp Gerontol 40:878–883
Lesch KP, Bengel D, Heils A et al (1996) Association of anxiety-related traits with a polymorphism in the serotonin transporter gene regulatory region. Science 274:1527–1531
Maier H, Gampe J, Jeune B, Robine JM, Vaupel JW (2010) Supercentenarians. Springer, Heidelberg
Mandawat A, Mandawat A, Mandawat MK, Tinetti ME (2012) Hospitalization rates and in-hospital mortality among centenarians. Arch Intern Med 172:1179–1180
Masui Y, Gondo Y, Inagaki H et al (2006) Do personality characteristics predict longevity? Finding from the Tokyo centenarian study. Age 28:353–361
Motta M, Bennati E, Ferlito L et al (2005) Successful aging in centenarians: myths and reality. Arch Gerontol Geriatr 40:241–251
Oeppen J, Vaupel JW (2002) Demography. Broken limits to life expectancy. Science 296:1029–1031
Perls TT (1995) The oldest old. Sci Am 272:50–55
Perls TT, Wilmoth J, Levenson R et al (2002) Life-long sustained mortality advantage of siblings of centenarians. Proc Natl Acad Sci U S A 99:8442–8447
Schoenhofen EA, Wyszynski DF, Andersen S et al (2006) Characteristics of 32 supercentenarians. J Am Geriatr Soc 54:1237–1240
Scoub WW (1994) Cytokine control of nutrition and metabolism in critical illness. Curr Probl Surg 340:448–454
Silver MH, Jilinskaia E, Perls TT (2001) Cognitive functional status of age-confirmed centenarians in a population-based study. J Gerontol B Psychol Sci Soc Sci 56:P134–P140
Stefan N, Vozarova B, Funahashi T et al (2002) Plasma adiponectin concentration is associated with skeletal muscle insulin receptor tyrosine phosphorylation, and low plasma concentration precedes a decrease in whole-body insulin sensitivity in humans. Diabetes 51:1884–1888
Suzuki M, Akisaka M, Ashitomi I et al (1995) Chronological study concerning ADL among Okinawan centenarians. Nippon Ronen Igakkai Zasshi 32:416–423
Takayama M, Hirose N, Arai Y et al (2007) Morbidity profile of Tokyo-area centenarians and its relationship to functional status. J Gerontol A Biol Sci Med Sci 62:774–782
Trujillo ME, Scherer PE (2006) Adipose tissue-derived factors: impact on health and disease. Endocr Rev 27:762–778
Vaupel JW (2010) Biodemography of human ageing. Nature 464:536–542

Wang Z, Al-Regaiey KA, Masternak MM, Bartke A (2006) Adipocytokines and lipid levels in Ames dwarf and calorie-restricted mice. J Gerontol A Biol Sci Med Sci 61:323–331

Willcox DC, Willcox BJ, Wang NC, He Q, Rosenbaum M, Suzuki M (2008) Life at the extreme limit: phenotypic characteristics of supercentenarians in Okinawa. J Gerontol A Biol Sci Med Sci 63:1201–1208

Chapter 3
Progeria and Genome Instability

Fanbiao Meng, Baohua Liu, and Zhongjun Zhou

Abstract Aging is a process of progressive decline in physiological functions. Cells and tissues in our body are constantly exposed to a variety of endogenous and exogenous assaults that cause, DNA damages. The accumulation of unrepaired/irreparable DNA damages results in a sustained DNA damage checkpoint response and induces cellular senescence, a permanent cell cycle arrest. Studies on human premature aging syndromes have suggested that accumulated damages might lead to exhaustion of resources that are required for replacement of the damaged tissues and thus accelerate aging. In this chapter, we summarize current knowledge on DNA damage repair machinery and evidence supporting the idea that defects in genomic maintenance are behind human premature aging syndromes. We put the emphasis on Hutchinson-Gilford Progeria Syndrome.

Keywords Premature aging • DNA damage • Lamin A • Progerin • Chromatin remodeling • Epigenetics

3.1 Introduction

Aging refers to the processes during which physiological functions necessary for fertility decline gradually accompanied by an increase in vulnerability, such as decreased fecundity, reduction of tissue repair capacity, increased propensity to infections and cancer predisposition (DiGiovanna 2000; Partridge and Mangel 1999). Although we are quite familiar with various changes of the bodies over times, aging processes are far more complicated than we could possibly explain with current knowledge, especially in higher vertebrate species, which occurs at

F. Meng • B. Liu
Health Science Center, Shenzhen University, Shenzhen, China
e-mail: ppliew@szu.edu.cn

Z. Zhou (✉)
Shenzhen Institute of Research and Innovation, The University of Hong Kong, Shenzhen, China

Department of Biochemistry, LKS Faculty of Medicine, The University of Hong Kong, Hong Kong
e-mail: zhongjun@hku.hk

© Springer Japan 2015

N. Mori, I. Mook-Jung (eds.), *Aging Mechanisms*,
DOI 10.1007/978-4-431-55763-0_3

multiple levels, e.g. molecular, cellular, tissue, and organismal levels. A variety of aging theories have been proposed during last few decades (Szilard 1959). For instance, the programmed theory proposes that aging is a consequence of a sequential switching on/off of certain genes. Supporting this hypothesis, a group of signaling pathways or genes is identified to accelerate aging or extend lifespan. Conserved from worms to mammals, mutations of insulin receptor or the phosphoinositide 3-kinase extend lifespan (Kenyon et al. 1993; Klass 1983; Kops et al. 1999; Morris et al. 1996). The reduction of food intake, i.e. calorie restriction (CR), extends healthspan and lifespan, attributable to amino acid concentration sensor-target of rapamycin (mTOR), AMP level sensor-AMPK and NAD+ level sensor-sirtuins (Houtkooper et al. 2010). mTOR exerts the effects of CR on protein synthesis, which is modulated by ribosomal protein S6 kinase (S6K) and translation initiation factor 4E-binding protein (4E-BP) (Hay and Sonenberg 2004). Sirtuins are NAD-dependent protein deacetylases conserved in metazoans. Ectopic sirtuins are able to extend lifespan in yeast, worms, flies and rodents (Herranz et al. 2010; Kanfi et al. 2012; Satoh et al. 2013). However, how significant these genes/signaling pathways contribute to human aging needs further investigation.

To date, the most direct evidence on the underlying mechanisms in human aging comes from studies in various human premature aging disorders, including Werner syndrome (WS), Bloom's syndrome (BS), Rothmund-Thomson syndrome (RTS), Cockayne syndrome (type A-CSA and type B-CSB), Xeroderma pigmentosum (XP), Trichothiodystrophy (TTD) and Hutchinson-Gilford progeria syndrome (HGPS). All of them except HGPS are caused by mutations in genes that are involved in DNA repair and genome maintenance, strongly supporting the DNA damage accumulation theory. Although gene involving in HGPS does not encode a protein directly participating in DNA repair, it was reported recently that HGPS patients suffer from genomic instability (Liu et al. 2005). In addition, mouse models that are deficient in essential DNA repair factors usually develop premature aging phenotypes, further supporting a direct correlation between genomic instability and aging (Hasty 2005; Hoeijmakers 2009). In this chapter, we summarize current knowledge on double strand break (DSB) repair pathways and discuss the evidences supporting the correlation between genomic instability and progeria.

3.2 Double Strand Break Repair

DNA is the fundamental genetic material that has to be transmitted to the next generation with high fidelity. Cells and tissues in our body are constantly exposed to a variety of endogenous and exogenous assaults, such as irradiation and oxygen radicals, leading to DNA damages (Maynard et al. 2009). The types of DNA damage include single-strand DNA gaps, nicks, base lesions, stalled replication forks and DSBs. Any of such damages will trigger a serial repair processes to maintain the genomic integrity through one or multiple repair mechanisms including homologous recombination (HR), non-homologous end-joining (NHEJ), nucleotide excision repair (NER), base excision repair (BER) and mismatch repair (MMR) (Fleck and Nielsen 2004). Among the various types of DNA damage, DSBs are the most severe assaults that could be lethal. Therefore, efficient repair of DSBs is essential for the maintenance of genome integrity. Failure in DSBs repair could lead to chromosomal deletions and/or translocations, which may eventually induce malignancy. DSBs could arise from either environmental challenges or endogenous stresses, such as reactive oxygen species and stalled DNA-replication forks (Osley et al. 2007; Pandita and Richardson 2009). Eukaryotic cells have evolved mainly two types of pathways to detect and eliminate DSBs, i.e. HR and NHEJ. HR requires a sister chromatid or chromosomal homologue as a template for the repair, and therefore preferentially occurs during late S-G2 phase when homologous chromosomes come in close vicinity (Osley et al. 2007). HR is a high-fidelity repair process. In contrast, NHEJ engages direct re-ligation of broken ends without regard to homologous template, is not restricted by cell cycle phase, and therefore is error-prone. As homologous sequences are not required, NHEJ is the predominant repair pathway in G1 and quiescent cells.

The broken DNA ends are immediately recognized by the Mre11-Rad50-Xrs2 (MRX) or Mre11-Rad50-Nbs1 (MRN) complex in mammals, and then processed by MRE11 to 3′ single-strand DNA (ssDNA) tails through a series of 5′→3′ strand resection activities (Karagiannis and El-Osta 2007; Osley and Shen 2006; van Attikum and Gasser 2005). The carboxyl terminus of NBS1 directly interacts with and recruits Ataxia-telangiectasia mutated (ATM), which belongs to phosphatidylinositol-3-like kinase-related kinase (PIKK) family (Lee and Paull 2004). In undamaged cells, ATM forms inactive dimer or multimer. Upon DSB, ATM is auto-phosphorylated at serine 1981, leading to its dissociation of dimer into activated monomers (Bakkenist and Kastan 2003). Histone variant-H2AX is then phosphorylated on its C-terminal Serine 139 by ATM, i.e. γ-H2AX, which then spread more than 1 kb region flanking the DSB as evidenced by immunofluorescence microscopy and Chip-seq (Iacovoni et al. 2010). γ-H2AX complex resides regions surrounding the DSB sites as irradiation-induced foci (IRIF) until DNA damage is completely repaired (Rogakou et al. 1998), serving as a platform for the recruitment of essential DNA repair proteins, such as 53BP1, Rad50, Rad51, Nbs1 and BRCA1 (Redon et al. 2002) and chromatin-remodeling complexes, such as INO80 and SWR1 etc. (Fernandez-Capetillo et al. 2004). For example, γ-H2AX recruits mediator of DNA-damage checkpoint 1 (MDC1), which further recruits the MRN complex in a positive feedback loop. Other substrates for ATM include SMC1, NBS1, CHK2, p53, BRCA1 and MDC1 etc. (Harper and Elledge 2007). Of these, phosphorylated Chk2 kinase and p53 inhibit cyclin-dependent kinase (CDK) activity and induce cell cycle arrest to allow sufficient time for DNA repair (Riley et al. 2008).

Followed is the recruitment of epistasis group proteins, including XRCC2, XRCC3, RAD51B, RAD51C and RAD51D etc., which further transmits DNA damage signals to downstream processing enzymes (Johnson et al. 1999). The single-strand overhangs are then rapidly bound by ssDNA-binding protein replication protein A (RPA), followed by recruitment of Rad51 and Rad52 to the damaged sites (Osley and Shen 2006). Loading of Rad51 onto the ssDNA tail results in the formation of ssDNA-Rad51 nucleoprotein filament, which then searches for its homologous counterpart in the corresponding intact sister chromatid. If the specific region of duplex DNA is found, strand invasion is initiated in the presence of another set of HR facilitating proteins (Rad54, Rad55, and Rad57) followed by strand exchange and joint molecule formation (Downs et al. 2007; Osley and Shen 2006). Once the Holliday junctions are resolved, distal broken ends are sealed via DNA polymerase mediated DNA synthesis, leading to an error-free repair event and preserving genetic contents (Osley et al. 2007).

In the case of NHEJ, the DSB ends are recognized and bound by end-binding heterodimers consisting of Ku70 and Ku80. Together with a DNA-PK catalytic subunit (DNA-PKcs), Ku protein forms a complex known as DNA-PK (van Attikum and Gasser 2005) to hold the two broken ends in close proximity for direct end-to-end religation by DNA ligase 4 (Lig4) and ligase-interacting factor 1 (Lif1) in yeast or XRCC4 in mammals. Alternatively, ends bound with Ku complex could be resected by MRX or MRN complex. Similarly, processed ends are joined directly by the action of Lig4-Lif1 (Lig4-XRCC4 in mammals)- mediated ligation, creating the repaired duplex DNA (Pandita and Richardson 2009).

Eukaryotic DNA does not disperse in nucleus, instead it wrap around the nucleosome containing two copies of each core histone (H2A, H2B, H3 and H4) in a hierarchical way (Davey et al. 2002). In addition to γ-H2AX, chromatin undergoes various dynamic modifications like methylation, acetylation, and phosphorylation in response to response to DNA damage. For instance, 53BP1 is recruited to DBSs through binding to di-methylated H4K20 via its tandem tudor domain (Botuyan et al. 2006). However, no change of H4K20 methylation is detected after DNA damage, indicating that other cooperation factor is required to facilitate the recruit 53BP1. Similarly, tri-methylation of H3K9 (H3K9me3) via SUV39H1 in human cells provides binding sites for HP1, and the mutant of Su(var) 3-9 Drosophila homolog resulted in spontaneous DNA damage particularly in heterochromatin. Recently, Ayrapetov et al. reported crosstalk between Tip60 and H3K9me3 surrounding DSBs. H3K9 methylation is activated by tip60 to facilitate the loading of a complex containing KAP-1, HP1, and the H3K9 methyltransferase suv39h1 through binding the chromodomain of HP1 at DSB. The presence of H3K9me3 is required for activation of Tip60 and consequent activation of ATM in response to DSB (Ayrapetov et al. 2014). Nucleosomes re-assembling after DNA repair mimics chromatin remodeling, which requires the acetylation of the globular core of H3 at lysine 56. H3K56 is acetylated in a new synthesized histone and disappear when entering into G2 phase of cell cycle in the absence of DNA damage. In response to DNA damage in S-phase, the H3K56 acetylation by Rtt109 in yeast and CBP/p300 in human cells is maintained during the checkpoint dependent mechanism (Das et al. 2009; Li et al. 2008). Tip60- mediated H4 acetylation results in the formation of open, flexible chromatin adjacent to the DNA break to facilitate the loading of BCRA1 and 53BP1 and DNA repair (Xu et al. 2012). The role of histones H2A and H2B in the DNA damage repair is to recruit the repair proteins such as BRCA1 through ubiquitination by E3 ubiquitin ligase RNF8 (RING finger protein 8) and RNF168 (Mailand et al. 2007). They accumulate at DSB sites and target E2 ubiquitin-conjugating enzyme UBC13 to ubiquitylate H2A, H2B and H2AX.

3.3 DNA Damage Accumulation Theory of Aging

It is estimated that one eukaryotic cell typically faces up to around 100,000 DNA lesions per day. While most of the DNA damages are successfully repaired by designated repair pathway(s) thus maintaining the integrity of genome, some of them remain unrepaired. The accumulation of unrepaired/irreparable DNA damages leads to a sustained DNA damage checkpoint response and consequently a stable cell cycle arrest, a state resembling cellular senescence (Halliwell and Whiteman 2004; Tanaka et al. 2006). This is supported by the finding that unrepaired/irreparable DNA damages positively stained with γ-H2AX accumulate in human senescent cells, in germ and somatic cells isolated from aged mice, and in dermal fibroblasts from aged nonhuman primates (Maslov and Vijg 2009). Further evidences come from increasing number of transgenic mouse models with premature aging phenotypes and human progeria syndromes harboring mutations in genes essential for DNA repair. For instance, progeria syndrome WS, BS and RTS are caused by defects in RecQ helicases, a group of highly conserved proteins involved in DNA replication, repair, recombination and gene transcription, while CS, XP and TTD share similar defects in NER (Balajee et al. 1999; Cooper et al. 2000; Lebel et al. 1999; Li and Comai 2000).

WS is an autosomal recessive genetic disorder, affecting about ten in one million (Multani and Chang 2007). Patients are usually born healthy with obvious growth retardation and other ageing-related features starting from second decade, including short stature, cataract, beaked nose, skin atrophy, hair loss, lipodystrophy, type 2 diabetes mellitus (T2DM), osteoporosis, arteriosclerosis, hypogonadism and neoplasia. Most of WS patients die of cardiovascular diseases or neoplasia in the fourth decade of life with an average life expectancy of 47 years. Due to the close resemblance to physiological aging, WS has been intensively investigated to understand the underlying molecular mechanism of normal aging. Skin fibroblasts derived from WS patient develop accelerated senescence with increased chromosome aberrations (Melcher et al. 2000; Salk et al. 1981). Before the identification of *LMNA* mutations in atypical WS, WRN was the only protein implicated in WS (Yu et al. 1996). WRN belongs to the family of RecQ helicases and is the only member with a specific exonuclease domain within the N-terminus (Gray et al. 1997; Huang et al. 1998). WRN interacts with proteins required for DNA replication, such as RPA, PCNA, FEN1 and DNA polymerase (Polδ) (Brosh et al. 1999, 2001; Kamath-Loeb et al. 2001; Lebel et al. 1999). Lebel and colleagues found that WRN is involved in the restoration of stalled replication forks (Cooper et al. 2000; Karmakar et al. 2002; Li and Comai 2000). WRN associates with telomeric repeat binding factor 1/2 (TRF1/2) and POT1 to maintain telomere integrity (Machwe et al. 2004; Multani and Chang 2007). Telomere shortening is one of the most important causes of replicative senescence (Blasco 2007). Kusumoto-Matsuo and colleagues showed that DNA-PKcs interacts with WRN thus stimulating its helicase activity while preventing the exonuclease digestion of telomeric D-loop (Kusumoto-Matsuo et al. 2010). In recent studies, *LMNA* mutations (A57P, R133L, L140R and E578V) are reported in autosomal dominant atypical WS, wherein patients presented with more severe phenotypes compared to those associated with WRN (Chen et al. 2003; Csoka et al. 2004; Fukuchi et al. 2004).

BS is a genetic disorder characterized by dwarfism, sun-induced erythema, T2DM, narrow face, prominent ears, infertility, and benign and malignant tumors. Individuals suffering from BS usually die from neoplasia before the age of 30. Gene mutation responsible for BS is the RecQ helicase BLM. In response to DNA damage, BLM forms discrete nucleoplasmic foci, which co-localize with RAD51 and BRCA1-associated genome surveillance complex (BASC) containing BRCA1, MLH1, MRN complex and ATM (Hickson 2003). BLM also modulates the correct localization and activation of topoisomerase IIIα (Wu et al. 2000). Deficiency of Blm in mice causes early embryonic lethality at E13.5. Mutant embryos exhibit an elevated frequency of sister chromatid exchange (Chester et al. 1998; Guo et al. 2004).

3.4 Hutchinson-Gilford Progeria Syndrome

Defective genome maintenance is also found to contribute to progeria features observed in HGPS as well (Liu et al. 2005). HGPS is extremely rare with a ratio of 1 to 4–8 million. HGPS was first reported by Drs. Jonathan Hutchinson and Hastings Gilford in 1886 and 1897, respectively (Hennekam 2006). Patients suffering from HGPS appear normal at birth, but develop to the characteristic symptoms of failure to thrive, hair loss, wrinkle skin and progressive lipodystrophy within 1 year. Joint mobility starts to decrease from second to third year, from knee to elbows and fingers. The average lifespan of HGPS patients is 13 years, and most of the patients die from progressive cardiovascular disease and stroke (Merideth et al. 2008). It is now clear that HGPS is predominantly caused by a de novo G608G mutation in *LMNA* gene.

Lamin A belongs to type V intermediate filament proteins and is the major constituents of the nuclear lamina and nuclear matrix. It is firstly synthesized as prelamin A with a carboxyl CAAX motif, which dictates a series of processing events including transient isoprenylation and two steps of proteolysis mediated by Zmpste24 metallopeptidase and Rce1 (Rusinol and Sinensky 2006). Albeit no amino acid substitution, the $LMNA^{G608G}$ mutation activates a cryptic splice site, resulting in a deletion of 50 amino acid within prelamin A protein (Eriksson et al. 2003). The 50-amino-acid truncation contains a cleavage site of Zmpste24, leading to an unprocessed, farnesylated and carboxyl-methylated prelamin A, namely progerin. Progerin accumulates in the nuclear sphere (Gordon et al. 2013), which leads to, by yet-to-be known mechanisms, the subsequent cellular and organismal disease symptoms. Interestingly, the wild-type allele of *LMNA* also undergoes alternative splicing at the same site and generates the 50-aa-truncated prelamin A but at low level (Scaffidi and Misteli 2006). The level of progerin and the number of cells expressing progerin increase along with aging (McClintock et al. 2007), and ageing-related telomere attrition correlates with the increased production of progerin in healthy individuals (Cao et al. 2011). These findings implicate that normal aging might share, partially if not all, similar molecular mechanism with HGPS.

Zmpste24, the mammalian orthologue of *Saccharomyces cerevisiae* Ste24p, is a zinc metalloproteinase and locates on the endoplasmic reticulum and the nuclear envelop. In yeast, Ste24p is responsible for maturation of mating pheromone α-factor (Fujimura-Kamada et al. 1997; Tam et al. 1998). In mammals, the nuclear prelamin A is so far the only reported substrate of Zmpste24 (Fong et al. 2004; Pendas et al. 2002; Varela et al. 2005). Mice lacking Zmpste24 are born indistinguishable from their littermates. However, progeroid features develop 2 months after, such as growth retardation, fat and hair loss, abnormal dentition, tissue fibrosis, dystrophic muscular fibers and decreased bone mineralization etc. Mutant mice are sterile and die around 4–6 months of age. Accumulation of unprocessed prelamin A in nuclear periphery and abnormal nuclear morphology with herniation, a hallmark of HGPS cells, were observed in cultured mouse embryonic fibroblasts (MEFs) derived from Zmpste24 deficient mice (Pendas et al. 2002). As prelamin A is so far the only identified substrate of Zmpste24, all the progeroid phenotypes observed in *Zmpste24* null mice are thought to be attributable to unprocessed prelamin A accumulation. Zmpste24 mutations are found to be associated with autosomal recessive restrictive dermopathy (RD), with severe progeroid features resembling HGPS (Denecke et al. 2006; Levy et al. 2005; Moulson et al. 2005; Navarro et al. 2004, 2005).

3.5 Genomic Instability and HGPS

Increasing evidences have shown that the repair machinery is dysfunctional and DNA damage accumulates in progeria cells. The *Zmpster24* knockout mice are vulnerable to DNA damage reagents like γ-irradiation (Liu et al. 2005). Increased micronuclei and aneuploidy are observed in *Zmpste24*$^{-/-}$ MEFs and HGPS dermal fibroblasts, indicating genomic instability. In line with this observation, the number of immunofluorescence foci containing 53BP1 and γ-H2AX are increased in *Zmpste24*$^{-/-}$ MEFs and HGPS cells. *Zmpste24*$^{-/-}$ MEFs are hyper-sensitive to DNA damaging agents, especially those can introduce DSBs. *Zmpste24*$^{-/-}$ MEFs and HGPS dermal fibroblasts showed delayed resolve of fragmented DNA upon massive DNA damage, attributable to delayed DNA-damage checkpoint response and defective DNA repair. Moreover, reduced homologous recombination

(HR) was detected in cells lacking Zmpste24 or with aberrant lamin A (unpublished data). It is further demonstrated that lamin A bind the SUV39H1, a H3K9 methyl transferase, and protects it from proteasomal degradation, while the prelamin A/progerin exhibits higher binding capacity to Suv39H1, causing increased levels of SUV39H1 and H3K9me3. Depletion of *Suv39H1* in *Zmpste24* null mice delays body weight loss, increases bone mineral density and extends lifespan by more than 30 % (Liu et al. 2013b). The presence of unprocessed prelamin A compromises ATM-KAP-1 pathway mediated chromatin remodeling, and depletion of KAP-1 rescues impaired chromatin remodeling, defective DNA repair and early senescence in cells lacking (Liu et al. 2013a). In addition, we examine the changes of histone modification in *Zmpste24* null mice, and found histone H4 was hypoacetylated at a lysine 16 (H4K16), which was caused by the reduced association of a histone acetyltransferase, Mof, to the nuclear matrix. When Mof was overexpressed, repair proteins were recruited to DNA damage sites together with the improvement of age syndromes. The lifespan of *Zmpste24* null mice extends if treated with Sodium Butyrate (NaB), a histone deacetylase inhibitor (Krishnan et al. 2011). The accumulation of progerin in the nuclear lamina compromise or change lamin A associated proteins, including DNA damage repair proteins, chromatin remodeling factors. For instance, lamin A interact with Sirt 1 via its C-terminal, while prelamin A or progerin also interact with Sirt1 in a weaker way, contributing to the mislocalization of Sirt1 in nuclear matrix (Liu et al. 2012). Another lamin A interacting Heterochromatin protein 1 (HP1), including HP1α, HP1β, and HP1γ, plays a versatile function in DNA replication, DNA damage repair, chromatin organization and nuclear architecture. The level of HP1α increased in *Zmpste24* null MEFs, and HP1α phosphorylation was compromised, leading to delayed formation of γ-H2AX foci in *Zmpste24* null MEFs following DNA damage (Liu et al. 2014).

In regard of NER, Xeroderma pigmentosum group A (XPA) protein recognizes DNA damage, recruits NER nuclease, and stabilizes repair intermediates (Sancar et al. 2004). Interestingly, as a specific protein in NER, XPA foci are not excluded from DSB sites in HGPS and RD fibroblasts, whereas other NER proteins, including XPC and replication protein A (RPA), does not co-localize in DSB sties in HGPS and RD cells. Further analysis reveals XPA-DSB co-localization in progeria cells that are resistant to repair. When XPA is depleted using RNA interference, recruitment of Rad50, Rad51 and Ku70 to γ-H2AX foci was partially restored, implicating that the XPA localized in DSB sites in HGPS cells interferes the recruitment of proteins required for normal DNA repair (Liu et al. 2008). The deterioration of normal repair in HGPS cells likely attributes to the accumulation of progerin, which disrupts the nuclear lamina and associated proteins, including the replicative proteins PCNA and polymerase δ. This would stall replication fork and causes the sustained presence of DSBs accessible to XPA, leading to cell arrest in S-phase (Liu et al. 2006). XPA was further shown to have higher affinity with ds-ssDNA junction in DNA damage process, which may explain the sequestration of PCNA at replication fork allow XPA bind the nascent ds-ssDNA junction (Yang et al. 2006).

3.6 Conclusions

Although the G608G mutation is responsible for HGPS, the underlying mechanism is rather complicated. Cells derived from HGPS patients and progeroid mouse models are vulnerable to DNA damage, are deficient in DSB repair response proteins, and harbors improperiate chromatin alteration, which together leads to the genome instability and accelerated senescence. Considering the similarity of progeria and normal old aging both in phenotypes and molecular mechanisms, investigation of progeria is beneficial for understanding the normal aging and the treatment of age-related diseases.

Acknowledgments B.L. is an Excellent Young Researcher of National Natural Science Foundation of China (81422016), an awardee of 1000-Young Talents Program (The Organization Department of the Communist Party of China Central Committee), and supported by the Major Research plan of NSFC (91439133). Works in Z.Z.'s laboratory are supported by a NSFC Key Project grant (81330009), a National Key Basic Research Program project (2011CB964700) from Ministry of Science and Technology of China, and a RGC grant (HKU2/CRF/13G) from Hong Kong SAR.

References

Ayrapetov MK, Gursoy-Yuzugullu O, Xu C, Xu Y, Price BD (2014) DNA double-strand breaks promote methylation of histone H3 on lysine 9 and transient formation of repressive chromatin. Proc Natl Acad Sci 111:9169–9174

Bakkenist CJ, Kastan MB (2003) DNA damage activates ATM through intermolecular autophosphorylation and dimer dissociation. Nature 421:499–506

Balajee AS, Machwe A, May A, Gray MD, Oshima J, Martin GM, Nehlin JO, Brosh R, Orren DK, Bohr VA (1999) The Werner syndrome protein is involved in RNA polymerase II transcription. Mol Biol Cell 10:2655–2668

Blasco MA (2007) Telomere length, stem cells and aging. Nat Chem Biol 3:640–649

Botuyan MV, Lee J, Ward IM, Kim J-E, Thompson JR, Chen J, Mer G (2006) Structural basis for the methylation state-specific recognition of histone H4-K20 by 53BP1 and Crb2 in DNA repair. Cell 127:1361–1373

Brosh RM Jr, Orren DK, Nehlin JO, Ravn PH, Kenny MK, Machwe A, Bohr VA (1999) Functional and physical interaction between WRN helicase and human replication protein A. J Biol Chem 274:18341–18350

Brosh RM Jr, von Kobbe C, Sommers JA, Karmakar P, Opresko PL, Piotrowski J, Dianova I, Dianov GL, Bohr VA (2001) Werner syndrome protein interacts with human flap endonuclease 1 and stimulates its cleavage activity. Embo J 20:5791–5801

Cao K, Blair CD, Faddah DA, Kieckhaefer JE, Olive M, Erdos MR, Nabel EG, Collins FS (2011) Progerin and telomere dysfunction collaborate to trigger cellular senescence in normal human fibroblasts. J Clin Invest 121:2833–2844

Chen L, Lee L, Kudlow BA, Dos Santos HG, Sletvold O, Shafeghati Y, Botha EG, Garg A, Hanson NB, Martin GM et al (2003) LMNA mutations in atypical Werner's syndrome. Lancet 362:440–445

Chester N, Kuo F, Kozak C, O'Hara CD, Leder P (1998) Stage-specific apoptosis, developmental delay, and embryonic lethality in mice homozygous for a targeted disruption in the murine Bloom's syndrome gene. Genes Dev 12:3382–3393

Cooper MP, Machwe A, Orren DK, Brosh RM, Ramsden D, Bohr VA (2000) Ku complex interacts with and stimulates the Werner protein. Genes Dev 14:907–912
Csoka AB, Cao H, Sammak PJ, Constantinescu D, Schatten GP, Hegele RA (2004) Novel lamin A/C gene (LMNA) mutations in atypical progeroid syndromes. J Med Genet 41:304–308
Das C, Lucia MS, Hansen KC, Tyler JK (2009) CBP/p300-mediated acetylation of histone H3 on lysine 56. Nature 459:113–117
Davey CA, Sargent DF, Luger K, Maeder AW, Richmond TJ (2002) Solvent mediated interactions in the structure of the nucleosome core particle at 1.9 Å resolution. J Mol Biol 319:1097–1113
Denecke J, Brune T, Feldhaus T, Robenek H, Kranz C, Auchus RJ, Agarwal AK, Marquardt T (2006) A homozygous ZMPSTE24 null mutation in combination with a heterozygous mutation in the LMNA gene causes Hutchinson-Gilford progeria syndrome (HGPS): insights into the pathophysiology of HGPS. Hum Mutat 27:524–531
DiGiovanna AG (2000) Human aging: biological perspectives. McGraw-Hill, New York
Downs JA, Nussenzweig MC, Nussenzweig A (2007) Chromatin dynamics and the preservation of genetic information. Nature 447:951–958
Eriksson M, Brown WT, Gordon LB, Glynn MW, Singer J, Scott L, Erdos MR, Robbins CM, Moses TY, Berglund P et al (2003) Recurrent de novo point mutations in lamin A cause Hutchinson-Gilford progeria syndrome. Nature 423:293–298
Fernandez-Capetillo O, Lee A, Nussenzweig M, Nussenzweig A (2004) H2AX: the histone guardian of the genome. DNA Repair 3:959–967
Fleck O, Nielsen O (2004) DNA repair. J Cell Sci 117:515–517
Fong LG, Ng JK, Meta M, Cote N, Yang SH, Stewart CL, Sullivan T, Burghardt A, Majumdar S, Reue K et al (2004) Heterozygosity for Lmna deficiency eliminates the progeria-like phenotypes in Zmpste24-deficient mice. Proc Natl Acad Sci U S A 101:18111–18116
Fujimura-Kamada K, Nouvet FJ, Michaelis S (1997) A novel membrane-associated metalloprotease, Ste24p, is required for the first step of NH2-terminal processing of the yeast a-factor precursor. J Cell Biol 136:271–285
Fukuchi K, Katsuya T, Sugimoto K, Kuremura M, Kim HD, Li L, Ogihara T (2004) LMNA mutation in a 45 year old Japanese subject with Hutchinson-Gilford progeria syndrome. J Med Genet 41:e67
Gordon LB, Rothman FG, López-Otín C, Misteli T (2013) Progeria: a paradigm for translational medicine. Cell 156:400–407
Gray MD, Shen JC, Kamath-Loeb AS, Blank A, Sopher BL, Martin GM, Oshima J, Loeb LA (1997) The Werner syndrome protein is a DNA helicase. Nat Genet 17:100–103
Guo G, Wang W, Bradley A (2004) Mismatch repair genes identified using genetic screens in Blm-deficient embryonic stem cells. Nature 429:891–895
Halliwell B, Whiteman M (2004) Measuring reactive species and oxidative damage in vivo and in cell culture: how should you do it and what do the results mean? Br J Pharmacol 142:231–255
Harper JW, Elledge SJ (2007) The DNA damage response: ten years after. Mol Cell 28:739–745
Hasty P (2005) The impact of DNA damage, genetic mutation and cellular responses on cancer prevention, longevity and aging: observations in humans and mice. Mech Ageing Dev 126:71–77
Hay N, Sonenberg N (2004) Upstream and downstream of mTOR. Genes Dev 18:1926–1945
Hennekam RCM (2006) Hutchinson–Gilford progeria syndrome: review of the phenotype. Am J Med Genet A 140A:2603–2624
Herranz D, Munoz-Martin M, Canamero M, Mulero F, Martinez-Pastor B, Fernandez-Capetillo O, Serrano M (2010) Sirt1 improves healthy ageing and protects from metabolic syndrome-associated cancer. Nat Commun 1:3
Hickson ID (2003) RecQ helicases: caretakers of the genome. Nat Rev Cancer 3:169–178
Hoeijmakers JH (2009) DNA damage, aging, and cancer. N Engl J Med 361:1475–1485
Houtkooper RH, Williams RW, Auwerx J (2010) Metabolic networks of longevity. Cell 142:9–14
Huang S, Li B, Gray MD, Oshima J, Mian IS, Campisi J (1998) The premature ageing syndrome protein, WRN, is a 3′-->5′ exonuclease. Nat Genet 20:114–116

Iacovoni JS, Caron P, Lassadi I, Nicolas E, Massip L, Trouche D, Legube G (2010) High resolution profiling of γH2AX around DNA double strand breaks in the mammalian genome. EMBO J 29:1446–1457

Johnson RD, Liu N, Jasin M (1999) Mammalian XRCC2 promotes the repair of DNA double-strand breaks by homologous recombination. Nature 401:397–399

Kamath-Loeb AS, Loeb LA, Johansson E, Burgers PM, Fry M (2001) Interactions between the Werner syndrome helicase and DNA polymerase delta specifically facilitate copying of tetraplex and hairpin structures of the d(CGG)n trinucleotide repeat sequence. J Biol Chem 276:16439–16446

Kanfi Y, Naiman S, Amir G, Peshti V, Zinman G, Nahum L, Bar-Joseph Z, Cohen HY (2012) The sirtuin SIRT6 regulates lifespan in male mice. Nature 483:218–221

Karagiannis TC, El-Osta A (2007) Chromatin modifications and DNA double-strand breaks: the current state of play. Leukemia 21:195–200

Karmakar P, Piotrowski J, Brosh RM Jr, Sommers JA, Miller SP, Cheng WH, Snowden CM, Ramsden DA, Bohr VA (2002) Werner protein is a target of DNA-dependent protein kinase in vivo and in vitro, and its catalytic activities are regulated by phosphorylation. J Biol Chem 277:18291–18302

Kenyon C, Chang J, Gensch E, Rudner A, Tabtiang R (1993) A C. elegans mutant that lives twice as long as wild type. Nature 366:461–464

Klass MR (1983) A method for the isolation of longevity mutants in the nematode Caenorhabditis elegans and initial results. Mech Ageing Dev 22:279–286

Kops GJPL, Ruiter ND, De Vries-Smits AMM, Powell DR, Bos JL, Burgering BMT (1999) Direct control of the Forkhead transcription factor AFX by protein kinase B. Nature 398:630–634

Krishnan V, Chow MZY, Wang Z, Zhang L, Liu B, Liu X, Zhou Z (2011) Histone H4 lysine 16 hypoacetylation is associated with defective DNA repair and premature senescence in Zmpste24-deficient mice. Proc Natl Acad Sci 108:12325–12330

Kusumoto-Matsuo R, Opresko PL, Ramsden D, Tahara H, Bohr VA (2010) Cooperation of DNA-PKcs and WRN helicase in the maintenance of telomeric D-loops. Aging 2:274–284

Lebel M, Spillare EA, Harris CC, Leder P (1999) The Werner syndrome gene product co-purifies with the DNA replication complex and interacts with PCNA and topoisomerase I. J Biol Chem 274:37795–37799

Lee JH, Paull TT (2004) Direct activation of the ATM protein kinase by the Mre11/Rad50/Nbs1 complex. Science 304:93–96

Levy N, Lopez-Otin C, Hennekam RC (2005) Defective prelamin A processing resulting from LMNA or ZMPSTE24 mutations as the cause of restrictive dermopathy. Arch Dermatol 141:1473–1474, author reply 1474

Li B, Comai L (2000) Functional interaction between Ku and the werner syndrome protein in DNA end processing. J Biol Chem 275:28349–28352

Li Q, Zhou H, Wurtele H, Davies B, Horazdovsky B, Verreault A, Zhang Z (2008) Acetylation of histone H3 lysine 56 regulates replication-coupled nucleosome assembly. Cell 134:244–255

Liu B, Wang J, Chan KM, Tjia WM, Deng W, Guan X, Huang J-d, Li KM, Chau PY, Chen DJ (2005) Genomic instability in laminopathy-based premature aging. Nat Med 11:780–785

Liu Y, Rusinol A, Sinensky M, Wang Y, Zou Y (2006) DNA damage responses in progeroid syndromes arise from defective maturation of prelamin A. J Cell Sci 119:4644–4649

Liu Y, Wang Y, Rusinol AE, Sinensky MS, Liu J, Shell SM, Zou Y (2008) Involvement of xeroderma pigmentosum group A (XPA) in progeria arising from defective maturation of prelamin A. FASEB J 22:603–611

Liu B, Ghosh S, Yang X, Zheng H, Liu X, Wang Z, Jin G, Zheng B, Kennedy BK, Suh Y (2012) Resveratrol rescues SIRT1-dependent adult stem cell decline and alleviates progeroid features in laminopathy-based progeria. Cell Metab 16:738–750

Liu B, Wang Z, Ghosh S, Zhou Z (2013a) Defective ATM-Kap-1-mediated chromatin remodeling impairs DNA repair and accelerates senescence in progeria mouse model. Aging Cell 12:316–318

Liu B, Wang Z, Zhang L, Ghosh S, Zheng H, Zhou Z (2013b) Depleting the methyltransferase Suv39h1 improves DNA repair and extends lifespan in a progeria mouse model. Nat Commun 4:1868

Liu J, Yin X, Liu B, Zheng H, Zhou G, Gong L, Li M, Li X, Wang Y, Hu J et al (2014) HP1alpha mediates defective heterochromatin repair and accelerates senescence in Zmpste24-deficient cells. Cell Cycle 13:1237–1247

Machwe A, Xiao L, Orren DK (2004) TRF2 recruits the Werner syndrome (WRN) exonuclease for processing of telomeric DNA. Oncogene 23:149–156

Mailand N, Bekker-Jensen S, Faustrup H, Melander F, Bartek J, Lukas C, Lukas J (2007) RNF8 ubiquitylates histones at DNA double-strand breaks and promotes assembly of repair proteins. Cell 131:887–900

Maslov AY, Vijg J (2009) Genome instability, cancer and aging. Biochim Biophys Acta 1790:963–969

Maynard S, Schurman SH, Harboe C, de Souza-Pinto NC, Bohr VA (2009) Base excision repair of oxidative DNA damage and association with cancer and aging. Carcinogenesis 30:2–10

McClintock D, Ratner D, Lokuge M, Owens DM, Gordon LB, Collins FS, Djabali K (2007) The mutant form of lamin A that causes Hutchinson-Gilford progeria is a biomarker of cellular aging in human skin. PLoS One 2:e1269

Melcher R, von Golitschek R, Steinlein C, Schindler D, Neitzel H, Kainer K, Schmid M, Hoehn H (2000) Spectral karyotyping of Werner syndrome fibroblast cultures. Cytogenet Cell Genet 91:180–185

Merideth MA, Gordon LB, Clauss S, Sachdev V, Smith ACM, Perry MB, Brewer CC, Zalewski C, Kim HJ, Solomon B et al (2008) Phenotype and course of Hutchinson–Gilford progeria syndrome. N Engl J Med 358:592–604

Morris JZ, Tissenbaum HA, Ruvkun G (1996) A phosphatidylinositol-3-OH kinase family member regulating longevity and diapause in Caenorhabditis elegans. Nature 382:536–539

Moulson CL, Go G, Gardner JM, van der Wal AC, Smitt JH, van Hagen JM, Miner JH (2005) Homozygous and compound heterozygous mutations in ZMPSTE24 cause the laminopathy restrictive dermopathy. J Invest Dermatol 125:913–919

Multani AS, Chang S (2007) WRN at telomeres: implications for aging and cancer. J Cell Sci 120:713–721

Navarro CL, De Sandre-Giovannoli A, Bernard R, Boccaccio I, Boyer A, Genevieve D, Hadj-Rabia S, Gaudy-Marqueste C, Smitt HS, Vabres P et al (2004) Lamin A and ZMPSTE24 (FACE-1) defects cause nuclear disorganization and identify restrictive dermopathy as a lethal neonatal laminopathy. Hum Mol Genet 13:2493–2503

Navarro CL, Cadinanos J, De Sandre-Giovannoli A, Bernard R, Courrier S, Boccaccio I, Boyer A, Kleijer WJ, Wagner A, Giuliano F et al (2005) Loss of ZMPSTE24 (FACE-1) causes autosomal recessive restrictive dermopathy and accumulation of Lamin A precursors. Hum Mol Genet 14:1503–1513

Osley MA, Shen X (2006) Altering nucleosomes during DNA double-strand break repair in yeast. Trends Genet 22:671–677

Osley MA, Tsukuda T, Nickoloff JA (2007) ATP-dependent chromatin remodeling factors and DNA damage repair. Mutat Res 618:65–80

Pandita TK, Richardson C (2009) Chromatin remodeling finds its place in the DNA double-strand break response. Nucleic Acids Res 37:1363–1377

Partridge L, Mangel M (1999) Messages from mortality: the evolution of death rates in the old. Trends Ecol Evol 14:438–442

Pendas AM, Zhou Z, Cadinanos J, Freije JM, Wang J, Hultenby K, Astudillo A, Wernerson A, Rodriguez F, Tryggvason K et al (2002) Defective prelamin A processing and muscular and adipocyte alterations in Zmpste24 metalloproteinase-deficient mice. Nat Genet 31:94–99

Redon C, Pilch D, Rogakou E, Sedelnikova O, Newrock K, Bonner W (2002) Histone H2A variants h2ax and h2az. Curr Opin Genet Dev 12:162–169

Riley T, Sontag E, Chen P, Levine A (2008) Transcriptional control of human p53-regulated genes. Nat Rev Mol Cell Biol 9:402–412

Rogakou EP, Pilch DR, Orr AH, Ivanova VS, Bonner WM (1998) DNA double-stranded breaks induce histone H2AX phosphorylation on serine 139. J Biol Chem 273:5858–5868

Rusinol AE, Sinensky MS (2006) Farnesylated lamins, progeroid syndromes and farnesyl transferase inhibitors. J Cell Sci 119:3265–3272

Salk D, Au K, Hoehn H, Martin GM (1981) Cytogenetics of Werner's syndrome cultured skin fibroblasts: variegated translocation mosaicism. Cytogenet Cell Genet 30:92–107

Sancar A, Lindsey-Boltz LA, Ünsal-Kaçmaz K, Linn S (2004) Molecular mechanisms of mammalian DNA repair and the DNA damage checkpoints. Annu Rev Biochem 73:39–85

Satoh A, Brace CS, Rensing N, Cliften P, Wozniak DF, Herzog ED, Yamada KA, Imai S (2013) Sirt1 extends life span and delays aging in mice through the regulation of Nk2 homeobox 1 in the DMH and LH. Cell Metab 18:416–430

Scaffidi P, Misteli T (2006) Lamin A-dependent nuclear defects in human aging. Science 312:1059–1063

Szilard L (1959) On the nature of the aging process. Proc Natl Acad Sci U S A 45:30–45

Tam A, Nouvet FJ, Fujimura-Kamada K, Slunt H, Sisodia SS, Michaelis S (1998) Dual roles for Ste24p in yeast a-factor maturation: NH2-terminal proteolysis and COOH-terminal CAAX processing. J Cell Biol 142:635–649

Tanaka T, Halicka HD, Huang X, Traganos F, Darzynkiewicz Z (2006) Constitutive histone H2AX phosphorylation and ATM activation, the reporters of DNA damage by endogenous oxidants. Cell Cycle 5:1940–1945

van Attikum H, Gasser SM (2005) The histone code at DNA breaks: a guide to repair? Nat Rev Mol Cell Biol 6:757–765

Varela I, Cadinanos J, Pendas AM, Gutierrez-Fernandez A, Folgueras AR, Sanchez LM, Zhou Z, Rodriguez FJ, Stewart CL, Vega JA et al (2005) Accelerated ageing in mice deficient in Zmpste24 protease is linked to p53 signalling activation. Nature 437:564–568

Wu L, Davies SL, North PS, Goulaouic H, Riou JF, Turley H, Gatter KC, Hickson ID (2000) The Bloom's syndrome gene product interacts with topoisomerase III. J Biol Chem 275:9636–9644

Xu Y, Ayrapetov MK, Xu C, Gursoy-Yuzugullu O, Hu Y, Price BD (2012) Histone H2A. Z controls a critical chromatin remodeling step required for DNA double-strand break repair. Mol Cell 48:723–733

Yang Z, Roginskaya M, Colis LC, Basu AK, Shell SM, Liu Y, Musich PR, Harris CM, Harris TM, Zou Y (2006) Specific and efficient binding of Xeroderma pigmentosum complementation group A to double-strand/single-strand DNA junctions with 3′-and/or 5′-ssDNA branches. Biochemistry 45:15921–15930

Yu CE, Oshima J, Fu YH, Wijsman EM, Hisama F, Alisch R, Matthews S, Nakura J, Miki T, Ouais S et al (1996) Positional cloning of the Werner's syndrome gene. Science 272:258–262

Part III
Stem Cells, Cultured Neurons, and Lower Animal Models

Chapter 4
Use of Induced Pluripotent Stem Cells in Aging Research

Ken-ichi Isobe

Abstract Human populations around the world are aging, which necessitates new ways to analyze age-related functional decline and disease mechanisms in elderly patients. Further, many elderly patients are waiting for new therapies for treatment of intractable age-related diseases such as Alzheimer's disease, Parkinson's disease, liver cirrhosis, lung failure and renal failure. The development of induced pluripotent stem cell (iPSC) technology opens the door for new approaches to the analysis of age-related disorders and to develop personalized cell therapy for age-related diseases. Progeria syndromes present early ageing characteristics. Human iPSCs (HiPSCs) have been established from progeria syndromes, including Hutchinson Gilford Progeria Syndrome (HGPS). Directed differentiation of HGPS-iPSCs to smooth muscle cells led to the appearance of premature senescence phenotypes associated with vascular ageing. Although the efficiency of establishing murine iPSCs from aged mice is lower than that from young mice, many HiPSCs from elderly humans have been established, including centenarians. HiPSCs from patients with age-related diseases have been established and are being used to elucidate the mechanism of disease progression. HiPSCs from Parkinson's disease (PD) patients were established soon after the discovery of HiPSCs technology. HiPSCs from PD patients have almost the same phenotype as HiPSCs from healthy people. However, long-term cultivation of dopamine neurons differentiated from Parkinson's-HiPSCs has revealed disease-specific characteristics, including fewer neurites and a significant increase in apoptotic cells. Recently, 3D culture has permitted the construction of tissue-specific organoids from HiPSCs. These form complete tissues in immune competent mice. iPSC technology will be used more widely for analysis of age-related disorders, age-related functional declines and future personalized therapy for age-related disorders.

Keywords iPS cells • Progeria • Stem cells • Aging • Age-related diseases

---

K.-i. Isobe (✉)
Department of Immunology, Nagoya University Graduate School of Medicine, Tsurumai, Showa-ku, Nagoya 466-8550, Japan
e-mail: kisobe@med.nagoya-u.ac.jp

© Springer Japan 2015

N. Mori, I. Mook-Jung (eds.), *Aging Mechanisms*,
DOI 10.1007/978-4-431-55763-0_4

4.1 Introduction

The discovery of iPSC technology by Yamanaka et al. has given researchers new approaches for the use of stem cells in regenerative medicine. In particular, iPSCs can be generated from the skin or blood of a patient with a serious illness. This technology may open the way to personalized medicine.

The human genome is heterogeneous. That is, no two people have the same DNA except for identical twins. When iPSCs are generated from one person, the DNA information in the cells is uniquely personalized. The iPSCs from a patient could be differentiated to tissue cells possessing the pathology in question. The cells could then be analyzed to better understand the disease characteristics. Then, the patient-specific iPSCs could be repaired by gene targeting to generate healthy iPSCs. The healthy autologous iPSCs could then be differentiated and transplanted back to the patient to effect repair of the tissue.

Here, I discuss two aspects of iPSCs and aging research. First, I discuss the use of iPSCs in studies of the mechanism of progeria and age-related disorders. Second, I discuss the possibility of using iPSCs for personalized regenerative medicine for age-related diseases, which include Alzheimer's, Parkinson's and other diseases.

4.2 HiPSCs from Progeria

Aging research has primarily been done by using cells and animals lower on the evolutionary tree, including budding yeast, nematodes and *Drosophila*. Recently, many studies have been published using murine models and extending the work done with lower animals. In contrast, it is difficult to conduct aging research using human subjects because the average life span of humans is relatively long. Thus, the diseases that show the phenotypes of an aged person in early life have been the target of researchers.

There are several genetic diseases characterized by early aging called progeria (Burtner and Kennedy 2010), including Hutchinson Gilford Progeria Syndrome (HGPS) (Kudlow et al. 2007), Werner syndrome (WS) (Goto et al. 2013; Mori et al. 2003) and Cockayne syndrome (Bertola et al. 2006). Whether these progeroid syndromes are really accelerated forms of human aging is one of the most important topics in aging research. Although many of the aspects of aging in progeria resemble natural aging (such as graying hair, cataracts, osteoporosis, and atherosclerosis), some phenotypes of aging are lacking in these progeria syndromes. For example, neurodegeneration, such as that which occurs in Alzheimer's and Parkinson's diseases, is seemingly not associated with WS (Goto et al. 2013; Mori et al. 2003).

In WS, cells express a mutant gene for ATP-dependent helicase (*WRN*). They possess critically short telomeres and defective synthesis in the lagging strand of sister telomeres, leading to premature senescence (Crabbe et al. 2004; Schulz

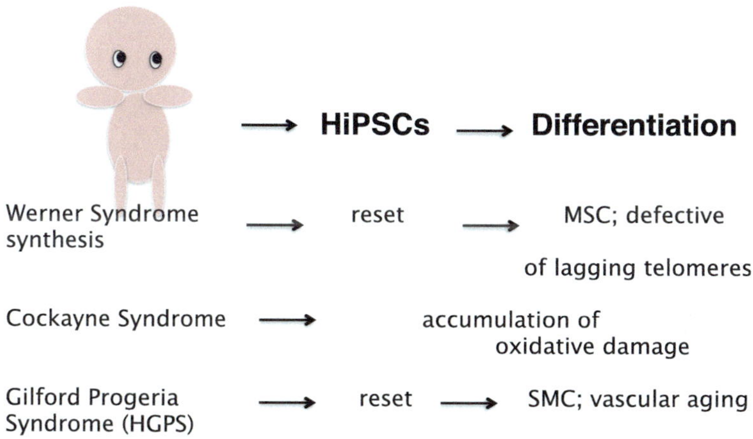

Fig. 4.1 HiPSCs from progeria syndromes. The three main progeria syndromes include Werner syndrome, Cockayne syndrome and Hutchinson Gilford Progeria Syndrome (HGPS). HiPSCs were established from the skin of these progeria patients. The aging characteristics of HiPSC from these syndromes were evaluated. HiPSCs from Werner syndrome and HGPS were reset, having the same characteristics as HiPSCs from normal donors. However, HiPSCs from Cockayne syndrome accumulated oxidative damage. Werner-HiPSCs were differentiated to mesenchymal stem cells (MSC), and they showed abnormal telomeres. HiPSCs from HGPS were differentiated to smooth muscle cells (SMCs), and they showed vascular aging (atherosclerosis)

et al. 1996). Cheung et al. established HiPSCs from WS fibroblasts. They found that WS-HiPSCs expressed pluripotency markers and showed characteristics of human embryonic stem cells (hESCs), including the stringent teratoma formation assay in SCID mice. WS-HiPSCs showed little deficit of telomere synthesis as revealed by telomere chromosome-orientation fluorescence in situ hybridization. Most of the cells could successfully synthesize the lagging strands of their telomeres despite the lack of WRN protein. These results indicated that WS-HiPSCs are reset to a pluripotent state. However, when WS-iPSCs were differentiated to mesenchymal stem cells (MSCs) or neural stem/progenitor cells (NPCs), they observed defective synthesis of lagging telomeres in MSCs but not in NPCs. These results indicate that WS has lineage-specific aging (Fig. 4.1).

Cockayne syndrome (CS) is an autosomal recessive human disorder characterized by neurodegeneration, retinal degeneration, cataracts, cachectic dwarfism and loss of subcutaneous fat. Patients are hypersensitive to sunlight (Filippi et al. 2008). CS is mainly caused by a defective excision-repair cross complementing group 6 (*ERCC6*) gene that encodes Cockayne syndrome group B (CSB) protein. CSB is involved in the removal and repair of oxidative DNA damage. Both defects in CSB and normal aging cause an accumulation of oxidative damage (Kyng et al. 2003; Yuan et al. 2007; Weidenheim et al. 2009). Andrade et al. established iPSCs from skin fibroblasts of a patient with CS. They found that the CSB-HiPSCs exhibited an elevated cell death rate and higher reactive oxygen species (ROS) production

(Andrade et al. 2012). These results indicate that CSB-HiPSCs show aging characteristics without differentiation (Fig. 4.1).

HGPS is a rare autosomal dominant syndrome that involves premature aging characterized by a small jaw, decreased subcutaneous fat, hair loss, etc. Such patients generally die around 13 years of age due to cardiovascular complications. Sporadic mutation of the lamin A (*LMNA*) gene leads to production of an abnormal protein (progerin) (Merideth et al. 2008). This change occurs at codon 608 of exon 11 from glycine GGC to glycine GGT, which induces abnormal splicing leading to aberrant removal of the 3′-terminal 150 nucleotides of this exon. The resulting Δ150 *LMNA* mRNA gives rise to a dominant gain-of-function lamin A isoform containing an internal deletion of 50 amino acids (Δ50 lamin A) (De Sandre-Giovannoli et al. 2003; Eriksson et al. 2003).

Does the same molecular mechanism that is responsible for HGPS also contribute to healthy aging? Scaffidi et al. examined fibroblast cell lines from elderly patients (81–96 years of age). Similar to HGPS cells, they found that aged fibroblasts showed downregulation of several nuclear proteins including heterochromatin protein HP1 and the LAP2 group of lamin A-associated proteins and low heterochromatin markers H3K9me3. Interestingly, all aged fibroblasts contained Δ150 LMNA DNA and Δ50 lamin A protein expression (Scaffidi and Misteli 2006). HGPS fibroblasts showed abnormal nuclear morphology, reduced expression of lamina components lamin B1 and LAP2b (also known as TMPO), loss of heterochromatin markers H3K9me3, HP1a (also known as CBX5) and HDAC1 and reduced expression of nuclear proliferation marker Ki67. These phenotypes of HGPS fibroblasts are due to the expression of progerin.

Liu et al. established HiPSCs from fibroblasts of HGPS patients in whom the LMNA mutation was confirmed. HiPSCs from HGPS fibroblasts lost the expression of progerin and reset the nuclear architecture and reset HGPS-specific gene expression. Progerin primarily accumulates in arterial smooth muscle cells (SMCs) of HGPS patients, and vascular SMC degeneration is one of the characteristics of HGPS-associated arteriosclerosis. They found that directed differentiation of HGPS-iPSCs to SMCs led to the appearance of premature senescence phenotypes associated with vascular ageing (Fig. 4.1). Thus, HGPS-iPSCs revealed not only pathologies of HGPS, but also pathogenesis of human premature and physiological vascular ageing (Liu et al. 2011). Recently, Miller et al. introduced progerin into PD HiPSC-derived dopamine neurons. They found that neuronal aging and PD phenotypes appeared in progerin-induced PD neurons such as pronounced dendrite degeneration, progressive loss of tyrosine hydroxylase (TH) expression, and enlarged mitochondria or Lewy-body-precursor inclusions (Miller et al. 2013). These results indicate that progerin speeds up the aging process and that the PD-specific aging characteristics are due to the introduction of progerin. Lamin A encodes the A-type nuclear lamins, intermediate filament proteins of the nuclear envelope. Mutation of *LMNA* causes human disorders called laminopathies (Worman et al. 2010).

4.3 Efficiency of Establishing iPSCs from Aged Mice and Elderly Humans

In order to use personalized therapy for the treatment of age-related diseases, it is necessary to evaluate the efficiency of iPSC production from aged individuals (Mahmoudi and Brunet 2012). Several studies have been done to compare the efficiency of establishing iPSCs. We compared the efficiency of iPSCs production by congenic C57BL/6 mice. We compared the efficiency to induce colonies of iPSCs from bone marrow cells between mice that were 23 months old and 2 months old. The colony number was lower and the time to appear as colonies was longer in aged B6 mice, although they could get iPSCs from aged mice (Cheng et al. 2011). Other studies also suggest that cells from aged mice tend to reprogram less efficiently than cells from young animals (Li et al. 2009; Wang et al. 2011). Wang compared the efficiency to establish iPSCs from muscle-derived fibroblasts (MuFs) of mdx mice, a strain that is a model of Duchenne muscular dystrophy. MuFs from 14-month-old (14 m) mdx mice showed low proliferative and reprogramming efficiency, gradually losing Nanog expression, and regressed in conventional ES medium during successive passages. Inhibition of TGFb and BMP signaling promoted self-renewal of 14m-MuF-iPSCs (Wang et al. 2011).

Although it is difficult to compare the efficiencies for establishing HiPSCs from different sources, many laboratories have succeeded in generating HiPSCs from the elderly. Somers was successful in establishing many HiPSCs from several age groups, ranging from neonates to 64-year-old normal and diseased donors (Somers et al. 2010). Boulting et al. established HiPSCs from 36- to 71-year-old healthy individuals and 29- to 82-year-old myotrophic lateral sclerosis (ALS; *SOD1* mutated) patients (Boulting et al. 2011). Further, Lapassset et al. established HiPSCs from a 74-year-old donor by transfecting *OSKM*, *NANOG* and *LIN28*. They showed that iPSCs generated from senescent and centenarian cells have reset their telomere size, gene expression profiles, oxidative stress response, and mitochondrial metabolism, and are indistinguishable from hESCs (Lapasset et al. 2011). Prigione et al. established iPSCs from dermal fibroblasts obtained from 56-, 80-, 82- and 84-year-old donors. They found chromosomal abnormalities. However, many aging characteristics were reset, including telomerase activity, high sensitivity to drug-induced apoptosis and low levels of oxidative stress and DNA damage (Prigione et al. 2011). Finally, iPSCs have been established from healthy centenarians (106- and 109-year-old donors) (Yagi et al. 2012) (Fig. 4.2).

4.4 HiPSCs and Age-Related Diseases

Aging in humans is accompanied by age-related diseases. However, studies of age-related diseases by sampling human tissues have some limitations. This situation has been changed by the development of iPSC technology. This approach

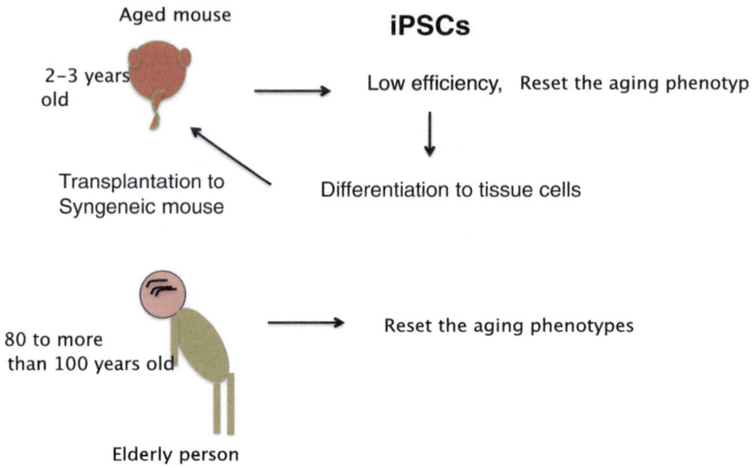

Fig. 4.2 iPSCs from aged mice and elderly humans. The efficiency of iPSC production was evaluated for iPSCs from aged mice. Although the efficiency to make iPSCs was lower in aged mice than that in young mice, it was still possible to establish iPSCs from old animals. The characteristics of aging were reset in aged-murine iPSCs. These iPSCs were used for model personalized therapy. HiPSCs have been established from many aged human donors. The characteristics of aging were reset in HiPSCs

offers an unlimited source of cells for in vitro study of age-related diseases. Here, I discuss the usefulness of iPSC technology for age-related research.

Soon after the development of human iPSC technology (Takahashi et al. 2007; Yu et al. 2007; Park et al. 2008a), HiPSCs from Parkinson's disease (PD) patients were established (Park et al. 2008b; Soldner et al. 2009). These developments were confirmed by several laboratories. PD is a common neurodegenerative disorder, characterized with rhythmical tremor, difficulty of walking and rigidity. The incidence of the disease rises with age, with a lifetime risk of developing the disease of 1.5 % (de Lau and Breteler 2006). PD is characterized by the degeneration of dopaminergic neurons in the substantia nigra of the midbrain and other brain regions. The cause of the degeneration of dopaminergic neurons is largely unknown. There have been major advances in the understanding of the mechanisms that contribute to dopaminergic cell death, including mitochondrial dysfunction, oxidative stress, altered protein handling, and inflammation (Dexter and Jenner 2013). PD is associated with a selective loss of dopaminergic neurons from the substantia nigra pars compacta, a region responsible for controlling body movement. Because patients have decreased levels of dopamine secretion from the substantia nigra, dopamine replacement therapy is common. Post mortem pathological examination has revealed the accumulation of Lewy bodies that were composed of α-synuclein. Around 90 % of PD cases are sporadic, while 10 % have a familial or genetic origin. Leucine-rich repeat kinase 2 (*LRRK2*) mutations (Gly2019Ser) are the most common cause of familial PD with a 1 % sporadic and a 4 % hereditary rate (Cookson 2010). African Arabs, Ashkenazi Jews and

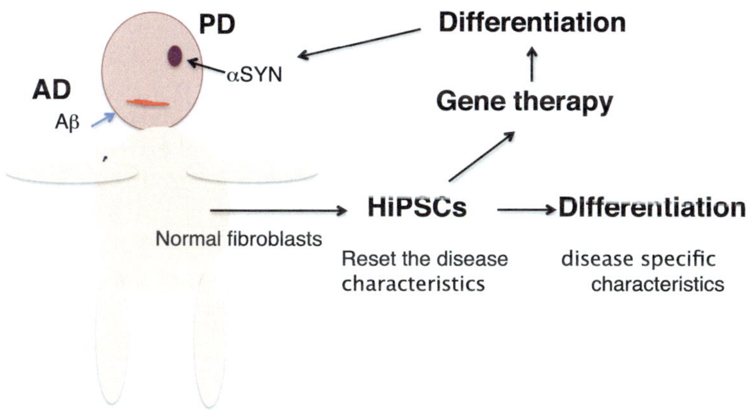

Fig. 4.3 HiPSCs from age-related diseases. HiPSCs were established from patients with age-related diseases such as Parkinson's disease (PD) and Alzheimer's disease (AD). The fibroblasts from these patients were converted to HiPSCs. PD-HiPSCs reset PD-specific characteristics except the expression of αSYN. AD-HiPSCs reset AD-specific disease characteristics. However, disease-specific characteristics appeared after differentiation of PD-HiPSCs or AD-HiPSCs to specific tissue cells

Portuguese people have a higher risk (Healy et al. 2008). *LRRK2* mutations correlate with alpha-synuclein (αSYN), the protein encoded by the *SNCA* gene, both related to Lewy body formation. *SNCA* was also identified as being mutated in PD (Spillantini et al. 1997).

HiPSCs from patients with *LRRK2* mutations have been established (Nguyen et al. 2011; Liu et al. 2012; Orenstein et al. 2013; Cooper et al. 2012; Sanchez-Danes et al. 2012; Reinhardt et al. 2013). *LRRK2* mutated-HiPSCs were differentiated to dopamine neurons. They developed neurodegeneration, including fewer neurites and a significant increase in apoptotic cells after 2.5 months culture. Interestingly, they showed decreased autophagic clearance (Sánchez-Danés et al. 2012). Neural stem cells differentiated from *LRRK2* mutated-HiPSCs had deterioration of nuclear architecture, which was one of characteristics of HGPS (Liu et al. 2012) (Fig. 4.3). HiPSCs have been established from other genetic disorders related to PD. A review of this subject was published recently (Jacobs 2014).

Alzheimer's disease (AD) is the most common age-related diseases in the world. Around 36,000,000 people are affected by AD or related dementia (World Alzheimer's Disease Report 2013). AD is a progressive neurodegenerative disease. AD patients exhibit memory loss and impairments in daily tasks. The two main pathological changes found in postmortem brains are amyloid plaques and neurofibrillary tangles, which are consisted with A-beta and Tau respectively. Diagnosis of AD before death is difficult and only symptomatic but not causative therapies are pursued at present. Although the majority of AD cases are late-onset sporadic form, there exist familial (genetic) cases associated with mutations in amyloid precursor protein (APP), presenilin-1 (PS1), presenilin-2 (PS2), Microtubule Associated

Protein Tau (MAPT; Tau), Apolipoprotein E4 (APOE4) and Triggering Receptor Expressed on Myeloid Cells 2 (TREM2). Israel et al. established H-iPSCs from two AD patients showing duplication of amyloid-b (APP^{Dp}), two sporadic AD (sAD) patients and two controls. All H-iPSCs generated from the three different types of patients could differentiate efficiently to form neurons and had the same synaptic functions. However, neurons from APP^{Dp} secreted significantly higher levels of amyloid-b(1–40) compared to control. Furthermore, neurons from patients APPDp and sAD had significantly higher aGSK-3b and p-tau(Thr 231) than control (Israel et al. 2012). Yagi et al. generated HiPSCs from fibroblasts of familial AD patients with mutations in PS1 (A246E) and PS2 (N141I). They found that differentiated neurons from AD-derived HiPSCs secreted higher amounts of Ab42 (Yagi et al. 2011). Muratore et al. established H-iPSCs from an AD patient having an amyloid precursor protein (APP) mutation (V717I). They found an increase in the levels of total and phosphorylated Tau in neurons with the APPV717I mutation (Muratore et al. 2014). Kondo et al. established HiPSCs from patients with the (APP)-E693D mutation and sporadic AD. Interestingly Ab-oligomers accumulated in cells of differentiated neurons of these patients. Ab-oligomers induced oxidative stresses and ER stresses that might destroy neurons (Kondo et al. 2013) (Fig. 4.3).

4.5 Toward Personalized Treatment of Age-Related Diseases by iPSCs

The generation of HiPSCs has opened the door for personalized stem cell therapy. Because iPSCs are developed from a patient's own cells, immune rejection is theoretically avoided. Although many papers have shown that iPSCs are capable of differentiating into tissue cells *in vitro*, they must engraft into a patient's own tissue to be of benefit. It has been questionable whether iPSCs from an aged patient can differentiate to tissue cells and can be transplanted back into his own tissue (Isobe et al. 2014). Type II diabetes is one of the most prevalent diseases in the elderly having diabetic polyneuropathy. Okawa tried to differentiate iPSCs to neural crest-like cells and transplanted them to a murine system mimicking diabetes. Specifically, the purified neural crest-like cells were transplanted to hind limb skeletal muscles of diabetes model B6 mice. *In vivo* neural crest-like cells differentiated into vascular smooth muscle cells or Schwann-like cells. The transplantation improved the impaired nerve and vascular functions (Okawa et al. 2013).

In order to pursue personalized therapy using a patient's own tissue, HiPSCs derived from a patient need to differentiate to cells for a specific organ or tissue. Two steps are needed. The first step is to generate an organ *in vitro*. The second step is to transplant the organ back into the patient. Watson et al. developed human intestinal organoids *in vitro* from HiPSCs, after which they transplanted them under the kidney capsule of immunocompromised nonobese diabetic severe combined immunodeficiency interleukin-2Rγnull (NSG) mice and allowed them to mature *in vivo*. Surprisingly, the entire human intestinal structure appeared in the transplanted organoid (Watson et al. 2014).

4.6 Future Directions

There are problems to be solved before iPSCs can be brought to the clinic for personalized cell therapy. These include the following. (1) Elimination of DNA duplications, rearrangement and mutations during the course of iPSC production. (2) Overcoming the potential immunogenicity of cells differentiated from iPSCs. (3) Reducing the risk of tumorigenicity. Substantial progress has been made in these areas and they are likely to be solved in near future. These topics were discussed in detail in another review (Isobe et al. 2014).

There remain unsolved questions related to aging. (1) Are aged somatic cells rejuvenated by iPSC production? (2) Do aged-iPSCs differentiate normally? With regard to the first issue, we showed some aspects of epigenetic rejuvenation of aged iPSCs (Cheng et al. 2011). However several reports showed incomplete epigenetic reprogramming (Rohanl et al. 2014). Studies of other ageing characteristics (such as telomere length, mitochondrial aging) have yielded different results (Isobe et al. 2012; Rohanl et al. 2014). The second question is more important. When we differentiated aged iPSCs to macrophages, they differentiated as expected. However, they soon died (unpublished results). However, when we differentiated the same aged-iPSCs to neural crest-like cells, they could grow for a considerable time (Okawa et al. 2013). It is evident that further studies are needed to bring iPSCs to the clinic for treatment of age-related diseases.

References

Andrade LN, Nathanson JL, Yeo GW, Menck CF, Muotri AR (2012) Evidence for premature aging due to oxidative stress in iPSCs from Cockayne syndrome. Hum Mol Genet 21:3825–3834

Bertola DR, Cao H, Albano LM, Oliveira DP, Kok F, Marques-Dias MJ, Kim CA, Hegele RA (2006) Cockayne syndrome type A: novel mutations in eight typical patients. J Hum Genet 51:701–705

Boulting GL, Kiskinis E, Croft GF, Amoroso MW, Oakley DH, Wainger BJ, Williams DJ, Kahler DJ, Yamaki M, Davidow L, Rodolfa CT, Dimos JT, Mikkilineni S, MacDermott AB, Woolf CJ, Henderson CE, Wichterle H, Eggan K (2011) A functionally characterized test set of human induced pluripotent stem cells. Nat Biotechnol 29:279–286

Burtner CR, Kennedy BK (2010) Progeria syndromes and ageing: what is the connection? Nat Rev Mol Cell Biol 11:567–578

Cheng Z, Ito S, Nishio N, Xiao H, Zhang R, Suzuki H, Okawa Y, Murohara T, Isobe K (2011) Establishment of induced pluripotent stem cells from aged mice using bone marrow-derived myeloid cells. J Mol Cell Biol 3:91–98

Cookson MR (2010) The role of leucine-rich repeat kinase 2 (LRRK2) in Parkinson's disease. Nat Rev Neurosci 11:791–797

Cooper O, Seo H, Andrabi S, Guardia-Laguarta C, Graziotto J, Sundberg M, McLean JR, Carrillo-Reid L, Xie Z, Osborn T, Hargus G, Deleidi M, Lawson T, Bogetofte H, Perez-Torres E, Clark L, Moskowitz C, Mazzulli J, Chen L, Volpicelli-Daley L, Romero N, Jiang H, Uitti RJ, Huang Z, Opala G, Scarffe LA, Dawson VL, Klein C, Feng J, Ross OA, Trojanowski JQ, Lee VM, Marder K, Surmeier DJ, Wszolek ZK, Przedborski S, Krainc D, Dawson TM, Isacson O (2012) Pharmacological rescue of mitochondrial deficits in iPSC-derived neural cells from patients with familial Parkinson's disease. Sci Transl Med 4:141ra90

Crabbe L, Verdun RE, Haggblom CI, Karlseder J (2004) Defective telomere lagging strand synthesis in cells lacking WRN helicase activity. Science 306:1951–1953

de Lau LM, Breteler MM (2006) Epidemiology of Parkinson's disease. Lancet Neurol 5:525–535

De Sandre-Giovannoli A, Bernard R, Cau P, Navarro C, Amiel J, Boccaccio I, Lyonnet S, Stewart CL, Munnich A, Le Merrer M, Lévy N (2003) Lamin a truncation in Hutchinson-Gilford progeria. Science 300:2055

Dexter DT, Jenner P (2013) Parkinson disease: from pathology to molecular disease mechanisms. Free Radic Biol Med 62:132–144

Eriksson M, Brown WT, Gordon LB, Glynn MW, Singer J, Scott L, Erdos MR, Robbins CM, Moses TY, Berglund P, Dutra A, Pak E, Durkin S, Csoka AB, Boehnke M, Glover TW, Collins FS (2003) Recurrent de novo point mutations in lamin A cause Hutchinson-Gilford progeria syndrome. Nature 423:293–298

Filippi S, Latini P, Frontini M, Palitti F, Egly JM, Proietti-De-Santis L (2008) CSB protein is (a direct target of HIF-1 and) a critical mediator of the hypoxic response. EMBO J 27:2545–2556

Goto M, Ishikawa Y, Sugimoto M, Furuichi Y (2013) Werner syndrome: a changing pattern of clinical manifestations in Japan (1917~2008). Biosci Trends 7:13–22

Healy DG, Falchi M, O'Sullivan SS, Bonifati V, Durr A, Bressman S, Brice A, Aasly J, Zabetian CP, Goldwurm S, Ferreira JJ, Tolosa E, Kay DM, Klein C, Williams DR, Marras C, Lang AE, Wszolek ZK, Berciano J, Schapira AH, Lynch T, Bhatia KP, Gasser T, Lees AJ, Wood NW, Consortium IL (2008) Phenotype, genotype, and worldwide genetic penetrance of LRRK2-associated Parkinson's disease: a case-control study. Lancet Neurol 7:583–590

Isobe K, Cheng Z, Ito S, Nishio N (2012) Aging in the mouse and perspectives of rejuvenation through induced pluripotent stem cells (iPSCs). Results Probl Cell Differ 55:413–427

Isobe K, Cheng Z, Nishio N, Suganya T, Tanaka Y, Ito S (2014) iPSCs, aging and age-related diseases. N Biotechnol 31:411–421

Israel MA, Yuan SH, Bardy C, Reyna SM, Mu Y, Herrera C, Hefferan MP, Van Gorp S, Nazor KL, Boscolo FS, Carson CT, Laurent LC, Marsala M, Gage FH, Remes AM, Koo EH, Goldstein LS (2012) Probing sporadic and familial Alzheimer's disease using induced pluripotent stem cells. Nature 482:216–220

Jacobs BM (2014) Stemming the hype: what can we learn from iPSC models of Parkinson's disease and how can we learn it? J Park Dis 4:15–27

Kondo T, Asai M, Tsukita K, Kutoku Y, Ohsawa Y, Sunada Y, Imamura K, Egawa N, Yahata N, Okita K, Takahashi K, Asaka I, Aoi T, Watanabe A, Watanabe K, Kadoya C, Nakano R, Watanabe D, Maruyama K, Hori O, Hibino S, Choshi T, Nakahata T, Hioki H, Kaneko T, Naitoh M, Yoshikawa K, Yamawaki S, Suzuki S, Hata R, Ueno S, Seki T, Kobayashi K, Toda T, Murakami K, Irie K, Klein WL, Mori H, Asada T, Takahashi R, Iwata N, Yamanaka S, Inoue H (2013) Modeling Alzheimer's disease with iPSCs reveals stress phenotypes associated with intracellular Aβ and differential drug responsiveness. Cell Stem Cell 12:487–496

Kudlow BA, Kennedy BK, Monnat RJ (2007) Werner and Hutchinson-Gilford progeria syndromes: mechanistic basis of human progeroid diseases. Nat Rev Mol Cell Biol 8:394–404

Kyng KJ, May A, Brosh RM, Cheng WH, Chen C, Becker KG, Bohr VA (2003) The transcriptional response after oxidative stress is defective in Cockayne syndrome group B cells. Oncogene 22:1135–1149

Lapasset L, Milhavet O, Prieur A, Besnard E, Babled A, Aït-Hamou N, Leschik J, Pellestor F, Ramirez JM, De Vos J, Lehmann S, Lemaitre JM (2011) Rejuvenating senescent and centenarian human cells by reprogramming through the pluripotent state. Genes Dev 25:2248–2253

Li H, Collado M, Villasante A, Strati K, Ortega S, Cañamero M, Blasco MA, Serrano M (2009) The Ink4/Arf locus is a barrier for iPS cell reprogramming. Nature 460:1136–1139

Liu GH, Barkho BZ, Ruiz S, Diep D, Qu J, Yang SL, Panopoulos AD, Suzuki K, Kurian L, Walsh C, Thompson J, Boue S, Fung HL, Sancho-Martinez I, Zhang K, Yates J, Izpisua Belmonte JC (2011) Recapitulation of premature ageing with iPSCs from Hutchinson-Gilford progeria syndrome. Nature 472:221–225

Liu GH, Qu J, Suzuki K, Nivet E, Li M, Montserrat N, Yi F, Xu X, Ruiz S, Zhang W, Wagner U, Kim A, Ren B, Li Y, Goebl A, Kim J, Soligalla RD, Dubova I, Thompson J, Yates J, Esteban CR, Sancho-Martinez I, Izpisua Belmonte JC (2012) Progressive degeneration of human neural stem cells caused by pathogenic LRRK2. Nature 491:603–607

Mahmoudi S, Brunet A (2012) Aging and reprogramming: a two-way street. Curr Opin Cell Biol 24:744–756

Merideth MA, Gordon LB, Clauss S, Sachdev V, Smith AC, Perry MB, Brewer CC, Zalewski C, Kim HJ, Solomon B, Brooks BP, Gerber LH, Turner ML, Domingo DL, Hart TC, Graf J, Reynolds JC, Gropman A, Yanovski JA, Gerhard-Herman M, Collins FS, Nabel EG, Cannon RO, Gahl WA, Introne WJ (2008) Phenotype and course of Hutchinson-Gilford progeria syndrome. N Engl J Med 358:592–604

Miller JD, Ganat YM, Kishinevsky S, Bowman RL, Liu B, Tu EY, Mandal PK, Vera E, Shim JW, Kriks S, Taldone T, Fusaki N, Tomishima MJ, Krainc D, Milner TA, Rossi DJ, Studer L (2013) Human iPSC-based modeling of late-onset disease via progerin-induced aging. Cell Stem Cell 13:691–705

Mori H, Tomiyama T, Maeda N, Ozawa K, Wakasa K (2003) Lack of amyloid plaque formation in the central nervous system of a patient with Werner syndrome. Neuropathology 23:51–56

Muratore CR, Rice HC, Srikanth P, Callahan DG, Shin T, Benjamin LN, Walsh DM, Selkoe DJ, Young-Pearse TL (2014) The familial Alzheimer's disease APPV717I mutation alters APP processing and Tau expression in iPSC-derived neurons. Hum Mol Genet 23:3523–3536

Nguyen HN, Byers B, Cord B, Shcheglovitov A, Byrne J, Gujar P, Kee K, Schüle B, Dolmetsch RE, Langston W, Palmer TD, Pera RR (2011) LRRK2 mutant iPSC-derived DA neurons demonstrate increased susceptibility to oxidative stress. Cell Stem Cell 8:267–280

Okawa T, Kamiya H, Himeno T, Kato J, Seino Y, Fujiya A, Kondo M, Tsunekawa S, Naruse K, Hamada Y, Ozaki N, Cheng Z, Kito T, Suzuki H, Ito S, Oiso Y, Nakamura J, Isobe K (2013) Transplantation of neural crest-like cells derived from induced pluripotent stem cells improves diabetic polyneuropathy in mice. Cell Transplant 22:1767–1783

Orenstein SJ, Kuo SH, Tasset I, Arias E, Koga H, Fernandez-Carasa I, Cortes E, Honig LS, Dauer W, Consiglio A, Raya A, Sulzer D, Cuervo AM (2013) Interplay of LRRK2 with chaperone-mediated autophagy. Nat Neurosci 16:394–406

Park IH, Arora N, Huo H, Maherali N, Ahfeldt T, Shimamura A, Lensch MW, Cowan C, Hochedlinger K, Daley GQ (2008a) Disease-specific induced pluripotent stem cells. Cell 134:877–886

Park IH, Zhao R, West JA, Yabuuchi A, Huo H, Ince TA, Lerou PH, Lensch MW, Daley GQ (2008b) Reprogramming of human somatic cells to pluripotency with defined factors. Nature 451:141–146

Prigione A, Hossini AM, Lichtner B, Serin A, Fauler B, Megges M, Lurz R, Lehrach H, Makrantonaki E, Zouboulis CC, Adjaye J (2011) Mitochondrial-associated cell death mechanisms are reset to an embryonic-like state in aged donor-derived iPS cells harboring chromosomal aberrations. PLoS One 6:e27352

Reinhardt P, Schmid B, Burbulla LF, Schöndorf DC, Wagner L, Glatza M, Höing S, Hargus G, Heck SA, Dhingra A, Wu G, Müller S, Brockmann K, Kluba T, Maisel M, Krüger R, Berg D, Tsytsyura Y, Thiel CS, Psathaki OE, Klingauf J, Kuhlmann T, Klewin M, Müller H, Gasser T, Schöler HR, Sterneckert J (2013) Genetic correction of a LRRK2 mutation in human iPSCs

links parkinsonian neurodegeneration to ERK-dependent changes in gene expression. Cell Stem Cell 12:354–367
Rohani L, Johnson AA, Arnold A, Stolzing A (2014) The aging signature: a hallmark of induced pluripotent stem cells? Aging Cell 13(1):2–7
Sánchez-Danés A, Richaud-Patin Y, Carballo-Carbajal I, Jiménez-Delgado S, Caig C, Mora S, Di Guglielmo C, Ezquerra M, Patel B, Giralt A, Canals JM, Memo M, Alberch J, López-Barneo J, Vila M, Cuervo AM, Tolosa E, Consiglio A, Raya A (2012) Disease-specific phenotypes in dopamine neurons from human iPS-based models of genetic and sporadic Parkinson's disease. EMBO Mol Med 4:380–395
Scaffidi P, Misteli T (2006) Lamin A-dependent nuclear defects in human aging. Science 312:1059–1063
Schulz VP, Zakian VA, Ogburn CE, McKay J, Jarzebowicz AA, Edland SD, Martin GM (1996) Accelerated loss of telomeric repeats may not explain accelerated replicative decline of Werner syndrome cells. Hum Genet 97:750–754
Soldner F, Hockemeyer D, Beard C, Gao Q, Bell GW, Cook EG, Hargus G, Blak A, Cooper O, Mitalipova M, Isacson O, Jaenisch R (2009) Parkinson's disease patient-derived induced pluripotent stem cells free of viral reprogramming factors. Cell 136:964–977
Spillantini MG, Schmidt ML, Lee VM, Trojanowski JQ, Jakes R, Goedert M (1997) Alpha-synuclein in Lewy bodies. Nature 388:839–840
Takahashi K, Tanabe K, Ohnuki M, Narita M, Ichisaka T, Tomoda K, Yamanaka S (2007) Induction of pluripotent stem cells from adult human fibroblasts by defined factors. Cell 131:861–872
Wang B, Miyagoe-Suzuki Y, Yada E, Ito N, Nishiyama T, Nakamura M, Ono Y, Motohashi N, Segawa M, Masuda S, Takeda S (2011) Reprogramming efficiency and quality of induced Pluripotent Stem Cells (iPSCs) generated from muscle-derived fibroblasts of mdx mice at different ages. PLoS Curr 3:RRN1274
Watson CL, Mahe MM, Múnera J, Howell JC, Sundaram N, Poling HM, Schweitzer JI, Vallance JE, Mayhew CN, Sun Y, Grabowski G, Finkbeiner SR, Spence JR, Shroyer NF, Wells JM, Helmrath MA (2014) An in vivo model of human small intestine using pluripotent stem cells. Nat Med 20:1310
Weidenheim KM, Dickson DW, Rapin I (2009) Neuropathology of Cockayne syndrome: evidence for impaired development, premature aging, and neurodegeneration. Mech Ageing Dev 130:619–636
World Alzheimer's Disease Report 2013 (2013) Available from: http://www.alz.co.uk/research/world-report-2013
Worman HJ, Ostlund C, Wang Y (2010) Diseases of the nuclear envelope. Cold Spring Harb Perspect Biol 2:a000760
Yagi T, Ito D, Okada Y, Akamatsu W, Nihei Y, Yoshizaki T, Yamanaka S, Okano H, Suzuki N (2011) Modeling familial Alzheimer's disease with induced pluripotent stem cells. Hum Mol Genet 20:4530–4539
Yagi T, Kosakai A, Ito D, Okada Y, Akamatsu W, Nihei Y, Nabetani A, Ishikawa F, Arai Y, Hirose N, Okano H, Suzuki N (2012) Establishment of induced pluripotent stem cells from centenarians for neurodegenerative disease research. PLoS One 7:e41572
Yu J, Vodyanik MA, Smuga-Otto K, Antosiewicz-Bourget J, Frane JL, Tian S, Nie J, Jonsdottir GA, Ruotti V, Stewart R, Slukvin II, Thomson JA (2007) Induced pluripotent stem cell lines derived from human somatic cells. Science 318:1917–1920
Yuan X, Feng W, Imhof A, Grummt I, Zhou Y (2007) Activation of RNA polymerase I transcription by cockayne syndrome group B protein and histone methyltransferase G9a. Mol Cell 27:585–595

Chapter 5
In Vitro Aging Revisited: The Longevity of Cultured Neurons

Nozomu Mori

Abstract Almost half a century ago, Leonard Hayflick reported that normal diploid fibroblasts can grow and age on cultured dishes in vitro and proposed that normal fibroblasts have a limited "replicative" lifespan, which was later termed the Hayflick limit. This limit was a paradigm of the cellular senescence of dividing cells. In contrast, it is not well-known how non-dividing cells, such as neurons and muscles, grow and age in vitro over a long time course in a culture dish. There is growing evidence that neurons dissociated from various brain areas, such as the cerebral neocortex or hippocampus, can survive for several months in culture dishes. Neurons initially proliferate, form mature synapses, and begin to exhibit aging associated with synaptic loss and neuronal elimination. The long-term cultured neurons in vitro seem to represent many similar aspects of physiological and possibly pathological aging that occur in vivo. The usefulness of this system as a new model for the investigation of non-replicative post-mitotic neuronal aging can be discussed.

Keywords Aging • Brain • Cellular aging • Longevity • Neuron • Primary culture

5.1 In Vitro Aging of Dividing Cells

It was nearly a half century ago when Leonard Hayflick reported a carefully controlled long-term culture system of human diploid fibroblasts and concluded that the normal human fibloblasts retain a finite "replicative" lifespan in vitro (Hayflick and Moorhead 1961; Hayflick 1965). Since this discovery, the so-called "cellular aging" paradigm, a model system of in vitro aging has been extensively studied during the following decades, and the system was later believed to represent a fundamental basis of replicative senescence that is processed in vivo (for reviews, see Cristofalo et al. 2004). The best-characterized hallmark of cellular aging is an accumulation of age pigment, i.e., lipofuscin (Gray and Woulfe 2005). This

N. Mori (✉)
Department of Anatomy and Neurobiology, Graduate School of Biomedical Sciences, Nagasaki University, 1-12-4 Sakamoto, Nagasaki 852-8523, Japan
e-mail: morinosm@nagasaki-u.ac.jp

© Springer Japan 2015

N. Mori, I. Mook-Jung (eds.), *Aging Mechanisms*,
DOI 10.1007/978-4-431-55763-0_5

age-related cellular deposit is also associated with the increased activity and/or expression of senescence-associated β-galactosidase (SAβG) (Cristofalo 2005). The cellular aging features in vitro are divided into three phases, i.e., Phase I, II, and III, in which Phase III cells are recognized as aged cells that are unable to divide and are senescent but not dead.

The in vitro aging system has provided a variety of opportunities to study the mechanisms of aging; for example, the phenomenon of telomere shortening and the identification of critical cell cycle machineries are among the key findings for understanding the intrinsic mechanisms of cell aging (Harley et al. 1990; Aubert and Lansdorp 2008; Deng et al. 2008). However, the cellular aging system has been applied to dividing cells, particularly to human diploid fibroblasts. Because organismal bodies consist of both dividing and non-dividing tissues and cells, it is important to investigate the aging properties of non-dividing cells in vitro. Therefore, it is important to determine how long non-dividing cells, such as neurons of the mammalian brain, can be maintained in culture and to examine how neurons age and whether neurons retain a defined limit to their lifespan in vitro.

It is well-known that neurons, particularly from embryonic and/or neonatal brains of rodents, can be cultured and maintained to develop synapses and to form neural networks (Banker and Cowan 1977, 1979; Kaech and Banker 2006). Numerous structural and functional studies have been performed thus far using cultured mature neurons from the hippocampus, cerebral cortex, and other brain regions (Mattson and Kater 1988; Brewer 1995, 1997; Aksenova et al. 1999; Lesuisse and Martin 2002); however, most studies did not examine older cultures over 2–3 months in vitro, possibly due to certain technical limitations.

5.2 In Vitro Aging of Post-Mitotic Neurons?

In our previous experiments using relatively "young" neuronal cultures addressing questions on the mechanisms of neuronal process outgrowth (Morii et al. 2006; Mori and Morii 2002) and synaptic maturation (Shiraishi et al. 2003), we sometimes noticed that the leftover neuronal cultures still survived, showing intact neuronal morphology. We therefore thought that neurons could possibly be maintained after relatively longer periods, even after 3–4 weeks in culture that is enriched with mature neurons, forming synapses and neural networks. To examine the optimal conditions for long-term neuronal cultures and to explore the nature of surviving or "aging" neurons, we initially tested the differential effects of plating densities, frequency and timing of media changes using the standard culture protocol of the so-called neurobasal medium supplemented with B27 cocktail (Brewer 1995). As this is a serum-free medium, it is suitable for retaining essentially the same culture conditions without experience experimental drift that may result from batch quality differences of serums.

The quest for cellular aging in vitro was a major topic in the 1960s and 1970s, as Leonard Hayflick first established the cell culture system of normal human diploid fibroblasts, which ultimately lead to the concept of 'replicative senescence' or 'cellular aging' (Cristofalo and Stanulis 1978; Cristofalo et al. 2004; Linskens et al. 1995). However, few studies have been performed to address how neurons (or other post-mitotic cells) grow, mature and age in vitro.

5.3 Experimental Trials for Long-Term Neuronal Cultures

Pettmann et al. first described the morphological and biochemical properties of dissociated neurons that were derived from chick embryos, including the telencephalon and other specific areas of the brain, in the absence of glial cells; however, the pure neuronal culture persisted for slightly over a week in vitro (Pettmann et al. 1979). Since this work and other pioneering work by Banker and Cowan (1977, 1979), numerous studies have accumulated, but most studies focused on neuronal differentiation and synaptic maturation using cultures for several weeks at most. Only a few studies explored how neurons age and die in long-term culture.

Among those were studies by Lesuisse and Martin (2002) and Mielke et al. (2005), in which they maintained and examined mouse cortical primary cultures and hippocampal slice cultures up to 60DIV (days in vitro) and 76DIV, respectively. These studies explored the nature of relatively long-term neural cultures, but the cultural dates achieved (days in vitro, i.e., DIV) were insufficient to study the aging and/or senescence stages.

In our preliminary report (Shiraishi-Yamaguchi and Mori 2008), we reported that rat hippocampal neurons could be maintained for at least 6 months, and a portion of neurons could survive over 1 year in normal serum-free culture conditions. It was observed that those neurons do age in vitro, which was associated with accumulation of various age-related markers. It seemed reasonable to consider that this in vitro neuronal culture system represents some features observed in the in vivo aging of neurons, and therefore, this in vitro neuronal aging system could be used as a new model to explore the molecular and cellular mechanisms of post-mitotic neuronal cell aging.

In our case, most of the rat hippocampal neurons survived easily for 5–6 months, i.e., 150–180DIV, and a significant portion reached to 1 year in vitro (Shiraishi-Yamaguchi and Mori 2008). In addition to hippocampal neurons, we also investigated the long-term neural cultures derived from cerebral cortex and hypothalamus of rat embryos (E18), and the results were largely the same, even though viability was the best in the case of the hippocampus (Okamoto et al. 2013). During these time courses, all neurons underwent progressive deteriorative changes in their morphology and biochemical protein expressions, which were initially normal (or physiological) and later abnormal (or pathological), which ultimately led to cell loss due to cell death. These stages could be arbitrarily sub-divided into five consecutive phases; the growing phase I (~1 M), the maturation phase II (1–2 M), the early aging phase III (3–6 M), the late aging (or middle senescence) phase IV

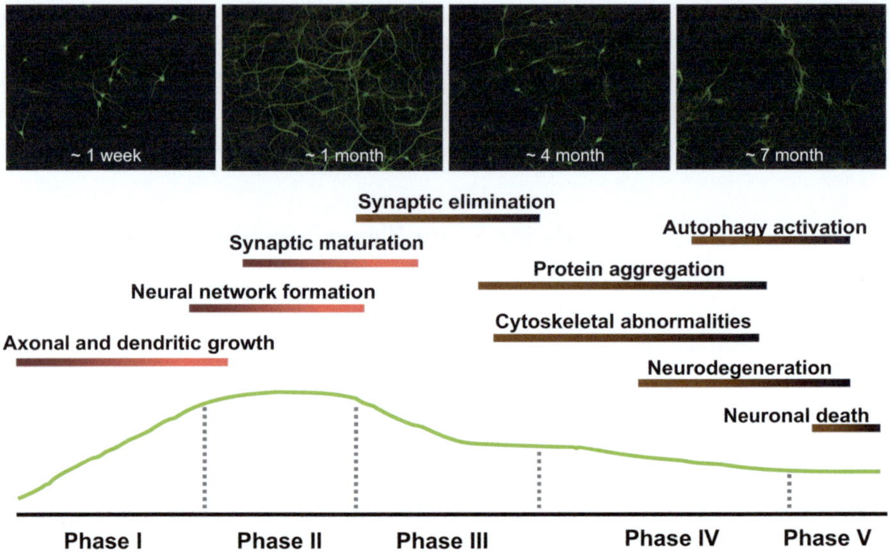

Fig. 5.1 Aging properties of cultured neurons in vitro

Table 5.1 Classification of neuronal aging in vitro

Stage	Age	Characteristics of neurons in culture	Cell numbers		Aging markers	
			Neurons	Astrocytes	SAβG	Lipofuscin
Phase I	~1 M	Neuronal growth and development	+~+++	+	+/−	−
		Networks to be formed approximately 2–3 W				
		Synaptic maturation occurs by 3–4 W				
Phase II	1–2 M	Synaptic maturation completed	+++	++	+	−
		Synaptic elimination and refinement				
Phase III	3–6 M	Early aging stage (physiological aging)	++	+++	++	+
Phase IV	7–9 M	Late aging stage (pathological aging)	++	++	+++	+++
		Fasciculation of neuronal fibers				
Phase V	10 M~	Senescent neurons	+	+/−	+++	+++

(7–9 M), and the late senescence phase V (10 M and over) (see Fig. 5.1 and Table 5.1).

It seemed that the hippocampal neurons of our culture system followed many aspects of aging in vivo. Although it is not clear at this stage whether this long-term in vitro neural culture model 'recapitulates' neuronal aging in vivo, the aged neurons on the dish eventually did exhibit a variety of features that are frequently observed in physiologically aged neurons in vivo, such as age pigment accumulation, neurotransmitter receptor reduction, dendritic actin disintegration, and synaptic loss. Thus, it seems reasonable to consider that an in vitro neuronal aging system could be used as a model to explore neuronal aging.

5.4 Lifespan of Long-Term Cultured Neurons

The viability of post-mitotic neurons in vitro was remarkable in contrast to the ordinary in vitro life span of human diploid fibroblasts, i.e., approximately 60–63 population doublings (PDLs), which correspond to approximately 250 DIV (Cristofalo and Stanulis 1978). Similarly, the viability of P19 cell-derived neurons is known to be limited to approximately 1 week (McBurney et al. 1988), and the in vitro life span of embryonic hippocampus-derived primary cultured neurons is notably extensive, possibly due to the stability of chromatin structures and/or components in comparison to the differentiation-enforced neurons. The longer survival and higher viability of those primary cultured neurons may also be due to intrinsic factors that support their life in the later stages after maturation. For example, the expression of SIRT1, a mammalian homologue of the yeast longevity regulator gene *Sir2*, gradually increased during in vitro aging. The aging neurons would have benefitted from increasing SIRT1 levels because accumulative evidence indicates that SIRT1 functions in neuroprotection through NAD-dependent transcriptional gene modulations (Araki et al. 2004; Qin et al. 2006; Kim et al. 2007; Gan and Mucke 2008). However, neuronal life seems to apparently depend upon the life of astrocytes and other glial cells, as well, because aged neurons (and their dendrites and axons) always stay along the remaining sheets of astrocytes. In this regard, the loss of glial cells in the late stage, e.g., approximately 8 months, could be due to the astrocyte's entrance into phase III in mitotic cell aging in vitro. Thus, the lifespan of postmitotic neurons may depend, at least in part, on the lifespan of mitotic glia. The potential influence of both intrinsic and extrinsic factors on the neuronal life in vitro should be determined in future experiments.

5.5 Does Neuronal Aging In Vitro Recapitulate Features of In Vivo Neuronal Aging?

We assumed that our long-term culture system of in vitro neuronal aging represented notably similar features of neuronal aging in vivo in the brain of organisms, but is it certainly so? The evaluation of the aging level by means of lipofuscin and SAβG revealed that both signals gradually increased after 2 months in culture. Lipofuscin in the intact hippocampus is known to accumulate in an age-dependent manner throughout the 2 years of life in rats (Kaur et al. 2001), suggesting that neuronal aging is similarly advanced in vitro and in vivo. Lipofuscin is a type of classical marker of aging with unknown roles in aging, but studies support the idea that it accumulates in aged cells in relation to autophagy, or lysosomal and/or proteasomal inhibition (Gray and Woulfe 2005; Kiselyov et al. 2007; Sulzer et al. 2008). Similarly to lipofuscin, the age-dependent alterations of synaptic components such as NMDA and AMPA receptor subunits and their anchoring molecules indicated that the in vitro system mimics the in vivo change (Clayton and Browning 2001; Clayton et al. 2002; Magnusson 2000; Magnusson et al. 2002; Wenk and Barnes 2000). The age-dependent loss of post synaptic components does not seem to be attributable to the loss of synaptic input because the expression of presynaptic markers did not decrease considerably during aging, at least by 5 months in vitro; however, it is uncertain whether those presynaptic terminals are always associated with spines or more general postsynaptic structures in the aged cultures. Changes in the expression of cytoskeletal components, for example, the loss of F-actin networks in dendritic spines and the temporal shift of Tau isoforms from the non-phosphorylated form to highly phosphorylated forms, are also consistent with in vivo brain aging, which is possibly directly associated with the hippocampal functional impairment of cognitive abilities (Rosenzweig and Barnes 2003; Burke and Barnes 2006; Dickstein et al. 2007).

All of these results, which occurred in phase II and phase III, seem consistent with in vivo changes; however, we are uncertain whether the abnormal morphological change of neuronal fibers, in other words, the fasciculation that was evident in phase IV neurons, indeed reflects in vivo aging. The neuronal fasciculation is an essential prerequisite for the proper functioning of axonal guidance and/or high conductance during motor neuron development (Bastiani et al. 1984; Hanson et al. 2008), but is also a prominent hallmark of motor neuron degeneration, as in Kennedy disease (Vucic and Kiernan 2007). Because the axonal fasciculation correlates with the increasing expression of neuronal cell adhesion molecules such as NCAM and TAG1 (Cremer et al. 1997; Wolfer et al. 1994), or extracellular protease (Oka et al. 2002), the expression of these proteins may be upregulated in phase IV neurons. During the early aging stage of phase III, cytoskeletal abnormalities might progress, as has been observed in the presence of increasing populations of slowly migrating bands of the microtubule-associated protein Tau, which is also observed in the early pathological conditions of Alzheimer's disease (Iqubal et al. 2005; Ueda et al. 1990). It is uncertain whether the expression changes of Tau may contribute to the morphological changes of neuronal axons, i.e., fasciculation; however, such changes in the structural components of neuronal synapses and spines during in vitro neuronal aging may be associated with the transition from normal to abnormal neuronal aging in vivo (Alvarez and Sabatini 2007; Dickstein et al. 2007). Thus, the transition between phase III and phase IV, i.e., 5–6 months in culture, may be the limit to differentiating the physiological and pathological aging of neurons in vitro.

The neurons in phase IV are aged neurons; however, they are, in other words, surviving (remaining) neurons. What would the difference be between the surviving neurons and other neurons that had to die during the transition stage of phase III to IV? There could be several explanations, including intrinsic and extrinsic mechanisms for those neurons; neurons might have some intrinsic age-counting machinery, similar to telomeres in replicating cells, or neurons might simply be affected by the deleterious conditions of supporting glia or dish coating. Because all the remaining neurons always stay on glial sheets, we assume that the aged neurons require physical and/or chemical influences from astrocytes and/or oligodendrocytes. It is necessary to determine the intrinsic and extrinsic factors that influence the lifespan of these long-term cultured neurons.

5.6 In Vitro Aging of Post-mitotic Neurons

The long-term culture system of hippocampal neurons offers a novel experimental system to precisely study the mechanisms of neuronal aging in vitro, and it has many advantages to complement many of the currently ongoing in vivo studies of neurons and brains of the aged organisms. Because we start culture every other week, we constantly maintain a reasonable number of 24-well plates. The cells on cover slips could be used any time for any comparison among 'young,' 'middle-aged' and 'aged' neurons. It is still time consuming because the rat neurons survive for up to 1 year, but it is less than the normal life span of the organism. Benefits of using the primary cultured neurons for aging studies also include the fact that environmental and/or genetic manipulations surrounding neurons could easily be accessible by bath application into the culture medium, by the transfection of DNAs and siRNAs, and/or by infecting viruses carrying the genes of interest such that the molecular factors and/or genes underlying neuronal aging may be relatively easily defined relative to the experiments in vivo. In addition, when starting with low-density plating, aged neurons are still dissociated, even if they form neural networks, and thus fine structures of spines and synapses can be visualized at a reasonable resolution to determine the subcellular localization of endogenous or exogenously expressed proteins for molecular imaging.

5.7 Conclusions

As we have learned considerably from the paradigm of replicative cell senescence (Campisi 2001; Jeyapalan and Sedivy 2008), we are now in a better position to explore the senescence of non-replicative cells. Many questions remain to be answered in the field of neurobiology of aging, including the following: What would be the primary neurobiological correlates of functional decline in normal aging? How many of these correlates would contribute to those in age-related neurodegenerative diseases (Morrison and Hof 1997)? How does the neuronal plasticity remain in the aged brain (Burke and Barnes 2006), and how do neurons in the brain contribute to determining the organismal lifespan (Mattson et al. 2002)? We believe that molecular mechanisms behind these questions could be investigated, at least in part, using the paradigm of the long-term aging culture of primary cultured neurons.

Acknowledgments The author would like to thank Drs. Y. Yamaguchi, K. Ohyama, and G. Matsumoto for their contribution to this work, and Ms. K. Onga and Ms. A. Okamoto for their excellent technical assistance. This work was supported by Grants-in-Aid for Scientific Research, Challenging Exploratory Research, and also in part by the Asian CORE Program of the Japan Society of Promotion of Science (JSPS): Asian Aging Core for Longevity.

References

Aksenova MV, Aksenov MY, Markesbery WR, Butterfield DA (1999) Aging in a dish: age-dependent changes of neuronal survival, protein oxidation, and creatine kinase BB expression in long-term hippocampal cell culture. J Neurosci Res 58:308–317

Alvarez VA, Sabatini BL (2007) Anatomical and physiological plasticity of dendritic spines. Annu Rev Neurosci 30:79–97

Araki T, Sasaki Y, Milbrandt J (2004) Increased nuclear NAD biosynthesis and SIRT1 activation prevent axonal degeneration. Science 305:1010–1013

Aubert G, Lansdorp PM (2008) Telomeres and aging. Physiol Rev 88:557–579

Banker GA, Cowan WM (1977) Rat hippocampal neurons in dispersed cell culture. Brain Res 126: 397–342

Banker GA, Cowan WM (1979) Further observations on hippocampal neurons in dispersed cell culture. J Comp Neurol 187:469–493

Bastiani M, Pearson KG, Goodman CS (1984) From embryonic fascicles to adult tracts: organization of neuropile from a developmental perspective. J Exp Biol 112:45–64

Brewer GJ (1995) Serum-free B27/neurobasal medium supports differentiated growth of neurons from the striatum, substantia nigra, septum, cerebral cortex, cerebellum, and dentate gyrus. J Neurosci Res 42:674–683

Brewer GJ (1997) Isolation and culture of adult rat hippocampal neurons. J Neurosci Methods 71: 143–155

Burke SN, Barnes CA (2006) Neural plasticity in the ageing brain. Nat Rev Neurosci 7:30–40

Campisi J (2001) From cells to organisms: can we learn about aging from cells in culture? Exp Gerontol 36:607–618

Clayton DA, Browning MD (2001) Deficits in the expression of the NR2B subunit in the hippocampus of aged Fisher 344 rats. Neurobiol Aging 22:165–168

Clayton DA, Grosshans DR, Browning MD (2002) Aging and surface expression of hippocampal NMDA receptors. J Biol Chem 277:14367–14369

Cremer H, Chazal G, Goridis C, Represa A (1997) NCAM is essential for axonal growth and fasciculation in the hippocampus. Mol Cell Neurosci 8:323–335

Cristofalo VJ (2005) SA beta Gal staining: biomarker or delusion. Exp Gerontol 40:836–838

Cristofalo VJ, Stanulis BM (1978) Cell aging: a model system approach. In: Behnke JA, Finch CE, Moment GB (eds) The biology of aging. Plenum Press, New York, pp 19–31

Cristofalo VJ, Lorenzini A, Allen RG, Torres C, Tresini M (2004) Replicative senescence: a critical review. Mech Ageing Dev 125:827–848

Deng Y, Chan SS, Chang S (2008) Telomere dysfunction and tumour suppression: the senescence connection. Nat Rev Cancer 8:450–458

Dickstein DL, Kabaso D, Rocher AB, Luebke JI, Wearne SL, Hof PR (2007) Changes in the structural complexity of the aged brain. Aging Cell 6:275–284
Gan L, Mucke L (2008) Paths of convergence: sirtuins in aging and neurodegeneration. Neuron 58(1):10–14
Gray DA, Woulfe J (2005) Lipofuscin and aging: a matter of toxic waste. Sci Aging Knowl Environ 2005(5):re1. doi:10.1126/sageke.2005.5.re1
Hanson MG, Milner LD, Landmesser LT (2008) Spontaneous rhythmic activity in early chick spinal cord influences distinct motor axon pathfinding decisions. Brain Res Rev 57:77–85
Harley CB, Futcher AB, Greider CW (1990) Telomeres shorten during ageing of human fibroblasts. Nature 345:458–460
Hayflick L (1965) The limited in vitro lifetime of human diploid cell strains. Exp Cell Res 37: 614–636
Hayflick L, Moorhead PS (1961) The serial cultivation of human diploid cell strains. Exp Cell Res 25:585–621
Iqbal K, Alonso Adel C, Chen S, Chohan MO, El-Akkad E, Gong CX, Khatoon S, Li B, Liu F, Rahman A, Tanimukai H, Grundke-Iqbal I (2005) Tau pathology in Alzheimer disease and other tauopathies. Biochim Biophys Acta 1739:198–210
Jeyapalan JC, Sedivy JM (2008) Cellular senescence and organismal aging. Mech Ageing Dev 129:467–474
Kaech S, Banker G (2006) Culturing hippocampal neurons. Nat Protoc 1:2406–2415
Kaur J, Sharma D, Singh R (2001) Acetyl-L-carnitine enhances Na(+), K(+)-ATPase glutathione-S-transferase and multiple unit activity and reduces lipid peroxidation and lipofuscin concentration in aged rat brain regions. Neurosci Lett 301:1–4
Kim D, Nguyen MD, Dobbin MM, Fischer A, Sananbenesi F, Rodgers JT, Delalle I, Baur JA, Sui G, Armour SM, Puigserver P, Sinclair DA, Tsai LH (2007) SIRT1 deacetylase protects against neurodegeneration in models for Alzheimer's disease and amyotrophic lateral sclerosis. EMBO J 26:3169–3179
Kiselyov K, Jennigs JJ Jr, Rbaibi Y, Chu CT (2007) Autophagy, mitochondria and cell death in lysosomal storage diseases. Autophagy 3:259–262
Lesuisse C, Martin LJ (2002) Long-term culture of mouse cortical neurons as a model for neuronal development, aging, and death. J Neurobiol 51:9–23
Linskens MH, Harley CB, West MD, Campisi J, Hayflick L (1995) Replicative senescence and cell death. Science 267:17
Magnusson KR (2000) Declines in mRNA expression of different subunits may account for differential effects of aging on agonist and antagonist binding to the NMDA receptor. J Neurosci 20:1666–1674
Magnusson KR, Nelson SE, Young AB (2002) Age-related changes in the protein expression of subunits of the NMDA receptor. Brain Res Mol Brain Res 99:40–45
Mattson MP, Kater SB (1988) Isolated hippocampal neurons in cryopreserved long-term cultures: development of neuroarchitecture and sensitivity to NMDA. Int J Dev Neurosci 6:439–452
Mattson MP, Duan W, Maswood N (2002) How does the brain control lifespan? Ageing Res Rev 1 (2):155–165
McBurney MW, Reuhl KR, Ally AI, Nasipuri S, Bell JC, Craig J (1988) Differentiation and maturation of embryonal carcinoma-derived neurons in cell culture. J Neurosci 8:1063–1073
Mielke JG, Comas T, Woulfe J, Monette R, Chakravarthy B, Mealing GA (2005) Cytoskeletal, synaptic, and nuclear protein changes associated with rat interface organotypic hippocampal slice culture development. Brain Res Dev Brain Res 160:275–286
Mori N, Morii H (2002) SCG10-related neuronal growth-associated proteins in neural development, plasticity, degeneration, and aging. J Neurosci Res 70:264–273
Morii H, Shiraishi-Yamaguchi Y, Mori N (2006) SCG10, a microtubule destabilizing factor, stimulates the neurite outgrowth by modulating microtubule dynamics in rat hippocampal primary cultured neurons. J Neurobiol 66:1101–1114
Morrison JH, Hof PR (1997) Life and death of neurons in the aging brain. Science 278:412–419

Oka T, Akisada M, Okabe A, Sakurai K, Shiosaka S, Kato K (2002) Extracellular serine protease neuropsin (KLK8) modulates neurite outgrowth and fasciculation of mouse hippocampal neurons in culture. Neurosci Lett 321:141–144

Okamoto A, Taguchi H, Onga K, Ohyama K, Mori N (2013) An in vitro aging model of long-term cultured hippocampal neurons. In: Cold Spring Harbor Asia Conference "Molecular basis of aging and disease", p 33

Pettmann B, Louis JC, Sensenbrenner M (1979) Morphological and biochemical maturation of neurones cultured in the absence of glial cells. Nature 281:378–380

Qin W, Yang T, Ho L, Zhao Z, Wang J, Chen L, Zhao W, Thiyagarajan M, MacGrogan D, Rodgers JT, Puigserver P, Sadoshima J, Deng H, Pedrini S, Gandy S, Sauve AA, Pasinetti GM (2006) Neuronal SIRT1 activation as a novel mechanism underlying the prevention of Alzheimer disease amyloid neuropathology by calorie restriction. J Biol Chem 281:21745–21754

Rosenzweig ES, Barnes CA (2003) Impact of aging on hippocampal function: plasticity, network dynamics, and cognition. Prog Neurobiol 69(3):143–179

Shiraishi Y, Mizutani A, Mikoshiba K, Furuichi T (2003) Coincidence in dendritic clustering and synaptic targeting of homer proteins and NMDA receptor complex proteins NR2B and PSD95 during development of cultured hippocampal neurons. Mol Cell Neurosci 22:188–201

Shiraishi-Yamaguchi Y, Mori N (2008) In vitro aging of the primary cultured hippocampal neurons. Biomed Gerontol (Tokyo) 32(3):27–30, in Japanese

Sulzer D, Mosharov E, Talloczy Z, Zucca FA, Simon JD, Zecca L (2008) Neuronal pigmented autophagic vacuoles: lipofuscin, neuromelanin, and ceroid as macroautophagic responses during aging and disease. J Neurochem 106:24–36

Ueda K, Masliah E, Saitoh T, Bakalis SL, Scoble H, Kosik KS (1990) Alz-50 recognizes a phosphorylated epitope of tau protein. J Neurosci 10:3295–3304

Vucic S, Kiernan MC (2007) Pathophysiologic insights into motor axonal function in Kennedy disease. Neurology 69:1828–1835

Wenk GL, Barnes CA (2000) Regional changes in the hippocampal density of AMPA and NMDA receptors across the lifespan of the rat. Brain Res 885:1–5

Wolfer DP, Henehan-Beatty A, Stoeckli ET, Sonderegger P, Lipp HP (1994) Distribution of TAG-1/axonin-1 in fibre tracts and migratory streams of the developing mouse nervous system. J Comp Neurol 345:1–32

Chapter 6
Cellular Longevity of Budding Yeast During Replicative and Chronological Aging

Kyung-Mi Choi and Cheol-Koo Lee

Abstract Mechanisms of aging and its retardation are evolutionarily conserved from unicellular to multicellular organisms. Several laboratory models, including budding yeast, have contributed to a better understanding of the complexity of aging and longevity. Budding yeast gradually loses the ability of producing daughter cells in rich media, and then loses viability in exhausted media during the aging process. According to these distinguishable losses, there are two measurable lifespans in budding yeast: replicative life span (RLS) and chronological life span (CLS). These two types of lifespans share common longevity-regulating pathways, such as Target of Rapamycin (TOR) signaling, and have non-overlapping pathways between chronologically long-lived mutants and replicatively long-lived mutants. CLS and RLS can be extended through genetic mutation, caloric restriction, and chemical treatment (e.g., rapamycin). We reviewed methodological properties of CLS and RLS, and discussed genes related to these lifespans. Particularly, we focused on two genes, Sir2 and Tor1, in context of their association with replicative and chronological cellular aging. We also described novel genes and primary biological processes responsible for cellular longevity from genome-wide studies.

Keywords Budding yeast • Replicative life span • Chronological life span • Target of Rapamycin (TOR) • Sirtuin • Genome-wide screening • Microarray

6.1 Introduction

A healthy and long lifespan is one of the long-cherished desires of humans. Scientists in aging research have been trying to discover ways to retard the aging process and achieve longevity for many centuries, which has resulted in several aging theories. One such theory is the free-radical theory of aging, which posits free radicals as major culprits that cause irreversible damage to macromolecules of the cell and, eventually, to organs and organisms (Harman 1992). In line with this,

K.-M. Choi • C.-K. Lee (✉)
Division of Biotechnology, College of Life Sciences and Biotechnology, Korea University, Seoul 136-701, Republic of Korea
e-mail: cklee2005@korea.ac.kr

© Springer Japan 2015

N. Mori, I. Mook-Jung (eds.), *Aging Mechanisms*,
DOI 10.1007/978-4-431-55763-0_6

extended lifespan by caloric restriction (CR) has been proposed via a decrease in the production of cellular reactive oxygen species (ROS) and an overall reduction of oxidative stress (Sohal and Weindruch 1996). Our recent studies of CR using a budding yeast system also showed reduction of ROS generation as an important factor for lifespan (Choi et al. 2011, 2013b, 2014; Choi and Lee 2013). However, other recent studies argue that CR actually increases levels of ROS, and that this mild oxidative stress is able to induce a secondary defense system that contributes to decreased oxidative damage (Ristow and Zarse 2010). This new concept was suggested as the mitohormesis theory, but the conflict with CR-mediated reduction of oxidative damage remains unclear. The rate of living theory, another biological theory of aging, was first suggested in the early 1900s, and postulates that metabolic rate is inversely correlated with lifespan. Although this hypothesis is not completely acceptable compared to other studies of lifespan, including CR mechanisms (Speakman et al. 2002; Hulbert et al. 2007), it is quite interesting in that lifespan is closely associated with cellular metabolism to secure energy. The disposable soma theory explains the aging process in terms of energy allocation (Kirkwood 2005). Organisms have a limited energy source that should be properly allocated for repair/maintenance and growth/reproduction to achieve optimum survival; however, these two processes compete against each other. The CR effect on lifespan could be explained by this theory. CR might force more energy allocation into somatic maintenance than growth/reproduction to achieve a long-lifespan (Shanley and Kirkwood 2000). The presence of many hypotheses in the field of aging research proves that the aging process and lifespan regulation are too complex to be explained by a single theory. With this complexity, the mechanism of cellular longevity might be accomplished via harmony among several pathways that lead to sustainable alteration of cellular systems over time.

Several model organisms have been used in the laboratory of aging research. The budding yeast *Saccharomyces cerevisiae*, a widely used model organism for aging, has several advantages. So far, yeast is the only organism that has a single gene deletion library constructed, in which each gene has been physically removed from the genome. In addition, DNA and protein tags have been successfully introduced into the genome with minimal damage to gene and protein function. Availability of these genetically modified strains allows us genome-wide observation using genomic, proteomic, and microfluidic approaches. Using these resources, important genes and pathways in the aging process and their regulation have been identified, many of which are conserved in higher organisms.

In this review, we describe cellular longevity based on two different yeast aging paradigms, replicative and chronological aging, and discuss characteristics of these two methods. In particular, we review two frequently overlapped genes in the aging process, SIR2 and TOR1, and their regulation, which are important in pathways involved in cellular longevity. We focus on the involvement of these two genes in replicative and chronological aging. Recently, we identified several biological processes that might have important roles in cellular longevity through the maintenance of optimal quiescence status (Choi et al. 2013a). Based on results from our and other's genome-wide screening, we suggest primary biological processes for achieving replicative and chronological longevity.

6.2 Replicative Aging and Chronological Aging

Measuring cellular lifespan is difficult; however, there are two different methods to measure budding yeast lifespan. They are replicative lifespan (RLS) and chronological lifespan (CLS). RLS measures the number of daughter cells that are produced from an isolated single mother cell under a microscope with a micropin. RLS is comparable to proliferative cells in higher organisms, such as stem cells (Mortimer and Johnston 1959; Egilmez and Jazwinski 1989). On the other hand, CLS measures the survival of cells with non-dividing status by counting the number of days these non-dividing cells remain viable in a batch culture. There are two assays to measure the CLS of cells, based on either colony forming ability or exclusion ability of non-permeable dye. CLS is comparable to post-mitotic cells, such as neurons and myocytes (Longo et al. 1996).

6.2.1 Assays to Measure Lifespan of Budding Yeast

RLS analysis uses microdissection by a micromanipulator equipped with a thin needle (Steffen et al. 2009; Sinclair 2013). An experienced technician picks a single yeast cell from a densely-populated area and isolates the cell in an addressed-empty area. Then, newly-born daughter cells (virgin cells) from the isolated mother cell are prepared as the starting point of the RLS measurement. Approximately 60–70 virgin cells per each experimental condition are arrayed in a row on the culture plate, and newborn daughter cells are removed manually until the virgin mother cell no longer produces buds. This method requires a skill and is labor-intensive. Indeed, RLS results are widely variable among laboratories (Huberts et al. 2014), and poor reproducibility of data from laboratory to laboratory is a problem. Therefore, more objective technical development is required, such as automatic assistance systems that use microfluidics to hold a mother cell and flow daughter cells for counting.

The recent development of technological applications, such as Mother Enrichment Program (MEP) (Lindstrom and Gottschling 2009), is promising in overcoming the above-mentioned RLS method. This approach uses genetically-modified yeast in which cell division is prevented in a daughter cell-specific manner. Two essential genes, UBC9 and CDC20, were disrupted only in daughter cells by Cre-*lox* recombination. The Cre recombinase is controlled by a daughter-specific promoter originating from SCW11, and is expressed only in daughter cells. Furthermore, the activity of Cre is dependent on estradiol, because the recombinase is fused with the estrogen-binding domain (EBD) of the murine estrogen receptor. In the presence of estradiol, the EBD translocates the modified Cre recombinase into the nucleus, where the Cre enzyme functions on the *loxP* sites. Thus, viability of the MEP strain in the presence of estradiol corresponds to RLS of the mother cells, and viability of the mother cells in liquid culture can be simply monitored by colony-based assays. Although there are several limitations to this technology that could permit the cells to avoid the selection strategy (estradiol resistance), including individual construction of MEP strains for target genes and natural mutations in the MEP strains, this method allows enrichment of mother cells at a specific age within the cell population by abolishing daughter cell proliferation. This could reduce the problem of cell loss caused by the purification step, and provide a number of cells for studying age-dependent phenotypes. A recent-developed microfluidic dissection platform allows for each mother cell to be separately trapped either chemically, through biotinylation of the mother cell surface (Xie et al. 2012), or mechanically, through geometric confinement (Lee et al. 2012; Zhang et al. 2012). The produced daughter cells can then be washed away by medium flow. Because these new tools retain mother cells throughout their lifespan, it is possible to automate RLS measurements and observe age-associated characteristics simultaneously.

To measure CLS, a colony-forming unit (CFU) assay is widely used as a gold standard (Hu et al. 2013). Non-dividing cells are harvested through chronological aging and plated onto solid media. After several days, yeast colonies on the plate are counted. This process is repeated until the CFU reaches below 1–10 %. Precision and reproducibility of the results depend on technical skill. This gold standard CFU assay is also a time-consuming process when multiple samples are tested. Therefore, other advanced approaches have been developed to enhance the efficiency of the CLS assay. The FUN-1 fluorescent probe is passively entered into a cell, and forms cylindrical intravacuolar structures that emit red fluorescence when cells are metabolically intact (Millard et al. 1997; Teng and Hardwick 2009). Propidium iodide (PI), a red-fluorescent nucleic acid probe, penetrates cells through a damaged plasma membrane. Thus, only dead or damaged cells show red fluorescence (Ocampo and Barrientos 2011). In addition, PI can be used in combination with other green-fluorescent dyes for counterstain, such as SYTO9 (Zhang and Fang 2004) or the acetoxymethyl ester (AM) of the esterase substrate 5-carboxy fluorescein diacetate (CFDA), abbreviated as CFDA-AM. After staining with these fluorescent dyes, the fluorescence signal can be detected with a flow cytometer or under a fluorescence microscope. When these dyes are used in a flow cytometer, consideration must be made toward the age-related autofluorescence (especially red fluorescence) that is produced by accumulation of oxidatively-damaged lipids and proteins in aged cells (Delori and Dorey 1998; Mesquita et al. 2010). This problem can be adjusted using a non-stained sample as a negative control for each measurement. Together, these fluorescence-based methods have several merits and exhibit high reproducibility from laboratory to laboratory.

Another high-throughput technique has been reported by the Kaeberlein lab (Murakami et al. 2008; Murakami and Kaeberlein 2009), in which CLS is determined from the growth curve of aged cells using an automated reader (Bioscreen C

MBR machine). In this system, chronologically-aging cells are diluted in fresh rich medium at each time-point, and the optical density (OD) at 600 nm is measured to plot a growth curve. The growth curve of aging cells is shifted to the right, and the time to reach a given OD is postponed compared to young cells. Based on the time delay between the growth curves of young and old cells, and the doubling time of the strain, a relative survival curve can be calculated. This method can also provide CLS data with better reproducibility and less effort than the other above-mentioned methods.

6.2.2 *Evolutionarily-Conserved Genes, SIR2 and TOR1, which Regulate RLS and CLS*

The pathways modulating cellular longevity are genetically conserved from yeast to mammals. The simplest model organism, budding yeast plays a pivotal role in identifying genes and pathways involved in the aging process and, thus, prevention of aging. Because SIR2 and TOR1 genes are key players in the lifespan of budding yeast, our review focuses on these two genes and their associated pathways in context of their effects on RLS and CLS.

6.2.2.1 Role of SIR2 in Replicative and Chronological Cell Aging

Yeast Silent Information Regulator 2 (SIR2), Sirtuin in mammals, is a heterochromatin protein that has nicotinamide adenine dinucleotide (NAD)-dependent histone deacetylase activity. Orthologs of SIR2 have been found in worms (SIR-2.1), flies (dSIR2), and mammals (SIRT1). This protein was first discovered as an anti-aging factor in yeast around the year 1999 (Fig. 6.1). SIR2 protein expression is reduced in an age-dependent manner (Dang et al. 2009). Deletion or inactivation of SIR2 leads to a shorter RLS than wild-type, whereas overexpression extends the RLS (Kaeberlein et al. 1999; Lin et al. 2000). In addition, several studies have reported that RLS extension by CR is partially mediated by SIR2 in a NAD^+-dependent way. For example, depletion of nicotinamide (NAM), a byproduct of NAD^+, was shown to activate SIR2, and this depletion was required for CR-mediated RLS extension (Anderson et al. 2003). Consistent with these results, overexpression of PNC1, encoding a nicotinamidase, that catalyzes NAM to nicotinic acid, increases RLS and prevents the SIR2-inhibition effect of NAM (Gallo et al. 2004). NAM is known to inhibit SIR2 activity non-competitively with NAD^+. Inhibition by NAM is suggested to occur through binding to a pocket near the NAD^+ binding site, thus blocking the binding and cleavage of NAD^+ (Bitterman et al. 2002; Zhao et al. 2004). However, another claim has proposed that NAM binds to the same pocket as NAD^+, and the non-competitive inhibition of NAM may be possible due to the low affinity of NAM to the binding pocket (Avalos et al. 2005). Moreover,

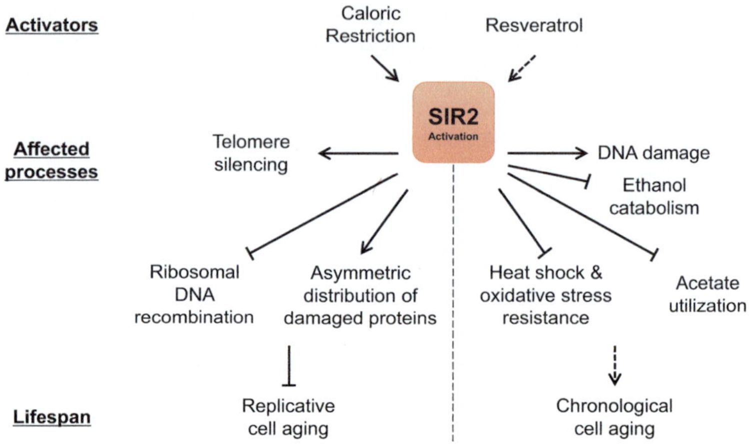

Fig. 6.1 Role of the SIR2 pathway in replicative and chronological aging processes. Caloric restriction (CR) and a CR mimetic, resveratrol, activate yeast SIR2. However, the role of resveratrol on activation of SIR2 is still under debate. SIR2 promotes telomere silencing and asymmetric distribution of damaged proteins, while it represses ribosomal DNA recombination. Consequently, SIR2 inhibits replicative cell aging. In contrast, SIR2 promotes chronological aging. The biological processes affected by activation of SIR2 related to chronological aging include reduction of ethanol catabolism, acetate utilization, and stress resistance, and an increase in DNA damage. *Dotted line*: in debate; *Arrow*: activation; *Bar*: inhibition

SIR2-mediated replicative longevity can be strengthened by other evidence that CR activates SIR2 by decreasing NADH, an inhibitor of SIR2 (Lin et al. 2004). Resveratrol, a type of polyphenol, has been suggested as a CR mimetic based on the mechanism by which the chemical activates SIR2, enhances DNA stability, and finally extends RLS (Howitz et al. 2003). However, follow-up studies had failed to repeat the results and concluded that SIR2-activation by resveratrol is a methodological artifact, and there is no effect on RLS (Kaeberlein et al. 2005a). Thus, the role of resveratrol on SIR2 and longevity is still under debate, and further research is needed.

The proposed mechanism by which SIR2 prolongs RLS is that the SIR2 protein represses recombination at the ribosomal DNA (rDNA) region and, consequently, decreases formation of toxic extrachromosomal rDNA circles (ERCs) (Kaeberlein et al. 1999). A recent study suggested that rDNA instability itself, instead of ERCs, is a primary cause of cellular senescence (Ganley et al. 2009). Along with this, other evidence has shown that SIR2 promotes RLS through maintenance of low acetylation levels on a histone H4K16 site at subtelomeric regions (Dang et al. 2009). Furthermore, overexpression of SIR2-orthologs in *Caenorhabditis elegans* (Tissenbaum and Guarente 2001; Schmeisser et al. 2013) and *Drosophila melanogaster* (Rogina and Helfand 2004; Hoffmann et al. 2013) extends their lifespan, while ERC accumulation was not observed in these organisms. Altogether, these results suggest that genomic stability, instead of ERC formation, is a key to maintaining cellular longevity. On the role of SIR2 in transcriptional silencing at rDNA (Gottlieb and Esposito 1989) and telomere loci (Strahl-Bolsinger et al. 1997; Armstrong et al. 2002), numerous studies have reported additional genes that act as a bridge between SIR2 and genome stability, which eventually influences lifespan. For instance, GAS1, encoding a beta-1,3-glucanosyltransferase required in cell wall assembly, interacts with SIR2 physically (Koch and Pillus 2009), and plays a negative role in rDNA stability in a SIR2-dependent manner (Ha et al. 2014). A recent report showed that TDH3, encoding a glyceraldehyde-3-phosphate dehydrogenase, also physically interacts with SIR2 and promotes transcriptional silencing at telomere and rDNA loci (Ringel et al. 2013). SGF73, encoding a component of the Spt-Ada-Gcn5-acetyltransferase (SAGA) complex, shows physical interaction with SIR2, and deletion of SGF73 extends RLS (McCormick et al. 2014). RLS extension was mediated by increased SIR2 action and was accompanied by enhanced genome stability, including decreased rDNA recombination and altered silencing at telomere regions. Another gene, RIF1, encoding RAP1 binding protein, is involved in transcriptional silencing (Hardy et al. 1992). The loss of RIF1 shortened RLS by limiting SIR2 localization on telomere regions, and this led to reduction of rDNA silencing by SIR2 (Salvi et al. 2013). Taken together, these findings show that SIR2-mediated RLS extension relies on genome stability, and might be a conserved longevity mechanism in other organisms.

The lifespan of newborn daughter cells from a mother cell was reset by the asymmetric cell division process. In this regard, several reports have suggested another role of SIR2 in rejuvenation by asymmetric distribution of damaged proteins. Aguilaniu et al. showed that carbonylated proteins are accumulated when cells age, however, the damaged proteins are not inherited into daughter cells (Aguilaniu et al. 2003). This asymmetric distribution of the damaged proteins failed when SIR2 was deleted (Aguilaniu et al. 2003; Erjavec et al. 2007). Overexpression of HSP104, encoding a heat shock protein that refolds denatured proteins, partially offset the defect in asymmetry and extended RLS in the *sir2*Δ mutant. This suggests that functional chaperones might contribute to the SIR2-dependent asymmetric process (Erjavec et al. 2007). UBP10, encoding a deubiquitinating enzyme, plays a role in genomic silencing mediated by SIR2 (Emre et al. 2005). Interestingly, a recent study showed that the SIR2-associated asymmetric segregation of oxidative damage was retained in cells lacking UBP10. This suggests that UBP10 is not needed in the asymmetric process and, thus, non-chromatin targets of SIR2 may be involved in the asymmetric retention of damaged proteins in mother cells (Orlandi et al. 2010).

In contrast with the above-mentioned evidence that SIR2 promotes replicative longevity through genomic stability and asymmetric processes, SIR2 either reduces CLS (Fabrizio et al. 2005; Casatta et al. 2013) or has almost no effect on CLS (Smith et al. 2007). Fabrizio et al. suggested possible mechanisms that block CLS extension by SIR2 (Fabrizio et al. 2005). Cells lacking SIR2 showed heat shock and oxidative stress resistance. Moreover, deletion of SIR2 in a long-lived mutant exhibited lower DNA mutation frequency. They also showed increased activity of an alcohol dehydrogenase, encoded by ADH2, in the *sir2*Δ mutant that leads to rapid removal of extracellular ethanol, a mediator of chronological aging. In addition, recent work has shown enhancement of acetate utilization in the *sir2Δ* mutant (Casatta et al. 2013). Acetate has been suggested to cause pro-aging signaling (Burtner et al. 2009). These findings show that SIR2 might play opposite roles in replicative and chronological aging processes, suggesting that Sirtuin might reflect different aspects of aging in isolated mammalian cells in culture and in cells of different tissues at other higher organisms.

6.2.2.2 Role of TOR in Replicative and Chronological Cell Aging

Another important factor in the aging process is Target of Rapamycin (TOR) signaling, which was first discovered as an aging modulator in worms (Vellai et al. 2003) in 2003 and flies (Kapahi et al. 2004) in 2004. TOR signaling also regulates the RLS (Kaeberlein et al. 2005b) and CLS (Powers et al. 2006; Bonawitz et al. 2007; Wei et al. 2008) of budding yeast (Fig. 6.2). In contrast with other higher organisms that have only one TOR, budding yeast have two paralogous TOR genes, TOR1 and TOR2. TOR Complex 1 (TORC1) contains either TOR1 or TOR2 with other components, such as KOG1, LST8, and TCO89 (Loewith et al. 2002; Reinke et al. 2004). In the case of TOR Complex 2 (TORC2), it consists of only TOR2, together with AVO1, AVO2, TSC11, LST8, BIT61, SLM1, and SLM2 (Loewith et al. 2002; Reinke et al. 2004; Fadri et al. 2005). TORC2 is insensitive to rapamycin because the rapamycin-FPR1p complex cannot bind to TOR2 (Loewith et al. 2002).

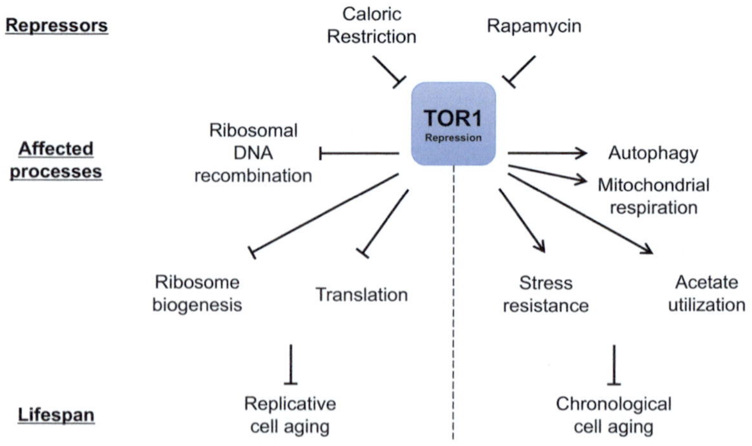

Fig. 6.2 Role of the TOR pathway in replicative and chronological aging. CR and rapamycin are known to repress activity of TORC1 (Target of Rapamycin Complex 1). Reduced TORC1 activity decreases ribosomal DNA (rDNA) recombination, ribosome biogenesis, and translation. As a result, inhibition of TORC1 retards replicative cellular senescence. Chronological cell aging is also repressed by inactivation of TORC1 through enhanced autophagy, mitochondrial respiration, acetate utilization, and stress resistance. *Dotted line*: in debate; *Arrow*: activation; *Bar*: inhibition

CR reduces TORC1 activity, and rapamycin treatment or deletion of the TOR1 gene extends both types of yeast lifespans (Kaeberlein et al. 2005b; Powers et al. 2006; Bonawitz et al. 2007; Wei et al. 2008). Regarding RLS extension by the inhibition of TOR signaling, a link between TOR and rDNA stability has been suggested. Deletion of TOR1 reduces rDNA recombination, while it does not affect rDNA silencing (Riesen and Morgan 2009). Medvedik et al. showed evidence that supports a model of how TOR modulates RLS in a SIR2-dependent way (Medvedik et al. 2007). Suppression of TOR by CR or rapamycin promotes localization of MSN2/4 into the nucleus, which leads to increased expression of PNC1. Then, SIR2 activated by PNC1 can stabilize the rDNA array and, consequently, contribute to RLS extension. A recent study also reported that inactivation of TORC1 by rapamycin enhances association between SIR2 and rDNA in a PNC1/NET1-dependent manner and increases rDNA stability (Ha and Huh 2011). These results suggest that TOR signaling may regulate RLS through downstream targets that are also affected by SIR2, along with Sirtuin-independent TOR pathways.

One of the TORC1-involved biological processes is regulation of cell growth by translation and ribosome biogenesis (Loewith et al. 2002). It was reported that reduction of 60S ribosomal subunits mediated by gene deletion or treatment with diazaborine, a chemical that disturbs pre-rRNA processing for the 60S ribosomal subunit, extends RLS, and this phenomenon is independent of SIR2 (Steffen et al. 2008). Instead of SIR2, GCN4 (a transcriptional activator of amino acid biosynthetic genes) is required for RLS extension by the reduction of large ribosomal subunits.

A lack of TOR1 extends CLS by enhancing mitochondrial respiration through increased translation of mitochondrial DNA-encoded oxidative phosphorylation (OXPHOS) subunits (Bonawitz et al. 2007). More precisely, reduced TOR signaling increases the abundance of OXPHOS subunits per mitochondrion without changing of the mitochondria number per cell (Pan and Shadel 2009). Furthermore, SCH9, a protein kinase that is an ortholog of the mammalian S6 kinase and resides downstream of TOR signaling, is responsible for the TOR-mitochondria link (Pan and Shadel 2009). Another report raised the possibility of carbon source substitution as a mediator of CLS extension by decreasing TOR/SCH9 signaling (Wei et al. 2009). They observed increased transcriptional expression of glycerol biosynthetic genes in *tor1Δ* and *sch9Δ* mutants. These mutants also accumulated extracellular glycerol, and the extended CLS of *sch9Δ* was reversed when glycerol biosynthetic genes were also deleted. Recently, Hu and colleagues showed that a deficiency in the TOR/SCH9 pathway exhibits rapid depletion of acetic acid in an ACH1 (a mitochondrial CoA transferase)-dependent manner (Hu et al. 2014). In addition, acetic acid in cells lacking SCH9 promotes the accumulation of reserve carbohydrate, trehalose. Taken together, these results suggest that metabolic switching toward mitochondrial respiration and accumulation of carbon sources may be primary mechanisms for CLS extension by TOR1 deficiency.

One of the many TOR-regulated biological processes is autophagy, which digests protein aggregates and allows for recycling of macromolecular elements, especially under starvation conditions (Noda and Ohsumi 1998). TORC1 is known to directly inhibit the autophagy process via phosphorylation of ATG13, causing a decrease in its affinity to the ATG1 complex and, as a result, inhibition of autophagy activation (Kamada et al. 2000, 2010). Recent studies suggest that induction of autophagy correlates with CLS extension (Eisenberg et al. 2009; Aris et al. 2013). In accordance with these findings, Alvers and coworkers showed that chronological longevity mediated by TORC1 inactivation requires autophagy that may contribute to recycling of cellular amino acids, including isoleucine, valine, and leucine (Alvers et al. 2009).

Stress resistance is another suggested mechanism that is responsible for CLS extension by decreased TOR signaling. It was reported that accumulation of storage carbohydrates, such as glycogen, glycerol, and trehalose, by TORC1 deficiency contributes to resistance to various types of stress, including heat shock, oxidative, and osmotic stress (Powers et al. 2006; Wei et al. 2009; Hu et al. 2014). Study of CR shows that activation of genetic pathways for the TOR-mediated stress response is well established, and includes the serine/threonine kinase RIM15 and the stress resistance transcription factor GIS1 (Wei et al. 2008). With regard to oxidative stress, early research suggested reduced mitochondrial ROS as a key pathway for chronological longevity (Harman 1956). However, recent studies have raised another possibility that adaptive mitochondrial ROS signaling during growth may participate in the TOR-associated longevity pathway (Pan et al. 2011). For instance, it was observed that elevated mitochondrial ROS during exponential growth in the *tor1*Δ mutant or rapamycin-treated cells is required for fully-extended CLS. These results were also supported by several reports showing that CR mediates an increase of hydrogen peroxide at the early stationary phase that activates superoxide dismutase and, consequently, reduces the production of superoxide anions (Mesquita et al. 2010; Weinberger et al. 2010). However, the concept of mitohormesis is still controversial for explaining all relationships between ROS and lifespan. Further research is required, especially with the application of methods for more precise measurement of each specific ROS, and might be helpful in progressing our understanding of ROS roles in longevity.

6.3 Biological Processes Required for Longevity: Insight from Genome-Wide Studies

Because of large-scale screening using yeast mutant libraries and microarray technology, our knowledge on the aging process has expanded rapidly (Fig. 6.3). In one of these approaches associated with replicative aging, 564 single-gene deletion mutants were used in a RLS assay, and 13 of them, including *bre5*Δ, *fob1*Δ, *idh2*Δ, *rei1*Δ, *rom2*Δ, *rpl31a*Δ, *rpl6b*Δ, *tor1*Δ, *ure2*Δ, *ybr238c*Δ, *ybr255w*Δ (*mtc4*Δ), *ybr266c*Δ (*slm6*Δ), and *yor135c*Δ (*irc14*Δ), were identified to increase RLS. Interestingly, TOR1, URE2, ROM2, YBR238C, RPL31A, and RPL6B reside in the TOR/SCH9 pathway, which suggests a link between RLS and

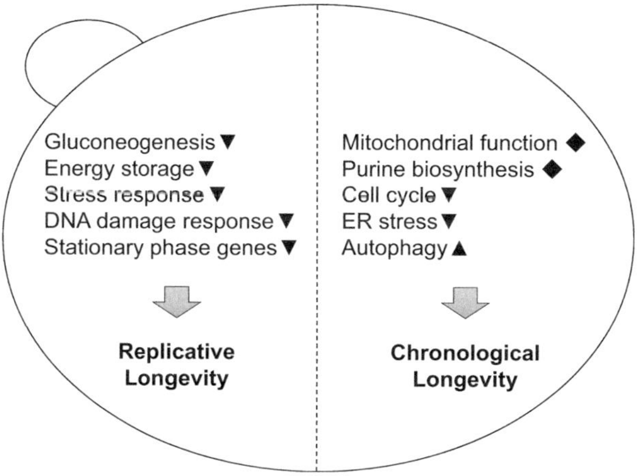

Fig. 6.3 Biological processes that determine replicative and chronological longevity. Based on genome-wide research, several biological processes lead to replicative and chronological longevity. Down-regulation of genes for gluconeogenesis, energy storage, response to environmental stress, DNA damage response, and the stationary phase is common in replicatively long-lived cells. For assurance of chronological longevity, proper alteration of gene expression for mitochondrial function and purine biosynthesis, down-regulation of genes for cell cycle and endoplasmic reticulum (ER) stress, and up-regulation of genes for autophagy are required

TOR/SCH9 signaling in budding yeast (Kaeberlein et al. 2005b). Another similar study using the deletion library showed that yeast lacking orthologs of worm aging genes, including AFG3, IDH2, INP53, RPL19A, RPL6B, RPL9A, SPT4, TIF1, TIF2, TIF4631, and TOR1, exhibit an extended RLS (Smith et al. 2008). This finding implies that the longevity pathway is evolutionarily conserved, and is related to regulation of ribosomal proteins and translation by TOR. Schleit and colleagues carried out a RLS assay for 166 single-gene deletion strains under control and CR conditions (Schleit et al. 2013). Their results revealed that vacuolar pH homeostasis, mitochondrial superoxide dismutase and mitochondrial proteostasis are required for CR-mediated RLS extension.

Several high-throughput methods using OD-based CLS measurements, barcode arrays detecting specific sequences of deletion mutants, and fluorescence-labeled yeast libraries have been applied in addition to large-scale individual CLS assays. In one study, 93 yeast deletion strains matched with candidate genes for human aging were investigated for CLS, and the results showed new longevity genes, including ALD4 (a mitochondrial aldehyde dehydrogenase responsible for ethanol metabolism), CYS4 (a cystathionine beta synthase required for cysteine biosynthesis), and PDX3 (a pyridoxine [pyridoxamine] phosphate oxidase involved in the pyridoxal phosphate salvage pathway) (Laschober et al. 2010). Another recent study reported the screening results of 550 single-gene deletion strains, and identified 34 genes whose deletion increased CLS (Burtner et al. 2011). Gene ontology (GO) analysis revealed GTPase activity to be significantly enriched in these long-lived mutants. In addition, reduction of acidification in culture medium was enriched in the CLS-extended mutants, based on a pH screen.

Based on the high correlation between OD and yeast viability, an OD-based CLS assay covering 4,800 single-gene deletion mutants was performed, and the importance of a decreased TOR pathway for CLS extension was identified (Powers et al. 2006). Two recent studies, which carried out CLS measurements using a barcode array after pooling deletion yeast mutants tagged with unique DNA barcode sequences, identified that autophagy and mitochondrial function are critical for maximum CLS (Matecic et al. 2010; Fabrizio et al. 2010). In addition to this, non-overlapping biological processes were also identified as a determinant of CLS. Matecic and colleagues showed that deletion of genes involved in purine biosynthesis or one-carbon metabolism extends lifespan (Matecic et al. 2010). Fabrizio et al. reported that short-lived strains lose genes for vacuolar protein sorting, and long-lived strains lose genes involved in fatty acid metabolism, cell signaling, and tRNA methylation (Fabrizio et al. 2010). Recently, a genome-wide CLS screen using a red fluorescent protein (RFP)-tagged yeast collection, including over 5,600 single- or double-deletion mutants, revealed that 14 % of the mutants are associated with cellular lifespan. In concordance with biological categories that were underscored in previous reports, deletion mutants involved in mitochondrial function, purine biosynthesis, and autophagy showed reduced CLS (Garay et al. 2014). This report also suggested novel longevity factors, SWR1, encoding a component of the histone exchange complex, and ARV1, encoding an endoplasmic reticulum (ER) protein required for lipid homeostasis (Garay et al. 2014). Cells lacking SWR1 lived longer than wild-type in nitrogen-rich media, but not in nitrogen-poor conditions, implying that SWR1 is partially involved in lifespan extension by dietary restriction. Furthermore, *arv1*Δ showed short CLS and impaired autophagy, which indicates a novel link between ARV1 and autophagy as a pro-longevity factor (Madeo et al. 2010).

Microarray is also a powerful tool for characterizing the longevity mechanism in an unbiased manner. In connection with replicative aging, several studies showed an altered gene expression profile of old cells compared to young cells. For instance, Lin and coworkers purified 7–8 generation old cells by magnetic sorting, and identified a metabolic shift toward gluconeogenesis and glucose storage in these old cells (Lin et al. 2001). In another study, microarray was performed on 1–3 generation young cells and 16–18 generation old cells, which were isolated by elutriation (Lesur and Campbell 2004). The results of this comparison study also revealed an increase in gene expression related to gluconeogenesis and glycogen production in the aged cells. Furthermore, this study identified increased expression of genes involved in stress response and DNA damage in the old cells, which suggests a response to genome instability by cellular senescence. Laun et al. obtained an old population, of which one third consisted of cells with approximately 15 bud scars, by centrifugal elutriation (Laun et al. 2005). The transcriptome analysis of these old cells relative to young cells showed induction of genes involved in DNA damage response, cell cycle, mitochondrial function, lipid metabolism and cell wall synthesis, which coincided with the transcriptome of *cdc48* and *orc2-1* mutants who partially lost their function and showed apoptotic phenotype at non-permissive temperatures (Laun et al. 2005). Other research on replicative senescence has used segmentalized magnetic sorting to obtain 1, 8, 12, and 18–20 generation old cells (Yiu et al. 2008). They found that global gene expression was significantly changed after the 12th generation, and the biological categories affected by replicative aging were consistent with previous observations, including increased aerobic metabolism, reduced ribosome gene expression, and a mixed change regarding environmental stress response. In addition, the replicative age-associated changes on transcripts included a decrease in gene expression of the methylation-related pathway and an increase in gene expression of nucleotide excision repair and the pseudostationary phase (Yiu et al. 2008). The pseudostationary phase is defined as a change in gene expression, similar to that of cells in the stationary phase, despite the cells having not actually entered the stationary phase. Furthermore, most regulatory subunits of GLC7 (a catalytic subunit of protein phosphatase 1) were up-regulated, which implies that protein phosphorylation cascades hold an important position in replicative aging (Yiu et al. 2008). Recent transcriptome data from an early stage of replicative aging (4, 7, and 11 generations old) showed up-regulation of genes for sugar metabolism, the tricarboxylic acid (TCA) cycle, and amino acid degradation, and down-regulation of genes for amino acid synthesis (Kamei et al. 2014). These results show coherence with additional metabolomic results that identified an increase of pyruvate and TCA cycle intermediates, and a decrease of amino acids (Kamei et al. 2014).

In a chronologically aging population, quiescent cells (dense and minimally-budded daughter cells) and non-quiescent cells (heterogeneous and replicatively-old cells) can be fractionated by density gradient centrifugation (Allen et al. 2006). By doing a gene expression comparison between the two fractions, 68 and 266 genes were highly expressed in either quiescent or non-quiescent cells, respectively. Gene ontology analysis showed that up-regulated genes in quiescent cells were enriched in those that play a role in response to water stress, energy metabolism, and response to ROS, while those of non-quiescent cells were enriched in genes involved in DNA recombination and transposition. Two independent studies for long-lived mutants, including *sch9*Δ, *ras2*Δ, and *tor1*Δ, showed transcriptional changes in a common direction, including reduced gene expression for the TCA cycle and oxidative phosphorylation, and increased gene expression for glycolysis, compared to wild-type (Cheng et al. 2007b; Wei et al. 2009). Furthermore, genes involved in ergosterol biosynthesis, glycerol biosynthesis, and stress response were up-regulated, whereas genes for the proteasome complex were down-regulated in these long-lived mutants (Wei et al. 2009). Cheng and coworkers also reported that transcriptional modification in these three mutants can be mediated by the stress response transcription factors MSN2, MSN4, and GIS1, and is possibly linked with other transcription factors, such as FHL1 and HSF1 (Cheng et al. 2007a).

When we apply high-throughput technology, such as a microarray experiment, it is important to assure which transcriptome profile should be obtained to provide an unbiased view for aging research. The microarray of the long-lived mutants discussed above was carried out at a single time-point (2.5 days after inoculation in SDC media) (Cheng et al. 2007a, b; Wei et al. 2009) to evaluate the effects of gene deletion. Recently, there was an additional attempt to characterize the *sch9Δ* mutant based on a time-course transcriptome profile from 12 to 120 h (Ge et al. 2010). Through the course of time, genes involved in stress response were increased, and genes encoding ribosomal RNA (rRNA) processing were decreased between 12 and 24 h. Moreover, mitochondrial function-related genes were downregulated after 24 h. In addition, from volatility analysis that provides information about variation of gene expression at a specific time, they selected 50 genes were selected that are most volatile in *sch9Δ*. By analyzing upstream sequences of the selected genes, three cis-regulatory elements, rRNA processing element (rRPE), polymerase A and C (PAC), and glucose response element (GRE), were found. To emphasize the importance of these elements, additional deletion of a GRE-binding transcription factor, AZF1, in *sch9Δ* abolished the extended CLS.

In our previous high-throughput studies, the transcriptome profile for several levels of CR strength at the exponential growth phase revealed 160 genes as CR markers, of which a significantly large number was mitochondria-related genes (Lee and Lee 2008). Most genes for mitochondrial function were gradually increased according to the CR strength (Lee and Lee 2008; Choi et al. 2011). However, our recent work after the diauxic shift phase, when CLS begins to increase under CR, showed down-regulation of genes for the TCA cycle and electron transport chain (Choi et al. 2013a). In addition, we observed that CR increased transcripts involved in transcriptional regulation, rRNA processing and nuclear genome stability, whereas CR decreased transcripts involved in general metabolic processes, ER function, stress response, and the cell cycle, from a longevity-assured time point. By analyzing the promoter regions of these longevity-related genes, AZF1, HSF1, and XBP1 were identified as the transcriptional factors causing these transcriptional changes (Choi et al. 2013a). Because timing of experiments affects the overall analysis, careful consideration on the cellular status for better understanding of the aging process is important.

6.4 Conclusion

Aging research in budding yeast has progressed in terms of chronological and replicative senescence. These two paradigms of aging are distinct, but partially overlapping, and contribute to an in-depth understanding of anti-aging genes and longevity pathways. Evolutionary conservation of the genetic network from yeast to metazoa enables us to find evidence of the genes and pathways involved in the aging process in different cell types. Although genome-wide approaches can produce false positives and negatives, these studies aid the identification of novel genes and pathways related to longevity that were previously unknown. Further technological improvements and sophistication of high-throughput techniques combined with various yeast transgenic libraries could accelerate the discovery of new genes and new biological processes. Understanding of replicative and chronological cell aging using the simplest aging model, budding yeast, might shed light on how organisms age at a cellular and molecular level and, ultimately, permits us to control the aging process in various mitotic and post-mitotic cell types of higher organisms, including humans.

Acknowledgments This work was supported by the "Cooperative Research Program for Agriculture Science & Technology Development (Project No. PJ01053302)" of the Rural Development Administration, Republic of Korea, and by a Korea University Grant.

References

Aguilaniu H, Gustafsson L, Rigoulet M, Nystrom T (2003) Asymmetric inheritance of oxidatively damaged proteins during cytokinesis. Science 299(5613):1751–1753. doi:10.1126/science.1080418

Allen C, Buttner S, Aragon AD, Thomas JA, Meirelles O, Jaetao JE, Benn D, Ruby SW, Veenhuis M, Madeo F, Werner-Washburne M (2006) Isolation of quiescent and nonquiescent cells from yeast stationary-phase cultures. J Cell Biol 174(1):89–100. doi:10.1083/jcb.200604072

Alvers AL, Fishwick LK, Wood MS, Hu D, Chung HS, Dunn WA Jr, Aris JP (2009) Autophagy and amino acid homeostasis are required for chronological longevity in Saccharomyces cerevisiae. Aging Cell 8(4):353–369. doi:10.1111/j.1474-9726.2009.00469.x

Anderson RM, Bitterman KJ, Wood JG, Medvedik O, Sinclair DA (2003) Nicotinamide and PNC1 govern lifespan extension by calorie restriction in Saccharomyces cerevisiae. Nature 423 (6936):181–185. doi:10.1038/nature01578

Aris JP, Alvers AL, Ferraiuolo RA, Fishwick LK, Hanvivatpong A, Hu D, Kirlew C, Leonard MT, Losin KJ, Marraffini M, Seo AY, Swanberg V, Westcott JL, Wood MS, Leeuwenburgh C, Dunn WA Jr (2013) Autophagy and leucine promote chronological longevity and respiration proficiency during calorie restriction in yeast. Exp Gerontol 48(10):1107–1119. doi:10.1016/j.exger.2013.01.006

Armstrong CM, Kaeberlein M, Imai SI, Guarente L (2002) Mutations in Saccharomyces cerevisiae gene SIR2 can have differential effects on in vivo silencing phenotypes and in vitro histone deacetylation activity. Mol Biol Cell 13(4):1427–1438. doi:10.1091/mbc.01-10-0482

Avalos JL, Bever KM, Wolberger C (2005) Mechanism of sirtuin inhibition by nicotinamide: altering the NAD(+) cosubstrate specificity of a Sir2 enzyme. Mol Cell 17(6):855–868. doi:10.1016/j.molcel.2005.02.022

Bitterman KJ, Anderson RM, Cohen HY, Latorre-Esteves M, Sinclair DA (2002) Inhibition of silencing and accelerated aging by nicotinamide, a putative negative regulator of yeast sir2 and human SIRT1. J Biol Chem 277(47):45099–45107. doi:10.1074/jbc.M205670200

Bonawitz ND, Chatenay-Lapointe M, Pan Y, Shadel GS (2007) Reduced TOR signaling extends chronological life span via increased respiration and upregulation of mitochondrial gene expression. Cell Metab 5(4):265–277. doi:10.1016/j.cmet.2007.02.009

Burtner CR, Murakami CJ, Kennedy BK, Kaeberlein M (2009) A molecular mechanism of chronological aging in yeast. Cell Cycle 8(8):1256–1270

Burtner CR, Murakami CJ, Olsen B, Kennedy BK, Kaeberlein M (2011) A genomic analysis of chronological longevity factors in budding yeast. Cell Cycle 10(9):1385–1396

Casatta N, Porro A, Orlandi I, Brambilla L, Vai M (2013) Lack of Sir2 increases acetate consumption and decreases extracellular pro-aging factors. Biochim Biophys Acta 1833 (3):593–601. doi:10.1016/j.bbamcr.2012.11.008

Cheng C, Fabrizio P, Ge H, Longo VD, Li LM (2007a) Inference of transcription modification in long-live yeast strains from their expression profiles. BMC Genomics 8:219. doi:10.1186/1471-2164-8-219

Cheng C, Fabrizio P, Ge H, Wei M, Longo VD, Li LM (2007b) Significant and systematic expression differentiation in long-lived yeast strains. PLoS One 2(10):e1095. doi:10.1371/journal.pone.0001095

Choi JS, Lee CK (2013) Maintenance of cellular ATP level by caloric restriction correlates chronological survival of budding yeast. Biochem Biophys Res Commun 439(1):126–131. doi:10.1016/j.bbrc.2013.08.014

Choi JS, Choi KM, Lee CK (2011) Caloric restriction improves efficiency and capacity of the mitochondrial electron transport chain in Saccharomyces cerevisiae. Biochem Biophys Res Commun 409(2):308–314. doi:10.1016/j.bbrc.2011.05.008

Choi KM, Kwon YY, Lee CK (2013a) Characterization of global gene expression during assurance of lifespan extension by caloric restriction in budding yeast. Exp Gerontol 48 (12):1455–1468. doi:10.1016/j.exger.2013.10.001

Choi KM, Lee HL, Kwon YY, Kang MS, Lee SK, Lee CK (2013b) Enhancement of mitochondrial function correlates with the extension of lifespan by caloric restriction and caloric restriction mimetics in yeast. Biochem Biophys Res Commun 441(1):236–242. doi:10.1016/j.bbrc.2013.10.049

Choi K, Kwon Y, Lee C (2014) Disruption of Snf3/Rgt2 glucose sensors decreases lifespan and caloric restriction effectiveness through Mth1/Std1 by adjusting mitochondrial efficiency in yeast. FEBS Lett. doi:10.1016/j.febslet.2014.12.020

Dang W, Steffen KK, Perry R, Dorsey JA, Johnson FB, Shilatifard A, Kaeberlein M, Kennedy BK, Berger SL (2009) Histone H4 lysine 16 acetylation regulates cellular lifespan. Nature 459 (7248):802–807. doi:10.1038/nature08085

Delori FC, Dorey CK (1998) In vivo technique for autofluorescent lipopigments. Methods Mol Biol 108:229–243. doi:10.1385/0-89603-472-0:229

Egilmez NK, Jazwinski SM (1989) Evidence for the involvement of a cytoplasmic factor in the aging of the yeast Saccharomyces cerevisiae. J Bacteriol 171(1):37–42

Eisenberg T, Knauer H, Schauer A, Buttner S, Ruckenstuhl C, Carmona-Gutierrez D, Ring J, Schroeder S, Magnes C, Antonacci L, Fussi H, Deszcz L, Hartl R, Schraml E, Criollo A, Megalou E, Weiskopf D, Laun P, Heeren G, Breitenbach M, Grubeck-Loebenstein B, Herker E, Fahrenkrog B, Frohlich KU, Sinner F, Tavernarakis N, Minois N, Kroemer G, Madeo F (2009) Induction of autophagy by spermidine promotes longevity. Nat Cell Biol 11 (11):1305–1314. doi:10.1038/ncb1975

Emre NC, Ingvarsdottir K, Wyce A, Wood A, Krogan NJ, Henry KW, Li K, Marmorstein R, Greenblatt JF, Shilatifard A, Berger SL (2005) Maintenance of low histone ubiquitylation by Ubp10 correlates with telomere-proximal Sir2 association and gene silencing. Mol Cell 17 (4):585–594. doi:10.1016/j.molcel.2005.01.007

Erjavec N, Larsson L, Grantham J, Nystrom T (2007) Accelerated aging and failure to segregate damaged proteins in Sir2 mutants can be suppressed by overproducing the protein aggregation-remodeling factor Hsp104p. Genes Dev 21(19):2410–2421. doi:10.1101/gad.439307

Fabrizio P, Gattazzo C, Battistella L, Wei M, Cheng C, McGrew K, Longo VD (2005) Sir2 blocks extreme life-span extension. Cell 123(4):655–667. doi:10.1016/j.cell.2005.08.042

Fabrizio P, Hoon S, Shamalnasab M, Galbani A, Wei M, Giaever G, Nislow C, Longo VD (2010) Genome-wide screen in Saccharomyces cerevisiae identifies vacuolar protein sorting, autophagy, biosynthetic, and tRNA methylation genes involved in life span regulation. PLoS Genet 6(7):e1001024. doi:10.1371/journal.pgen.1001024

Fadri M, Daquinag A, Wang S, Xue T, Kunz J (2005) The pleckstrin homology domain proteins Slm1 and Slm2 are required for actin cytoskeleton organization in yeast and bind phosphatidylinositol-4,5-bisphosphate and TORC2. Mol Biol Cell 16(4):1883–1900. doi:10.1091/mbc.E04-07-0564

Gallo CM, Smith DL Jr, Smith JS (2004) Nicotinamide clearance by Pnc1 directly regulates Sir2-mediated silencing and longevity. Mol Cell Biol 24(3):1301–1312
Ganley AR, Ide S, Saka K, Kobayashi T (2009) The effect of replication initiation on gene amplification in the rDNA and its relationship to aging. Mol Cell 35(5):683–693. doi:10.1016/j.molcel.2009.07.012
Garay E, Campos SE, Gonzalez de la Cruz J, Gaspar AP, Jinich A, Deluna A (2014) High-resolution profiling of stationary-phase survival reveals yeast longevity factors and their genetic interactions. PLoS Genet 10(2):e1004168. doi:10.1371/journal.pgen.1004168
Ge H, Wei M, Fabrizio P, Hu J, Cheng C, Longo VD, Li LM (2010) Comparative analyses of time-course gene expression profiles of the long-lived sch9Delta mutant. Nucleic Acids Res 38 (1):143–158. doi:10.1093/nar/gkp849
Gottlieb S, Esposito RE (1989) A new role for a yeast transcriptional silencer gene, SIR2, in regulation of recombination in ribosomal DNA. Cell 56(5):771–776
Ha CW, Huh WK (2011) Rapamycin increases rDNA stability by enhancing association of Sir2 with rDNA in Saccharomyces cerevisiae. Nucleic Acids Res 39(4):1336–1350. doi:10.1093/nar/gkq895
Ha CW, Kim K, Chang YJ, Kim B, Huh WK (2014) The beta-1,3-glucanosyltransferase Gas1 regulates Sir2-mediated rDNA stability in Saccharomyces cerevisiae. Nucleic Acids Res 42 (13):8486–8499. doi:10.1093/nar/gku570
Hardy CF, Sussel L, Shore D (1992) A RAP1-interacting protein involved in transcriptional silencing and telomere length regulation. Genes Dev 6(5):801–814
Harman D (1956) Aging: a theory based on free radical and radiation chemistry. J Gerontol 11 (3):298–300
Harman D (1992) Free radical theory of aging. Mutat Res 275(3–6):257–266
Hoffmann J, Romey R, Fink C, Yong L, Roeder T (2013) Overexpression of Sir2 in the adult fat body is sufficient to extend lifespan of male and female Drosophila. Aging 5(4):315–327
Howitz KT, Bitterman KJ, Cohen HY, Lamming DW, Lavu S, Wood JG, Zipkin RE, Chung P, Kisielewski A, Zhang LL, Scherer B, Sinclair DA (2003) Small molecule activators of sirtuins extend Saccharomyces cerevisiae lifespan. Nature 425(6954):191–196. doi:10.1038/nature01960
Hu J, Wei M, Mirisola MG, Longo VD (2013) Assessing chronological aging in Saccharomyces cerevisiae. Methods Mol Biol 965:463–472. doi:10.1007/978-1-62703-239-1_30
Hu J, Wei M, Mirzaei H, Madia F, Mirisola M, Amparo C, Chagoury S, Kennedy B, Longo VD (2014) Tor-Sch9 deficiency activates catabolism of the ketone body-like acetic acid to promote trehalose accumulation and longevity. Aging Cell 13(3):457–467. doi:10.1111/acel.12202
Huberts DH, Gonzalez J, Lee SS, Litsios A, Hubmann G, Wit EC, Heinemann M (2014) Calorie restriction does not elicit a robust extension of replicative lifespan in Saccharomyces cerevisiae. Proc Natl Acad Sci U S A 111(32):11727–11731. doi:10.1073/pnas.1410024111
Hulbert AJ, Pamplona R, Buffenstein R, Buttemer WA (2007) Life and death: metabolic rate, membrane composition, and life span of animals. Physiol Rev 87(4):1175–1213. doi:10.1152/physrev.00047.2006
Kaeberlein M, McVey M, Guarente L (1999) The SIR2/3/4 complex and SIR2 alone promote longevity in Saccharomyces cerevisiae by two different mechanisms. Genes Dev 13 (19):2570–2580
Kaeberlein M, McDonagh T, Heltweg B, Hixon J, Westman EA, Caldwell SD, Napper A, Curtis R, DiStefano PS, Fields S, Bedalov A, Kennedy BK (2005a) Substrate-specific activation of sirtuins by resveratrol. J Biol Chem 280(17):17038–17045. doi:10.1074/jbc.M500655200
Kaeberlein M, Powers RW 3rd, Steffen KK, Westman EA, Hu D, Dang N, Kerr EO, Kirkland KT, Fields S, Kennedy BK (2005b) Regulation of yeast replicative life span by TOR and Sch9 in response to nutrients. Science 310(5751):1193–1196. doi:10.1126/science.1115535
Kamada Y, Funakoshi T, Shintani T, Nagano K, Ohsumi M, Ohsumi Y (2000) Tor-mediated induction of autophagy via an Apg1 protein kinase complex. J Cell Biol 150(6):1507–1513

Kamada Y, Yoshino K, Kondo C, Kawamata T, Oshiro N, Yonezawa K, Ohsumi Y (2010) Tor directly controls the Atg1 kinase complex to regulate autophagy. Mol Cell Biol 30 (4):1049–1058. doi:10.1128/MCB.01344-09

Kamei Y, Tamada Y, Nakayama Y, Fukusaki E, Mukai Y (2014) Changes in transcription and metabolism during the early stage of replicative cellular senescence in budding yeast. J Biol Chem 289(46):32081–32093. doi:10.1074/jbc.M114.600528

Kapahi P, Zid BM, Harper T, Koslover D, Sapin V, Benzer S (2004) Regulation of lifespan in Drosophila by modulation of genes in the TOR signaling pathway. Curr Biol: CB 14 (10):885–890. doi:10.1016/j.cub.2004.03.059

Kirkwood TB (2005) Understanding the odd science of aging. Cell 120(4):437–447. doi:10.1016/j.cell.2005.01.027

Koch MR, Pillus L (2009) The glucanosyltransferase Gas1 functions in transcriptional silencing. Proc Natl Acad Sci U S A 106(27):11224–11229. doi:10.1073/pnas.0900809106

Laschober GT, Ruli D, Hofer E, Muck C, Carmona-Gutierrez D, Ring J, Hutter E, Ruckenstuhl C, Micutkova L, Brunauer R, Jamnig A, Trimmel D, Herndler-Brandstetter D, Brunner S, Zenzmaier C, Sampson N, Breitenbach M, Frohlich KU, Grubeck-Loebenstein B, Berger P, Wieser M, Grillari-Voglauer R, Thallinger GG, Grillari J, Trajanoski Z, Madeo F, Lepperdinger G, Jansen-Durr P (2010) Identification of evolutionarily conserved genetic regulators of cellular aging. Aging Cell 9(6):1084–1097. doi:10.1111/j.1474-9726.2010.00637.x

Laun P, Ramachandran L, Jarolim S, Herker E, Liang P, Wang J, Weinberger M, Burhans DT, Suter B, Madeo F, Burhans WC, Breitenbach M (2005) A comparison of the aging and apoptotic transcriptome of Saccharomyces cerevisiae. FEMS Yeast Res 5(12):1261–1272. doi:10.1016/j.femsyr.2005.07.006

Lee YL, Lee CK (2008) Transcriptional response according to strength of calorie restriction in Saccharomyces cerevisiae. Mol Cells 26(3):299–307

Lee SS, Avalos Vizcarra I, Huberts DH, Lee LP, Heinemann M (2012) Whole lifespan microscopic observation of budding yeast aging through a microfluidic dissection platform. Proc Natl Acad Sci U S A 109(13):4916–4920. doi:10.1073/pnas.1113505109

Lesur I, Campbell JL (2004) The transcriptome of prematurely aging yeast cells is similar to that of telomerase-deficient cells. Mol Biol Cell 15(3):1297–1312. doi:10.1091/mbc.E03-10-0742

Lin SJ, Defossez PA, Guarente L (2000) Requirement of NAD and SIR2 for life-span extension by calorie restriction in Saccharomyces cerevisiae. Science 289(5487):2126–2128

Lin SS, Manchester JK, Gordon JI (2001) Enhanced gluconeogenesis and increased energy storage as hallmarks of aging in Saccharomyces cerevisiae. J Biol Chem 276(38):36000–36007. doi:10.1074/jbc.M103509200

Lin SJ, Ford E, Haigis M, Liszt G, Guarente L (2004) Calorie restriction extends yeast life span by lowering the level of NADH. Genes Dev 18(1):12–16. doi:10.1101/gad.1164804

Lindstrom DL, Gottschling DE (2009) The mother enrichment program: a genetic system for facile replicative life span analysis in Saccharomyces cerevisiae. Genetics 183(2):413–422. doi:10.1534/genetics.109.106229, 411SI-413SI

Loewith R, Jacinto E, Wullschleger S, Lorberg A, Crespo JL, Bonenfant D, Oppliger W, Jenoe P, Hall MN (2002) Two TOR complexes, only one of which is rapamycin sensitive, have distinct roles in cell growth control. Mol Cell 10(3):457–468

Longo VD, Gralla EB, Valentine JS (1996) Superoxide dismutase activity is essential for stationary phase survival in Saccharomyces cerevisiae. Mitochondrial production of toxic oxygen species in vivo. J Biol Chem 271(21):12275–12280

Madeo F, Tavernarakis N, Kroemer G (2010) Can autophagy promote longevity? Nat Cell Biol 12 (9):842–846. doi:10.1038/ncb0910-842

Matecic M, Smith DL, Pan X, Maqani N, Bekiranov S, Boeke JD, Smith JS (2010) A microarray-based genetic screen for yeast chronological aging factors. PLoS Genet 6(4):e1000921. doi:10.1371/journal.pgen.1000921

McCormick MA, Mason AG, Guyenet SJ, Dang W, Garza RM, Ting MK, Moller RM, Berger SL, Kaeberlein M, Pillus L, La Spada AR, Kennedy BK (2014) The SAGA histone deubiquitinase module controls yeast replicative lifespan via Sir2 interaction. Cell Rep 8(2):477–486. doi:10.1016/j.celrep.2014.06.037

Medvedik O, Lamming DW, Kim KD, Sinclair DA (2007) MSN2 and MSN4 link calorie restriction and TOR to sirtuin-mediated lifespan extension in Saccharomyces cerevisiae. PLoS Biol 5(10):e261. doi:10.1371/journal.pbio.0050261

Mesquita A, Weinberger M, Silva A, Sampaio-Marques B, Almeida B, Leao C, Costa V, Rodrigues F, Burhans WC, Ludovico P (2010) Caloric restriction or catalase inactivation extends yeast chronological lifespan by inducing H2O2 and superoxide dismutase activity. Proc Natl Acad Sci U S A 107(34):15123–15128. doi:10.1073/pnas.1004432107

Millard PJ, Roth BL, Thi HP, Yue ST, Haugland RP (1997) Development of the FUN-1 family of fluorescent probes for vacuole labeling and viability testing of yeasts. Appl Environ Microbiol 63(7):2897–2905

Mortimer RK, Johnston JR (1959) Life span of individual yeast cells. Nature 183 (4677):1751–1752

Murakami C, Kaeberlein M (2009) Quantifying yeast chronological life span by outgrowth of aged cells. J Vis Exp 27. doi:10.3791/1156. http://www.jove.com/index/Details.stp?ID=1156

Murakami CJ, Burtner CR, Kennedy BK, Kaeberlein M (2008) A method for high-throughput quantitative analysis of yeast chronological life span. J Gerontol A Biol Sci Med Sci 63 (2):113–121

Noda T, Ohsumi Y (1998) Tor, a phosphatidylinositol kinase homologue, controls autophagy in yeast. J Biol Chem 273(7):3963–3966

Ocampo A, Barrientos A (2011) Quick and reliable assessment of chronological life span in yeast cell populations by flow cytometry. Mech Ageing Dev 132(6–7):315–323. doi:10.1016/j.mad.2011.06.007

Orlandi I, Bettiga M, Alberghina L, Nystrom T, Vai M (2010) Sir2-dependent asymmetric segregation of damaged proteins in ubp10 null mutants is independent of genomic silencing. Biochim Biophys Acta 1803(5):630–638. doi:10.1016/j.bbamcr.2010.02.009

Pan Y, Shadel GS (2009) Extension of chronological life span by reduced TOR signaling requires down-regulation of Sch9p and involves increased mitochondrial OXPHOS complex density. Aging 1(1):131–145

Pan Y, Schroeder EA, Ocampo A, Barrientos A, Shadel GS (2011) Regulation of yeast chronological life span by TORC1 via adaptive mitochondrial ROS signaling. Cell Metab 13 (6):668–678. doi:10.1016/j.cmet.2011.03.018

Powers RW 3rd, Kaeberlein M, Caldwell SD, Kennedy BK, Fields S (2006) Extension of chronological life span in yeast by decreased TOR pathway signaling. Genes Dev 20 (2):174–184. doi:10.1101/gad.1381406

Reinke A, Anderson S, McCaffery JM, Yates J 3rd, Aronova S, Chu S, Fairclough S, Iverson C, Wedaman KP, Powers T (2004) TOR complex 1 includes a novel component, Tco89p (YPL180w), and cooperates with Ssd1p to maintain cellular integrity in Saccharomyces cerevisiae. J Biol Chem 279(15):14752–14762. doi:10.1074/jbc.M313062200

Riesen M, Morgan A (2009) Calorie restriction reduces rDNA recombination independently of rDNA silencing. Aging Cell 8(6):624–632. doi:10.1111/j.1474-9726.2009.00514.x

Ringel AE, Ryznar R, Picariello H, Huang KL, Lazarus AG, Holmes SG (2013) Yeast Tdh3 (glyceraldehyde 3-phosphate dehydrogenase) is a Sir2-interacting factor that regulates transcriptional silencing and rDNA recombination. PLoS Genet 9(10):e1003871. doi:10.1371/journal.pgen.1003871

Ristow M, Zarse K (2010) How increased oxidative stress promotes longevity and metabolic health: the concept of mitochondrial hormesis (mitohormesis). Exp Gerontol 45(6):410–418. doi:10.1016/j.exger.2010.03.014

Rogina B, Helfand SL (2004) Sir2 mediates longevity in the fly through a pathway related to calorie restriction. Proc Natl Acad Sci U S A 101(45):15998–16003. doi:10.1073/pnas.0404184101

Salvi JS, Chan JN, Pettigrew C, Liu TT, Wu JD, Mekhail K (2013) Enforcement of a lifespan-sustaining distribution of Sir2 between telomeres, mating-type loci, and rDNA repeats by Rif1. Aging Cell 12(1):67–75. doi:10.1111/acel.12020

Schleit J, Johnson SC, Bennett CF, Simko M, Trongtham N, Castanza A, Hsieh EJ, Moller RM, Wasko BM, Delaney JR, Sutphin GL, Carr D, Murakami CJ, Tocchi A, Xian B, Chen W, Yu T, Goswami S, Higgins S, Holmberg M, Jeong KS, Kim JR, Klum S, Liao E, Lin MS, Lo W, Miller H, Olsen B, Peng ZJ, Pollard T, Pradeep P, Pruett D, Rai D, Ros V, Singh M, Spector BL, Vander Wende H, An EH, Fletcher M, Jelic M, Rabinovitch PS, MacCoss MJ, Han JD, Kennedy BK, Kaeberlein M (2013) Molecular mechanisms underlying genotype-dependent responses to dietary restriction. Aging Cell 12(6):1050–1061. doi:10.1111/acel.12130

Schmeisser K, Mansfeld J, Kuhlow D, Weimer S, Priebe S, Heiland I, Birringer M, Groth M, Segref A, Kanfi Y, Price NL, Schmeisser S, Schuster S, Pfeiffer AF, Guthke R, Platzer M, Hoppe T, Cohen HY, Zarse K, Sinclair DA, Ristow M (2013) Role of sirtuins in lifespan regulation is linked to methylation of nicotinamide. Nat Chem Biol 9(11):693–700. doi:10.1038/nchembio.1352

Shanley DP, Kirkwood TB (2000) Calorie restriction and aging: a life-history analysis. Evol Int J Org Evol 54(3):740–750

Sinclair DA (2013) Studying the replicative life span of yeast cells. Methods Mol Biol 1048:49–63. doi:10.1007/978-1-62703-556-9_5

Smith DL Jr, McClure JM, Matecic M, Smith JS (2007) Calorie restriction extends the chronological lifespan of Saccharomyces cerevisiae independently of the Sirtuins. Aging Cell 6 (5):649–662. doi:10.1111/j.1474-9726.2007.00326.x

Smith ED, Tsuchiya M, Fox LA, Dang N, Hu D, Kerr EO, Johnston ED, Tchao BN, Pak DN, Welton KL, Promislow DE, Thomas JH, Kaeberlein M, Kennedy BK (2008) Quantitative evidence for conserved longevity pathways between divergent eukaryotic species. Genome Res 18(4):564–570. doi:10.1101/gr.074724.107

Sohal RS, Weindruch R (1996) Oxidative stress, caloric restriction, and aging. Science 273 (5271):59–63

Speakman JR, Selman C, McLaren JS, Harper EJ (2002) Living fast, dying when? The link between aging and energetics. J Nutr 132(6 Suppl 2):1583S–1597S

Steffen KK, MacKay VL, Kerr EO, Tsuchiya M, Hu D, Fox LA, Dang N, Johnston ED, Oakes JA, Tchao BN, Pak DN, Fields S, Kennedy BK, Kaeberlein M (2008) Yeast life span extension by depletion of 60s ribosomal subunits is mediated by Gcn4. Cell 133(2):292–302. doi:10.1016/j.cell.2008.02.037

Steffen KK, Kennedy BK, Kaeberlein M (2009) Measuring replicative life span in the budding yeast. J Vis Exp 28. doi:10.3791/1209. http://www.jove.com/index/Details.stp?ID=1209

Strahl-Bolsinger S, Hecht A, Luo K, Grunstein M (1997) SIR2 and SIR4 interactions differ in core and extended telomeric heterochromatin in yeast. Genes Dev 11(1):83–93

Teng X, Hardwick JM (2009) Reliable method for detection of programmed cell death in yeast. Methods Mol Biol 559:335–342. doi:10.1007/978-1-60327-017-5_23

Tissenbaum HA, Guarente L (2001) Increased dosage of a sir-2 gene extends lifespan in Caenorhabditis elegans. Nature 410(6825):227–230. doi:10.1038/35065638

Vellai T, Takacs-Vellai K, Zhang Y, Kovacs AL, Orosz L, Muller F (2003) Genetics: influence of TOR kinase on lifespan in C. elegans. Nature 426(6967):620. doi:10.1038/426620a

Wei M, Fabrizio P, Hu J, Ge H, Cheng C, Li L, Longo VD (2008) Life span extension by calorie restriction depends on Rim15 and transcription factors downstream of Ras/PKA, Tor, and Sch9. PLoS Genet 4(1):e13. doi:10.1371/journal.pgen.0040013

Wei M, Fabrizio P, Madia F, Hu J, Ge H, Li LM, Longo VD (2009) Tor1/Sch9-regulated carbon source substitution is as effective as calorie restriction in life span extension. PLoS Genet 5(5): e1000467. doi:10.1371/journal.pgen.1000467

Weinberger M, Mesquita A, Caroll T, Marks L, Yang H, Zhang Z, Ludovico P, Burhans WC (2010) Growth signaling promotes chronological aging in budding yeast by inducing superoxide anions that inhibit quiescence. Aging 2(10):709–726

Xie Z, Zhang Y, Zou K, Brandman O, Luo C, Ouyang Q, Li H (2012) Molecular phenotyping of aging in single yeast cells using a novel microfluidic device. Aging Cell 11(4):599–606. doi:10.1111/j.1474-9726.2012.00821.x

Yin G, McCord A, Wise A, Jindal R, Hardee J, Kuo A, Shimogawa MY, Cahoon L, Wu M, Kloke J, Hardin J, Mays Hoopes LL (2008) Pathways change in expression during replicative aging in Saccharomyces cerevisiae. J Gerontol A Biol Sci Med Sci 63(1):21–34

Zhang T, Fang HH (2004) Quantification of Saccharomyces cerevisiae viability using BacLight. Biotechnol Lett 26(12):989–992

Zhang Y, Luo C, Zou K, Xie Z, Brandman O, Ouyang Q, Li H (2012) Single cell analysis of yeast replicative aging using a new generation of microfluidic device. PLoS One 7(11):e48275. doi:10.1371/journal.pone.0048275

Zhao K, Harshaw R, Chai X, Marmorstein R (2004) Structural basis for nicotinamide cleavage and ADP-ribose transfer by NAD(+)-dependent Sir2 histone/protein deacetylases. Proc Natl Acad Sci U S A 101(23):8563–8568. doi:10.1073/pnas.0401057101

Chapter 7
Oxidative Stress and *C. elegans* Models

Naoaki Ishii, Takamasa Ishii, and Philip S. Hartman

Abstract Oxidative stress is thought to be an important contributor to cellular and organismal aging. While there are many reports that support this notion, some recent evidence using transgenic animals indicates that oxidative defense systems, including antioxidant enzymes, may not affect life extension. This leads to speculation that oxidative stress does not play a major role in aging. However, it is difficult to ascertain the role of oxidative stress on aging under complex mechanisms of ROS production and the defense systems in normal cells that maintain a favorable redox balance. The nematode *Caenorhabditis elegans* has gained widespread favor for the study of many biological processes, including aging. Several lines of *C. elegans* research relating to oxidative stress and aging are discussed in this review, including the use of transgenic organisms with altered superoxide dismutase levels as well as studies that focus on mitochondrial mutations.

Keywords *Caenorhabditis elegans* • Aging • Oxidative stress • Reactive oxygen species • Electron transport

7.1 Oxidative Stress and Aging

Aging is controlled by a complex interplay of both genetic and environmental factors. Investigators have been examining these factors from a wide variety of viewpoints. Much attention has focused on the hypothesis that oxidative damage plays an important determinative role in cellular and organismal aging (Harman 1956; Jazwinski 1996; Holiday 1997; Liochev 2013; Clancy and Birdsall 2013) (Fig. 7.1). Reactive oxygen species (ROS) such as superoxide anion (O_2^-), hydrogen peroxide (H_2O_2), and hydroxyl radicals (OH) can readily attack a wide variety of cellular entities, resulting in damage that compromises cellular integrity and

N. Ishii (✉) • T. Ishii
Department of Molecular Life Science, Tokai University School of Medicine, Isehara, Kanagawa 259-1193, Japan
e-mail: nishii@is.icc.u-tokai.ac.jp; ishiit@tokai.ac.jp

P.S. Hartman
Department of Biology, Texas Christian University, Fort Worth, TX 76129, USA
e-mail: P.Hartman@tcu.edu

© Springer Japan 2015
N. Mori, I. Mook-Jung (eds.), *Aging Mechanisms*,
DOI 10.1007/978-4-431-55763-0_7

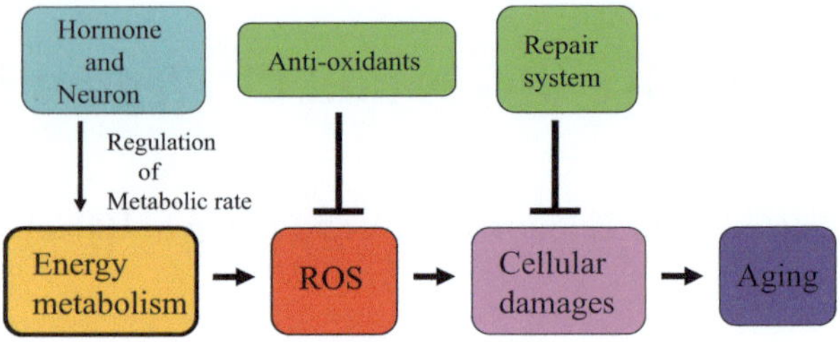

Fig. 7.1 An aging theory that relates to energy metabolism

function (Chance et al. 1979; Vuillaume 1987; Collins et al. 1997). This can cause or at least contribute to a variety of pathologies, including some in humans (Cross et al. 1987; Reddy and Beal 2005; Abou-Sleiman et al. 2006; Valko et al. 2006). To lessen the consequences of this damage, cells have evolved complex defense mechanisms, including enzymatic ones (e.g., superoxide dismutase (SOD) and catalase) as well as various non-enzymatic antioxidants (e.g., vitamins C and E and glutathione) that act to detoxify the offending molecules (Chance et al. 1979; Rajendran et al. 2014). Oxidative damage resulting from an unfavorable balance between oxidative stress and antioxidant defenses may determine the individual aging rate of living things.

Genetic approaches using the small, free-living nematode *Caenorhabditis elegans* have helped elucidate various mechanisms of aging (Houthoofd and Vanfleteren 2007; Kenyon 2010). This review focuses specifically on the role oxidative stress plays in *C. elegans* aging.

7.2 *C. elegans* as a Model System for Aging Research

C. elegans can easily grow in petri plates on a simple diet of *Escherichia coli* and reproduces with a rapid life cycle of approximately 3.5 days at 20 °C. Embryonic development is rapid, taking only 13 h at 20 °C (Sulston 1988; Wood 1988a). The cell lineage has been traced from single-celled zygote to adult, and the entire cell lineage has been determined (Sulston and Horvitz 1977; Sulson et al. 1983; Sulston 1988). After hatching, larval development proceeds through four molts. About 10 % of the cells in the first larval stage undergo further cell divisions during larval development, contributing to the hypodermis, neurons, musculature, and somatic gonadal structures. Adults contain fewer than 1,000 somatic cells.

C. elegans offers several distinct advantages for aging research, which include a short maximum lifespan of approximately 30 days. In addition, *C. elegans* has received much attention as a genetic model, in part because its hermaphroditic mode of reproduction makes for ready isolation of mutants and allows rapid inbreeding. Males, which arise via meiotic non-disjunction can be used to construct stocks and map mutations (Brenner 1974; Wood 1988a, 1988b; Epstein and Shakes 1995; Riddle et al. 1997). Literally thousands of mutants have been isolated that affect virtually all biological processes. These genetic approaches have been useful in identifying and mapping genes that regulate aging (Guarente and Kenyon 2000; Finkel and Holbrook 2000; Kenyon 2010). Mutations can be readily analyzed at the molecular level, thus providing specific insights to the various biochemical and physiological elements of lifespan determination. Techniques include the process of germ-line transformation, which is readily accomplished through microinjection, and enables the creation of transgenics, including those with reporter genes. *C. elegans* has also proven susceptible to the phenomenon of RNAi. By exposing animals to double-stranded RNA, an organismal response is triggered that can mimic the null phenotype of the gene corresponding to that particular RNA (Fraser et al. 2000). Because virtually all putative genes have been identified *in silico*, RNAi provides the opportunity to at least crudely determine the phenotypes resulting from inactivation of many *C. elegans* genes. In addition to these advantages, an adult soma consisting of fewer than 1,000 cells, all of which are postmitotic, offers the ability to detect cumulative age-related cellular alterations (Hosokawa et al. 1994; Adachi et al. 1998; Ishii et al. 2002).

7.3 Oxidative Stress and Defense Systems

There are numerous reports dealing with oxidative stress defense systems and their relationship to lifespan. For example, SOD activity is positively correlated in several organs with the maximum lifespan for various animal species, including primates (Tolmasoff et al. 1980). Similar correlations were observed for several other radical scavengers, including plasma urate, carotenoids, and vitamin E (Cutler 1985). There are some good recent articles that review defense systems against oxidative stress and aging of *C. elegans* (Gems and Doonan 2009; Back et al. 2012; Honda et al. 2010). Some evidence indicates that the defense systems, including antioxidant enzymes, may not affect life extension, which leads to speculation that oxidative stress does not play a major role in aging. For example, *sod-1* knockdown/out or transgenic *C. elegans* do not particularly affect lifespan (Back et al. 2012). Specifically, despite oxidative stress increases (Yang et al. 2007; Raamsdonk and Hekimi 2009), MnSOD knockdown/out mutants did not reduce (Yang et al. 2007; Doonan et al. 2008; Honda et al. 2008; Yen et al. 2009) or extended their lifespans (Raamsdonk and Hekimi 2009; Yang and Hekimi 2010; Dingley et al. 2010). Overexpression of MnSOD and Cu/ZnSOD extended lifespan but did not decrease oxidative damage (Cabreiro at al. 2011). Conversely, overexpression of catalase unexpectedly reduced lifespan (Doonan et al. 2008). These results do not support the oxidative stress hypothesis. However, there are some complexities inherent to interpreting these experiments. Specifically, superoxide anion converts superoxide anion to H_2O_2, which can be catalyzed to H_2O and oxygen by catalase or peroxidase. However, H_2O_2 also produces ·OH in response to superoxide and metal ions (Fig. 7.2). The amount and mixture of ROS produced in cells depends on site (cytoplasm)/subcellular organelle (mitochondria or endoplasmic reticulum) and the balance of anti-oxidative enzymes that are present. Thus, overexpressing one

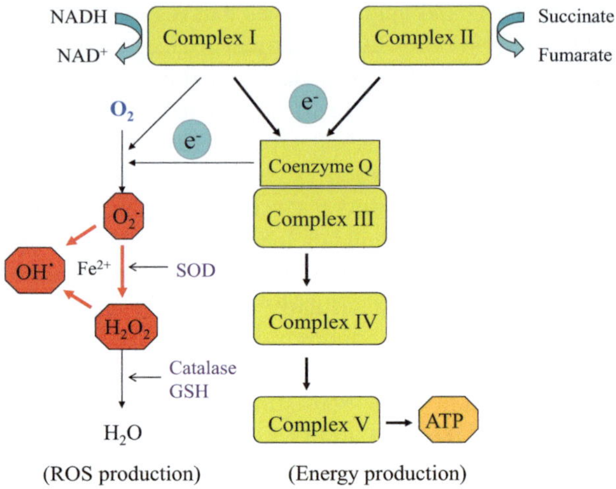

Fig. 7.2 Energy and ROS production from the electron transport chain; *SOD* superoxide dismutase, *GSH*: reduced glutathione

enzyme might have more complex effects than simply depleting its target substrate. Further, the redox balance of reduction and oxidation are finely tuned in healthy cells. Consequently, the redox imbalance in the SOD transgenics may lead to metabolic change including in energy metabolism. In animals which only one defense gene specific to one kind of ROS is knocked down or out or over-expressed, other ROS may be overproduced or imbalance of redox state may result. Just as Orr and Sohal (1994) demonstrated that lifespan was significantly longer in transgenic flies carrying both SOD and catalase genes, experiments using combinations of anti-oxidation enzymes may be necessary to clear up the relation between oxidative stress and aging. In addition, it is critical to determine the cellular amount and distribution of ROS in these transgenic animals.

7.4 Mitochondrial Oxidative Stress

The major endogenous source of reactive oxygen species ROS derives from the electron transport system in mitochondria (Nohl and Hegner 1978; Chance at al. 1979) (Fig. 7.2). It has been estimated that generation of O_2^- and its dismutated product H_2O_2 may constitute as much as 1–2 % of total electron flow, although others have placed this value at 0.1–0.2 % (Tahara et al. 2009). It is known that oxygen is initially converted to O_2^- by electrons leaked from complexes I and mainly complex III (Turrens 1997; Lenaz 1998; Finkel and Holbrook 2000; Raha and Robinson 2000).

Energy metabolism in aerobic organisms is almost exclusively the result of glycolysis, the Krebs cycle and electron transport. With respect to electron transport, five membrane-bound complexes within mitochondria form the respiratory chain that sequentially transfers electrons through a series of donor/acceptors, with oxygen (O_2) as the final acceptor (Wallace 1999; Leonard and Schapira 2000). The eukaryotic mitochondrial electron transport system is composed of more than 80 subunits and requires more than 100 additional genes for its assembly (Attardi and Schatz 1988). The *C. elegans* electron transport is composed of about 70 nuclear and 12 mitochondrial genes products. The metabolism and structure of the *C. elegans* electron transport closely parallel its mammalian counterpart, and its mitochondrial DNA (mtDNA) is similar in size and gene content to the human mtDNA (Murfitt et al. 1976; Okimoto et al. 1992).

The nuclear gene *gas-1* (*fc21*) encodes a homologue of the Ip49 kDa iron protein, a subunit of complex I of the electron transport system (Kayser et al. 1999). The 49-kDa iron protein is abundantly expressed in multiple tissues, including neurons and body wall muscle in *C. elegans* (Kayser et al. 2001). The subunit seems to be implicated in binding CoQ and, therefore, it is believed to be essential for the core function of complex I (Xu et al. 1992; Anderson and Trgovcich-Zacok 1995). Mutations in *gas-1* were isolated based upon their ability to confer hypersensitivity to volatile anesthetics such as halothane or diethyl ether (Morgan and Sedensky 1994; Kayser et al. 1999). The mitochondria isolated from *gas-1* mutant reduced complex I enzymatic activities and increased complex II-dependent metabolism (Kayser et al. 2001). We have determined that *gas-1* mutants are hypersensitive to hyperoxia in a temperature-dependent fashion (Hartman et al. 2001). Specifically, the *gas-1* mutant was three times more sensitive than wild type at 15 °C and over six times more sensitive at 25 °C. These temperatures are typically employed as the permissive and restrictive temperatures for *C. elegans*. In these experiments, survival was defined as the ability of first-stage larvae to complete development. The hypersensitivity of *gas-1* was not restricted to larval development, as the ability to complete embryogenesis was also strongly influenced by oxygen concentration. The *gas-1* mutation also caused a dramatic decrease in lifespan upon exposure to hyperoxia. This was most dramatically observed when animals were reared under atmospheric oxygen and shifted to 60 % oxygen upon sexual maturity.

The question that is most directly related to the scope of this review is as follows: what is the mechanism by which a complex I defect creates hypersensitivity to ROS? This is at least partially answered by the observation that superoxide anion levels in sub-mitochondrial particles are more than two times greater than in wild-type (Ishii and Ishii, unpublished data). Thus, the hypersensitivity and precocious aging caused by the *gas-1* mutation is likely due to excess ROS production.

We have also demonstrated that O_2^- is produced from complex II in a genetic background that compromises complex II functionality (Ishii et al. 2006, 2013). Namely, the *mev-1* mutant is oxygen and ROS-generating chemical methyl viologen (paraquat) hypersensitive with respect to both development and aging (Ishii et al. 1990). The *mev-1*(*kn-1*) mutation, which results in an amino acid substitution at the 71st position from glycine to glutamate (G71E), has been identified as residing in the putative gene *cyt-1* (a human SDHC gene homologue), which is homologous to the succinate dehydrogenase (SDH) cytochrome b large subunit in complex II (Ishii et al. 1998). The mutation results in a greater than 80 %

reduction in complex II activity in the mitochondrial membrane fraction. Complex II catalyzes electron transport from succinate to ubiquinone and contains the citric cycle enzyme succinate dehydrogenase (SDH), which is composed of the flavin protein (Fp), the iron-sulfur protein (Ip) and two other subunits (a small subunit of cytochrome b and a large subunit of cytochrome b encoded by *cyt-1*) (Cecchini 2003; Cecchini et al. 2003; Maklashina and Cecchini 2010). The cytochrome b large subunit is essential for electron transport to ubiquinone in complex III. Based upon its position, the mutation site in *mev-1* may affect the domain binding to ubiquinone.

The mean and maximum lifespans of both the wild type and *mev-1* mutant were influenced by oxygen (Honda et al. 1993). Wild-type lifespans were not affected by oxygen concentrations between 2 % and 40 %. On the other hand, the mean and maximum lifespans of the *mev-1* mutant under atmospheric conditions (21 % oxygen) were shorter than wild type (Honda et al. 1993). Fluorescent materials (lipofuscin) and protein carbonyl derivatives are formed in vivo as a result of metal-catalyzed oxidation and accumulate during aging in disparate model systems (Strehler et al. 1959; Spoerri et al. 1974; Stadman and Oliver 1991; Stadman 1992). The presence of fluorescent materials and protein carbonyl modifications can be a specific indicator of oxidized lipid and protein. The *mev-1* mutants accumulated fluorescent materials and protein-carbonyl derivatives at significantly higher rates than did their wild-type cohorts (Hosokawa et al. 1994; Adachi et al. 1998). Thus, the aging process in *mev-1* animals approximates that of wild type except for its precocious nature. The biochemical pathologies of *mev-1* include elevated ROS. Specificity, O_2^- levels in both intact mitochondria and sub-mitochondrial particles were approximately two times greater in *mev-1* mutants as compared to wild type (Senoo-Matsuda et al. 2001). Given that most O_2^- generation is thought to occur around complex III, this means that the *mev-1* mutation either exacerbates O_2^- production at this location or, in some indirect way, increases O_2^- production at another point in electron transport, even at complex II. Another of the biochemical pathologies is that of reduced glutathione concentration in *mev-1* animals (Senoo-Matsuda et al. 2001). The *mev-1* mutation also caused supernumerary embryonic apoptosis especially under hyperoxia (Senoo-Matsuda et al. 2003). The abnormal apoptosis was suppressed by mutations in either *ced-3* or *ced-4*, indicating that the inappropriate signal in *mev-1* embryos stimulated induction of the normal *ced-9*/*ced-3*/*ced-4* apoptotic pathway in *C. elegans* (Senoo-Matsuda et al. 2003). Furthermore, the *mev-1;ced-3* double mutant lived longer than *mev-1*, which suggests that the supernumerary apoptosis contributed to the phenotype of life shortening in *mev-1* (Senoo-Matsuda et al. 2003). In addition, the oxidative stress by hyperoxia in *mev-1* animals rendered them hypermutable to nuclear mutations (Hartman et al. 2001). Finally, a number of biochemical pathologies likely derive from the role played by succinate dehydrogenase in the citric cycle. First, the ratio of lactate to pyruvate is significantly higher in *mev-1* mutants, suggesting that a metabolic imbalance known as lactate acidosis occurs in these animals. Second, a number of citric cycle intermediates are present at abnormal concentrations in *mev-1* mutants.

Conversely, ATP levels are normal in *mev-1* mutants. This was initially surprising but may suggest that *mev-1* animals rely more heavily on glycolysis for energy acquisition, thus explaining the elevated lactate levels. However, it is also possible that ATP consumption is decreased in *mev-1* because of some sort of global decrease in the metabolic rate that acts to counterbalance the compromised ATP generation in *mev-1* (Senoo-Matsuda et al. 2001). These results suggest that age-related complex II deterioration might also produce O_2^- and consequently accelerate aging.

In a similar fashion, Lemire and colleagues constructed transgenic *C. elegans* strains with a series of mutations in the succinate dehydrogenase iron-sulfur subunit (SDHB-1) (a human SDHB homologue) (Huang and Lemire 2009). They also resulted in reduced lifespans. These strains are also more sensitive to oxygen and paraquat. They overproduced superoxide anion with decreased succinate–cytochrome c reductase activity compared to the control strain. Thus, they recapitulate the phenotypes of the *mev-1* mutant. On the other hand, the *gas-1* and *mev-1* mutants do display some divergent phenotypes. For example, mutations of *isp-1* (*qm150*) and *lrs-2* genes, which encode iron sulfur protein of complex III and mitochondrial leucyl-tRNA synthetase, respectively, increased lifespan (Feng et al. 2001; Lee et al. 2003).

7.5 The Mitochondrial Paradox

While indispensable as a source of ATP generation, mitochondria are also the major endogenous source of ROS. Most of this occurs at complex III, although we have provided evidence that ROS can be generated at complexes I and II. In either case, these ROS can then attack all components of the electron transport system, damaging complexes that then leads to the production of even more ROS. The net result of this cascade is cellular and organismal aging. The metabolic abnormality affects electron flow and leads to a decreased mitochondrial membrane potential ($\Delta\Psi_m$). This may ultimately disrupt the mitochondrial structure and function. It is thought that this metabolic abnormality and ROS generation causes degenerative disease and aging. On the other hand, the reduction of energy metabolism may actually reduce ROS generation from mitochondria and consequently extend lifespan. In addition to the *isp-1* (*qm150*) and lrs-2 mutants described above, for example, RNAi treatment of *atp-3* (a subunit of complex V), *nuo-2* (a subunit of complex I), *cyc-1* (a subunit of complex III) and *cco-1* (a subunit of complex IV) genes resulted in adult animals with reduced ATP levels and prolonged lifespans (Dillin et al. 2002). In addition, a *clk-1* mutant [defective in demethoxy ubiquinone (DMQ)], whose gene encodes hydoxylase, exhibit a longer life than wild type (Lakowski and Hekimi 1996). CoQ biosynthesis is dramatically altered in *clk-1* animals such that mitochondria lack detectable levels of CoQ_9, and instead contain DMQ_9 (Miyadera et al. 2001). Furthermore, Larsen and Clarke (2002) showed that CoQ-less diets, which are the result of growing nematodes on a bacterial strain

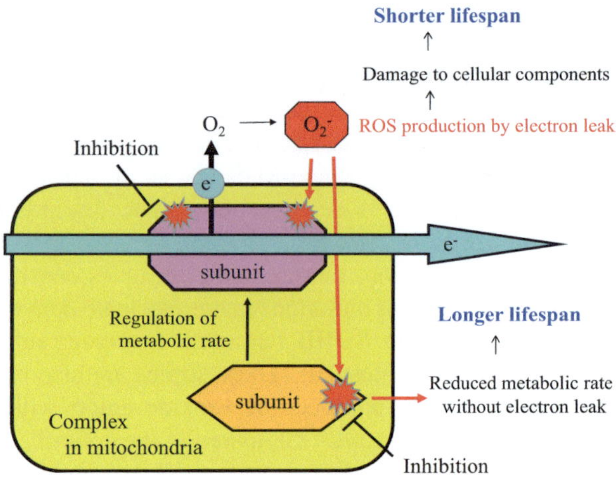

Fig. 7.3 Effects on lifespan by inhibition to subunits of the electron transport complexes (Ishii and Hartman 2003)

lacking CoQ, increase the lifespan of wild type. They also postulated that CoQ-deficient diet may affect aerobic respiration such that less superoxide anion is generated. In the case of the RNAi experiments (Dillin et al. 2002), this is somewhat akin to the effects of caloric restriction. The two contrary results (that is, the reduced lifespan with compromised complex II activity versus the increased lifespan with compromised complex I, III, IV and V activities) may depend on different functionalities of each subunit in the complexes. As described above, the *cyt-1* (= *mev-1*) mutation reduced lifespan and plays a direct role in electron flow from complex II to CoQ. Indeed, this subunit has a binding site to CoQ. Conversely, RNAi of *atp-3*, *nuo-2*, *cyc-1* and *cco-1* gene yielded animals with longer lifespans (Dillin et al. 2002). These gene functions may not affect electron flow directly but instead lower metabolic rate without electron leakage (Fig. 7.3). In addition, the presence of other isoforms may be partially compensatory. Indeed, there are such candidates in the genome (e.g., ceSHDA in complex II). In either case, avoiding electron leakage from electron transport and the resultant ROS production seems to be essential for a normal lifespan.

7.6 Conclusion

Since Harman (1956) postulated in his free-radical theory of aging, much attention has focused on the hypothesis that oxidative damage plays an important determinative role in cellular and organismal aging. In the 50-plus years since then, there are many results in support of this hypothesis, while certain observations could be interpreted as contradicting it. Normal cells, which produce several types of ROS as byproduct of energy metabolism and remove them by the complex defense systems, maintain a balance between reduction and oxidation states. Artificially changing (by drugs or transgenic over- and under-gene expression) can lead to an imbalanced redox state and then to metabolic changes including energy metabolism via mitochondria, which could ultimately impact cellular and organismal wellbeing. In addition, even if the activity or amount of an antioxidation enzyme or antioxidant are changed, cells have a complex systems to compensate for it. On the balance, the evidence points to the fact that ROS can and do impose considerable damage throughout an organism's lifespan. Thus, the evidence supports. The evidence includes studies using the nematode *C. elegans* in which mutants and transgenics have been subjected to a variety of analyses

References

Abou-Sleiman PM, Muqit MM, Wood NW (2006) Expanding insights of mitochondrial dysfunction in Parkinson's disease. Nat Rev Neurosci 7:207–219

Adachi H, Fujiwara Y, Ishii N (1998) Effects of oxygen on protein carbonyl and aging in *Caenorhabditis elegans* mutants with long (*age-1*) and short (*mev-1*) life spans. J Gerontol A Biol Sci Med Sci 53:B240–B244

Anderson WM, Trgovcich-Zacok D (1995) Carbocyanine dyes with long alkyl side-chains: broad spectrum inhibitors of mitochondrial electron transport chain activity. Biochem Pharmacol 49:1303–1131

Attardi G, Schatz G (1988) Biogenesis of mitochondria. Annu Rev Cell Biol 4:289–333

Back P, Braeckman BP, Matthijssens F (2012) ROS in aging *Caenorhabditis elegans*: damage or signaling? Oxidative Med Cell Longev. Vol 2012, Article ID 608478, p. 14

Brenner S (1974) The genetics of *Caenorhabditis elegans*. Genetics 77:71–94

Cabreiro F, Ackerman D, Doonan R, Araiz C, Back P, Papp D, Braeckman BP, Gems D (2011) Increased life span from overexpression of superoxide dismutase in *Caenorhabditis elegans* is not caused by decreased oxidative damage. Free Radic Biol Med 51:1575–1582

Cecchini G (2003) Function and structure of complex II of the respiratory chain. Annu Rev Biochem 72:77–109

Cecchini G, Maklashina E, Yankovskaya V, Iverson TM, Iwata S (2003) Variation in proton donor/acceptor pathways in succinate: quinone oxidoreductases. FEBS Lett 545:31–38

Chance B, Sies H, Boveris A (1979) Hydroperoxide metabolism in mammalian organs. Physiol Rev 59:527–589

Clancy D, Birdsall J (2013) Flies, worms and the free radical theory of ageing. Ageing Res Rev 12:404–412

Collins AR, Duthie SJ, Fillion L, Gedik CM, Vaughan N, Wood SG (1997) Oxidative DNA damage in human cells: the influence of antioxidants and DNA repair. Biochem Soc Trans 25:326–331

Cross CE, Halliwell B, Borish ET, Pryor WA, Ames BN, Saul RL, McCord JM, Harman D (1987) Oxygen radicals and diseases. Ann Intern Med 107:526–545

Cutler RG (1985) Antioxidants and longevity of mammalian species. In: Woodhead AD, Blackett AD, Hollaender A (eds) Molecular biology of aging. Plenum Press, New York/London, pp 15–73

Dillin A, Hsu AL, Arantes-Oliveira N, Lehrer-Graiwer J, Hsin H, Fraser AG, Kamath RS, Ahringer J, Kenyon C (2002) Rates of behavior and aging specific by mitochondrial function during development. Science 298:2398–2401

Dingley S, Polyak E, Lightfoot R, Ostrovsky J, Rao M, Greco T, Ischiropoulos H, Falk MJ (2010) Mitochondrial respiratory chain dysfunction variably increases oxidant stress in *Caenorhabditis elegans*. Mitochondrion 10:125–136

Doonan R, McElwee JJ, Matthijssens F, Walker GA, Houthoofd K, Back P, Matscheski A, Vanfleteren JR, Gems D (2008) Against the oxidative damage theory of aging: superoxide dismutases protect against oxidative stress but have little or no effect on life span in *Caenorhabditis elegans*. Genes Dev 22:3236–3241

Epstein HF, Shakes DC (1995) Methods in cell biology. Academic, San Diego

Feng J, Bussiere F, Hekimi S (2001) Mitochondrial electron transport is a key determinant of life span in *Caenorhabditis elegans*. Dev Cell 1:633–644

Finkel T, Holbrook NJ (2000) Oxidants, oxidative stress and the biology of ageing. Nature 408:239–247

Fraser AG, Kamath RS, Zipperlen P, Martinez-Campos M, Sohrmann M, Ahringer J (2000) Functional genomic analysis of *C. elegans* chromosome I by systematic RNA interference. Nature 408:325–330

Gems D, Doonan R (2009) Antioxidant defense and aging in *C. elegans*: is the oxidative damage theory of aging wrong? Cell Cycle 8:1681–1697

Guarente L, Kenyon C (2000) Genetic pathways that regulate ageing in model organisms. Nature 408:255–262

Harman D (1956) Aging: a theory based on free radical and radiation chemistry. J Gerontol 11:298–300

Hartman PS, Ishii N, Kayser EB, Morgan PG, Sedensky MM (2001) Mitochondrial mutations differentially affect aging, mutability and anesthetic sensitivity in *Caenorhabditis elegans*. Mech Ageing Dev 122:1187–1201

Holiday R (1997) Understanding aging. Philos Trans R Soc Lond B Biol Sci 352:1793–1797

Honda S, Ishii N, Suzuki K, Matsuo M (1993) Oxygen-dependent perturbation of life span and aging rate in the nematode. J Gerontol Ser A Biol Sci Med Sci 48:B57–B61

Honda Y, Tanaka M, Honda S (2008) Modulation of longevity and diapause by redox regulation mechanisms under the insulin-like signaling control in *Caenorhabditis elegans*. Exp Gerontol 43:520–529

Honda Y, Tanaka M, Honda S (2010) Redox regulation, gene expression and longevity. Geriatr Gerontol Int 10(Suppl 1):S59–S69

Hosokawa H, Ishii N, Ishida H, Ichimori K, Nakazawa H, Suzuki K (1994) Rapid accumulation of fluorescent material with aging in an oxygen-sensitive mutant *mev-1* of *Caenorhabditis elegans*. Mech Ageing Dev 74:161–170

Houthoofd K, Vanfleteren JR (2007) Public and private mechanisms of life extension in *Caenorhabditis elegans*. Mol Genet Genomics 277:601–617

Huang J, Lemire BD (2009) Mutations in the *C. elegans* succinate dehydrogenase iron–sulfur subunit promote superoxide generation and premature aging. J Mol Biol 387:559–569

Ishii N, Hartman PS (2003) Electron transport and life span in *C. elegans*. In: Mattson MP (ed) Energy metabolism and lifespan determination. Vol 14, Elsevier, Baltimore, pp 177–195

Ishii N, Takahashi T, Tomita S, Keino T, Honda S, Yoshino K, Suzuki K (1990) A methyl viologen-sensitive mutant of the nematode *Caenorhabditis elegans*. Mutat Res 237:165–171

Ishii N, Fujii M, Hartman PS, Tsuda M, Yasuda K, Senoo-Matsuda N, Yanase S, Ayusawa D, Suzuki K (1998) A mutation in succinate dehydrogenase cytochrome b causes oxidative stress and ageing in nematodes. Nature 394:694–697

Ishii N, Goto S, Hartman PS (2002) Protein oxidation during aging of the nematode *Caenorhabditis elegans*. Free Radic Biol Med 33:1021–1025

Ishii N, Ishii T, Hartman PS (2006) The role of the electron transport gene SDHC on lifespan and cancer. Exp Gerontol 41:952–956

Ishii T, Miyazawa M, Onouchi H, Yasuda K, Hartman PS, Ishii N (2013) Model animals for the study of oxidative stress from complex II. Biochim Biophys Acta 1827:588–597

Jazwinski SM (1996) Longevity, genes, and aging. Science 273:54–59

Kayser EB, Morgan PG, Sedensky MM (1999) GAS-1: a mitochondrial protein controls sensitivity to volatile anesthetics in the nematode *Caenorhabditis elegans*. Anesthesiology 90:545–554

Kayser EB, Morgan PG, Hoppel CL, Sedensky MM (2001) Mitochondrial expression and function of GAS-1 in *Caenorhabditis elegans*. J Biol Chem 276:20551–20558

Kenyon CJ (2010) The genetics of ageing. Nature 464:504–512

Lakowski B, Hekimi S (1996) Determination of life-span in *Caenorhabditis elegans* by four clock genes. Science 272:1010–1013

Larsen PL, Clarke CF (2002) Extension of life-span in *Caenorhabditis elegans* by a diet lacking coenzyme Q. Science 295:120–123

Lee SS, Lee RYN, Fraser AG, Kamath RS, Ahringer J, Ruvkun G (2003) A systematic RNAi screening identifies a critical role for mitochondria in *C. elegans* longevity. Nat Genet 33:40–48

Lenaz G (1998) Role of mitochondria in oxidative stress and ageing. Biochim Biophys Acta 1366:53–67

Leonard JV, Schapira AH (2000) Mitochondrial respiratory chain disorders: I. mitochondrial DNA defects. Lancet 355:299–304

Liochev SI (2013) Reactive oxygen species and the free radical theory of aging. Free Radic Biol Med 60:1–4

Maklashina E, Cecchini G (2010) The quinone-binding and catalytic site of complex II. Biochim Biophys Acta 1797:1877–1882

Miyadera H, Amino H, Hiraishi A, Taka H, Murayama K, Miyoshi H, Sakamoto K, Ishii N, Hekimi S, Kita K (2001) Altered quinone biosynthesis in the long-lived *clk-1* mutants of *Caenorhabditis elegans*. J Biol Chem 276:7713–7716

Morgan PG, Sedensky MM (1994) Mutations conferring new patterns of sensitivity to volatile anesthetics in *Caenorhabditis elegans*. Anesthesiology 81:888–898

Murfitt RR, Vogel K, Sanadi DR (1976) Characterization of the mitochondria of the free-living nematode, *Caenorhabditis elegans*. Comp Biochem Physiol B 53:423–430

Nohl H, Hegner D (1978) Do mitochondria produce oxygen radicals in vivo? Eur J Biochem 82:563–567

Okimoto R, Macfarlane JL, Clary DO, Wolstenholme DR (1992) The mitochondrial genomes of two nematodes, *Caenorhabditis elegans* and *Ascaris suum*. Genetics 130:471–498

Orr WC, Sohal RS (1994) Extension of life span by over expression of superoxide dismutase and catalase in *Drosophila melanogaster*. Science 263:1128–1130

Raamsdonk JM, Van Hekimi S (2009) Deletion of the mitochondrial superoxide dismutase sod-2 extends lifespan in *Caenorhabditis elegans*. PLoS Genet 5(2), e1000361

Raha S, Robinson BH (2000) Mitochondria, oxygen free radicals, disease and ageing. Trends Biochem Sci 25:502–508

Rajendran P, Nandakumar N, Rengarajan T, Palaniswami R, Gnanadhas EN, Lakshminarasaiah U, Gopas J, Nishigaki I (2014) Antioxidants and human diseases. Clin Chim Acta 436:332–347

Reddy PH, Beal MF (2005) Are mitochondria critical in the pathogenesis of Alzheimer's disease? Brain Res Rev 49:618–632

Riddle DL, Blumenthal T, Mayer BJ, Priess JR (1997) C. elegans II. Cold Spring Harbor Laboratory, New York

Senoo-Matsuda N, Yasuda K, Tsuda M, Ohkubo T, Yoshimura S, Nakazawa H, Hartman PS, Ishii N (2001) A defect in the cytochrome b large subunit in complex II causes both superoxide anion overproduction and abnormal energy metabolism in *Caenorhabditis elegans*. J Biol Chem 276:41553–41558

Senoo-Matsuda N, Hartman PS, Akatsuka A, Yoshimura S, Ishii N (2003) A complex II defect affects mitochondrial structure, leading to *ced-3*- and *ced-4*-dependent apoptosis and aging. J Biol Chem 278:22031–22036

Spoerri PE, Glass P, Ghazzawi PE (1974) Accumulation of lipofuscin in the myocardium of senile guinia pigs; dissolution and removal of lipofuscin following dimethylaminoethyl p-chloroohenoxyacetate administration. An electron microscopy study. Mech Ageing Dev 3:311–321

Stadman ER (1992) Protein oxidation and aging. Science 257:1220–1224

Stadman ER, Oliver CN (1991) Metal-catalyzed oxidation of proteins. J Biol Chem 266:2005–2008

Strehler BL, Mark DD, Mildvan AS, Gee MV (1959) Rate and magnitude of age pigment accumulation in the human myocardium. J Gerontol 14:257–264

Sulston JE (1988) Cell lineage. In: Wood WB (ed) The nematode *Caenorhabditis elegans*. Cold Spring Harbor Laboratory, New York, pp 123–155

Sulston JE, Horvitz HR (1977) Post embryonic cell lineages of the nematode *Caenorhabditis elegans*. Dev Biol 56:110–156

Sulston JE, Schiernberg E, White JG, Thomson JN (1983) The embryonic cell lineage of the nematode *Caenorhabditis elegans*. Dev Biol 100:64–119

Tahara EB, Navarete FDT, Kowaltowski AJ (2009) Tissue-, substrate-, and site-specific characteristics of mitochondrial reactive oxygen species generation. Free Radic Biol Med 46:1283–1297

Tolmasoff JM, Ono T, Cutler RG (1980) Superoxide dismutase: correlation with life-span and specific metabolic rate in primate species. Proc Natl Acad Sci 77:2777–2781

Turrens JF (1997) Superoxide production by the mitochondrial respiratory chain. Biosci Rep 17:3–8

Valko M, Rhodes CJ, Moncol J, Izakovic M, Mazur M (2006) Free radicals, metals and antioxidants in oxidative stress-induced cancer. Chem Biol Interact 160:1–40

Vuillaume M (1987) Reduced oxygen species, mutation, induction and cancer initiation. Mutat Res 186:43–72

Wallace DC (1999) Mitochondrial diseases in man and mouse. Science 283:1482–1488

Wood WB (1988a) Embryology. In: Wood WB (ed) The nematode *Caenorhabditis elegans*. Cold Spring Harbor Laboratory, New York, pp 215–241

Wood WB (1988b) Aging of *C. elegans*: mosaics and mechanisms. Cell 95:147–150

Xu X, Matsuno-Yagi A, Yagi T (1992) Gene cluster of the energy-transducing NADH-quinone oxidoreductase of Paracoccus denitrificans: characterization of four structural gene products. Biochemistry 31:6925–6932

Yang W, Hekimi S (2010) A mitochondrial superoxide signal triggers increased longevity in *Caenorhabditis elegans*. PLoS Biol 8(12), e1000556

Yang W, Li JJ, Hekimi S (2007) A measurable increase in oxidative damage due to reduction in superoxide detoxification fails to shorten the life span of long-lived mitochondrial mutants of *Caenorhabditis elegans*. Genetics 177:2063–2074

Yen K, Patel HB, Lublin AL, Mobbs CV (2009) SOD isoforms play no role in lifespan in ad lib or dietary restricted conditions, butmutational inactivation of SOD-1 reduces life extension by cold. Mech Ageing Dev 130:173–178

Chapter 8
Genes and Pathways That Influence Longevity in *Caenorhabditis elegans*

Yujin Lee, Seon Woo A. An, Murat Artan, Mihwa Seo, Ara B. Hwang, Dae-Eun Jeong, Heehwa G. Son, Wooseon Hwang, Dongyeop Lee, Keunhee Seo, Ozlem Altintas, Sangsoon Park, and Seung-Jae V. Lee

Abstract The roundworm *Caenorhabditis elegans* is one of the most popular model organisms for research on aging because of its short lifespan and genetic tractability. Studies using *C. elegans* have identified many genes and pathways that regulate aging, several of which are conserved in other species, including mammals. In this chapter, we describe longevity-regulatory pathways including insulin/IGF-1 (insulin-like

---

Y.L., S.W.A.A., M.A., M.S., A.B.H., D.-E.J., H.G.S., W.H., D.L., and K.S. contributed equally.

Y. Lee • S.W.A. An • A.B. Hwang • D.-E. Jeong • H.G. Son • W. Hwang • D. Lee • K. Seo • S. Park
Department of Life Sciences, Pohang University of Science and Technology, Pohang, Gyeongbuk, South Korea

M. Artan
Information Technology Convergence Engineering, Pohang University of Science and Technology, Pohang, Gyeongbuk, South Korea

M. Seo
School of Interdisciplinary Bioscience and Bioengineering, Pohang University of Science and Technology, Pohang, Gyeongbuk, South Korea

Center for Plant Aging Research, Institute for Basic Science, Daegu Gyeongbuk Institute of Science and Technology, Daegu, South Korea

Department of New Biology, Daegu Gyeongbuk Institute of Science and Technology, Daegu, South Korea

O. Altintas
School of Interdisciplinary Bioscience and Bioengineering, Pohang University of Science and Technology, Pohang, Gyeongbuk, South Korea

S.-J.V. Lee (✉)
Department of Life Sciences, Pohang University of Science and Technology, Pohang, Gyeongbuk, South Korea

Information Technology Convergence Engineering, Pohang University of Science and Technology, Pohang, Gyeongbuk, South Korea

School of Interdisciplinary Bioscience and Bioengineering, Pohang University of Science and Technology, Pohang, Gyeongbuk, South Korea
e-mail: seungjaelee@postech.ac.kr

© Springer Japan 2015

N. Mori, I. Mook-Jung (eds.), *Aging Mechanisms*,
DOI 10.1007/978-4-431-55763-0_8

growth factor 1) signaling, TOR (target of rapamycin) signaling, autophagy, mitochondrial respiration, and HIF-1 (hypoxia-inducible factor 1) pathways. We also review the effects of dietary restriction, a key environmental factor that influences aging, on longevity-regulatory genetic factors. In addition, we illustrate the roles of two important *C. elegans* tissues, those of the sensory neural and reproductive systems, in regulating longevity at the molecular level. For each of the subtopics, we explain how changes in the expression of genes involved in each pathway and system alter longevity. We also speculate on the evolutionary significance of the genes and pathways that affect longevity. Given the conserved nature of longevity regulation, the dissection of the roles of these genetic factors in determining the *C. elegans* lifespan will provide important clues for understanding the secrets of human aging.

Keywords *C. elegans* • Aging • Insulin/IGF-1 • Target of rapamycin • Dietary restriction • Autophagy • Hypoxia-inducible factor • Mitochondria • Sensory neurons • Reproductive system

8.1 Introduction

For a long time, the lifespans of living organisms were believed to be limited by the passive, age-dependent degeneration of tissues, which eventually leads to death. However, the findings of studies conducted in the past two decades tell us otherwise. Scientists have shown that organismal lifespans are subjected to active regulation by many genes and pathways. Although the exact mechanisms remain unclear, we now know that genetic factors influence the rate of organismal aging, in response to changes in environmental signals as well as physiologic inputs.

The small roundworm *Caenorhabditis elegans* has been exploited as a fundamental tool for research on aging, revealing crucial lifespan-regulatory pathways. One of the best advantages of *C. elegans* as a model for research on aging is that the *C. elegans* lifespan is only a few weeks. In addition, *C. elegans* undergoes clear age-dependent physiologic and behavioral changes and possesses many lifespan-regulatory pathways that are conserved across phyla. In fact, many evolutionarily conserved genes and pathways that affect organismal longevity were first identified in *C. elegans*.

The first lifespan-regulatory pathway identified in *C. elegans* was the insulin/insulin-like growth factor 1 (IGF-1) signaling (IIS) pathway. Subsequent research led to the identification of target of rapamycin (TOR) signaling, dietary restriction (DR), steroid signaling, autophagy, reduced mitochondrial respiration, the hypoxia inducible factor 1 (HIF-1) pathway, and the sensory and reproductive systems as major lifespan-regulatory pathways in *C. elegans*. The identification of genes acting in those pathways and systems shed light on ensuing research in more complex organisms by revealing that many of the pathways are indeed evolutionarily conserved. In this chapter, we will review the roles and mechanisms by which key genes in lifespan-regulatory pathways modulate *C. elegans* lifespan. We include a table with an extensive list of *C. elegans* longevity-influencing genes, many of which are not described in the text due to space limits (Table 8.1). Furthermore, we speculate regarding the physiologic natures of the lifespan-regulatory pathways and possible reasons why these pathways modulate aging in *C. elegans*.

Table 8.1 *C. elegans* genes that affect lifespan via acting in representative longevity pathways

C. elegans gene	Human ortholog	Gene function/Domains	Pathway(s)	Effects on lifespan: Overexpression/Gain of function	Effects on lifespan: Loss of function	References
aak-2	*PRKAA2*	Catalytic alpha subunit of AMP-activated protein kinase (AMPK)	TOR, DR, Mit	Increase[OE]	S.L.[m] of *rsks-1(−)*, *DR*, *isp-1(−)*, *clk-1(−)*, *sir-2.1*[OE], ROS Decrease[m]	Apfeld et al. (2004), Curtis et al. (2005), Greer et al. (2007), Selman et al. (2009), Seo et al. (2013), Hwang et al. (2014)
aakb-2	*PRKAB2*	Regulatory beta subunit of AMPK	Mit	ND	S.L.[m] of ROS Decrease[m]	Hwang et al. (2014)
aakg-4	*PRKAG1*	Regulatory gamma subunit of AMPK	IIS	ND	S.L.[m, i] of *daf-2(−)*	Tullet et al. (2014)
aap-1	*PIK3R3*	Phosphoinositide 3-kinase (PI3K) adaptor subunit	IIS	ND	Increase[m] (25.5 °C)	Wolkow et al. (2002)
age-1	*RIK3CA*	Phosphoinositide 3-kinase (PI3K)	IIS	ND	Increase[m, i]	Friedman and Johnson (1988)
aha-1	*HIFβ*	Aryl hydrocarbon receptor nuclear translocator	Mit, HIF	ND	S.L.[i] of *isp-1(−)*, *clk-1(−)*, *tpk-1(−)*	Lee et al. (2010), Khan et al. (2013)
akt-1	*AKT1*	Serine/threonine kinase Akt/PKB	IIS	ND	Increase[m, i]	Hertweck et al. (2004), Hamilton et al. (2005), Tullet et al. (2008), Zhang et al. (2008), Shen et al. (2012)
akt-2	*AKT3*	Serine/threonine kinase Akt/PKB	IIS	ND	Increase[i]	Hertweck et al. (2004), Tullet et al. (2008)
aqp-1/dod-4	*AQP10*	Aquaporin	IIS, Mit	ND	S.L.[i] of *clk-1(−)* Decrease[m]	Cristina et al. (2009), Lee et al. (2009)
atg-18	*WIPI1/2*	WD40 repeat-containing protein	IIS, TOR, Atg, Mit, Rpd	ND	S.L.[m] of *daf-2(−)*, *let-363 (RNAi)*, *clk-1(−)*, *atp-3(RNAi)* S.L.[i] of g. c.(−), *rsks-1(−)* Decrease[m]	Toth et al. (2008), Lapierre et al. (2011), Lapierre et al. (2013a)

(continued)

Table 8.1 (continued)

C. elegans gene	Human ortholog	Gene function/Domains	Pathway(s)	Effects on lifespan: Overexpression/Gain of function	Effects on lifespan: Loss of function	References
atg-4.1	*ATG4A*	Cysteine protease ATG4A	DR, Atg	ND	S.L.[i] of *mir-34(−)*	Yang et al. (2013)
atg-7	*ATG7*	E1 ubiquitin-activating-like enzyme	DR, Atg	ND	S.L.[i] of DR; Increase[i]	Jia and Levine (2007), Hashimoto et al. (2009)
atg-9	*ATG9A*	Autophagy-related protein 9A	DR, Atg	ND	S.L.[i] of *mir-34(−), daf-2(−)*; Decrease[i]	Toth et al. (2008), Yang et al. (2013)
atp-2	*ATP5B*	F1 portion of ATP synthase	Mit	ND	Increase[m]	Tsang et al. (2001)
atp-3	*ATP5O*	ATP synthase subunit	Mit	ND	Increase[i]	Dillin et al. (2002b)
bar-1	*JUP*	Beta-catenin	IIS	ND	Decrease[m]	Essers et al. (2005)
bec-1	*BECN1*	Class III phosphatidylinositol 3-kinase complex	IIS, TOR, DR, Atg, Rpd	ND	S.L. [m, i] of *daf-2(−)*, DR, *daf-15/+*, g.c.(−), *mir-34(−), let-363(RNAi), atp-3(RNAi), frh-1(RNAi)*; Decrease[m, i]; Increase[i]	Melendez et al. (2003), Jia and Levine (2007), Hansen et al. (2008), Toth et al. (2008), Hashimoto et al. (2009), Lapierre et al. (2011), Schiavi et al. (2013), Yang et al. (2013)
cbp-1	*CREBBP*	CREB binding protein	IIS, DR	ND	S.L.[i] of DR, *daf-2(−), mir-80 (−)*; Decrease[i]	Zhang et al. (2009a), Vora et al. (2013)
cco-1	*COX5B*	Cytochrome C oxidase	Mit	ND	Increase[i]	Dillin et al. (2002b)
ced-3	*CASP2*	Caspase, a cysteine-aspartate protease, CASP9-like	Mit	ND	S.L.[m] *isp-1(−), nuo-6(−)*, ROS	Yee et al. (2014)
ced-4	*APAF1*	Apoptotic peptidase activating factor 1-like	Mit	ND	S.L.[m] *isp-1(−), nuo-6(−)*, ROS	Yee et al. (2014)
ced-9	*BCL2L2*	Bcl-2-like	Mit	ND	S.L.[m] *isp-1(−), nuo-6(−)*, ROS	Yee et al. (2014)
ced-13	–	BH3 domain	Mit	ND	S.L.[m] of *isp-1(−), nuo-6(−)*	Yee et al. (2014)
ceh-23	*EMX2*	Transcription factor	Mit	Increase[OE]	S.L.[m] of *isp-1(−), clk-1(−)*	Walter et al. (2011)

cep-1	*p53*	p53-like transcription factor	Mit	ND	S.L.[m] of *isp-1(−)*, *nuo-6(−)*	Torgovnick et al. (2010), Baruah et al. (2014)
che-2	*IFT80*	Intraflagellar transport	SN	ND	Increase[m]	Apfeld and Kenyon (1999)
che-3	*DNAH9*	Dynein heavy chain (DHC)	SN	ND	Increase[m]	Apfeld and Kenyon (1999)
che-11	*IFT140*	Intraflagellar transport	SN	ND	Increase[m]	Apfeld and Kenyon (1999)
che-13	*IFT57*	Intraflagellar transport	SN	ND	Increase[m]	Apfeld and Kenyon (1999)
clk-1	*COQ7/ CAT5*	Demethoxyubiquinone (DMQ) hydroxylase	Mit	ND	Increase[m]	Wong et al. (1995), Braeckman et al. (1999)
col-10[#]	*COL3A1*	Collagen	IIS, TOR, DR, Rpd	Increase[OE]	S.L.[i] of *daf-2(−)*(15 °C), TOR, DR, g.c.(−)	Ewald et al. (2014)
col-13[#]	*COL3A1*	Collagen	IIS, TOR, DR, Rpd	Increase[OE]	S.L.[i] of *daf-2(−)*(15 °C), TOR, DR, g.c.(−)	Ewald et al. (2014)
					Decrease[i] in *daf-2(−)* (20 °C)	
col-120[#]	*COL10A1*	Collagen	IIS, TOR, DR, Rpd	Increase[OE]	S.L.[i] of *daf-2(−)*(15 °C), TOR, DR, g.c.(−)	Ewald et al. (2014)
cst-1	*Serine/ threonine protein kinase 3*	Serine/threonine protein kinase	IIS	Increase[OE]	S.L.[i] of *daf-2(−)*	Lehtinen et al. (2006)
cul-1	*CUL1*	Scaffolding protein	IIS	ND	S.L.[i] of *daf-2(−)*	Ghazi et al. (2007)
					Decrease[i]	
cup-4	*CHRNB1*	non-alpha ligand-gated ion channel	DR	Decrease[OE]	S.L.[m, i] of DR	Park et al. (2010)
					Decrease[m, i]	
cyc-1	*CYC1*	Cytochrome C reductase	Mit	ND	Increase[i]	Dillin et al. (2002b)
daf-2	*INSR/IGF-1R*	Insulin receptor	IIS	ND	Increase[m, i]	Kenyon et al. (1993), Dillin et al. (2002a)
daf-6	*PTCHD3*	Patched domain- containing protein	SN	ND	Increase[m]	Apfeld and Kenyon (1999)
daf-7[#]	*GDF11*	Transforming growth factor beta superfamily	IIS	ND	Increase[m]	Shaw et al. (2007)

(continued)

Table 8.1 (continued)

C. elegans gene	Human ortholog	Gene function/Domains	Pathway(s)	Effects on lifespan: Overexpression/ Gain of function	Effects on lifespan: Loss of function	References
daf-9[*#]	*CYP2U1*	Cytochrome P450	IIS, DR, SN, Rpd,	No change[OE]	S.L.[m] of *daf-2(−)*, DR, g.c.(−)	Gerisch et al. (2001), Jia et al. (2002), Gerisch et al. (2007), Lee and Kenyon (2009), Thondamal et al. (2014)
					Decrease[m]	
					Increase[m]	
daf-10	*IFT122*	Intraflagellar transport	SN	ND	Increase[m]	Apfeld and Kenyon (1999)
daf-11	–	Guanylyl cylase	SN	ND	Increase[m]	Hahm et al. (2009)
daf-12[*]	*NR1H3*	Nuclear receptor subfamily 1	IIS, SN, Rpd	Increase[gf] (20 °C)	S.S.[m] of thermosensory mutants (25 °C)	Larsen et al. (1995), Gems et al. (1998), Hsin and Kenyon (1999), Fisher and Lithgow (2006), Berman and Kenyon (2006), Lee and Kenyon (2009)
					S.L.[m, i] of g.c.(−), weak *daf-2 (−)* alleles	
					Decrease[m]	
daf-15	*RAPTOR*	Regulatory associated protein of mTOR	IIS, TOR, DR	ND	Increase[m/+, i] (22.5 °C)	Jia et al. (2004), Ching et al. (2010), Seo et al. (2013)
daf-16	*FOXO*	Transcription factor	IIS, TOR, DR, SN, Rpd,	Increase[OE]	S.L.[m] of *daf-2(−)*, *ifg-1 (RNAi)*, *ragc-1(RNAi)*, *rsks-1 (−)*, *daf-2(−); rsks-1(−)*, *mir-80(−)*, sensory mutants, g.c. (−), DR, *daf-15(+/−)*	Kenyon et al. (1993), Apfeld and Kenyon (1999), Hsin and Kenyon (1999), Lin et al. (2001), Boehm and Slack (2005), Berman and Kenyon (2006), Greer et al. (2007) Hansen et al. (2007), Steinkraus et al. (2008), Zhang et al. (2009a), de Lencastre et al. (2010), Yang et al. (2011), Robida-Stubbs et al. (2012), Chen et al. (2013b), Seo et al. (2013), Vora et al. (2013), Riera et al. (2014)
					S.L.[i] of *daf-2(−)*, *ife-2(−)*, *lin-14(−)*, *mir-239(−)*, g.c.(−)	
					Decrease[m, i]	

daf-18	*PTEN*	Lipid phosphatase	IIS	Increase^{OE}	S.L.^{m} of *daf-2(−)* Decrease^{m}	Dorman et al. (1995), Larsen et al. (1995), Mihaylova et al. (1999), Masse et al. (2005), Brisbin et al. (2009)
daf-19	*RFX2*	Transcription factor	SN	ND	Increase^{m}	Apfeld and Kenyon (1999)
daf-28	insulin	Beta-type insulin	IIS	ND	Increase$^{i,\,dn}$	Malone et al. (1996), Okuyama et al. (2010)
daf-36	*RFESDP1*	Rieske oxygenase	Rpd	ND	S.L.^{m} of g.c.(−) Decrease^{m}	Rottiers et al. (2006), Gerisch et al. (2007)
ddl-1	*CCD53*	Coiled-coil protein	IIS	ND	Increase^{i}	Hansen et al. (2005), Chiang et al. (2012)
ddl-2	*WASH2P*	Proline-rich domains	IIS	ND	Increase^{i}	Hansen et al. (2005), Chiang et al. (2012)
dhs-16	*SDR9C7*	Short chain dehydrogenase/ reductase family	Rpd	ND	S.L.^{m} of g.c.(−)	Wollam et al. (2012)
dox-1	*KIN17*	DNA and RNA binding protein	IIS	ND	Decrease^{i} S.L.^{i} of *daf-2(−)*	Henis-Korenblit et al. (2010)
drp-1	*DRP1*	Dynamin-related protein	IIS	ND	Increase^{m,i} in *daf-2(−)*	Yang et al. (2011)
drr-2	*eIF4H*	Eukaryotic translation initiation factor	TOR, DR	S.L.^{OE} of *let-363(RNAi); daf-15(RNAi)*, DR No change^{OE}	Increase^{i}	Hansen et al. (2005), Ching et al. (2010)
dve-1	*SATB1*	SATB homeobox 1 transcription factor	Mit, DR	ND	S.L.^{i} of DR, *isp-1(−)* Decrease^{i}	Zhang et al. (2009a), Durieux et al. (2011)
eak-3	*RNF11*	RING finger protein 11	IIS	ND	Increase ^{m} in *daf-2(−)*	Zhang et al. (2008)
eak-7	*TLDC1*	TBC/LysM-associated domain containing 1	IIS	ND	Increase$^{m,\,i}$	Samuelson et al. (2007), Alam et al. (2010)
eat-2	*CHRNA7*	Ligand-gated ion channel subunit	DR	ND	Increase^{m}	Lakowski and Hekimi (1998)
egl-9	*EGLN1*	Proline hydroxylase	DR, HIF	ND	S.L.^{m} of DR Increase$^{m,\,i}$	Chen et al. (2009), Mehta et al. (2009), Lee et al. (2010)
egl-27	*MTA1*	Metastasis-associated protein	IIS	Increase^{OE}	S.L.^{i} of *daf-2(−)* Decrease^{m}	Xu and Kim (2012)

(continued)

Table 8.1 (continued)

C. elegans gene	Human ortholog	Gene function/Domains	Pathway(s)	Effects on lifespan: Overexpression/ Gain of function	Effects on lifespan: Loss of function	References
elt-3	*GATA3*	GATA transcription factor 3	IIS	ND	S.L.$^{m, i}$ of *daf-2(−)* Decrease$^{m, i}$	Budovskaya et al. (2008), Kim et al. (2013) but see Tonsaker et al. (2012)
ets-4	*SPDEF*	ETS class transcription factor (highly similar to SAM pointed domain-containing ETS transcription factor)	IIS	ND	Increase$^{m, i}$	Thyagarajan et al. (2010)
faah-1	*FAAH*	Fatty acid amide hydrolase	DR	Increase^{OE}		Lucanic et al. (2011)
fard-1	*FAR1*	Fatty acyl CoA reductase 1	Rpd	ND	S.L.^{i} of g.c.(−)	McCormick et al. (2012)
fat-6/fat-7	*SCD*	Stearoyl-CoA desaturase (delta-9 fatty acid desaturase)	Rpd	ND	S.L.^{m} of g.c.(−)	Goudeau et al. (2011), Brock et al. (2006)
flcn-1	*FLCN*	Folliculin	HIF	ND	Increase$^{m, i}$	Gharbi et al. (2013) but see Possik et al. (2014)
frh-1	*FXN*	Frataxin	Mit	ND	Increase^{i}	Ventura et al. (2005) but see Zarse et al. (2007), Schiavi et al. (2013)
fstr-1/fstr-2/gfi-1	*Mucin-5 AC (Fragment)*	–	Mit	ND	S.L.^{i} of *clk-1(−)*	Cristina et al. (2009)
ftn-1/ftn-2	*FTH1*	Ferritin	Mit	ND	S.L.^{i} of *isp-1(−), nuo-6(−)* S.S.^{i} of *isp-1(−); hif-1(−)*	Baruah et al. (2014), Hwang et al. (2014)
ftt-2	*YWHAZ*	Tyrosine 3-monooxygenase/ tryptophan 5-monooxygenase activation protein	IIS, Rpd	Increase^{OE}	S.L.^{i} of *daf-2(−)*,g.c.(−) Decrease^{i}	Berdichevsky et al. (2006), Wang et al. (2006), Araiz (2008), McCormick et al. (2012), Li et al. (2007a)
gcn-2	*EIF2AK4*	Eukaryotic translation initiation factor 2 alpha kinase 4	Mit	ND	S.L.^{i} of *clk-1(−)*	Baker et al. (2012)
glp-1	*NOTCH1*	N-glycosylated transmembrane protein	Rpd	Decrease^{gf}	Increase$^{m, i}$	Arantes-Oliveira et al. (2002), Curran and Ruvkun

glp-4	–	–	Rpd	ND	Increase^{m}	Arantes-Oliveira et al. (2002)
gpa-1	*GNAO1*	Guanine nucleotide binding protein (G protein), alpha activating activity polypeptide O	SN	Decrease^{OE}	Increase^{m}	Lans and Jansen (2007)
gpa-2	*GNAO1*	Guanine nucleotide binding protein (G protein), alpha activating activity polypeptide O	SN	Increase^{OE}	Decrease^{m}	Lans and Jansen (2007)
gpa-3	*GNAO1*	Guanine nucleotide binding protein (G protein), alpha activating activity polypeptide O	SN	Increase^{gf}	Decrease^{m}	Lans and Jansen (2007), Hahm et al. (2009)
				No change^{gf}		
gpa-5	–	G protein	SN	No change^{OE}	Increase^{m}	Lans and Jansen (2007)
gpa-9	*GNAI2*	Guanine nucleotide binding protein (G protein), alpha inhibiting activity polypeptide 2	SN	Decrease^{OE}	Increase^{m}	Lans and Jansen (2007)
gpa-10	*GNAT3*	Guanine nucleotide binding protein, alpha transducing 3	SN	ND	Decrease^{m}	Lans and Jansen (2007)
gpa-11	*GNAT3*	Guanine nucleotide binding protein, alpha transducing 3	SN	Increase^{OE}		Lans and Jansen (2007)
gpc-1	*GNG7*	Gγ subunit	SN	ND	Increase^{m}	Lans and Jansen (2007)
hcf-1	*HCFC1*	Human host cell factor 1	IIS	ND	Increase$^{m, i}$	Li et al. (2008), Rizki et al. (2011)
hif-1$^{\#}$	*HIF1A*	Hipoxia-inducible factor 1, transcription factor	DR, Mit, HIF	Increase$^{OE, gf}$	S.L.^{m} of *vhl-1(−)*, *egl-9(−)*	Chen et al. (2009), Mehta et al. (2009) Zhang et al. (2009b), Lee et al. (2010), Leiser et al. (2011), Hwang et al. (2014)
					S.L.$^{m, i}$ of *isp-1(−)*, *clk-1(−)*, *daf-15(+/−)*, *rsks-1(−)*	
					S.L.^{m} of ROS	
					Increase^{m} (25 °C)	

(continued)

Table 8.1 (continued)

C. elegans gene	Human ortholog	Gene function/Domains	Pathway(s)	Effects on lifespan: Overexpression/Gain of function	Effects on lifespan: Loss of function	References
hlh-30	*TFEB*	Basic helix-loop-helix (bHLH) transcription factor	IIS, DR, Atg, Mit, Rpd	Increase^{OE}	S.L.^{m} of *mxl-3(−)*	O'Rourke and Ruvkun (2013). Lapierre et al. (2013a)
					S.L.^{i} of DR, *daf-2(−)*, *clk-1(−)*, *rsks-1(−)*, g.c.(−)	
					Decrease^{m}	
hsf-1	*HSF1*	Heat-shock transcription factor	IIS, TOR, DR	Increase^{OE}	S.L.^{m} of *daf-2(−)*, *rsks-1(−)*, *daf-2(−); rsks-1(−)*, *rps-6(RNAi)*, *rps-15(RNAi)*, *mir-80(−)*, *ddl-1(RNAi)*, *ddl-2(RNAi)*, DR	Hsu et al. (2003), Morley and Morimoto (2004), Steinkraus et al. (2008), Lee and Kenyon (2009), Zhang et al. (2009a), Chiang et al. (2012), Seo et al. (2013), Vora et al. (2013)
					S.L.^{i} of *daf-2(−)*, *age-1(−)*	
					Decrease$^{m, i}$	
hsp-16/16.1/16.2/16.11/16.49	*HSPB5*	Heat shock protein	IIS, TOR	ND	Decrease^{i}	Hsu et al. (2003), Morley and Morimoto, (2004), Seo et al. (2013)
					S.L.^{i} of *daf-2(−)*, *age-1(−)*, *rsks-1(−)*, *hsf-1*^{OE}	
ife-2	*eIF4E*	Translation initiation factor 4E (eIF4E) family	IIS, TOR	ND	Increase$^{m, i}$	Hansen et al. (2007), Syntichaki et al. (2007)
					S.L.^{i} of *daf-2(−)*, *age-1(−)*	
ifg-1	*eIF4G*	Translation initiation factor 4G (eIF4G) family	IIS, TOR	ND	Increase^{i}	Curran and Ruvkun (2007), Hansen et al. (2007), Pan et al. (2007)
					S.L.^{i} of *daf-2(−)*	
iftb-1	*eIF2S2*	eIF2beta translation initiation factor	IIS	ND	S.L.^{i} of *daf-2(−)*	Hansen et al. (2007)
					Increase^{i}	
ins-1	–	Insulin/IGF-like peptides	IIS	Increase^{OE} (26 °C)	Increase^{m}	Pierce et al. (2001)
ins-7	–	Insulin/IGF-like peptides	IIS	Decrease^{OE}	Increase$^{m, i}$	Murphy et al. (2003), Murphy et al. (2007), Matsunaga et al. (2012)
ins-18	–	Insulin/IGF-like peptides	IIS	Increase^{OE} (25 °C)	S.L.^{m} of *daf-2(−)*, *ins-7(−)*	Matsunaga et al. (2012)

ire-1	*ERN1*	Serine/threonine-protein kinase/endoribonuclease IRE1	IIS	ND	Decrease$^{m, i}$ S.L.$^{m, i}$ of *daf-2(−)*	Henis-Korenblit et al. (2010)
isw-1	*SMARCA1*	SWI/SNF related, matrix associated, actin dependent regulator of chromatin, subfamily a, member 1	IIS	ND	S.L.^{i} of *daf-2(−)*	Curran et al. (2009)
isp-1	*UQCRFS1*	Rieske iron sulphur protein	Mit	ND	Increase$^{m, i}$	Feng et al. (2001), Rea et al. (2007)
jnk-1	*MAPK10*	Serine/threonine kinase	IIS	Increase^{OE}	Decrease^{m}	Oh et al. (2005)
kin-29	*SIK3*	Serine/threonine-protein kinase SIK3	SN	ND	Increase^{m}	Lanjuin and Sengupta (2002)
klf-1	*KLF4*	Krueppel-like factor 4 transcription factor	DR	Increase^{OE}	S.L.^{i} of DR	Carrano et al. (2014)
kri-1	*KRIT1*	Krev interaction trapped/cerebral cavernous malformation 1	Rpd	Increase^{OE}	S.L.$^{m, i}$ of g.c.(−)	Berman and Kenyon (2006)
let-363	*MTOR*	Mechanistic target of rapamycin, serine/threonine kinase	TOR, DR	ND	Increase$^{m, i}$	Vellai et al. (2003), Hansen et al. (2007)
let-60	*KRAS*	GTPase KRas	IIS	Decrease^{gf}	S.L.^{m} of *daf-2(−)* Decrease^{m}	Nanji et al. (2005)
lgg-1	*GABARAP*	LC3, GABARAP and GATE-16 family Gamma-aminobutyric acid receptor-associated protein	Atg, Rpd	ND	S.L.^{i} of *daf-2(−)*, g.c.(−) Decrease^{i}	Toth et al. (2008), Lapierre et al. (2011)
lin-4	–	MicroRNA	IIS	Increase^{OE}	S.L.^{m} of *daf-2(−)* Decrease^{m}	Boehm and Slack (2005)
lin-14	–	ND	IIS, Rpd	Decrease^{gf}	S.S.$^{m, i}$ of *lin-4(−)* S.S.^{i} of *mir-84(−); mir-241 (−)*; g.c.(−) Increase$^{m, i}$	Boehm and Slack (2005), Shen et al. (2012)

(continued)

Table 8.1 (continued)

C. elegans gene	Human ortholog	Gene function/Domains	Pathway(s)	Effects on lifespan: Overexpression/Gain of function	Effects on lifespan: Loss of function	References
lipl-4	*LIPM*	Lipase member M	Atg, Rpd	Increase[OE]	S.L.[i] of g.c.(−)	Wang et al. (2008), Lapierre et al. (2011)
lips-17	–	Triacylglycerol lipase	Rpd	ND	S.L.[i] of g.c.(−)	McCormick et al. (2012)
mct-1/2	*MCT9*	Monocarboxylate transporter	SN	Increase[OE]	S.L.[i] of *daf-10(−)*	Gaglia et al. (2012)
mdl-1	*MXI1*	Basic helix-loop-helix (bHLH) protein	IIS		Increase[m]	Johnson et al. (2014)
mec-8	*RBPMS2*	RNA binding	SN	ND	Increase[m]	Apfeld and Kenyon (1999)
mekk-3	–	Kinase	DR	ND	Increase[i]	Chamoli et al. (2014)
mes-1	*TYK2*	Receptor tyrosine kinase-like protein	IIS, Rpd	ND	Increase[m]	Arantes-Oliveira et al. (2002)
mes-4	*WHSC1*	SET domain-containing protein that also contains three plant homeodomain (PHD) fingers	IIS	ND	S.L.[i] of *daf-2(−)*	Curran et al. (2009)
					Increase[i]	
mir-34	–	MicroRNA	DR, Atg	ND	S.L.[m] of DR	Yang et al. (2013)
					Increase[m]	
mir-71	–	MicroRNA	IIS, DR, Rpd	Increase[OE]	S.L.[m] of *daf-2(−)*,g.c.(−)	de Lencastre et al. (2010), Boulias and Horvitz (2012), Smith-Vikos et al. (2014)
					Decrease[m]	
mir-80	–	MicroRNA	IIS, DR	ND	Increase[m]	Vora et al. (2013)
mir-84/ mir-241	–	MicroRNAs	Rpd	No change[OE]	S.L.[m] of g.c.(−)	Shen et al. (2012)
mir-228	–	MicroRNA	DR	Decrease[OE]	Increase[m]	Smith-Vikos et al. (2014)
mir-239.1		MicroRNA	IIS	Decrease[OE]	Increase[m]	de Lencastre et al. (2010)
mml-1	*MLXIP*	MLX interacting protein	IIS	ND	S.L.[i] of *daf-2(−)*	Johnson et al. (2014)
					Decrease[m]	
mrpl-1	*MRPL1*	Mitochondrial ribosomal protein, large	Mit	ND	Increase[i]	Houtkooper et al. (2013)

mrpl-2	*MRPL2*	Mitochondrial ribosomal protein, large	Mit	ND	Increase^{i}	Houtkooper et al. (2013)
mrpl-37	*MRPL37*	Mitochondrial ribosomal protein, large	Mit	ND	Increase^{i}	Houtkooper et al. (2013)
mrps-5	*MRPS5*	Mitochondrial ribosomal protein, small	Mit	ND	Increase^{i}	Houtkooper et al. (2013)
mxl-1	*MAX*	Basic helix-loop-helix protein	IIS, DR	ND	Increase^{m}	Johnson et al. (2014)
mxl-2	*MLX*	Basic helix-loop-helix protein	IIS, DR	ND	S.L.^{m} of *daf-2(−)*, DR	Johnson et al. (2014)
		MAX dimerization protein			S.L.^{i} of *mdl-1(−)*, *mxl-1(−)*, *daf-2(−)*, DR	
					Decrease^{m}	
nhr-8	*NR1H*	Nuclear hormone receptor	DR	ND	S.L.^{m} of DR	Thondamal et al. (2014), Magner et al. (2013)
					Decrease^{m}	
nhr-49	*NR2A1*	Transcription factor	Mit, Rpd	Increase^{OE}	Decrease$^{m, i}$	Van Gilst et al. (2005), Khan et al. (2013), Ratnappan et al. (2014)
					S.L.^{m}of g.c.(−)	
					S.L.^{i} of *isp-1(−)*, *g.c.(−)*	
nhr-62	*HNF4A*	Transcription factor	DR	ND	S.L.$^{m, i}$ of DR	Heestand et al. (2013)
nhr-80	*HNF4A*	Hepatocyte nuclear factor 4-alpha-3	Rpd	No change^{OE}	S.L.^{m} of g.c.(−)	Brock et al. (2006), Goudeau et al. (2011)
					Slight decrease^{m}	
nkcc-1	*NKCC2*	Na-K-Cl cotransporter	Mit	ND	Increase^{i}	Houtkooper et al. (2013)
nlp-7	–	Neuropeptide-like protein	DR	Decrease^{OE}	S.L.^{m,i} of DR	Park et al. (2010)
					Decrease^{m,i}	
nmur-1$^{\$}$	*NMUR2*	Neuromedin-U receptor 2	SN	ND	Increase^{m}	Maier et al. (2010)
nuo-1	*NDUFV1*	NADH ubiquinone oxidoreductase	Mit	ND	Increase^{m}	Tsang et al. (2001)
nuo-2	*NDUFS3*	NADH ubiquinone oxidoreductase	Mit	ND	Increase^{i}	Dillin et al. (2002b)
nuo-6	*NDUFB4/ B15*	NADH ubiquinone oxidoreductase	Mit	ND	Increase^{m}	Yang and Hekimi (2010a)

(continued)

Table 8.1 (continued)

C. elegans gene	Human ortholog	Gene function/Domains	Pathway(s)	Effects on lifespan: Overexpression/Gain of function	Effects on lifespan: Loss of function	References
ocr-2	*TRPV6*	Transient receptor potential channel, vanilloid subfamily	SN	ND	Increase[m]	Lee and Ashrafi (2008), Riera et al. (2014)
odr-2	–	Membrane-associated protein	SN	ND	Increase[m]	Alcedo and Kenyon (2004)
odr-3	*GNAT3*	G protein	SN	ND	Increase[m]	Alcedo and Kenyon (2004), Lans and Jansen (2007)
odr-7	*NR2E1*	Olfactory-specific member of the nuclear receptor superfamily	SN	ND	Increase[m]	Alcedo and Kenyon (2004)
oga-1	*MGEA5*	O-GlcNAc selective N-Acetyl-beta-D-glucosaminidase (O-GlcNAcase)	IIS	ND	Increase[m]	Rahman et al. (2010)
ogt-1	*OGT*	12 N-terminal tetratricopeptide (TPR) domains and a C-terminal putative catalytic domain	IIS	ND	S.L.[m] of *age-1(−)*, *sgk-1(−)*, *daf-2(−)*	Rahman et al. (2010)
					Decrease[m]	
osm-1	*IFT172*	Intraflagellar transport	SN	ND	Increase[m]	Apfeld and Kenyon (1999)
osm-3	*KIF17*	Kinesin-2 family member	SN	ND	Increase[m]	Apfeld and Kenyon (1999)
osm-5	*IFT88*	Intraflagellar transport	SN	ND	Increase[m]	Apfeld and Kenyon (1999)
osm-6	*IFT52*	Intraflagellar transport	SN	ND	Increase[m]	Apfeld and Kenyon (1999)
par-4	*LKB1*/*STK11*	Serine/threonine protein kinase	Mit	ND	S.L.[m] of ROS	Hwang et al. (2014)
					Decrease[m]	
par-5	*YWHAZ*	One of two *C. elegans* 14-3-3 proteins	IIS	Increase[OE]	Decrease[i]	Berdichevsky et al. (2006), Wang et al. (2006), Li et al. (2007a)
pdk-1	*PDPK1*	3-phosphoinositide-dependent kinase 1	IIS	Decrease[gf]	Increase[m]	Paradis et al. (1999)
pgl-1	*TAF15*	RNA-binding protein	IIS	ND	S.L.[i] of *daf-2(−)*	Curran et al. (2009)
pgl-2	–	P granule abnormality protein	IIS	ND	S.L.[i] of *daf-2(−)*	Curran et al. (2009)

pgl-3	–	P granule abnormality protein	IIS	ND	S.L.i of *daf-2(−)*	Curran et al. (2009)
pha-4	*FOXA1*	Forkhead box protein A1 transcription factor	TOR, DR, Rpd	Increase^{OE}	S.L.^{i} of *rsks-1(−)*, DR, *mir-228(−)*, g.c.(−) Decrease^{i}	Panowski et al. (2007), Sheaffer et al. (2008), Lapierre et al. (2011), Smith-Vikos et al. (2014)
phi-62	*RNASEK*	Ribonuclease kappa	DR, Mit, Rpd	ND	S.L.^{i} of *isp-1(−)*, DR, g.c.(−) Decrease^{i}	McCormick et al. (2012)
pie-1	–	Zinc-finger protein	IIS	ND	S.L.^{i} of *daf-2(−)*	Curran et al. (2009)
pnc-1	–	pyrazinamidase/ nicotinamidase	DR	ND	S.L.^{m} by DR Decrease^{i}	van der Horst et al. (2007), Moroz et al. (2014)
pptr-1	*PPP2R5E*	PP2A holoenzyme regulatory subunit	IIS	Increase^{OE}	S.L.^{i} of *daf-2(−)*	Padmanabhan et al. (2009)
pqm-1	–	C2H2-type zinc finger and leucine zipper-containing protein	IIS	Decrease^{OE}	S.L.$^{m, i}$ of *daf-2(−)*	Tepper et al. (2013)
prmt-1	*PRMT1*	Type I protein arginine methyltransferase	IIS	No change^{OE}	S.L.^{m} of *daf-2(−)* Decrease^{m}	Takahashi et al. (2011)
rab-10	*RAB10*	Rab-like GTPase	DR	ND	Increase^{i}	Hansen et al. (2005)
raga-1	*RagA*	Ras-related GTP binding protein A	IIS, TOR, DR	Decrease^{gf}, Increase^{dn}	Increase$^{m, i}$	Schreiber et al. (2010), Robida-Stubbs et al. (2012)
ragc-1	*RagC*	Ras-related GTP binding protein C	IIS, TOR	ND	Increase^{i}	Robida-Stubbs et al. (2012), Seo et al. (2013)
rheb-1	*Rheb*	Rheb GTPase	TOR, DR	ND	Increase^{i}	Honjoh et al. (2009)
rict-1$^{\$}$	*RICTOR*	Rapamycin insensitive companion of mTOR	TOR	ND	Decrease^{m}	Soukas et al. (2009)
rle-1	*RC3H1*	E3 ubiquitin ligase	IIS	ND	Increase^{m}	Li et al. (2007b)
rpl-4	*RPL4*	Ribosomal subunit	TOR	ND	Increase^{i}	Hansen et al. (2007)
rpl-6	*RPL6*	Ribosomal subunit	TOR	ND	Increase^{i}	Hansen et al. (2007)
rpl-9	*RPL9*	Ribosomal subunit	TOR	ND	Increase^{i}	Hansen et al. (2007)
rpl-19	*RPL19*	Ribosomal subunit	TOR	ND	Increase^{i}	Hansen et al. (2007)
rpl-30	*RPL30*	Ribosomal subunit	TOR	ND	Increase^{i}	Hansen et al. (2007)

(continued)

Table 8.1 (continued)

C. elegans gene	Human ortholog	Gene function/Domains	Pathway(s)	Effects on lifespan: Overexpression/ Gain of function	Effects on lifespan: Loss of function	References
rps-3	*RPS3*	Ribosomal subunit	TOR	ND	Increase[i]	Curran and Ruvkun (2007)
rps-5	*RPS5*	Ribosomal subunit	TOR	ND	Increase[i]	Kim and Sun (2007)
rps-6	*RPS6*	Ribosomal subunit	TOR	ND	Increase[i]	Hansen et al. (2007), Seo et al. (2013)
rps-8	*RPS8*	Ribosomal subunit	TOR	ND	Increase[i]	Curran and Ruvkun (2007)
rps-10	*RPS10*	Ribosomal subunit	TOR	ND	Increase[i]	Hansen et al. (2007)
rps-11	*RPS11*	Ribosomal subunit	TOR	ND	Increase[i]	Curran and Ruvkun (2007), Hansen et al. (2007)
rps-15	*RPS15*	Ribosomal subunit	TOR	ND	Increase[i]	Hansen et al. (2007), Seo et al. (2013)
rps-22	*RPS22*	Ribosomal subunit	TOR	ND	Increase[i]	Hansen et al. (2007)
rps-23	*RPS23*	Ribosomal subunit	TOR	ND	Increase[i]	Kim and Sun (2007)
rps-26	*RPS26*	Ribosomal subunit	TOR	ND	Increase[i]	Hansen et al. (2007)
rsks-1	*RPS6KB*	Ribosomal subunit 6 kinase	TOR	ND	Increase[m, i]	Hansen et al. (2007), Pan et al. (2007), Seo et al. (2013)
sams-1	*MAT1A*	S-adenosyl methionine synthetase	DR	ND	Increase[i]	Hansen et al. (2005)
sea-2	*EP400*	Functions to interpret the X: A chromosomal ratio	IIS	ND	Increase[m, i] Increase[m] in *daf-2(−)*	Huang et al. (2011)
sem-5	*GRB2*	Src homology (SH) domain 2 and 3-containing protein	IIS	ND	S.L.[m] of *daf-2(−)*	Nanji et al. (2005)
ser-3	*ADRA1A*	Octopamine receptor	SN	ND	Increase[m]	Petrascheck et al. (2007)
ser-4	*HTR1B*	Serotonin receptor	SN	ND	Increase[m]	Petrascheck et al. (2007)

sgk-1	*SGK1*	Serine/threonine protein kinase	IIS, TOR	Increase^{gf}, Decrease^{OE}, Increase^{OE} (intestine)	S.L.^{m} of *akt-1(−)*, *trpa-1*^{OE} Decrease^{m} Increase^{i}	Hertweck et al. (2004), Tullet et al. (2008), Soukas et al. (2009), Alam et al. (2010), Rahman et al. (2010), Chen et al. (2013a), Xiao et al. (2013)
shc-1	*SHC1*	Src Homology domain C-terminal adaptor homolog	IIS	No change^{OE}	S.L.^{m} of *daf-2(−)* Decrease^{m}	Neumann-Haefelin et al. (2008)
sir-2.1	*SIRT1*	NAD-dependent protein deacetylase	IIS, DR	Increase^{OE}	S.L.^{m} of DR Decrease$^{m, i}$	Tissenbaum and Guarente (2001), Wang and Tissenbaum (2006), Berdichevsky et al. (2006), Rizki et al. (2011), Mouchiroud et al. (2013), Schmeisser et al. (2013) but see Burnett et al. (2011), Viswanathan and Guarente (2011)
				No change^{OE}	Increase^{i}	
skn-1	*NFE2L2*	bZip transcription factor	IIS, TOR, DR	Increase^{OE}	S.L.^{m} of *daf-2(−)*, DR, *ife-2 (RNAi)*, *ragc-1(RNAi)* S.L.^{i} of *mir-228(−)* Decrease$^{m, i}$	Bishop and Guarente (2007), Tullet et al. (2008), Wang et al. (2010), Okuyama et al. (2010), Robida-Stubbs et al. (2012), Smith-Vikos et al. (2014)
skr-1/2	*SKP1*	SKP1-related (ubiquitin ligase complex component)	IIS	ND	S.L.^{i} of *daf-2(−)* Decrease^{i}	Ghazi et al. (2007)
smk-1	*SMEK1*	Mammalian and Dictyostelium discoideum SMEK (suppressor of MEK null)	IIS, DR, Rpd	ND	Decrease^{i} S.L.^{i} of *daf-2(−)*, DR, g.c.(−)	Wolff et al. (2006), Panowski et al. (2007)

(continued)

Table 8.1 (continued)

C. elegans gene	Human ortholog	Gene function/Domains	Pathway(s)	Effects on lifespan: Overexpression/ Gain of function	Effects on lifespan: Loss of function	References
sod-2	*SOD2*	Mitochondrial superoxide dismutase	Mit	Increase^{OE}	S.L.^{m} of *isp-1(−)*	Yang et al. (2007), Van Raamsdonk and Hekimi (2009), Cabreiro et al. (2011)
					Increase^{m} of *clk-1(−)*, DR, g. c.(−)	
sod-3	*SOD3*	Mitochondrial superoxide dismutase	Mit	ND	S.L.^{m} of ROS	Yee et al. (2014)
sos-1	*SOS1/2*	Ras-activating guanine nucleotide exchange factor (GEF)	IIS	ND	S.L.^{m} of *daf-2(−)*	Nanji et al. (2005)
str-2	–	G protein-coupled receptor	SN	ND	Increase^{i}	Alcedo and Kenyon (2004)
taf-4	*TAF4B*	Component of transcription factor TFIID complex	Mit	ND	S.L.^{i} of *isp-1(−)*, *clk-1(−)*, *tpk-1(−)*	Khan et al. (2013)
tax-2$^{\#}$	*CNGB3*	Cyclic nucleotide-gated channel β-subunit	SN	ND	Increase^{m} (15 °C, 20 °C)	Apfeld and Kenyon (1999), Lee and Kenyon (2009)
					Decrease^{m} (25 °C)	
tax-4$^{\#}$	*CNGA1*	Cyclic nucleotide-gated channel α-subunit	SN	ND	Increase^{m} (15 °C, 20 °C)	Apfeld and Kenyon (1999), Lee and Kenyon (2009)
					Decrease^{m} (25 °C)	
tcer-1	*TCERG1*	Transcription elongation regulator	Rpd	Increase^{OE}	S.L.$^{\mathrm{m,\,i}}$ of g.c.(−)	Ghazi et al. (2009)
tpk-1	*TPK1*	Thiamine pyrophosphokinase	Mit	ND	Increase^{m}	de Jong et al. (2004)
trpa-1$^{\#}$	*TRPA1*	cold-sensitive transient receptor potential ion channel subfamily A	SN, IIS	Increase^{OE} (20 °C and 15 °C)	Decrease^{m} (20 °C and 15 °C)	Xiao et al. (2013)
				No change^{OE} (25 °C)		

tts-1	–	transcribed telomeric sequence 1/long noncoding RNA (lncRNA)	IIS, Mit	ND	S.L.^{i} of *daf-2(−)*, and *clk-1(−)*	Essers et al. (2015)
ttx-1$^{\#}$	*OTX1*	Homeodomain transcription factor	SN	ND	Decrease^{m} (25 °C)	Lee and Kenyon (2009)
tub-1	*TUB*	Tubby bipartite transcription factor	IIS	ND	Increase^{m}	Mukhopadhyay et al. (2005)
ubc-18	*UBE2L3*	E2 ubiquitin-conjugating enzyme	DR	No change^{OE}	S.L.^{i} of DR	Carrano et al. (2009)
					Decrease^{m}	
ubl-5	*UBL5*	Coactivator of DVE-1	Mit	ND	S.L.^{i} of *isp-1(−), clk-1(−)*	Durieux et al. (2011), Taylor and Dillin (2013)
					Decrease^{i}	
unc-31	*CADPS*	Pleckstrin homology (PH) domain-containing protein	IIS	ND	Increase^{m}	Ailion et al. (1999)
unc-51	*ULK2*	Serine/threonine protein kinase orthologous to *S. cerevisiae* Atg1p	TOR, DR, Atg, Mit, Rpd	ND	S.L.^{m} of TOR, DR, *atp-3 (RNAi)*	Toth et al. (2008), Lapierre et al. (2011)
					S.L.^{i} of g.c.(−)	
					Decrease^{m}	
unc-64	*STX1A*	Syntaxin, a plasma membrane receptor	IIS	ND	Increase^{m}	Ailion et al. (1999)
utx-1	*KDM6A*	Lysine (K)-specific H3K27 demethylase	IIS	ND	Increase$^{m/+, i}$	Jin et al. (2011), Maures et al. (2011)
vang-1	*VANGL1*	Wnt-directed planar cell polarity (PCP) protein orthologous to *Drosophila* VAN GOGH	IIS	ND	Increase$^{m, i}$	Honnen et al. (2012)
vhl-1	*VHL*	Substrate-recognition subunit of E3 ligase	HIF	ND	Increase$^{m, i}$	Mehta et al. (2009), Muller et al. (2009), Hwang et al. (2014)

(continued)

Table 8.1 (continued)

C. elegans gene	Human ortholog	Gene function/Domains	Pathway(s)	Effects on lifespan: Overexpression/ Gain of function	Effects on lifespan: Loss of function	References
vps-34	*PIK3C3*	Phosphoinositide 3-kinase	DR, Atg, Rpd	ND	S.L.^{i} of DR, g.c.(−), *lipl-4*^{OE}, *rab-10(−)*	Hansen et al. (2008), Lapierre et al. (2011)
wwp-1	*WWP1*	HECT E3 ubiquitin ligase	DR	Increase^{OE}	S.L.$^{m, i}$ of DR	Carrano et al. (2009)
xbp-1	*XBP1*	X-box binding protein/bZIP transcription factor	IIS	Increase^{OE} (neurons, intestine) Decrease^{OE} (muscle)	S.L.^{m} of *daf-2(−)* Decrease^{m,i}	Henis-Korenblit et al. (2010), Taylor and Dillin (2013)

These are selected *C. elegans* lifespan-regulatory genes that act in the pathways described in the text

Notes: descriptions regarding loss of function mutations or RNAi of genes, which did not cause lifespan changes, were omitted. Genes that affect lifespan via DAF-16/FOXO were included as IIS genes for simplicity

IIS insulin/IGF-1 signaling, *TOR* target of rapamycin, *DR* dietary restriction, *Atg* autophagy, *Mit* mitochondria, *HIF* hypoxia-inducible factor 1, *ROS* reactive oxygen species, *SN* sensory neurons, *Rpd* reproduction, *g.c.(−)* germ cell ablation. Overexpression (^{OE}), gain of function (^{gf}), dominant negative mutation (^{dn}), loss of function mutations (^{m}), heterozygote loss of function mutations (m/+), and RNAi knockdown (^{i}) of genes, which affect lifespan in *C. elegans*. *C. elegans* lifespan phenotypes that show: *, allele dependency; #, temperature dependency; $, food dependency; decrease or increase, decrease or increase in lifespan; S.L., suppression of longevity; S.S., suppression of short lifespan. ND, not determined

8.2 Longevity-Regulatory Pathways in *C. elegans*

8.2.1 The Insulin/Insulin-Like Growth Factor 1 Signaling Pathway

The insulin/IGF-1 signaling (IIS) pathway is one of the most highly characterized and evolutionarily conserved pathways that regulate aging (Fig. 8.1). In *C. elegans*, IIS is presumably initiated by the modulation of the activity of DAF-2, an insulin/

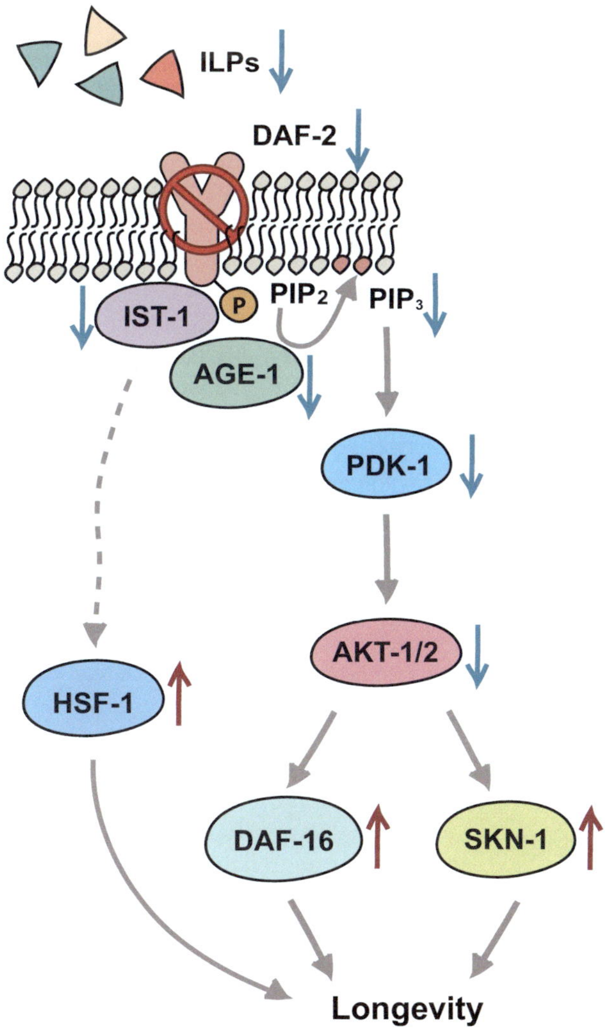

Fig. 8.1 Lifespan regulation by the insulin/IGF-1 signaling pathway in *C. elegans*. Insulin-like peptides (ILPs) bind to insulin/IGF-1 receptor DAF-2 to regulate its phosphorylation. The reduction of DAF-2 activity leads to decreased binding of IST-1, the insulin receptor substrate, resulting in the inactivation of phosphoinositide-3 kinase (AGE-1), which is responsible for the conversion of PI(4,5)P_2 to PI(3,4,5)P_3. This event decreases the activities of phosphoinositide-dependent kinase 1 (PDK-1) and the AKT-1 and AKT-2 (AKT-1/2) kinases. This reduces the phosphorylation of the DAF-16/FOXO and SKN-1/NRF2 transcription factors, increasing their activities. The inhibition of DAF-2 also increases the activity of heat shock transcription factor 1 (HSF-1). The activation of DAF-16, SKN-1, and HSF-1 leads to longevity by transcriptionally regulating the expression of downstream longevity genes

IGF-1 receptor homolog, upon the binding of insulin-like peptides (ILPs). *C. elegans* expresses 40 ILPs, some of which are predicted to be DAF-2 agonists (e.g., *ins-7*) or antagonists (e.g., *ins-18*) (Kawano et al. 2000; Murphy et al. 2003, 2007; Fernandes de Abreu et al. 2014). The up-regulation of DAF-2 leads to the activation of a phosphoinositide 3-kinase (PI3K) cascade, which in turn regulates several transcription factors that affect lifespan (reviewed in Kaletsky and Murphy 2010; Murphy and Hu 2013).

Many components of IIS regulate lifespan. For example, mutations in *daf-2* can double the lifespan (Kenyon et al. 1993). The importance of DAF-2 as a longevity-regulatory factor is highlighted by the findings that mouse and human DAF-2 homologs are associated with longevity (reviewed in Kenyon 2010). Mutations in *age-1*, which encodes the catalytic subunit of PI3K (Morris et al. 1996), greatly extend lifespan as well (Friedman and Johnson 1988). Other genes in the IIS pathway whose genetic perturbation extends lifespan include *ins-7* and *daf-28*, two ILP genes; *ist-1* (insulin receptor substrate homolog); *pdk-1*, which encodes a phosphoinositide-dependent kinase (PDK); and *akt-1* and *akt-2*, two *C. elegans* Akt/PKB homologs (Murphy et al. 2003; Paradis and Ruvkun 1998; Malone et al. 1996; Paradis et al. 1999; Wolkow et al. 2002). The transcription factors that act downstream of IIS, including DAF-16/FOXO, heat shock factor 1 (HSF-1) and SKN-1/NRF2, are required for longevity in animals with reduced IIS (Kenyon et al. 1993; Lin et al. 1997; Ogg et al. 1997; Hsu et al. 2003; Morley and Morimoto 2004; Tullet et al. 2008). Among those, DAF-16 is the best characterized longevity-promoting transcription factor, and its human homolog FOXO3A is also linked to longevity (reviewed in Kenyon 2010). DAF-16 is inactivated by phosphorylation via AKT-1 and AKT-2, and the dephosphorylation of DAF-16 leads to DAF-16 activation via nuclear localization (Lin et al. 2001; Henderson and Johnson 2001; Lee et al. 2001). The activation of DAF-16 leads to the induction of various genes that are crucial for longevity, including molecular chaperones, antioxidants, anti-microbials, and stress resistance genes (reviewed in Kaletsky and Murphy 2010; Murphy and Hu 2013). Many other longevity-regulatory genes involved in IIS have been identified using genetic, genomic, and proteomic approaches (Table 8.1). It will be important to characterize the relationships among those many factors to ascertain their impact on lifespan regulation.

What is the physiologic interpretation of the effects of IIS on longevity? IIS influences not only longevity but also other physiologic processes including larval development, fat metabolism, immunity, and stress resistance (reviewed in Kaletsky and Murphy 2010; Murphy and Hu 2013). Increased resistance to various stresses, including heat (Lithgow et al. 1995), osmotic stress (Lamitina and Strange 2005), reactive oxygen species (ROS) (Honda and Honda 1999), hypoxia (Scott et al. 2002) and endoplasmic reticulum (ER) stress (Henis-Korenblit et al. 2010), and proteotoxicity (Morley et al. 2002; Hsu et al. 2003) will confer survival benefits to *C. elegans* in general. At the cellular level, the stress resistance phenotypes might contribute to longevity by promoting cellular maintenance capabilities. In addition, animals with reduced IIS have higher chances to survive infection by various pathogens because of enhanced innate immunity (Garsin et al. 2003). This appears to lengthen lifespan in *C. elegans*, because the main cause of death in aged *C. elegans* in laboratory is infection by *E. coli*, the worm's bacterial food (Garigan et al. 2002). Moreover, somatic cells in *C. elegans* with reduced IIS have the characteristics of germline stem cells, which are robustly protected from various stresses (Curran et al. 2009). Overall, *C. elegans* appears to employ endocrine IIS to enhance cellular protection and maintenance in the face of harsh environmental conditions, which may lead to longevity.

8.2.2 Lifespan-Regulating Genes in the Target of Rapamycin (TOR) Signaling Pathway

The TOR pathway is another evolutionarily well-conserved signaling pathway that influences aging (Fig. 8.2). TOR is a serine/threonine kinase that plays various physiologic roles, including involvement in cellular growth, metabolism, protein synthesis and autophagy, and aging, in response to changes in nutrient status (Stanfel et al. 2009; Kapahi et al. 2010; Evans et al. 2011; Laplante and Sabatini 2012; Johnson et al. 2013). TOR interacts with other proteins such as the regulatory-associated protein raptor and the rapamycin-insensitive companion rictor. Those interactions determine the formation of TOR complex 1 (TORC1) and TOR complex 2 (TORC2), respectively. Although both complexes are linked to longevity, the signaling pathway of TORC1 is characterized in more detail than that

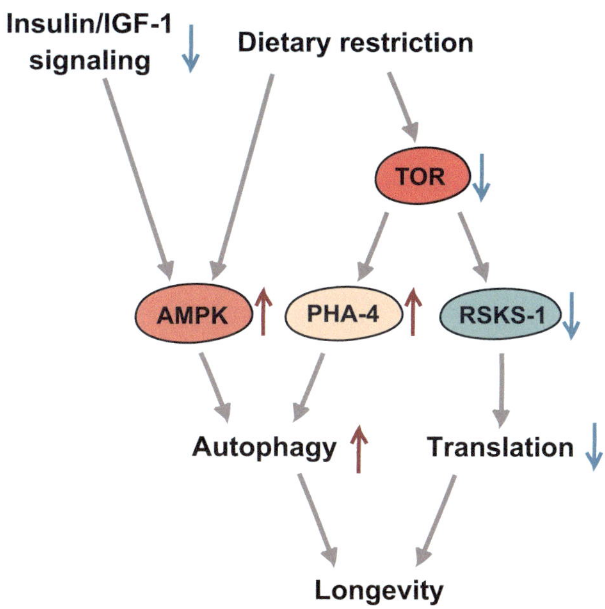

Fig. 8.2 Dietary restriction increases lifespan through AMPK, TOR signaling, and autophagy. Dietary restriction extends lifespan through energy-sensing pathways in general. Dietary restriction increases the activity of AMP-activated protein kinase (AMPK) while down-regulating target of rapamycin (TOR) kinase. Activated AMPK extends lifespan partly by increasing autophagy. Reduced TOR signaling leads to the activation of autophagy via PHA-4/FOXA and the reduction of translation via the down-regulation of ribosomal S6 kinase 1 (RSKS-1)

of TORC2. TORC1 exerts its effects by regulating downstream targets, including ribosomal protein S6 kinase, which promotes protein synthesis via the phosphorylation of ribosomal protein subunit 6.

The inhibition of various TOR pathway genes extends the lifespan of *C. elegans*. The RNAi knockdown or mutation of *let-363*/TOR, *daf-15*/RAPTOR, *rsks-1* (ribosomal protein S6 kinase), ribosomal subunits, or translational initiation factors increases the lifespan of *C. elegans* (Vellai et al. 2003; Jia et al. 2004; Ching et al. 2010; Chen et al. 2013b; Seo et al. 2013; Hansen et al. 2007; Pan et al. 2007; Syntichaki et al. 2007; Curran and Ruvkun 2007). In addition, treatment with rapamycin, the inhibitor of TOR, extends lifespan in *C. elegans* (Robida-Stubbs et al. 2012). Reduced translation underlies the longevity caused by the inhibition of TOR signaling (Hansen et al. 2007; Pan et al. 2007). The inhibition of the translation initiation factors *ife-2*/eIF4E and *ifg-1*/eIF4G, which are predicted to be regulated by TOR, extends lifespan (Pan et al. 2007; Wang et al. 2010; Rogers et al. 2011; Hansen et al. 2007; Syntichaki et al. 2007; Curran and Ruvkun 2007). Several longevity-promoting transcription factors including *pha-4*/FoxA (Sheaffer et al. 2008), HSF-1 (Seo et al. 2013), SKN-1 (Robida-Stubbs et al. 2012), and DAF-16 (Seo et al. 2013; Robida-Stubbs et al. 2012; Hansen et al. 2007) mediate the longevity caused by reduced TOR signaling. In addition, AMP-activated protein kinase (AMPK), a nutrient-sensing and longevity-promoting kinase, is required for the extended lifespan of *rsks-1* mutants (Selman et al. 2009). Thus, reduced TORC1 signaling appears to lead to decreased translation, which in turn up-regulates various downstream longevity factors.

How does TOR signaling modulate longevity by influencing mRNA translation rates? Because protein synthesis requires large amounts of energy and metabolic resources such as ATP and amino acids, thrifty usage of proteins may be a cost-effective way for organisms to use resources for maintenance. Additionally, a slowed rate of protein synthesis may give organisms a chance to increase overall protein quality, because protein repair and degradation systems can be efficiently act on a relatively small amount of proteins. In addition to reducing translation, the inhibition of TOR signaling enhances autophagy-related processes (reviewed in Green et al. 2014). This in turn removes and/or recycles damaged proteins and organelles by selectively transporting them to lysosomes, which can promote healthy cellular environments. Thus, the inhibition of TOR signaling may benefit longevity by enhancing protein quality and reducing proteotoxicity during aging.

8.2.3 Genes That Mediate Dietary Restriction-Induced Longevity

Dietary restriction, which is defined as the restriction of food intake without malnutrition, extends lifespan in various species (reviewed in Piper et al. 2011). In *C. elegans*, diverse DR regimens have been used, such as genetic mutations that decrease feeding rates, the dilution or deprivation of food (bacteria) concentrations, intermittent fasting, and culture in axenic media that contain sparse nutrients (summarized in Greer and Brunet 2009). Interestingly, different genes and pathways mediate the effects of diverse DR regimens on longevity in *C. elegans* (Fig. 8.2, Table 8.1). The most notable genetic factors that mediate the longevity conferred by DR are cellular energy sensors. Dietary restriction decreases ATP levels and subsequently increases the AMP/ATP ratio, and this in turn activates the energy sensor AMPK (Hardie 2014). The catalytic α subunit of AMPK in *C. elegans*, AAK-2, is necessary and sufficient for DR-induced longevity (Greer and Brunet 2009; Greer et al. 2007). Target of rapamycin (TOR), another major energy sensor, is also implicated as a mediator of DR-induced longevity (Hansen et al. 2007). The inhibition of *let-363*, *C. elegans* TOR, extends lifespan, perhaps by mimicking DR. The lifespan extension conferred by intermittent fasting is mediated by RHEB-1 (GTPase), an activator of TOR (Honjoh et al. 2009). One possible mechanism by which reduced TOR levels mediate DR-induced longevity is the up-regulation of autophagy (reviewed in Green et al. 2014), which helps cellular maintenance during DR. Thus, DR appears to increase lifespan by regulating energy sensors to maximize cellular maintenance under conditions where energy is scarce.

DR alters the activities of longevity-promoting transcription factors. The AMPK activated upon DR up-regulates DAF-16 to mediate longevity (Greer et al. 2007). Dietary restriction reduces TOR signaling, which leads to changes in the activity of transcription factors including PHA-4 (Panowski et al. 2007; Sheaffer et al. 2008), hypoxia-inducible factor 1 (HIF-1) (Chen et al. 2009), HSF-1 (Seo et al. 2013; Steinkraus et al. 2008), DAF-16 (Robida-Stubbs et al. 2012; Seo et al. 2013), and SKN-1 (Bishop and Guarente 2007; Robida-Stubbs et al. 2012). SIR-2.1 (sirtuin: NAD-dependent protein deacetylase) (Wang and Tissenbaum 2006) and PNC-1 (a key component of the NAD^+ salvage pathway) (Moroz et al. 2014) mediate DR-induced longevity. NHR-62 (nuclear receptor) mediates the lifespan-increasing effects of DR by controlling fat metabolism and autophagy (Heestand et al. 2013). WWP-1 (E3 ligase) and UBC-18 (E2 conjugating enzyme), components of the ubiquitin system, extend lifespan upon DR by degrading kruppel-like factor 1 (KLF-1) (Carrano et al. 2014; Carrano et al. 2009).

Under DR, organisms decrease their rates of growth and reproduction to preserve resources for survival until food conditions become more favorable. These maintenance responses appear to promote stress resistances and eventually lead to longevity. In *C. elegans*, various factors, including cellular energy sensors, in diverse longevity pathways mediate DR-induced longevity in a diet regimen-dependent manner. Thus, these factors may sense various DR cues and transmit longevity signals to different pathways. It will be important to dissect the mechanisms by which these diverse factors or pathways interact with each other for lifespan extension in response to DR.

8.2.4 Autophagy-Related Genes Required for Longevity

Autophagy is a process that promotes the degradation of cellular components to recycle macromolecules and organelles (reviewed in Levine and Klionsky 2004). The appropriate clearance of damaged cellular components mediated by autophagy is one of the crucial requirements for lifespan extension in *C. elegans*. Autophagy was first identified in mammals, and subsequent genetic studies using yeast identified many autophagy-related genes (ATGs). Autophagy begins with the induction of membrane changes (regulated by *unc-51/ATG1*), vesicle nucleation (regulated by *bec-1/ATG6*, *vps-34/VPS34*), vesicle expansion (regulated by *atg-7/ATG7*, *lgg-1/ATG8*, *lgg-3/ATG12*), and eventually retrieval (regulated by *atg-18/ATG18*). Those ATGs are well-conserved across species including *C. elegans*.

Many lifespan-regulating factors such as IIS, TOR, DR, and reproductive pathways have been shown to modulate autophagy in *C. elegans*. The role of autophagy in promoting longevity was first shown for reduced IIS (Melendez et al. 2003). *daf-2* mutants display increased levels of autophagy and require autophagy-related genes, including *bec-1/ATG6* and *lgg-3/APG12*, for longevity (Melendez et al. 2003; Hars et al. 2007). Acting downstream of IIS (Apfeld et al. 2004), AMPK contributes to the up-regulation of autophagy in *daf-2* mutants (Egan et al. 2011). Although DAF-16 is not required for increased autophagy in *daf-2* mutants (Hansen et al. 2008), overexpression of the DAF-16 is sufficient to induce autophagy (Jia et al. 2009).

Dietary restriction induces autophagy, and essential autophagy genes are required for the longevity caused by genetic mimesis of DR (Jia and Levine 2007; Hansen et al. 2008). Dietary restriction appears to reduce TOR signaling (reviewed in Johnson et al. 2013) and up-regulate the transcription factors PHA-4 and TFEB/*hlh-30* to mediate autophagy-induced longevity (Hansen et al. 2008; Lapierre et al. 2013a). Overall, increased autophagy is required for the longevity caused by multiple signaling pathways, most if not all of which are sensitive to nutrient conditions. Thus, autophagy may provide the nutrients and energy required for longevity pathways. Because evidence supporting the hypothesis that enhanced autophagy *per se* is sufficient for longevity is scarce, it seems likely that autophagy is a limiting factor for longevity.

8.2.5 Longevity Caused by Reduced Mitochondrial Function

Mitochondria are crucial for many physiologic processes including energy production. Interestingly, a mild inhibition of the mitochondrial electron transport chain (ETC) generally promotes longevity in *C. elegans* (Fig. 8.3) (reviewed in Hwang et al. 2012). One of the first long-lived mitochondrial ETC mutants that were identified was the *clk-1* (demethoxyubiquinone hydroxylase) mutant (Wong et al. 1995; Braeckman et al. 1999; Ewbank et al. 1997). Other long-lived ETC

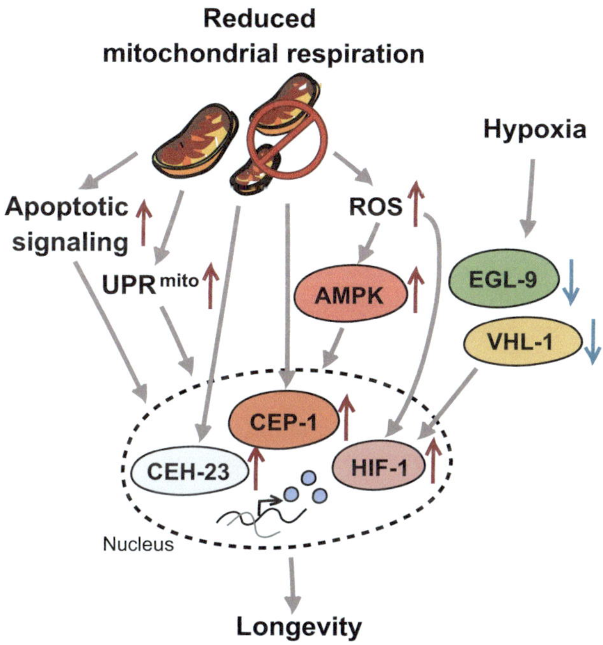

Fig. 8.3 Lifespan-regulatory pathways of reduced mitochondrial respiration and HIF-1. Impaired mitochondrial respiration extends lifespan by causing a global change in gene expression via non-mitochondrial mediators. These mediators include nuclear transcription factors, AMPK (AMP-activated protein kinase), the apoptotic signaling pathway, and the mitochondrial unfolded protein response (UPR^{mito}). In addition, increased reactive oxygen species (ROS) levels mediate longevity in respiratory mutants through AMPK and HIF-1. HIF-1 is also stabilized upon hypoxia via EGL-9 (proline hydroxylase) and VHL-1 (ubiquitin E3 ligase)

mutants include the *isp-1* (iron-sulfur protein) mutants, which are defective in mitochondrial complex III (Feng 2001), and the *nuo-6* (NADH ubiquinone oxidoreductase) mutants, which are defective in mitochondrial complex I (Yang and Hekimi 2010b). In addition, RNAi knockdown of many genes that encode mitochondrial ETC components leads to longevity (Dillin et al. 2002b; Lee et al. 2003; Hansen et al. 2005).

Several key downstream factors and signaling pathways are known to mediate the longevity of mitochondrial ETC mutants. These include FSTR-1/-2, AMPK, HIF-1, CEH-23 (homeobox domain transcription factor), the mitochondrial unfolded protein response (UPR^{mito}) genes, CEP-1 (p53 homolog), components of the apoptotic signaling pathway, and the TAF-4/TFIID complex (Baruah et al. 2014; Yee et al. 2014; Khan et al. 2013; Walter et al. 2011; Durieux et al. 2011; Lee et al. 2010; Ventura et al. 2009; Cristina et al. 2009; Curtis et al. 2006; Hwang et al. 2014). The inhibition of the ETC appears to modulate signaling from the mitochondria to other cellular organelles including the nucleus. For example, reduced mitochondrial respiration leads to changes in global gene expression, which contribute to a long lifespan (Yee et al. 2014; Cristina et al. 2009). The global changes in gene expression appear to be mediated by transcription factors such as HIF-1 and CEP-1 (Hwang et al. 2014; Baruah et al. 2014).

Many long-lived ETC mutants display increased ROS levels, which actually contribute to longevity (Hwang et al. 2014; Lee et al. 2010; Yang and Hekimi 2010a; Van Raamsdonk and Hekimi 2012). Long-lived mitochondrial mutants have increased mitochondrial ROS levels (Yang and Hekimi 2010a; Hwang et al. 2014; Lee et al. 2010), and antioxidant treatment suppresses this longevity (Yang and Hekimi 2010a; Van Raamsdonk and Hekimi 2012). Thus, increased ROS levels seem to contribute to the long lifespan of ETC mutants. Furthermore, increased HIF-1 and AMPK activities in response to elevated ROS levels mediate this ROS-induced longevity (Hwang et al. 2014; Lee et al. 2010). These findings invite a revision of the free radical theory of aging (Harman 1956, 1972), which proposes that ROS cause aging and therefore shorten lifespan.

Another key parallel signaling pathway required for the longevity of ETC mutants is the UPR^{mito}. The UPR^{mito} is a stress response that relays signals from the mitochondria to the nucleus to induce mitochondrial chaperon proteins (reviewed in Haynes et al. 2013). Impaired ETC function in one tissue (e.g., neurons) activates the UPR^{mito} and relays yet unidentified longevity signals to other tissues (e.g., intestinal cells) to extend lifespan (Durieux et al. 2011). However, the activation of the UPR^{mito} is not sufficient to promote longevity (Bennett et al. 2014).

Since the first long-lived mitochondrial respiratory *clk-1* mutants were identified, numerous studies have been conducted to reveal the molecular mechanisms underlying this lifespan regulation. Only recently, scientists started to understand the paradox of how reduced ETC delays aging and increases lifespan. Interestingly, simple and small animal species that have high respiration rates tend to live shorter lives, whereas complex and large species with low respiration rates tend to live longer lives (reviewed in Kenyon 2010). Perhaps the longevity displayed by *C. elegans* ETC mutants mimics the evolution of longevity among species.

8.2.6 The Regulation of Lifespan by the Hypoxia-Inducible Factor 1-Regulatory Pathway

Hypoxia-inducible factor 1 (HIF-1) is a key transcription factor that regulates responses to conditions of low oxygen (Fig. 8.3) (reviewed in Powell-Coffman 2010; Semenza 2012). Under normal oxygen conditions, HIF-1 is hydroxylated by the proline hydroxylase EGL-9 and ubiquitinated by von Hippel-Lindau-1 (VHL-1), an E3 ligase component. Under conditions of low oxygen, EGL-9 cannot hydroxylate HIF-1, leading to the stabilization of HIF-1 and the induction of HIF-1 target genes. HIF-1 modulates various biological processes, including lifespan- and aging-related processes in *C. elegans*.

The up-regulation of HIF-1, by the genetic inhibition of VHL-1 or EGL-9 (Lee et al. 2010; Mehta et al. 2009; Muller et al. 2009), or by the overexpression of *hif-1*, increases lifespan (Zhang et al. 2009). The activation of HIF-1 also contributes to the longevity conferred by mitochondrial ROS in a positive-feedback fashion and through the modulation of iron-metabolism genes (Lee et al. 2010; Hwang et al. 2014). Interestingly, HIF-1 also regulates lifespan in a temperature-dependent manner, possibly through IIS (Lee et al. 2010; Leiser et al. 2011; Chen et al. 2009; Zhang et al. 2009), and mediates DR-induced longevity (Chen et al. 2009). Overall, HIF-1 appears to act as a sensor and mediator for various lifespan-regulatory signals such as oxygen concentration, mitochondrial ROS, temperature changes, and nutrient levels.

Because HIF-1 is one of the recently identified factors that regulate aging in *C. elegans*, the mechanisms by which HIF-1 increases lifespan remain elusive. Different from vertebrate models, the availability of viable *hif-1, vhl-1*, and *egl-9* mutants has made *C. elegans* a unique and important model organism to study the role of HIF-1 in aging. Future studies regarding HIF-1, including tissue-specific roles and the functional characterization of upstream and downstream factors, will provide mechanistic insights into how this evolutionarily conserved transcription factor exerts its effects on longevity.

8.2.7 Sensory Neuronal Regulation of Longevity

C. elegans is equipped with a sensory nervous system that perceives environmental changes. Intriguingly, sensory neurons modulate lifespan in *C. elegans* (Fig. 8.4), and this phenomenon is also observed in *Drosophila* and mice (Linford et al. 2011; Jeong et al. 2012; Riera et al. 2014). Structural perturbations of a subset of ciliated sensory neurons, including the genetic disruption of *che-2/IFT80*, *daf-10/IFT122*, *daf-19/RFX2*, or *osm-5/IFT88*, increase lifespan in *C. elegans* (Apfeld and Kenyon 1999; Alcedo and Kenyon 2004). Many *C. elegans* mutants that have defects in sensory signal transduction also live long. The genetic inhibition of *str-2*, a putative sensory G protein-coupled receptor, or of *kin-29* (SIK3 kinase), which regulates the expression of subsets of neuronal sensory receptors, lengthens lifespan (Lanjuin and Sengupta 2002; Alcedo and Kenyon 2004). The inhibition of G proteins that act downstream of sensory receptors such as *gpa-1*, *gpa-5*, *gpa-9*, and *odr-3* extends lifespan as well (Lans and Jansen 2007; Alcedo and Kenyon 2004). The genetic modulation of downstream cation channels, including *tax-4* (cyclic nucleotide-gated channel subunit), *ocr-2* and *osm-9* (neuronal transient receptor potential vanilloid (TRPV) channels), and cold-sensitive *trpa-1* (TRPA channel), can increase lifespan (Apfeld and Kenyon 1999; Lee and Ashrafi 2008; Lee and Kenyon 2009; Xiao et al. 2013; Riera et al. 2014). Thus, the inhibition of the sensory neural structure or function generally increases lifespan in *C. elegans*.

The longevity caused by sensory impairment appears to be mediated at least partly by the IIS pathway. Defects in sensory neurons promote the nuclear

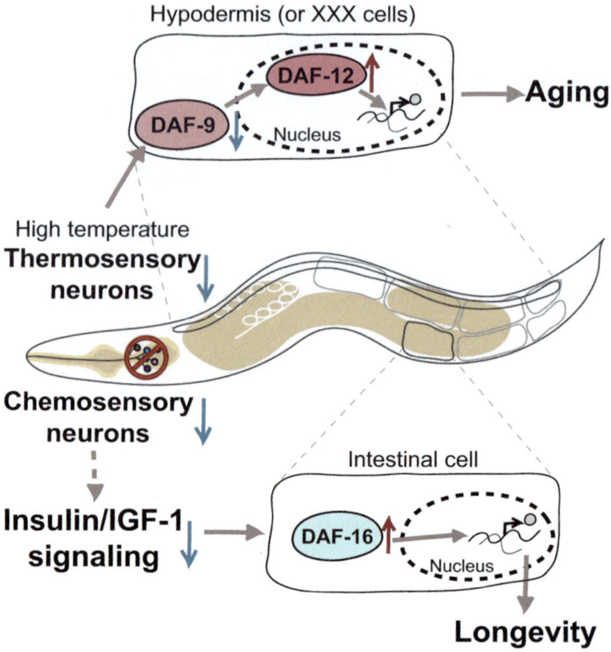

Fig. 8.4 Pathways that act downstream of sensory neurons to modulate lifespan. The inhibition of thermosensory neurons decreases *C. elegans* lifespan at high temperature (25 °C) by decreasing the expression of DAF-9 (cytochrome P450) in distal tissues such as hypodermis and XXX cells, resulting in the activation of DAF-12 (nuclear receptor). The perturbation of chemosensory neurons presumably decreases insulin/IGF-1 signaling, which promotes the nuclear localization and activation of DAF-16/FOXO to enhance longevity

localization and transcriptional activation of DAF-16 (Lin et al. 2001; Xiao et al. 2013; Gaglia et al. 2012). In addition, the long lifespans caused by sensory mutations are largely suppressed by *daf-16* mutations (Apfeld and Kenyon 1999; Hahm et al. 2009; Lanjuin and Sengupta 2002; Lee and Ashrafi 2008; Xiao et al. 2013; Alcedo and Kenyon 2004; Lans and Jansen 2007). Thus, the disruption of sensory neurons increases lifespan in *C. elegans* through the activation of DAF-16. In addition, the long lifespan of sensory *daf-10/IFT122* mutants requires the induction of *mct-1*, a putative monocarboxylate transporter (Gaglia et al. 2012). This suggests that the transportation of hormones or small molecules modulates lifespan by acting downstream of the sensory perturbation. In contrast to these long-lived sensory mutants, which mostly have chemosensory defects, mutants that have defects in thermosensory AFD neurons are short lived at high temperatures (25 °C) (Lee and Kenyon 2009). This lifespan regulation is mediated by steroid signaling, composed of DAF-9 (cytochrome P450) and DAF-12 (nuclear receptor) in multiple tissues, including hypodermis and endocrine XXX cells (Lee and Kenyon 2009). How these various sensory modalities affect lifespan by employing different downstream factors is currently unclear.

It is intriguing that the inhibition of a small number of sensory neurons can have a large effect on the organismal lifespan. Several signaling pathways that regulate aging, including those involved in IIS, TOR, DR, and autophagy, are concerned with nutrient and food availability. Because foods have smells and tastes as well as nutrients, the sensory neurons may be an intrinsic factor that acts at the upstream end of longevity signaling pathways that are linked with food availability. In fact, sensory cues can directly influence lifespan via sensory neurons in *C. elegans* and *Drosophila* (Libert et al. 2007; Maier et al. 2010). Hence, one can speculate that sensory neurons monitor environmental changes such as food availability and temperature fluctuations and modulate physiologic processes that eventually affect lifespan.

8.2.8 Lifespan Regulation by the Reproductive System

Organismal longevity is frequently associated with reduced reproduction. In *C. elegans*, the removal of the germline promotes longevity (Fig. 8.5) (Hsin and Kenyon 1999). This phenomenon does not result from a simple trade-off between longevity and reproduction, because the removal of the somatic gonad together with germline does not result in longevity (Hsin and Kenyon 1999). Instead, when the

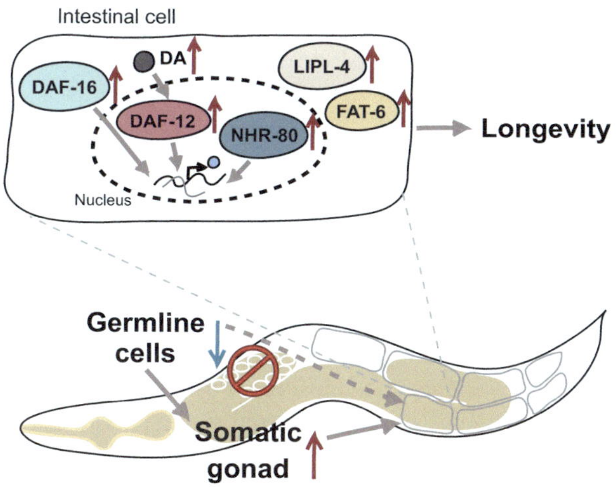

Fig. 8.5 The components of the reproductive system, which regulates longevity. The removal of germline cells increases lifespan by transmitting longevity signals from the somatic gonad to intestinal cells in *C. elegans*. This gonadal signaling increases the synthesis of dafachronic acid (DA), which results in the activation of DAF-12 (nuclear receptor). In addition, signals from the gonad enhance the nuclear localization and transcriptional activity of DAF-16/FOXO. Moreover, the gonadal signaling modulates fat metabolism by up-regulating the NHR-49, 80 (nuclear receptors: NHRs) and LIPL-4 (lipase) to promote longevity

germline is removed, the somatic gonad actively promotes longevity by sending signals that modulate steroid signaling, DAF-16 activities, and fat metabolism.

The DAF-12 is one of the key components in steroid signaling (Antebi et al. 2000) and is required for longevity in germline-ablated worms (Hsin and Kenyon 1999). The activity of DAF-12 is regulated by dafachronic acids (DAs), which are bile acid-like steroid ligands (Motola et al. 2006). DAs are synthesized from cholesterol by multiple enzymatic components such as DAF-36 (Rieske-like oxygenase), DHS-16 (3-hydroxysteroid dehydrogenase), and DAF-9 (Rottiers et al. 2006; Wollam et al. 2012; Jia et al. 2002; Gerisch et al. 2001). Those components, as well as DAs, contribute to the longevity induced by the lack of the germline resulting from laser ablation or *glp-1* (germ line proliferation 1) mutations. For example, mutations in *daf-36*, *dhs-16*, or *daf-9* suppress the long lifespan induced by the lack of germ cells (Rottiers et al. 2006; Wollam et al. 2012; Gerisch et al. 2001). Dafachronic acids are ligands of DAF-12 that promote lifespan extension in animals lacking germline cells (Gerisch et al. 2007; Yamawaki et al. 2010; Mahanti et al. 2014), although treatment with DAs is not sufficient to increase lifespan in wild-type worms (Gerisch et al. 2007; Yamawaki et al. 2010). Thus, the loss of the germline leads to the production of high levels of DAs in the somatic gonad, which activate DAF-12 and promote longevity.

Another component that mediates the longevity conferred by germline loss is DAF-16, which is activated by germline loss and is required for the lifespan extension associated with germline loss (Hsin and Kenyon 1999; Lin et al. 2001). Upon germline removal, intestinal DAF-16 translocates from the cytosol to the nucleus (Lin et al. 2001). This process is mediated by DAF-9, DAF-12, and KRI-1/KRIT1/CCM1, independently of IIS (Berman and Kenyon 2006). Moreover, the transcriptional activity of nuclear DAF-16 is regulated by several factors such as TCER-1/TCERG-1, PHI-62 (a predicted RNA-binding protein), and FTT-2/14-3-3 (Ghazi et al. 2009; McCormick et al. 2012). Thus, the loss of the germline enhances the transcriptonal activity of DAF-16 to induce longevity genes and increases lifespan.

Fat metabolism also plays key roles in the regulation of lifespan by the reproductive system. Oil red O fat staining and Coherent Anti-Stokes Raman Scattering (CARS) microscopy indicate that germline loss increases fat storage (O'Rourke et al. 2009; Lapierre et al. 2013b). Moreover, several factors that regulate fat metabolism are required for the longevity conferred by germline loss, including NHR-49 and NHR-80, nuclear receptors that regulate fat metabolism (Goudeau et al. 2011; Ratnappan et al. 2014). The gonadal signaling is also mediated by the induction of *lipl-4* (a triglyceride lipase) that functions to increase lifespan (Wang et al. 2008; Lapierre et al. 2011). Thus, changes in fat metabolism contribute to the extension of lifespan by germline loss.

In *C. elegans*, the somatic gonad seems to relay longevity signals to other body parts upon sensing the loss of the germline. When the germ cells are compromised, the somatic gonad sends signals that may help the survival of the soma, and the animals may resume reproduction under conditions that favor reproduction. This may help the animals balance the whole system between the maintenance of somatic health and the promotion of reproduction. Interestingly, the regulation of lifespan by the reproductive system is also observed in other species, including *Drosophila* (Flatt et al. 2008) and mice (Cargill et al. 2003). Thus, the elucidation of the mechanisms by which the reproductive system regulates longevity will provide useful information regarding how reproduction and aging have evolved in an interlocked manner in complex organisms such as humans.

8.3 Conclusions

In this chapter, we reviewed representative pathways and genes that influence longevity in *C. elegans*. More than 20 years of research using *C. elegans* has provided invaluable information about the genetics of aging. Importantly, many of the genes that regulate aging in *C. elegans* are implicated in the longevity of mammals, including humans. For example, the identification of IIS as a longevity pathway in *C. elegans* has led the way for the discovery of *FOXO3A* variants in long-lived humans. In addition, the dissection of TOR signaling as a target of anti-aging medicine has helped the emergence of rapamycin as a lifespan-extending drug in mice (Harrison et al. 2009). Overall, it is indisputable that the aging research using *C. elegans* has provided pivotal clues to the basis for slowing human aging and delaying the onset of age-related diseases.

Organismal lifespan is highly plastic and subject to changes in environmental and internal conditions. Under normal conditions, lifespan-regulatory pathways support growth, reproduction, and other essential cellular functions such as translation and energy production. However, under harsh conditions, including low food availability and the presence of various stressors, those pathways appear to shift from growth and reproduction to protective states, which eventually lead to longevity. For example, a reduction in food availability alters the function of multiple lifespan-regulatory pathways, including TOR, IIS, and autophagy, to promote longevity. The loss of the germline extends lifespan, likely by sending longevity signals from the reproductive organs to other body parts to support the health of somatic tissues. Decreases in sensory perception can extend lifespan, probably by transmitting cues for environmental stresses, including low food availability, into internal longevity signals. These examples can help us to interpret how organisms switch their physiologic status between growth/reproduction and maintenance/protection at the molecular level upon changes in extrinsic and intrinsic conditions.

Many of the genes and pathways described in this chapter have emerged as promising targets for anti-aging medicine. However, most of these pathways are tightly linked to one another. Hence, it will be difficult to predict the exact physiologic outcome of intervention in one pathway, which may lead to alterations in the function of other vital pathways. In this regard, a major challenge in aging research may be to unravel the complex network among the lifespan-regulatory pathways. In addition, some factors that increase longevity come with expenses including reduced fitness. Therefore, it will be important to uncouple the longevity from adverse side effects by elucidating precise mechanisms. Because many of the pathways that regulate aging are evolutionarily conserved, findings regarding *C. elegans* longevity genes will likely impact on aging research in mammals as well. Thus, deciphering the whole network among these pathways and genes in *C. elegans* will eventually help us to achieve longer and healthier lifespans in humans.

References

Ailion M, Inoue T, Weaver CI, Holdcraft RW, Thomas JH (1999) Neurosecretory control of aging in *Caenorhabditis elegans*. Proc Natl Acad Sci U S A 96(13):7394–7397

Alam H, Williams TW, Dumas KJ, Guo C, Yoshina S, Mitani S, Hu PJ (2010) EAK-7 controls development and life span by regulating nuclear DAF-16/FoxO activity. Cell Metab 12(1):30–41. doi:10.1016/j.cmet.2010.05.004

Alcedo J, Kenyon C (2004) Regulation of *C. elegans* longevity by specific gustatory and olfactory neurons. Neuron 41(1):45–55

Antebi A, Yeh WH, Tait D, Hedgecock EM, Riddle DL (2000) *daf-12* encodes a nuclear receptor that regulates the Dauer diapause and developmental age in *C. elegans*. Genes Dev 14 (12):1512–1527

Apfeld J, Kenyon C (1999) Regulation of lifespan by sensory perception in *Caenorhabditis elegans*. Nature 402(6763):804–809. doi:10.1038/45544

Apfeld J, O'Connor G, McDonagh T, DiStefano PS, Curtis R (2004) The AMP-activated protein kinase AAK-2 links energy levels and insulin-like signals to lifespan in *C. elegans*. Genes Dev 18(24):3004–3009

Araiz C, Château MT, Galas S (2008) 14-3-3 regulates life span by both DAF-16-dependent and-independent mechanisms in *Caenorhabditis elegans*. Exp Gerontol 43(6):505–519. doi:10.1016/j.exger.2008.03.001

Arantes-Oliveira N, Apfeld J, Dillin A, Kenyon C (2002) Regulation of life-span by germ-line stem cells in *Caenorhabditis elegans*. Science 295(5554):502–505. doi:10.1126/science.1065768

Baker BM, Nargund AM, Sun T, Haynes CM (2012) Protective coupling of mitochondrial function and protein synthesis via the eIF2alpha kinase GCN-2. PLoS Genet 8(6), e1002760. doi:10.1371/journal.pgen.1002760

Baruah A, Chang H, Hall M, Yuan J, Gordon S, Johnson E, Shtessel LL, Yee C, Hekimi S, Derry WB, Lee SS (2014) CEP-1, the *Caenorhabditis elegans* p53 homolog, mediates opposing longevity outcomes in mitochondrial electron transport chain mutants. PLoS Genet 10(2), e1004097. doi:10.1371/journal.pgen.1004097

Bennett CF, Vander Wende H, Simko M, Klum S, Barfield S, Choi H, Pineda VV, Kaeberlein M (2014) Activation of the mitochondrial unfolded protein response does not predict longevity in *Caenorhabditis elegans*. Nat Commun 5:3483. doi:10.1038/ncomms4483

Berdichevsky A, Viswanathan M, Horvitz HR, Guarente L (2006) *C. elegans* SIR-2.1 interacts with 14-3-3 proteins to activate DAF-16 and extend life span. Cell 125(6):1165–1177. doi:10.1016/j.cell.2006.04.036

Berman JR, Kenyon C (2006) Germ-cell loss extends *C. elegans* life span through regulation of DAF-16 by *kri-1* and lipophilic-hormone signaling. Cell 124(5):1055–1068. doi:10.1016/j.cell.2006.01.039

Bishop NA, Guarente L (2007) Two neurons mediate diet-restriction-induced longevity in *C. elegans*. Nature 447(7144):545–549. doi:10.1038/nature05904

Boehm M, Slack F (2005) A developmental timing microRNA and its target regulate life span in *C. elegans*. Science 310(5756):1954–1957. doi:10.1126/science.1115596

Boulias K, Horvitz HR (2012) The *C. elegans* microRNA mir-71 acts in neurons to promote germline-mediated longevity through regulation of DAF-16/FOXO. Cell Metab 15(4):439–450. doi:10.1016/j.cmet.2012.02.014
Braeckman BP, Houthoofd K, De Vreese A, Vanfleteren JR (1999) Apparent uncoupling of energy production and consumption in long-lived *Clk* mutants of *Caenorhabditis elegans*. Curr Biol 9 (9):493–496
Brisbin S, Liu J, Boudreau J, Peng J, Evangelista M, Chin-Sang I (2009) A role for *C. elegans* Eph RTK signaling in PTEN regulation. Dev Cell 17(4):459–469. doi:10.1016/j.devcel.2009.08.009
Brock TJ, Browse J, Watts JL (2006) Genetic regulation of unsaturated fatty acid composition in *C. elegans*. PLoS Genet 2(7), e108. doi:10.1371/journal.pgen.0020108
Budovskaya YV, Wu K, Southworth LK, Jiang M, Tedesco P, Johnson TE, Kim SK (2008) An elt-3/elt-5/elt-6 GATA transcription circuit guides aging in *C. elegans*. Cell 134(2):291–303. doi:10.1016/j.cell.2008.05.044
Burnett C, Valentini S, Cabreiro F, Goss M, Somogyvari M, Piper MD, Hoddinott M, Sutphin GL, Leko V, McElwee JJ, Vazquez-Manrique RP, Orfila AM, Ackerman D, Au C, Vinti G, Riesen M, Howard K, Neri C, Bedalov A, Kaeberlein M, Soti C, Partridge L, Gems D (2011) Absence of effects of Sir2 overexpression on lifespan in *C. elegans* and *Drosophila*. Nature 477 (7365):482–485. doi:10.1038/nature10296
Cabreiro F, Ackerman D, Doonan R, Araiz C, Back P, Papp D, Braeckman BP, Gems D (2011) Increased life span from overexpression of superoxide dismutase in *Caenorhabditis elegans* is not caused by decreased oxidative damage. Free Radic Biol Med 51(8):1575–1582. doi:10.1016/j.freeradbiomed.2011.07.020
Cargill SL, Carey JR, Muller HG, Anderson G (2003) Age of ovary determines remaining life expectancy in old ovariectomized mice. Aging Cell 2(3):185–190
Carrano AC, Liu Z, Dillin A, Hunter T (2009) A conserved ubiquitination pathway determines longevity in response to diet restriction. Nature 460(7253):396–399. doi:10.1038/nature08130
Carrano AC, Dillin A, Hunter T (2014) A Kruppel-like factor downstream of the E3 ligase WWP-1 mediates dietary-restriction-induced longevity in *Caenorhabditis elegans*. Nat Commun 5:3772. doi:10.1038/ncomms4772
Chamoli M, Singh A, Malik Y, Mukhopadhyay A (2014) A novel kinase regulates dietary restriction-mediated longevity in *Caenorhabditis elegans*. Aging Cell 13(4):641–655. doi:10.1111/acel.12218
Chen D, Thomas EL, Kapahi P (2009) HIF-1 modulates dietary restriction-mediated lifespan extension via IRE-1 in *Caenorhabditis elegans*. PLoS Genet 5(5), e1000486. doi:10.1371/journal.pgen.1000486
Chen AT, Guo C, Dumas KJ, Ashrafi K, Hu PJ (2013a) Effects of *Caenorhabditis elegans sgk-1* mutations on lifespan, stress resistance, and DAF-16/FoxO regulation. Aging Cell 12(5):932–940. doi:10.1111/acel.12120
Chen D, Li PW, Goldstein BA, Cai W, Thomas EL, Chen F, Hubbard AE, Melov S, Kapahi P (2013b) Germline signaling mediates the synergistically prolonged longevity produced by double mutations in *daf-2* and *rsks-1* in *C. elegans*. Cell Rep 5(6):1600–1610. doi:10.1016/j.celrep.2013.11.018
Chiang WC, Ching TT, Lee HC, Mousigian C, Hsu AL (2012) HSF-1 regulators DDL-1/2 link insulin-like signaling to heat-shock responses and modulation of longevity. Cell 148(1–2):322–334. doi:10.1016/j.cell.2011.12.019
Ching TT, Paal AB, Mehta A, Zhong L, Hsu AL (2010) *drr-2* encodes an eIF4H that acts downstream of TOR in diet-restriction-induced longevity of *C. elegans*. Aging Cell 9 (4):545–557. doi:10.1111/j.1474-9726.2010.00580.x
Cristina D, Cary M, Lunceford A, Clarke C, Kenyon C (2009) A regulated response to impaired respiration slows behavioral rates and increases lifespan in *Caenorhabditis elegans*. PLoS Genet 5(4), e1000450. doi:10.1371/journal.pgen.1000450

Curran SP, Ruvkun G (2007) Lifespan regulation by evolutionarily conserved genes essential for viability. PLoS Genet 3(4), e56. doi:10.1371/journal.pgen.0030056

Curran SP, Wu X, Riedel CG, Ruvkun G (2009) A soma-to-germline transformation in long-lived *Caenorhabditis elegans* mutants. Nature 459(7250):1079–1084. doi:10.1038/nature08106

Curtis R, O'Connor G, DiStefano PS (2006) Aging networks in *Caenorhabditis elegans*: AMP-activated protein kinase (*aak-2*) links multiple aging and metabolism pathways. Aging Cell 5 (2):119–126. doi:10.1111/j.1474-9726.2006.00205.x

de Jong L, Meng Y, Dent J, Hekimi S (2004) Thiamine pyrophosphate biosynthesis and transport in the nematode *Caenorhabditis elegans*. Genetics 168(2):845–854. doi:10.1534/genetics.104.028605

de Lencastre A, Pincus Z, Zhou K, Kato M, Lee SS, Slack FJ (2010) MicroRNAs both promote and antagonize longevity in *C. elegans*. Curr Biol 20(24):2159–2168. doi:10.1016/j.cub.2010.11.015

Dillin A, Crawford DK, Kenyon C (2002a) Timing requirements for insulin/IGF-1 signaling in *C. elegans*. Science 298(5594):830–834. doi:10.1126/science.1074240

Dillin A, Hsu AL, Arantes-Oliveira N, Lehrer-Graiwer J, Hsin H, Fraser AG, Kamath RS, Ahringer J, Kenyon C (2002b) Rates of behavior and aging specified by mitochondrial function during development. Science 298(5602):2398–2401. doi:10.1126/science.1077780

Dorman JB, Albinder B, Shroyer T, Kenyon C (1995) The *age-1* and *daf-2* genes function in a common pathway to control the lifespan of *Caenorhabditis elegans*. Genetics 141(4):1399–1406

Durieux J, Wolff S, Dillin A (2011) The cell-non-autonomous nature of electron transport chain-mediated longevity. Cell 144(1):79–91. doi:10.1016/j.cell.2010.12.016

Egan DF, Shackelford DB, Mihaylova MM, Gelino S, Kohnz RA, Mair W, Vasquez DS, Joshi A, Gwinn DM, Taylor R, Asara JM, Fitzpatrick J, Dillin A, Viollet B, Kundu M, Hansen M, Shaw RJ (2011) Phosphorylation of ULK1 (hATG1) by AMP-activated protein kinase connects energy sensing to mitophagy. Science 331(6016):456–461. doi:10.1126/science.1196371

Essers MA, de Vries-Smits LM, Barker N, Polderman PE, Burgering BM, Korswagen HC (2005) Functional interaction between beta-catenin and FOXO in oxidative stress signaling. Science 308(5725):1181–1184. doi:10.1126/science.1109083

Essers Paul B, Nonnekens J, Goos Yvonne J, Betist Marco C, Viester Marjon D, Mossink B, Lansu N, Korswagen Hendrik C, Jelier R, Brenkman Arjan B, MacInnes Alyson W (2015) A long noncoding RNA on the ribosome is required for lifespan extension. Cell Rep 10(3):339–345. doi:10.1016/j.celrep.2014.12.029

Evans DS, Kapahi P, Hsueh WC, Kockel L (2011) TOR signaling never gets old: aging, longevity and TORC1 activity. Ageing Res Rev 10(2):225–237. doi:10.1016/j.arr.2010.04.001

Ewald CY, Landis JN, Abate JP, Murphy CT, Blackwell TK (2014) Dauer-independent insulin/IGF-1-signalling implicates collagen remodelling in longevity. Nature. doi:10.1038/nature14021

Ewbank JJ, Barnes TM, Lakowski B, Lussier M, Bussey H, Hekimi S (1997) Structural and functional conservation of the *Caenorhabditis elegans* timing gene *clk-1*. Science 275 (5302):980–983

Feng J, Bussiere F, Hekimi S (2001) Mitochondrial electron transport is a key determinant of life span in *Caenorhabditis elegans*. Dev Cell 1(5):633–644

Fernandes de Abreu DA, Caballero A, Fardel P, Stroustrup N, Chen Z, Lee K, Keyes WD, Nash ZM, Lopez-Moyado IF, Vaggi F, Cornils A, Regenass M, Neagu A, Ostojic I, Liu C, Cho Y, Sifoglu D, Shen Y, Fontana W, Lu H, Csikasz-Nagy A, Murphy CT, Antebi A, Blanc E, Apfeld J, Zhang Y, Alcedo J, Ch'ng Q (2014) An insulin-to-insulin regulatory network orchestrates phenotypic specificity in development and physiology. PLoS Genet 10(3), e1004225. doi:10.1371/journal.pgen.1004225

Fisher AL, Lithgow GJ (2006) The nuclear hormone receptor DAF-12 has opposing effects on *Caenorhabditis elegans* lifespan and regulates genes repressed in multiple long-lived worms. Aging Cell 5(2):127–138. doi:10.1111/j.1474-9726.2006.00203.x

Flatt T, Min KJ, D'Alterio C, Villa-Cuesta E, Cumbers J, Lehmann R, Jones DL, Tatar M (2008) *Drosophila* germ-line modulation of insulin signaling and lifespan. Proc Natl Acad Sci U S A 105(17):6368–6373. doi:10.1073/pnas.0709128105

Friedman DB, Johnson TE (1988) Three mutants that extend both mean and maximum life span of the nematode, *Caenorhabditis elegans*, define the *age-1* gene. J Gerontol 43(4):B102–B109

Gaglia MM, Jeong DE, Ryu EA, Lee D, Kenyon C, Lee SJ (2012) Genes that act downstream of sensory neurons to influence longevity, Dauer formation, and pathogen responses in *Caenorhabditis elegans*. PLoS Genet 8(12), e1003133. doi:10.1371/journal.pgen.1003133

Garigan D, Hsu AL, Fraser AG, Kamath RS, Ahringer J, Kenyon C (2002) Genetic analysis of tissue aging in *Caenorhabditis elegans*: a role for heat-shock factor and bacterial proliferation. Genetics 161(3):1101–1112

Garsin DA, Villanueva JM, Begun J, Kim DH, Sifri CD, Calderwood SB, Ruvkun G, Ausubel FM (2003) Long-lived *C. elegans daf-2* mutants are resistant to bacterial pathogens. Science 300 (5627):1921. doi:10.1126/science.1080147

Gems D, Sutton AJ, Sundermeyer ML, Albert PS, King KV, Edgley ML, Larsen PL, Riddle DL (1998) Two pleiotropic classes of *daf-2* mutation affect larval arrest, adult behavior, reproduction and longevity in *Caenorhabditis elegans*. Genetics 150(1):129–155

Gerisch B, Weitzel C, Kober-Eisermann C, Rottiers V, Antebi A (2001) A hormonal signaling pathway influencing *C. elegans* metabolism, reproductive development, and life span. Dev Cell 1(6):841–851

Gerisch B, Rottiers V, Li D, Motola DL, Cummins CL, Lehrach H, Mangelsdorf DJ, Antebi A (2007) A bile acid-like steroid modulates *Caenorhabditis elegans* lifespan through nuclear receptor signaling. Proc Natl Acad Sci U S A 104(12):5014–5019. doi:10.1073/pnas.0700847104

Gharbi H, Fabretti F, Bharill P, Rinschen MM, Brinkkotter S, Frommolt P, Burst V, Schermer B, Benzing T, Muller RU (2013) Loss of the Birt-Hogg-Dube gene product folliculin induces longevity in a hypoxia-inducible factor-dependent manner. Aging Cell 12(4):593–603. doi:10.1111/acel.12081

Ghazi A, Henis-Korenblit S, Kenyon C (2007) Regulation of *Caenorhabditis elegans* lifespan by a proteasomal E3 ligase complex. Proc Natl Acad Sci U S A 104(14):5947–5952. doi:10.1073/pnas.0700638104

Ghazi A, Henis-Korenblit S, Kenyon C (2009) A transcription elongation factor that links signals from the reproductive system to lifespan extension in *Caenorhabditis elegans*. PLoS Genet 5 (9), e1000639. doi:10.1371/journal.pgen.1000639

Goudeau J, Bellemin S, Toselli-Mollereau E, Shamalnasab M, Chen Y, Aguilaniu H (2011) Fatty acid desaturation links germ cell loss to longevity through NHR-80/HNF4 in *C. elegans*. PLoS Biol 9(3), e1000599. doi:10.1371/journal.pbio.1000599

Green DR, Galluzzi L, Kroemer G (2014) Metabolic control of cell death. Science 345 (6203):1250256

Greer EL, Brunet A (2009) Different dietary restriction regimens extend lifespan by both independent and overlapping genetic pathways in *C. elegans*. Aging Cell 8(2):113–127. doi:10.1111/j.1474-9726.2009.00459.x

Greer EL, Dowlatshahi D, Banko MR, Villen J, Hoang K, Blanchard D, Gygi SP, Brunet A (2007) An AMPK-FOXO pathway mediates longevity induced by a novel method of dietary restriction in *C. elegans*. Curr Biol 17(19):1646–1656. doi:10.1016/j.cub.2007.08.047

Hahm JH, Kim S, Paik YK (2009) Endogenous cGMP regulates adult longevity via the insulin signaling pathway in *Caenorhabditis elegans*. Aging Cell 8(4):473–483. doi:10.1111/j.1474-9726.2009.00495.x

Hamilton B, Dong Y, Shindo M, Liu W, Odell I, Ruvkun G, Lee SS (2005) A systematic RNAi screen for longevity genes in *C. elegans*. Genes Dev 19(13):1544–1555. doi:10.1101/gad.1308205

Hansen M, Hsu AL, Dillin A, Kenyon C (2005) New genes tied to endocrine, metabolic, and dietary regulation of lifespan from a *Caenorhabditis elegans* genomic RNAi screen. PLoS Genet 1(1):119–128. doi:10.1371/journal.pgen.0010017

Hansen M, Taubert S, Crawford D, Libina N, Lee SJ, Kenyon C (2007) Lifespan extension by conditions that inhibit translation in *Caenorhabditis elegans*. Aging Cell 6(1):95–110. doi:10.1111/j.1474-9726.2006.00267.x

Hansen M, Chandra A, Mitic LL, Onken B, Driscoll M, Kenyon C (2008) A role for autophagy in the extension of lifespan by dietary restriction in *C. elegans*. PLoS Genet 4(2), e24. doi:10.1371/journal.pgen.0040024

Hardie DG (2014) AMPK-sensing energy while talking to other signaling pathways. Cell Metab 20(6):939–952. doi:10.1016/j.cmet.2014.09.013

Harman D (1956) Aging: a theory based on free radical and radiation chemistry. J Gerontol 11 (3):298–300

Harman D (1972) The biologic clock: the mitochondria? J Am Geriatr Soc 20(4):145–147

Harrison DE, Strong R, Sharp ZD, Nelson JF, Astle CM, Flurkey K, Nadon NL, Wilkinson JE, Frenkel K, Carter CS, Pahor M, Javors MA, Fernandez E, Miller RA (2009) Rapamycin fed late in life extends lifespan in genetically heterogeneous mice. Nature 460(7253):392–395. doi:10.1038/nature08221

Hars ES, Qi H, Ryazanov AG, Jin S, Cai L, Hu C, Liu LF (2007) Autophagy regulates ageing in *C. elegans*. Autophagy 3(2):93–95

Hashimoto Y, Ookuma S, Nishida E (2009) Lifespan extension by suppression of autophagy genes in *Caenorhabditis elegans*. Genes Cells 14(6):717–726. doi:10.1111/j.1365-2443.2009.01306.x

Haynes CM, Fiorese CJ, Lin YF (2013) Evaluating and responding to mitochondrial dysfunction: the mitochondrial unfolded-protein response and beyond. Trends Cell Biol 23(7):311–318. doi:10.1016/j.tcb.2013.02.002

Heestand BN, Shen Y, Liu W, Magner DB, Storm N, Meharg C, Habermann B, Antebi A (2013) Dietary restriction induced longevity is mediated by nuclear receptor NHR-62 in *Caenorhabditis elegans*. PLoS Genet 9(7), e1003651. doi:10.1371/journal.pgen.1003651

Henderson ST, Johnson TE (2001) *daf-16* integrates developmental and environmental inputs to mediate aging in the nematode *Caenorhabditis elegans*. Curr Biol 11(24):1975–1980

Henis-Korenblit S, Zhang P, Hansen M, McCormick M, Lee SJ, Cary M, Kenyon C (2010) Insulin/IGF-1 signaling mutants reprogram ER stress response regulators to promote longevity. Proc Natl Acad Sci U S A 107(21):9730–9735. doi:10.1073/pnas.1002575107

Hertweck M, Gobel C, Baumeister R (2004) *C. elegans* SGK-1 is the critical component in the Akt/PKB kinase complex to control stress response and life span. Dev Cell 6(4):577–588

Honda Y, Honda S (1999) The *daf-2* gene network for longevity regulates oxidative stress resistance and Mn-superoxide dismutase gene expression in *Caenorhabditis elegans*. FASEB J 13(11):1385–1393

Honjoh S, Yamamoto T, Uno M, Nishida E (2009) Signalling through RHEB-1 mediates intermittent fasting-induced longevity in *C. elegans*. Nature 457(7230):726–730. doi:10.1038/nature07583

Honnen SJ, Buchter C, Schroder V, Hoffmann M, Kohara Y, Kampkotter A, Bossinger O (2012) *C. elegans* VANG-1 modulates life span via insulin/IGF-1-like signaling. PLoS One 7(2), e32183. doi:10.1371/journal.pone.0032183

Houtkooper RH, Mouchiroud L, Ryu D, Moullan N, Katsyuba E, Knott G, Williams RW, Auwerx J (2013) Mitonuclear protein imbalance as a conserved longevity mechanism. Nature 497 (7450):451–457. doi:10.1038/nature12188

Hsin H, Kenyon C (1999) Signals from the reproductive system regulate the lifespan of *C. elegans*. Nature 399(6734):362–366. doi:10.1038/20694

Hsu AL, Murphy CT, Kenyon C (2003) Regulation of aging and age-related disease by DAF-16 and heat-shock factor. Science 300(5622):1142–1145. doi:10.1126/science.1083701

Huang X, Zhang H, Zhang H (2011) The zinc-finger protein SEA-2 regulates larval developmental timing and adult lifespan in *C. elegans*. Development 138(10):2059–2068. doi:10.1242/dev.057109

Hwang AB, Jeong DE, Lee SJ (2012) Mitochondria and organismal longevity. Curr Genomics 13 (7):519–532. doi:10.2174/138920212803251427

Hwang AB, Ryu EA, Artan M, Chang HW, Kabir MH, Nam HJ, Lee D, Yang JS, Kim S, Mair WB, Lee C, Lee SS, Lee SJ (2014) Feedback regulation via AMPK and HIF-1 mediates ROS-dependent longevity in *Caenorhabditis elegans*. Proc Natl Acad Sci U S A 111(42):E4458–E4467. doi:10.1073/pnas.1411199111

Jeong DE, Artan M, Seo K, Lee SJ (2012) Regulation of lifespan by chemosensory and thermosensory systems: findings in invertebrates and their implications in mammalian aging. Front Genet 3:218. doi:10.3389/fgene.2012.00218

Jia K, Albert PS, Riddle DL (2002) DAF-9, a cytochrome P450 regulating *C. elegans* larval development and adult longevity. Development 129(1):221–231

Jia K, Chen D, Riddle DL (2004) The TOR pathway interacts with the insulin signaling pathway to regulate *C. elegans* larval development, metabolism and life span. Development 131 (16):3897–3906. doi:10.1242/dev.01255

Jia K, Levine B (2007) Autophagy is required for dietary restriction-mediated life span extension in *C. elegans*. Autophagy 3(6):597–599

Jia K, Thomas C, Akbar M, Sun Q, Adams-Huet B, Gilpin C, Levine B (2009) Autophagy genes protect against Salmonella typhimurium infection and mediate insulin signaling-regulated pathogen resistance. Proc Natl Acad Sci U S A 106(34):14564–14569. doi:10.1073/pnas.0813319106

Jin C, Li J, Green CD, Yu X, Tang X, Han D, Xian B, Wang D, Huang X, Cao X, Yan Z, Hou L, Liu J, Shukeir N, Khaitovich P, Chen CD, Zhang H, Jenuwein T, Han JD (2011) Histone demethylase UTX-1 regulates *C. elegans* life span by targeting the insulin/IGF-1 signaling pathway. Cell Metab 14(2):161–172. doi:10.1016/j.cmet.2011.07.001

Johnson SC, Rabinovitch PS, Kaeberlein M (2013) mTOR is a key modulator of ageing and age-related disease. Nature 493(7432):338–345. doi:10.1038/nature11861

Johnson DW, Llop JR, Farrell SF, Yuan J, Stolzenburg LR, Samuelson AV (2014) The *Caenorhabditis elegans* Myc-Mondo/Mad complexes integrate diverse longevity signals. PLoS Genet 10(4), e1004278. doi:10.1371/journal.pgen.1004278

Kaletsky R, Murphy CT (2010) The role of insulin/IGF-like signaling in *C. elegans* longevity and aging. Dis Model Mech 3(7–8):415–419. doi:10.1242/dmm.001040

Kapahi P, Chen D, Rogers AN, Katewa SD, Li PW, Thomas EL, Kockel L (2010) With TOR, less is more: a key role for the conserved nutrient-sensing TOR pathway in aging. Cell Metab 11 (6):453–465. doi:10.1016/j.cmet.2010.05.001

Kawano T, Ito Y, Ishiguro M, Takuwa K, Nakajima T, Kimura Y (2000) Molecular cloning and characterization of a new insulin/IGF-like peptide of the nematode *Caenorhabditis elegans*. Biochem Biophys Res Commun 273(2):431–436. doi:10.1006/bbrc.2000.2971

Kenyon C, Chang J, Gensch E, Rudner A, Tabtiang R (1993) A *C. elegans* mutant that lives twice as long as wild type. Nature 366(6454):461–464. doi:10.1038/366461a0

Kenyon CJ (2010) The genetics of ageing. Nature 464(7288):504–512. doi:10.1038/nature08980

Khan MH, Ligon M, Hussey LR, Hufnal B, Farber R 2nd, Munkacsy E, Rodriguez A, Dillow A, Kahlig E, Rea SL (2013) TAF-4 is required for the life extension of *isp-1*, *clk-1* and *tpk-1* Mit mutants. Aging 5(10):741–758

Kim SK, Budovskaya YV, Johnson TE (2013) Reconciliation of *daf-2* suppression by *elt-3* in *Caenorhabditis elegans* from Tonsaker et al. (2012) and Kim et al. (2012). Mech Ageing Dev 134(1–2):64–65. doi:10.1016/j.mad.2012.12.004

Kim Y, Sun H (2007) Functional genomic approach to identify novel genes involved in the regulation of oxidative stress resistance and animal lifespan. Aging Cell 6(4):489–503. doi:10.1111/j.1474-9726.2007.00302.x

Lamitina ST, Strange K (2005) Transcriptional targets of DAF-16 insulin signaling pathway protect *C. elegans* from extreme hypertonic stress. Am J Physiol Cell Physiol 288(2):C467–C474. doi:10.1152/ajpcell.00451.2004

Lanjuin A, Sengupta P (2002) Regulation of chemosensory receptor expression and sensory signaling by the KIN-29 Ser/Thr kinase. Neuron 33(3):369–381

Lans H, Jansen G (2007) Multiple sensory G proteins in the olfactory, gustatory and nociceptive neurons modulate longevity in *Caenorhabditis elegans*. Dev Biol 303(2):474–482. doi:10.1016/j.ydbio.2006.11.028

Lapierre LR, Gelino S, Melendez A, Hansen M (2011) Autophagy and lipid metabolism coordinately modulate life span in germline-less *C. elegans*. Curr Biol 21(18):1507–1514. doi:10.1016/j.cub.2011.07.042

Lapierre LR, De Magalhaes Filho CD, McQuary PR, Chu CC, Visvikis O, Chang JT, Gelino S, Ong B, Davis AE, Irazoqui JE, Dillin A, Hansen M (2013a) The TFEB orthologue HLH-30 regulates autophagy and modulates longevity in *Caenorhabditis elegans*. Nat Commun 4:2267. doi:10.1038/ncomms3267

Lapierre LR, Silvestrini MJ, Nunez L, Ames K, Wong S, Le TT, Hansen M, Melendez A (2013b) Autophagy genes are required for normal lipid levels in *C. elegans*. Autophagy 9(3):278–286. doi:10.4161/auto.22930

Laplante M, Sabatini DM (2012) mTOR signaling in growth control and disease. Cell 149(2):274–293. doi:10.1016/j.cell.2012.03.017

Larsen PL, Albert PS, Riddle DL (1995) Genes that regulate both development and longevity in *Caenorhabditis elegans*. Genetics 139(4):1567–1583

Lakowski B, Hekimi S (1998) The genetics of caloric restriction in *Caenorhabditis elegans*. Proc Natl Acad Sci 95(22):13091–13096

Lee BH, Ashrafi K (2008) A TRPV channel modulates *C. elegans* neurosecretion, larval starvation survival, and adult lifespan. PLoS Genet 4(10), e1000213. doi:10.1371/journal.pgen.1000213

Lee SJ, Kenyon C (2009) Regulation of the longevity response to temperature by thermosensory neurons in *Caenorhabditis elegans*. Curr Biol 19(9):715–722. doi:10.1016/j.cub.2009.03.041

Lee RY, Hench J, Ruvkun G (2001) Regulation of *C. elegans* DAF-16 and its human ortholog FKHRL1 by the *daf-2* insulin-like signaling pathway. Curr Biol 11(24):1950–1957

Lee SS, Lee RY, Fraser AG, Kamath RS, Ahringer J, Ruvkun G (2003) A systematic RNAi screen identifies a critical role for mitochondria in *C. elegans* longevity. Nat Genet 33(1):40–48. doi:10.1038/ng1056

Lee SJ, Murphy CT, Kenyon C (2009) Glucose shortens the life span of *C. elegans* by downregulating DAF-16/FOXO activity and aquaporin gene expression. Cell Metab 10 (5):379–391. doi:10.1016/j.cmet.2009.10.003

Lee SJ, Hwang AB, Kenyon C (2010) Inhibition of respiration extends *C. elegans* life span via reactive oxygen species that increase HIF-1 activity. Curr Biol 20(23):2131–2136. doi:10.1016/j.cub.2010.10.057

Lehtinen MK, Yuan Z, Boag PR, Yang Y, Villen J, Becker EB, DiBacco S, de la Iglesia N, Gygi S, Blackwell TK, Bonni A (2006) A conserved MST-FOXO signaling pathway mediates oxidative-stress responses and extends life span. Cell 125(5):987–1001. doi:10.1016/j.cell.2006.03.046

Leiser SF, Begun A, Kaeberlein M (2011) HIF-1 modulates longevity and healthspan in a temperature-dependent manner. Aging Cell 10(2):318–326. doi:10.1111/j.1474-9726.2011.00672.x

Levine B, Klionsky DJ (2004) Development by self-digestion: molecular mechanisms and biological functions of autophagy. Dev Cell 6(4):463–477

Li J, Tewari M, Vidal M, Lee SS (2007a) The 14-3-3 protein FTT-2 regulates DAF-16 in *Caenorhabditis elegans*. Dev Biol 301(1):82–91

Li W, Gao B, Lee SM, Bennett K, Fang D (2007b) RLE-1, an E3 ubiquitin ligase, regulates *C. elegans* aging by catalyzing DAF-16 polyubiquitination. Dev Cell 12(2):235–246. doi:10.1016/j.devcel.2006.12.002

Li J, Ebata A, Dong Y, Rizki G, Iwata T, Lee SS (2008) *Caenorhabditis elegans* HCF-1 functions in longevity maintenance as a DAF-16 regulator. PLoS Biol 6(9), e233. doi:10.1371/journal.pbio.0060233

Libert S, Zwiener J, Chu X, Vanvoorhies W, Roman G, Pletcher SD (2007) Regulation of *Drosophila* life span by olfaction and food-derived odors. Science 315(5815):1133–1137. doi:10.1126/science.1136610

Lin K, Dorman JB, Rodan A, Kenyon C (1997) *daf-16*: An HNF-3/forkhead family member that can function to double the life-span of *Caenorhabditis elegans*. Science 278(5341):1319–1322

Lin K, Hsin H, Libina N, Kenyon C (2001) Regulation of the *Caenorhabditis elegans* longevity protein DAF-16 by insulin/IGF-1 and germline signaling. Nat Genet 28(2):139–145. doi:10.1038/88850

Linford NJ, Kuo TH, Chan TP, Pletcher SD (2011) Sensory perception and aging in model systems: from the outside in. Annu Rev Cell Dev Biol 27:759–785. doi:10.1146/annurev-cellbio-092910-154240

Lithgow GJ, White TM, Melov S, Johnson TE (1995) Thermotolerance and extended life-span conferred by single-gene mutations and induced by thermal stress. Proc Natl Acad Sci U S A 92 (16):7540–7544

Lithgow GJ, Kapahi P (2011) Life span extension via eIF4G inhibition is mediated by posttranscriptional remodeling of stress response gene expression in *C. elegans*. Cell Metab 14(1):55–66. doi:10.1016/j.cmet.2011.05.010

Lucanic M, Held JM, Vantipalli MC, Klang IM, Graham JB, Gibson BW, Lithgow GJ, Gill MS (2011) N-acylethanolamine signalling mediates the effect of diet on lifespan in *Caenorhabditis elegans*. Nature 473(7346):226–229. doi:10.1038/nature10007

Magner DB, Wollam J, Shen Y, Hoppe C, Li D, Latza C, Rottiers V, Hutter H, Antebi A (2013) The NHR-8 nuclear receptor regulates cholesterol and bile acid homeostasis in *C. elegans*. Cell Metab 18(2):212–224. doi:10.1016/j.cmet.2013.07.007

Mahanti P, Bose N, Bethke A, Judkins JC, Wollam J, Dumas KJ, Zimmerman AM, Campbell SL, Hu PJ, Antebi A, Schroeder FC (2014) Comparative metabolomics reveals endogenous ligands of DAF-12, a nuclear hormone receptor, regulating *C. elegans* development and lifespan. Cell Metab 19(1):73–83. doi:10.1016/j.cmet.2013.11.024

Maier W, Adilov B, Regenass M, Alcedo J (2010) A neuromedin U receptor acts with the sensory system to modulate food type-dependent effects on *C. elegans* lifespan. PLoS Biol 8(5), e1000376. doi:10.1371/journal.pbio.1000376

Malone EA, Inoue T, Thomas JH (1996) Genetic analysis of the roles of *daf-28* and *age-1* in regulating *Caenorhabditis elegans* Dauer formation. Genetics 143(3):1193–1205

Masse I, Molin L, Billaud M, Solari F (2005) Lifespan and Dauer regulation by tissue-specific activities of *Caenorhabditis elegans* DAF-18. Dev Biol 286(1):91–101. doi:10.1016/j.ydbio.2005.07.010

Matsunaga Y, Gengyo-Ando K, Mitani S, Iwasaki T, Kawano T (2012) Physiological function, expression pattern, and transcriptional regulation of a *Caenorhabditis elegans* insulin-like peptide, INS-18. Biochem Biophys Res Commun 423(3):478–483. doi:10.1016/j.bbrc.2012.05.145

Maures TJ, Greer EL, Hauswirth AG, Brunet A (2011) The H3K27 demethylase UTX-1 regulates *C. elegans* lifespan in a germline-independent, insulin-dependent manner. Aging Cell 10 (6):980–990. doi:10.1111/j.1474-9726.2011.00738.x

McCormick M, Chen K, Ramaswamy P, Kenyon C (2012) New genes that extend *Caenorhabditis elegans*' lifespan in response to reproductive signals. Aging Cell 11(2):192–202. doi:10.1111/j.1474-9726.2011.00768.x

Mehta R, Steinkraus KA, Sutphin GL, Ramos FJ, Shamieh LS, Huh A, Davis C, Chandler-Brown D, Kaeberlein M (2009) Proteasomal regulation of the hypoxic response modulates aging in *C. elegans*. Science 324(5931):1196–1198. doi:10.1126/science.1173507

Melendez A, Talloczy Z, Seaman M, Eskelinen EL, Hall DH, Levine B (2003) Autophagy genes are essential for Dauer development and life-span extension in *C. elegans*. Science 301 (5638):1387–1391. doi:10.1126/science.1087782

Mihaylova VT, Borland CZ, Manjarrez L, Stern MJ, Sun H (1999) The PTEN tumor suppressor homolog in *Caenorhabditis elegans* regulates longevity and Dauer formation in an insulin receptor-like signaling pathway. Proc Natl Acad Sci U S A 96(13):7427–7432

Morley JF, Morimoto RI (2004) Regulation of longevity in *Caenorhabditis elegans* by heat shock factor and molecular chaperones. Mol Biol Cell 15(2):657–664. doi:10.1091/mbc.E03-07-0532

Morley JF, Brignull HR, Weyers JJ, Morimoto RI (2002) The threshold for polyglutamine-expansion protein aggregation and cellular toxicity is dynamic and influenced by aging in *Caenorhabditis elegans*. Proc Natl Acad Sci U S A 99(16):10417–10422. doi:10.1073/pnas.152161099

Moroz N, Carmona JJ, Anderson E, Hart AC, Sinclair DA, Blackwell TK (2014) Dietary restriction involves NAD(+) -dependent mechanisms and a shift toward oxidative metabolism. Aging Cell 13(6):1075–1085. doi:10.1111/acel.12273

Morris JZ, Tissenbaum HA, Ruvkun G (1996) A phosphatidylinositol-3-OH kinase family member regulating longevity and diapause in *Caenorhabditis elegans*. Nature 382(6591):536–539. doi:10.1038/382536a0

Motola DL, Cummins CL, Rottiers V, Sharma KK, Li T, Li Y, Suino-Powell K, Xu HE, Auchus RJ, Antebi A, Mangelsdorf DJ (2006) Identification of ligands for DAF-12 that govern Dauer formation and reproduction in *C. elegans*. Cell 124(6):1209–1223. doi:http://dx.doi.org/10.1016/j.cell.2006.01.037

Mouchiroud L, Houtkooper RH, Moullan N, Katsyuba E, Ryu D, Canto C, Mottis A, Jo YS, Viswanathan M, Schoonjans K, Guarente L, Auwerx J (2013) The NAD(+)/Sirtuin pathway modulates longevity through activation of mitochondrial UPR and FOXO signaling. Cell 154 (2):430–441. doi:10.1016/j.cell.2013.06.016

Mukhopadhyay A, Deplancke B, Walhout AJ, Tissenbaum HA (2005) *C. elegans* tubby regulates life span and fat storage by two independent mechanisms. Cell Metab 2(1):35–42. doi:10.1016/j.cmet.2005.06.004

Muller RU, Fabretti F, Zank S, Burst V, Benzing T, Schermer B (2009) The von Hippel Lindau tumor suppressor limits longevity. J Am Soc Nephrol 20(12):2513–2517. doi:10.1681/asn.2009050497

Murphy CT, Hu PJ (2013) Insulin/insulin-like growth factor signaling in C. elegans. WormBook:1–43. doi:10.1895/wormbook.1.164.1

Murphy CT, Lee SJ, Kenyon C (2007) Tissue entrainment by feedback regulation of insulin gene expression in the endoderm of *Caenorhabditis elegans*. Proc Natl Acad Sci U S A 104 (48):19046–19050. doi:10.1073/pnas.0709613104

Murphy CT, McCarroll SA, Bargmann CI, Fraser A, Kamath RS, Ahringer J, Li H, Kenyon C (2003) Genes that act downstream of DAF-16 to influence the lifespan of *Caenorhabditis elegans*. Nature 424(6946):277–283. doi:10.1038/nature01789

Nanji M, Hopper NA, Gems D (2005) LET-60 RAS modulates effects of insulin/IGF-1 signaling on development and aging in *Caenorhabditis elegans*. Aging Cell 4(5):235–245. doi:10.1111/j.1474-9726.2005.00166.x

Neumann-Haefelin E, Qi W, Finkbeiner E, Walz G, Baumeister R, Hertweck M (2008) SHC-1/p52Shc targets the insulin/IGF-1 and JNK signaling pathways to modulate life span and stress response in *C. elegans*. Genes Dev 22(19):2721–2735. doi:10.1101/gad.478408

Ogg S, Paradis S, Gottlieb S, Patterson GI, Lee L, Tissenbaum HA, Ruvkun G (1997) The Fork head transcription factor DAF-16 transduces insulin-like metabolic and longevity signals in *C. elegans*. Nature 389(6654):994–999. doi:10.1038/40194

Oh SW, Mukhopadhyay A, Svrzikapa N, Jiang F, Davis RJ, Tissenbaum HA (2005) JNK regulates lifespan in *Caenorhabditis elegans* by modulating nuclear translocation of forkhead transcription factor/DAF-16. Proc Natl Acad Sci U S A 102(12):4494–4499. doi:10.1073/pnas.0500749102

Okuyama T, Inoue H, Ookuma S, Satoh T, Kano K, Honjoh S, Hisamoto N, Matsumoto K, Nishida E (2010) The ERK-MAPK pathway regulates longevity through SKN-1 and insulin-like signaling in *Caenorhabditis elegans*. J Biol Chem 285(39):30274–30281. doi:10.1074/jbc.M110.146274

O'Rourke EJ, Ruvkun G (2013) MXL-3 and HLH-30 transcriptionally link lipolysis and autophagy to nutrient availability. Nat Cell Biol 15(6):668–676. doi:10.1038/ncb2741

O'Rourke EJ, Soukas AA, Carr CE, Ruvkun G (2009) *C. elegans* major fats are stored in vesicles distinct from lysosome-related organelles. Cell Metab 10(5):430–435. doi:10.1016/j.cmet.2009.10.002

Padmanabhan S, Mukhopadhyay A, Narasimhan SD, Tesz G, Czech MP, Tissenbaum HA (2009) A PP2A regulatory subunit regulates *C. elegans* insulin/IGF-1 signaling by modulating AKT-1 phosphorylation. Cell 136(5):939–951. doi:10.1016/j.cell.2009.01.025

Paek J, Lo JY, Narasimhan SD, Nguyen TN, Glover-Cutter K, Robida-Stubbs S, Suzuki T, Yamamoto M, Blackwell TK, Curran SP (2012) Mitochondrial SKN-1/Nrf mediates a conserved starvation response. Cell Metab 16(4):526–537. doi:10.1016/j.cmet.2012.09.007

Pan KZ, Palter JE, Rogers AN, Olsen A, Chen D, Lithgow GJ, Kapahi P (2007) Inhibition of mRNA translation extends lifespan in *Caenorhabditis elegans*. Aging Cell 6(1):111–119. doi:10.1111/j.1474-9726.2006.00266.x

Panowski SH, Wolff S, Aguilaniu H, Durieux J, Dillin A (2007) PHA-4/Foxa mediates diet-restriction-induced longevity of *C. elegans*. Nature 447(7144):550–555. doi:10.1038/nature05837

Paradis S, Ailion M, Toker A, Thomas JH, Ruvkun G (1999) A PDK1 homolog is necessary and sufficient to transduce AGE-1 PI3 kinase signals that regulate diapause in *Caenorhabditis elegans*. Genes Dev 13(11):1438–1452

Paradis S, Ruvkun G (1998) *Caenorhabditis elegans* Akt/PKB transduces insulin receptor-like signals from AGE-1 PI3 kinase to the DAF-16 transcription factor. Genes Dev 12(16):2488–2498

Park SK, Link CD, Johnson TE (2010) Life-span extension by dietary restriction is mediated by NLP-7 signaling and coelomocyte endocytosis in *C. elegans*. FASEB J 24(2):383–392. doi:10.1096/fj.09-142984, fj.09-142984 [pii]

Petrascheck M, Ye X, Buck LB (2007) An antidepressant that extends lifespan in adult *Caenorhabditis elegans*. Nature 450(7169):553–556

Pierce SB, Costa M, Wisotzkey R, Devadhar S, Homburger SA, Buchman AR, Ferguson KC, Heller J, Platt DM, Pasquinelli AA, Liu LX, Doberstein SK, Ruvkun G (2001) Regulation of DAF-2 receptor signaling by human insulin and *ins-1*, a member of the unusually large and diverse *C. elegans* insulin gene family. Genes Dev 15(6):672–686. doi:10.1101/gad.867301

Piper MD, Partridge L, Raubenheimer D, Simpson SJ (2011) Dietary restriction and aging: a unifying perspective. Cell Metab 14(2):154–160. doi:10.1016/j.cmet.2011.06.013

Possik E, Jalali Z, Nouet Y, Yan M, Gingras MC, Schmeisser K, Panaite L, Dupuy F, Kharitidi D, Chotard L, Jones RG, Hall DH, Pause A (2014) Folliculin regulates ampk-dependent autophagy and metabolic stress survival. PLoS Genet 10(4), e1004273. doi:10.1371/journal.pgen.1004273

Powell-Coffman JA (2010) Hypoxia signaling and resistance in *C. elegans*. Trends Endocrinol Metab 21(7):435–440. doi:10.1016/j.tem.2010.02.006

Rahman MM, Stuchlick O, El-Karim EG, Stuart R, Kipreos ET, Wells L (2010) Intracellular protein glycosylation modulates insulin mediated lifespan in *C.elegans*. Aging 2(10):678–690

Ratnappan R, Amrit FR, Chen SW, Gill H, Holden K, Ward J, Yamamoto KR, Olsen CP, Ghazi A (2014) Germline signals deploy NHR-49 to modulate fatty-acid beta-oxidation and desaturation in somatic tissues of *C. elegans*. PLoS Genet 10(12), e1004829. doi:10.1371/journal.pgen.1004829

Rea SL, Ventura N, Johnson TE (2007) Relationship between mitochondrial electron transport chain dysfunction, development, and life extension in *Caenorhabditis elegans*. PLoS Biol 5 (10), e259. doi:10.1371/journal.pbio.0050259

Riera CE, Huising MO, Follett P, Leblanc M, Halloran J, Van Andel R, de Magalhaes Filho CD, Merkwirth C, Dillin A (2014) TRPV1 pain receptors regulate longevity and metabolism by neuropeptide signaling. Cell 157(5):1023–1036. doi:10.1016/j.cell.2014.03.051

Rizki G, Iwata TN, Li J, Riedel CG, Picard CL, Jan M, Murphy CT, Lee SS (2011) The evolutionarily conserved longevity determinants HCF-1 and SIR-2.1/SIRT1 collaborate to regulate DAF-16/FOXO. PLoS Genet 7(9), e1002235. doi:10.1371/journal.pgen.1002235

Robida-Stubbs S, Glover-Cutter K, Lamming DW, Mizunuma M, Narasimhan SD, Neumann-Haefelin E, Sabatini DM, Blackwell TK (2012) TOR signaling and rapamycin influence longevity by regulating SKN-1/Nrf and DAF-16/FoxO. Cell Metab 15(5):713–724. doi:10.1016/j.cmet.2012.04.007

Rogers AN, Chen D, McColl G, Czerwieniec G, Felkey K, Gibson BW, Hubbard A, Melov S, Lithgow GJ, Kapahi P (2011) Life span extension via eIF4G inhibition is mediated by posttranscriptional remodeling of stress response gene expression in *C. elegans*. Cell Metab 14(1):55–66. doi:10.1016/j.cmet.2011.05.010

Rottiers V, Motola DL, Gerisch B, Cummins CL, Nishiwaki K, Mangelsdorf DJ, Antebi A (2006) Hormonal control of *C. elegans* dauer formation and life span by a Rieske-like oxygenase. Dev Cell 10(4):473–482. doi:10.1016/j.devcel.2006.02.008

Samuelson AV, Klimczak RR, Thompson DB, Carr CE, Ruvkun G (2007) Identification of *Caenorhabditis elegans* genes regulating longevity using enhanced RNAi-sensitive strains. Cold Spring Harb Symp Quant Biol 72:489–497. doi:10.1101/sqb.2007.72.068

Schiavi A, Torgovnick A, Kell A, Megalou E, Castelein N, Guccini I, Marzocchella L, Gelino S, Hansen M, Malisan F, Condo I, Bei R, Rea SL, Braeckman BP, Tavernarakis N, Testi R, Ventura N (2013) Autophagy induction extends lifespan and reduces lipid content in response to frataxin silencing in *C. elegans*. Exp Gerontol 48(2):191–201. doi:10.1016/j.exger.2012.12.002

Schmeisser K, Mansfeld J, Kuhlow D, Weimer S, Priebe S, Heiland I, Birringer M, Groth M, Segref A, Kanfi Y, Price NL, Schmeisser S, Schuster S, Pfeiffer AF, Guthke R, Platzer M, Hoppe T, Cohen HY, Zarse K, Sinclair DA, Ristow M (2013) Role of sirtuins in lifespan regulation is linked to methylation of nicotinamide. Nat Chem Biol 9(11):693–700. doi:10.1038/nchembio.1352

Schreiber MA, Pierce-Shimomura JT, Chan S, Parry D, McIntire SL (2010) Manipulation of behavioral decline in *Caenorhabditis elegans* with the Rag GTPase *raga-1*. PLoS Genet 6(5), e1000972. doi:10.1371/journal.pgen.1000972

Scott BA, Avidan MS, Crowder CM (2002) Regulation of hypoxic death in *C. elegans* by the insulin/IGF receptor homolog DAF-2. Science 296(5577):2388–2391. doi:10.1126/science.1072302

Selman C, Tullet JM, Wieser D, Irvine E, Lingard SJ, Choudhury AI, Claret M, Al-Qassab H, Carmignac D, Ramadani F, Woods A, Robinson IC, Schuster E, Batterham RL, Kozma SC, Thomas G, Carling D, Okkenhaug K, Thornton JM, Partridge L, Gems D, Withers DJ (2009) Ribosomal protein S6 kinase 1 signaling regulates mammalian life span. Science 326 (5949):140–144. doi:10.1126/science.1177221

Semenza GL (2012) Hypoxia-inducible factors in physiology and medicine. Cell 148(3):399–408. doi:10.1016/j.cell.2012.01.021

Seo K, Choi E, Lee D, Jeong DE, Jang SK, Lee SJ (2013) Heat shock factor 1 mediates the longevity conferred by inhibition of TOR and insulin/IGF-1 signaling pathways in *C. elegans*. Aging Cell 12(6):1073–1081. doi:10.1111/acel.12140

Shaw WM, Luo S, Landis J, Ashraf J, Murphy CT (2007) The *C. elegans* TGF-beta dauer pathway regulates longevity via insulin signaling. Curr Biol 17(19):1635–1645. doi:10.1016/j.cub.2007.08.058

Sheaffer KL, Updike DL, Mango SE (2008) The target of rapamycin pathway antagonizes pha-4/FoxA to control development and aging. Curr Biol 18(18):1355–1364. doi:10.1016/j.cub.2008.07.097

Shen Y, Wollam J, Magner D, Karalay O, Antebi A (2012) A steroid receptor-microRNA switch regulates life span in response to signals from the gonad. Science 338(6113):1472–1476. doi:10.1126/science.1228967

Smith-Vikos T, de Lencastre A, Inukai S, Shlomchik M, Holtrup B, Slack FJ (2014) MicroRNAs mediate dietary-restriction-induced longevity through PHA-4/FOXA and SKN 1/Nrf transcription factors. Curr Biol 24(19):2238–2246. doi:10.1016/j.cub.2014.08.013

Soukas AA, Kane EA, Carr CE, Melo JA, Ruvkun G (2009) Rictor/TORC2 regulates fat metabolism, feeding, growth, and life span in *Caenorhabditis elegans*. Genes Dev 23(4):496–511. doi:10.1101/gad.1775409

Stanfel MN, Shamieh LS, Kaeberlein M, Kennedy BK (2009) The TOR pathway comes of age. Biochim Biophys Acta 1790(10):1067–1074. doi:10.1016/j.bbagen.2009.06.007

Steinkraus KA, Smith ED, Davis C, Carr D, Pendergrass WR, Sutphin GL, Kennedy BK, Kaeberlein M (2008) Dietary restriction suppresses proteotoxicity and enhances longevity by an *hsf-1*-dependent mechanism in *Caenorhabditis elegans*. Aging Cell 7(3):394–404. doi:10.1111/j.1474-9726.2008.00385.x

Syntichaki P, Troulinaki K, Tavernarakis N (2007) eIF4E function in somatic cells modulates ageing in *Caenorhabditis elegans*. Nature 445(7130):922–926. doi:10.1038/nature05603

Takahashi Y, Daitoku H, Hirota K, Tamiya H, Yokoyama A, Kako K, Nagashima Y, Nakamura A, Shimada T, Watanabe S, Yamagata K, Yasuda K, Ishii N, Fukamizu A (2011) Asymmetric arginine dimethylation determines life span in *C. elegans* by regulating forkhead transcription factor DAF-16. Cell Metab 13(5):505–516. doi:10.1016/j.cmet.2011.03.017

Taylor RC, Dillin A (2013) XBP-1 is a cell-nonautonomous regulator of stress resistance and longevity. Cell 153(7):1435–1447. doi:10.1016/j.cell.2013.05.042

Tepper RG, Ashraf J, Kaletsky R, Kleemann G, Murphy CT, Bussemaker HJ (2013) PQM-1 complements DAF-16 as a key transcriptional regulator of DAF-2-mediated development and longevity. Cell 154(3):676–690. doi:10.1016/j.cell.2013.07.006

Thondamal M, Witting M, Schmitt-Kopplin P, Aguilaniu H (2014) Steroid hormone signalling links reproduction to lifespan in dietary-restricted *Caenorhabditis elegans*. Nat Commun 5:4879. doi:10.1038/ncomms5879

Thyagarajan B, Blaszczak AG, Chandler KJ, Watts JL, Johnson WE, Graves BJ (2010) ETS-4 is a transcriptional regulator of life span in *Caenorhabditis elegans*. PLoS Genet 6(9), e1001125. doi:10.1371/journal.pgen.1001125

Tissenbaum HA, Guarente L (2001) Increased dosage of a *sir-2* gene extends lifespan in *Caenorhabditis elegans*. Nature 410(6825):227–230. doi:10.1038/35065638

Tonsaker T, Pratt RM, McGhee JD (2012) Re-evaluating the role of ELT-3 in a GATA transcription factor circuit proposed to guide aging in *C. elegans*. Mech Ageing Dev 133(1):50–53. doi:10.1016/j.mad.2011.09.006

Torgovnick A, Schiavi A, Testi R, Ventura N (2010) A role for p53 in mitochondrial stress response control of longevity in *C. elegans*. Exp Gerontol 45(7–8):550–557. doi:10.1016/j.exger.2010.02.007

Toth ML, Sigmond T, Borsos E, Barna J, Erdelyi P, Takacs-Vellai K, Orosz L, Kovacs AL, Csikos G, Sass M, Vellai T (2008) Longevity pathways converge on autophagy genes to regulate life span in *Caenorhabditis elegans*. Autophagy 4(3):330–338

Tsang WY, Sayles LC, Grad LI, Pilgrim DB, Lemire BD (2001) Mitochondrial respiratory chain deficiency in *Caenorhabditis elegans* results in developmental arrest and increased life span. J Biol Chem 276(34):32240–32246. doi:10.1074/jbc.M103999200

Tullet JM, Araiz C, Sanders MJ, Au C, Benedetto A, Papatheodorou I, Clark E, Schmeisser K, Jones D, Schuster EF, Thornton JM, Gems D (2014) DAF-16/FoxO directly regulates an atypical AMP-activated protein kinase gamma isoform to mediate the effects of insulin/IGF-

1 signaling on aging in *Caenorhabditis elegans*. PLoS Genet 10(2), e1004109. doi:10.1371/journal.pgen.1004109

Tullet JM, Hertweck M, An JH, Baker J, Hwang JY, Liu S, Oliveira RP, Baumeister R, Blackwell TK (2008) Direct inhibition of the longevity-promoting factor SKN-1 by insulin-like signaling in *C. elegans*. Cell 132(6):1025–1038. doi:10.1016/j.cell.2008.01.030

van der Horst A, Schavemaker JM, Pellis-van Berkel W, Burgering BM (2007) The *Caenorhabditis elegans* nicotinamidase PNC-1 enhances survival. Mech Ageing Dev 128 (4):346–349. doi:10.1016/j.mad.2007.01.004

Van Gilst MR, Hadjivassiliou H, Jolly A, Yamamoto KR (2005) Nuclear hormone receptor NHR-49 controls fat consumption and fatty acid composition in *C. elegans*. PLoS Biol 3(2), e53. doi:10.1371/journal.pbio.0030053

Van Raamsdonk JM, Hekimi S (2009) Deletion of the mitochondrial superoxide dismutase *sod-2* extends lifespan in *Caenorhabditis elegans*. PLoS Genet 5(2), e1000361. doi:10.1371/journal.pgen.1000361

Van Raamsdonk JM, Hekimi S (2012) Superoxide dismutase is dispensable for normal animal lifespan. Proc Natl Acad Sci U S A 109(15):5785–5790. doi:10.1073/pnas.1116158109

Vellai T, Takacs-Vellai K, Zhang Y, Kovacs AL, Orosz L, Muller F (2003) Genetics: influence of TOR kinase on lifespan in *C. elegans*. Nature 426(6967):620. doi:10.1038/426620a

Ventura N, Rea S, Henderson ST, Condo I, Johnson TE, Testi R (2005) Reduced expression of frataxin extends the lifespan of *Caenorhabditis elegans*. Aging Cell 4(2):109–112

Ventura N, Rea SL, Schiavi A, Torgovnick A, Testi R, Johnson TE (2009) p53/CEP-1 increases or decreases lifespan, depending on level of mitochondrial bioenergetic stress. Aging Cell 8 (4):380–393. doi:10.1111/j.1474-9726.2009.00482.x

Viswanathan M, Guarente L (2011) Regulation of *Caenorhabditis elegans* lifespan by sir-2.1 transgenes. Nature 477(7365):E1–E2. doi:10.1038/nature10440

Vora M, Shah M, Ostafi S, Onken B, Xue J, Ni JZ, Gu S, Driscoll M (2013) Deletion of microRNA-80 activates dietary restriction to extend *C. elegans* healthspan and lifespan. PLoS Genet 9(8), e1003737. doi:10.1371/journal.pgen.1003737

Walter L, Baruah A, Chang HW, Pace HM, Lee SS (2011) The homeobox protein CEH-23 mediates prolonged longevity in response to impaired mitochondrial electron transport chain in *C. elegans*. PLoS Biol 9(6), e1001084. doi:10.1371/journal.pbio.1001084

Wang J, Robida-Stubbs S, Tullet JM, Rual JF, Vidal M, Blackwell TK (2010) RNAi screening implicates a SKN-1-dependent transcriptional response in stress resistance and longevity deriving from translation inhibition. PLoS Genet 6(8). doi:10.1371/journal.pgen.1001048

Wang MC, O'Rourke EJ, Ruvkun G (2008) Fat metabolism links germline stem cells and longevity in *C. elegans*. Science 322(5903):957–960. doi:10.1126/science.1162011

Wang Y, Tissenbaum HA (2006) Overlapping and distinct functions for a *Caenorhabditis elegans* SIR2 and DAF-16/FOXO. Mech Ageing Dev 127(1):48–56. doi:10.1016/j.mad.2005.09.005

Wang Y, Oh SW, Deplancke B, Luo J, Walhout AJ, Tissenbaum HA (2006) *C. elegans* 14-3-3 proteins regulate life span and interact with SIR-2.1 and DAF-16/FOXO. Mech Ageing Dev 127(9):741–747. doi:10.1016/j.mad.2006.05.005

Wolff S, Ma H, Burch D, Maciel GA, Hunter T, Dillin A (2006) SMK-1, an essential regulator of DAF-16-mediated longevity. Cell 124(5):1039–1053. doi:10.1016/j.cell.2005.12.042

Wolkow CA, Munoz MJ, Riddle DL, Ruvkun G (2002) Insulin receptor substrate and p55 orthologous adaptor proteins function in the *Caenorhabditis elegans* daf-2/insulin-like signaling pathway. J Biol Chem 277(51):49591–49597. doi:10.1074/jbc.M207866200

Wollam J, Magner DB, Magomedova L, Rass E, Shen Y, Rottiers V, Habermann B, Cummins CL, Antebi A (2012) A novel 3-hydroxysteroid dehydrogenase that regulates reproductive development and longevity. PLoS Biol 10(4), e1001305. doi:10.1371/journal.pbio.1001305

Wong A, Boutis P, Hekimi S (1995) Mutations in the *clk-1* gene of *Caenorhabditis elegans* affect developmental and behavioral timing. Genetics 139(3):1247–1259

Xiao R, Zhang B, Dong Y, Gong J, Xu T, Liu J, Xu XZ (2013) A genetic program promotes *C. elegans* longevity at cold temperatures via a thermosensitive TRP channel. Cell 152(4):806–817. doi:10.1016/j.cell.2013.01.020

Xu X, Kim SK (2012) The GATA transcription factor *egl-27* delays aging by promoting stress resistance in *Caenorhabditis elegans*. PLoS Genet 8(12), e1003108. doi:10.1371/journal.pgen.1003108

Yamawaki TM, Berman JR, Suchanek-Kavipurapu M, McCormick M, Gaglia MM, Lee SJ, Kenyon C (2010) The somatic reproductive tissues of *C. elegans* promote longevity through steroid hormone signaling. PLoS Biol 8(8). doi:10.1371/journal.pbio.1000468

Yang CC, Chen D, Lee SS, Walter L (2011) The dynamin-related protein DRP-1 and the insulin signaling pathway cooperate to modulate *Caenorhabditis elegans* longevity. Aging Cell 10 (4):724–728. doi:10.1111/j.1474-9726.2011.00711.x

Yang J, Chen D, He Y, Melendez A, Feng Z, Hong Q, Bai X, Li Q, Cai G, Wang J, Chen X (2013) MiR-34 modulates *Caenorhabditis elegans* lifespan via repressing the autophagy gene atg9. Age (Dordr) 35(1):11–22. doi:10.1007/s11357-011-9324-3

Yang W, Hekimi S (2010a) A mitochondrial superoxide signal triggers increased longevity in *Caenorhabditis elegans*. PLoS Biol 8(12), e1000556. doi:10.1371/journal.pbio.1000556

Yang W, Hekimi S (2010b) Two modes of mitochondrial dysfunction lead independently to lifespan extension in *Caenorhabditis elegans*. Aging Cell 9(3):433–447, doi:ACE571 [pii]

Yee C, Yang W, Hekimi S (2014) The Intrinsic Apoptosis Pathway Mediates the Pro-Longevity Response to Mitochondrial ROS in *C. elegans*. Cell 157(4):897–909. doi:10.1016/j.cell.2014.02.055

Zarse K, Schulz TJ, Birringer M, Ristow M (2007) Impaired respiration is positively correlated with decreased life span in *Caenorhabditis elegans* models of Friedreich Ataxia. FASEB J 21 (4):1271–1275. doi:10.1096/fj.06-6994com

Zhang Y, Xu J, Puscau C, Kim Y, Wang X, Alam H, Hu PJ (2008) *Caenorhabditis elegans* EAK-3 inhibits dauer arrest via nonautonomous regulation of nuclear DAF-16/FoxO activity. Dev Biol 315(2):290–302. doi:10.1016/j.ydbio.2007.12.032

Zhang M, Poplawski M, Yen K, Cheng H, Bloss E, Zhu X, Patel H, Mobbs CV (2009a) Role of CBP and SATB-1 in aging, dietary restriction, and insulin-like signaling. PLoS Biol 7(11), e1000245. doi:10.1371/journal.pbio.1000245

Zhang Y, Shao Z, Zhai Z, Shen C, Powell-Coffman JA (2009b) The HIF-1 hypoxia-inducible factor modulates lifespan in *C. elegans*. PLoS One 4(7), e6348. doi:10.1371/journal.pone.0006348

Chapter 9
Aging-Related Neurodegenerative Diseases in *Caenorhabditis elegans*

Dong Kyu Kim and Seung-Jae Lee

Abstract Neurodegenerative diseases, such as Alzheimer's disease and Parkinson's disease, are aging-dependent. As aged population grows, neurodegenerative diseases have become the major threats to human health and serious social burdens worldwide. Recently, research in neurodegenerative diseases has entered the Renaissance age, as several genes responsible for the diseases were identified. In this chapter, we explain why *Caenorhabditis elegans* (*C. elegans*) is an excellent animal model for the studies on neurodegenerative diseases and describe how this animal model has contributed to understanding of the mechanism of the diseases. Several features of *C. elegans* make it a particularly useful model for searching for genetic modifiers of disease phenotypes and for investigating the role of aging in disease development and progression.

Keywords Neurodegenerative diseases • *Caenorhabditis elegans* • Aging • Protein aggregation • Alzheimer's disease • Parkinson's disease • Huntington's disease • Amyotrophic lateral sclerosis

9.1 Introduction

Neurodegenerative diseases are mostly aging-dependent and characterized by accumulation of protein aggregates that are specific to particular diseases. These protein aggregates include amyloid β aggregates in the senile plaques and tau aggregates in the neurofibrillary tangles, both in Alzheimer's disease (AD), α-synuclein aggregates in Lewy bodies in Parkinson's disease (PD), huntingtin

D.K. Kim
Neuroscience Research Institute and Department of Medicine, Seoul National University College of Medicine, Seoul 110-799, Korea

Department of Biomedical Science and Technology, Konkuk University, Seoul 143-701, Korea

S.-J. Lee (✉)
Neuroscience Research Institute and Department of Medicine, Seoul National University College of Medicine, Seoul 110-799, Korea
e-mail: Sjlee66@snu.ac.kr

© Springer Japan 2015

N. Mori, I. Mook-Jung (eds.), *Aging Mechanisms*,
DOI 10.1007/978-4-431-55763-0_9

aggregates with expanded polyglutamine (polyQ) in Huntington's disease (HD), and TDP-43 aggregates in amyotrophic lateral sclerosis (ALS). Genomic mutations that cause these diseases generally result in increased aggregation of the corresponding proteins, suggesting that protein aggregation is probably the critical step in disease pathogenesis.

The protein aggregates usually occur in distinct brain regions at the early stages of disease and spread to larger areas as the disease progresses. Although not perfectly consistent, the patterns of protein aggregate spreading tend to be highly sequential, and some studies suggested that the aggregate spreading is the underlying principle for the progressive diversification of clinical symptoms. Therefore, understanding of the mechanism of specific protein aggregation and the spreading of the aggregates might be of foremost importance for unraveling how the disease initiates and progresses.

Numerous model systems have been established to study the mechanism of protein aggregates and their roles in neurodegeneration. The most widely used *in vivo* models have been constructed in rodents. Although the rodent models have been very useful in recapitulating some of the major features of neurodegenerative diseases, the results obtained in these models have largely been correlative, due to limitations the rodent models have, which includes anatomical complexity and difficulties of genetic modification. In addition, relatively long incubation period in rodents makes it difficult to assess the role of aging process in disease pathogenesis.

C. elegans (*Caenorhabditis elegans*) have several advantages as an in vivo model, complementing these limitations that rodents have. *C. elegans* is a small nematode, which is inexpensive to maintain in Nematode Growth medium (NGM) plate, and has a transparent body that allows for visualization of anatomical structures in live (Brenner 1974). Short generation cycle and high number of progeny, as well as short lifespan of 2–3 weeks make it an ideal model for studying age-related diseases. The entire body plan was mapped out, and there is massive pool of resources, such as lines with specific gene mutations. Most importantly, genetic and epigenetic manipulations of *C. elegans* are so simple and convenient that genome-wide screening for genetic modifiers can be done with much less effort and expense than in rodents (Sulston et al. 1983). In *C. elegans*, homologues about 60–80 % of human genes exist (Lai et al. 2000), and these include several orthologous genes involved in neurodegenerative diseases (Culetto and Sattelle 2000). In this chapter, we briefly introduce the methodology, by which *C. elegans* can be used in the study of neurological disorders, and review what we have learned about human neurodegenerative diseases from the studies conducted using this model.

9.2 Methods for Genetic Modification in *C. elegans* Model

Transformation methods have been commonly used for ectopic gene expression (Mello et al. 1991). In these transgenic models, transgenes are expressed in selected cells using specific *C. elegans* promoters. Stable transgenic lines are produced by integrating transgene into the genome of *C. elegans* with UV or gamma irradiation (Riddle et al. 1997).

Screening for genetic modifiers has largely conducted with two methodologies, ethyl methane sulfate (EMS) mutagenesis (Brenner 1974) and RNA interference (RNAi) (Fire et al. 1998). EMS is a useful tool for forward genetics, generating random mutations by nucleotide transitions of G/C to A/T. On the other hand, RNAi knocks down gene functions by reducing expression of specific genes. There are three ways of delivering vectors for RNAi to *C. elegans*: injection (Fire et al. 1998), soaking (Tabara et al. 1998), and feeding (Timmons and Fire 1998). Among these methods, soaking and feeding have been widely used for high throughput screening.

9.3 Nervous System of *C. elegans*

C. elegans has a simple nervous system comprised of 302 neurons (White et al. 1986), and a complete composition of synaptic connections has been mapped out, including 5000 chemical synapses and 600 gap junctions (Sulston 1983). Neuronal types as well as trafficking and release of synaptic vesicles (Nonet et al. 1998) in *C. elegans* were highly conserved with mammals. Several neurotransmitters were identified, and they include dopamine (Sulston et al. 1975), γ-aminobutyric acid (Mclintire et al. 1993), acetylcholine, serotonin (Waggoner et al. 1998), and glutamate (Hart et al. 1995; Maricq et al. 1995). There are totally eight dopaminergic neurons: four cephalic neurons (CEP), two anterior deirid neurons (ADE) and two posterior deirid neurons (PDE) (White et al. 1986). Additionally, there are 26 GABAergic neurons divided into six classes (McIntire et al. 1993; White et al. 1986), 100 cholinergic neurons (Hall et al. 2006) and 34 glutamatergic neurons (Lee et al. 1999) in *C. elegans*.

9.4 Daf-2 Signaling Pathway in Aging

The most extensively investigated pathway for regulating aging in *C. elegans* is the insulin/insulin-like growth factor-1 (IGF-1) signaling (IIS) pathway (Kenyon 2010) (Fig. 9.1). DAF-2 gene, encoding an insulin/IGF-1 receptor, inhibits the forkhead (FOXO) transcription factor, encoded by DAF-16 gene, in *C. elegans* (Lee et al. 2001). The daf-2/daf-16 pathway regulates longevity in such a way that mutations in daf-2 gene extend life span while those in daf-16 reduce life span.

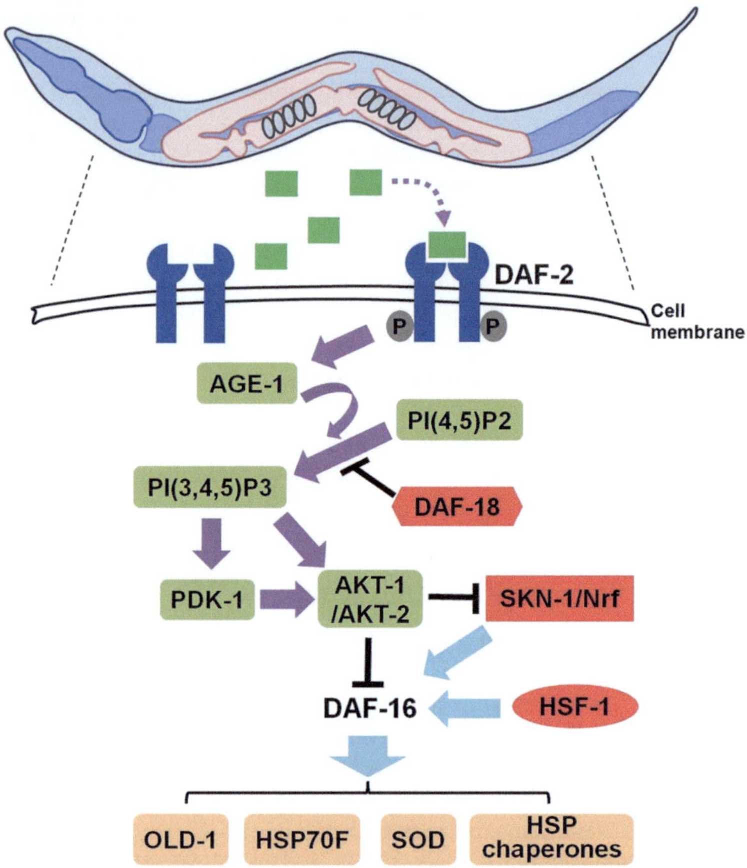

Fig. 9.1 **The IIS pathway in *C. elegans*.** Binding of insulin-like peptide to DAF-2 induces self-phosphorylation and dimerization of DAF-2 and leads to activation of the phosphatidylinositol 3-kinase (AGE-1). Activation of AGE-1 results in conversion of phosphatidylinositol-4,5-bisphosphate (PI(4,5)P2) to phosphatidylinositol-3,4,5-trisphosphate (PI(3,4,5)P3) and then causes activation of PDK-1 and AKT-1/AKT-2. Activated AKT inhibits the FOXO transcription factor, DAF-16, by phosphorylating this protein, thereby blocking the expression of its target genes, which mediate longevity and stress resistance signaling. On the other hand, the phosphatase, DAF-18 (a PTEN orthologue), blocks turnover of PI(4,5)P2 into PI(3,4,5)P3, and SKN-1 and Heat-shock factor 1 (HSF-1) activate DAF-16, turning on the longevity and stress resistance genes, such as OLD-1, HSP70F, sod-3 and small heat-shock protein (HSP) chaperones

The daf-2/daf-16 pathway also regulates stress-resistance and induction of heat shock genes, such as *OLD-1*, HSP70F, *sod-3* and small heat-shock protein (HSP) chaperones.

Removal of germ-line cells in *C. elegans* (Hsin and Kenyon 1999) and dietary restriction (McCay et al. 1989) have Extended life span. In contrast, inhibition of autophagy by mutations of Atg genes, such as *atg-1*, *atg-7* and *bec-1*, reduced life span of *C. elegans* (Toth et al. 2008). Germ-line removal and dietary restriction induced autophagy activation by triggering activation of DAF-16 and inhibition of IIS (Kenyon 2010).

9.5 Modeling Age-Related Neurodegenerative Diseases in *C. elegans*

Aging is the major risk factor for neurodegenerative diseases. Taking advantage of short life span, complete understanding of anatomical structures, and well-established genetic pathways of aging, *C. elegans* has been one of the favorite models for the study of aging-related neurodegenerative diseases (Fig. 9.2).

9.5.1 Alzheimer's Disease (AD)

Alzheimer's disease (AD) is characterized by the amyloid plaques containing β-amyloid (Aβ) peptides and neurofibrillary tangles (NFTs) containing hyperphosphorylated tau proteins. Transgenic animals that overexpress Aβ42 peptide in body wall muscle using *unc-54* promoter exhibited intracellular cytoplasmic Aβ inclusions, toxicity of Aβ aggregates, and progressive, age-dependent paralysis phenotype (Link 1995, 2001). Aβ amyloid aggregates and disease phenotypes were alleviated by reducing the activity of the IIS pathway or by knockdown of daf-2 gene (Cohen et al. 2006). Consistently, Aβ aggregation and toxicity were increased

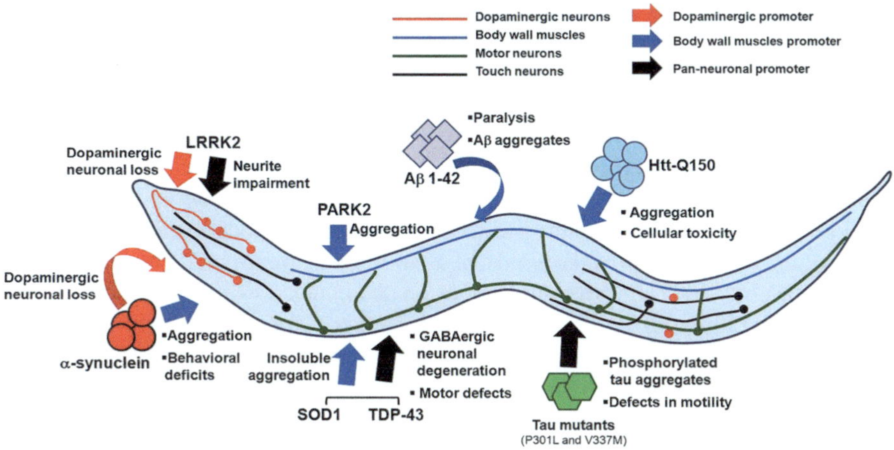

Fig. 9.2 Age-related neurodegenerative diseases in *C. elegans*. Summary of transgenic *C. elegans* neurodegenerative models expressing major disease-causing proteins under the specific promoters. *Arrows* indicate promoters used in the individual studies, and *lines* indicate cell types where transgenes were expressed.

by knockdown of *daf-16* or *hsf-1*, downstream genes of daf-2 in the IIS pathway. In a tauopathy model, transgenic worms expressing P301L and V337M mutants under the *aex-3* promoter, a pan-neuronal promoter, exhibited accumulation of phosphorylated tau aggregates and severe defects in motility (Kraemer et al. 2003).

9.5.2 Parkinson Disease (PD)

Pathological features of Parkinson disease (PD) are include selective loss of dopaminergic neurons in the substantia nigra pars compacta and intracellular inclusion bodies known as Lewy bodies and Lewy neurites, which are composed mainly of amyloid fibril forms of α-synuclein. While six PD genes were identified in *C. elegans*, including *pdr-1* (PARK2), *ubh-1* (PARK5), *pink-1* (PARK6), *djr-1.1/1.2* (PARK7), *lrk-1* (PARK8), *catp-6* (PARK9), there is no α-synuclein gene (PARK1) in *C. elegans* (Harrington et al. 2010). Nevertheless, transgenic overexpression of human α-synuclein in *C. elegans* has generated useful models for synucleinopathies. Transgenic *C. elegans* expressing human α-synuclein-GFP in body wall muscles exhibited accumulation of protein aggregates with aging and induced proteotoxicity (Hamamichi et al. 2008; van Ham et al. 2008; Springer et al. 2005). Additionally, overexpression of wild-type human α-synuclein in dopaminergic neurons of *C. elegans* resulted in age-dependent neuronal loss (Cao et al. 2005). A genome-wide RNAi screening identified several genetic modifiers of α-synuclein aggregation, which include genes involved in inclusion formation, vesicular transport, lipid metabolism and aging-pathways (van Ham et al. 2008).

Other PD genes have been mutated or overexpressed to generate PD models in *C. elegans*. Worms with mutations in the *pdr-1*, an ortholog of parkin (PARK2), changed solubility of the parkin protein and exhibited α-synuclein aggregation (Springer et al. 2005). Transgenic models overexpressing human wild-type Leucine Rich Repeat Kinase 2 (LRRK2) using the *snb-1* promoter (pan-neuronal expression) resulted in dopaminergic neuronal loss (Saha et al. 2009). Worms in the absence of *lrk-1*, an ortholog of human LRRK2 (PARK8) were generated by deletion of the kinase or GTPase domain (Sakaguchi-Nakashima et al. 2007; Saha et al. 2009). These mutant lines exhibited aberrant localization of synaptic proteins in the presynaptic terminals of neurons and motor defects (Sakaguchi-Nakashima et al. 2007). Furthermore, deletion and knock down of *lrk-1* reduced life span and mitochondrial abnormalities by rotenone treatment (Saha et al. 2009). On the other hand, transgenic lines expressing PD-linked LRRK2 mutants, such as LRRK2 G2019S (increased kinase activity) and LRRK2 R1441C (decreased GTPase activity), using *dat-1* promoter (dopaminergic neuronal expression) displayed severe dopaminergic neuronal loss and progressive behavioral deficits (Yao et al. 2010).

9.5.3 Huntington's Disease (HD)

Huntington's disease (HD), an autosomal dominant genetic disease, is characterized by neuronal accumulation of insoluble protein aggregates composed of huntingtin proteins with expanded polyglutamine (polyQ) motif. The longer the length of the expansion, the more severe and faster disease onset and progression. *C. elegans* does not contain a huntingtin ortholog gene, however, has been widely used to generate HD models, which have been useful in elucidating the mechanism of the disease and identifying disease-modifying gene candidates. Transgenic animals that were engineered to produce polyQ protein aggregates exhibited proteostasis impairment (Satyal et al. 2000). Transgenic expression of expanded polyQ proteins in body wall muscle resulted in protein aggregates and cellular toxicity with aging, and these phenotypes were alleviated by autophagy activation (Jia et al. 2007). Furthermore, reduction of IIS by mutation delayed polyQ protein aggregation, reduced proteotoxicity, and extended life span (Morley et al. 2002).

9.5.4 Amyotrophic Lateral Sclerosis (ALS)

ALS is characterized by progressive degeneration of motor neurons in the brain and the spinal cord, and some of the familial ALS is caused by mutations in genes, such as Cu/Zn superoxide dismutase 1 (SOD1) and TAR DNA-binding protein-43 (TDP-43). The *C. elegans* models expressing human SOD1 mutants consistently exhibited SOD1 aggregation and cell death. Expression of mutant SOD1 (A4V, G37R, and G93A) under the control of *myo-3* promoter resulted in aggregation of SOD1 mutants in the muscle (Oeda et al. 2001). Transgenic worms expressing SOD1 mutants (G85R or G85R-YFP) using *snb-1* promoter (pan-neuronal expression) also exhibited insoluble SOD1 aggregates in neurons (Wang et al. 2009). Likewise, transgenic expression of YFP-tagged wild-type and mutant SOD1 (G85R, G93A, G127X) resulted in SOD1 aggregation in the body wall muscle (Gidalevitz et al. 2009). Expression of G93A SOD1 induced age-dependent aggregates and motor defects (Li et al. 2013). Consistent with the idea that aging plays an important role, the insoluble protein aggregates and motor defects were alleviated by reducing daf-2 activity in G85R and G93A SOD1 worms (Boccitto et al. 2012; Li et al. 2013).

TDP 43-associated ALS is characterized by pathological inclusions containing ubiquitinated, hyperphosphorylated TDP-43. The transgenic worms expressing human TDP-43 under the control of *snb-1* promoter exhibited abnormal motor symptom (Ash et al. 2010). Expression of human TDP-43 and mutants (G290A, A315T, or M337V) in *C. elegans* caused motor dysfunction with aging (Liachko et al. 2010). Furthermore, expression of TDP-43 mutants caused degeneration of GABAergic processes (Liachko et al. 2010). Expression of YFP-tagged wild-type or the C-terminal fragment of TDP-43 also resulted in insoluble aggregates and motor defects (Zhang et al. 2011). TDP-43 aggregation and abnormal motor behavior were alleviated by mutation of daf-2, while deletion of daf-16 accelerated neurotoxicity and aggregation (Zhang et al. 2011).

9.6 Conclusion

C. elegans has been a powerful model system for investigation of the mechanisms underlying the pathogenesis of neurodegenerative diseases. This model has been particularly useful to analyze protein aggregation and the pathophysiological consequences of protein aggregation, especially in the context of aging. Screenings for genetic modifiers of protein aggregation revealed importance of maintaining proteostasis and vesicle trafficking systems. Small molecule screenings in *C. elegans* models might be a promising approach to identify drug-like molecules that interfere with protein aggregation and pathogenic actions of aggregates. In the future, we anticipate that *C. elegans* model will also be useful in understanding the mechanism of protein aggregate propagation and disease progression.

Acknowledgements This work was supported by the National Research Foundation (NRF) grant funded by the Korean Government (MEST) (No. 2010-0015188), and the Korea Health Technology R&D Project, Ministry of Health & Welfare, Republic of Korea (HI14C0093).

References

Ash PE, Zhang YJ, Roberts CM, Saldi T, Hutter H, Buratti E, Petrucelli L, Link CD (2010) Neurotoxic effects of TDP-43 overexpression in C. elegans. Hum Mol Genet 19(16):3206–3218

Boccitto M, Lamitina T, Kalb RG (2012) Daf-2 signaling modifies mutant SOD1 toxicity in C. elegans. PLoS One 7(3):e33494

Brenner S (1974) The genetics of Caenorhabditis elegans. Genetics 77(1):71–94

Cao S, Gelwix CC, Caldwell KA, Caldwell GA (2005) Torsin-mediated protection from cellular stress in the dopaminergic neurons of Caenorhabditis elegans. J Neurosci 25(15):3801–3812. doi:10.1523/JNEUROSCI.5157-04.2005

Cohen E, Bieschke J, Perciavalle RM, Kelly JW, Dillin A (2006) Opposing activities protect against age-onset proteotoxicity. Science 313(5793):1604–1610. doi:10.1126/science.1124646

Culetto E, Sattelle DB (2000) A role for Caenorhabditis elegans in understanding the function and interactions of human disease genes. Hum Mol Genet 9(6):869–877

Fire A, Xu S, Montgomery MK, Kostas SA, Driver SE, Mello CC (1998) Potent and specific genetic interference by double-stranded RNA in Caenorhabditis elegans. Nature 391 (6669):806–811. doi:10.1038/35888

Gidalevitz T, Krupinski T, Garcia S, Morimoto RI (2009) Destabilizing protein polymorphisms in the genetic background direct phenotypic expression of mutant SOD1 toxicity. PLoS Genet 5(3):e1000399

Hall DH, Lints R, Altun Z (2006) Nematode neurons: anatomy and anatomical methods in Caenorhabditis elegans. Int Rev Neurobiol 69:1–35. doi:10.1016/S0074-7742(05)69001-0

Hamamichi S, Rivas RN, Knight AL, Cao S, Caldwell KA, Caldwell GA (2008) Hypothesis-based RNAi screening identifies neuroprotective genes in a Parkinson's disease model. Proc Natl Acad Sci U S A 105(2):728–733. doi:10.1073/pnas.0711018105

Harrington AJ, Hamamichi S, Caldwell GA, Caldwell KA (2010) C. elegans as a model organism to investigate molecular pathways involved with Parkinson's disease. Dev Dyn: Off Publ Am Assoc Anat 239(5):1282–1295. doi:10.1002/dvdy.22231

Hart AC, Sims S, Kaplan JM (1995) Synaptic code for sensory modalities revealed by C. elegans GLR-1 glutamate receptor. Nature 378(6552):82–85. doi:10.1038/378082a0

Hsin H, Kenyon C (1999) Signals from the reproductive system regulate the lifespan of C. elegans. Nature 399(6734):362–366. doi:10.1038/20694

Jia K, Hart AC, Levine B (2007) Autophagy genes protect against disease caused by polyglutamine expansion proteins in Caenorhabditis elegans. Autophagy 3(1):21–25

Kenyon CJ (2010) The genetics of ageing. Nature 464(7288):504–512. doi:10.1038/nature08980

Kraemer BC, Zhang B, Leverenz JB, Thomas JH, Trojanowski JQ, Schellenberg GD (2003) Neurodegeneration and defective neurotransmission in a Caenorhabditis elegans model of tauopathy. Proc Natl Acad Sci U S A 100(17):9980–9985. doi:10.1073/pnas.1533448100

Lai CH, Chou CY, Ch'ang LY, Liu CS, Lin W (2000) Identification of novel human genes evolutionarily conserved in Caenorhabditis elegans by comparative proteomics. Genome Res 10(5):703–713

Lee JH, Daud AN, Cribbs LL, Lacerda AE, Pereverzev A, Klockner U, Schneider T, Perez-Reyes E (1999) Cloning and expression of a novel member of the low voltage-activated T-type calcium channel family. J Neurosci 19(6):1912–1921

Lee RY, Hench J, Ruvkun G (2001) Regulation of C. elegans DAF-16 and its human ortholog FKHRL1 by the daf-2 insulin-like signaling pathway. Curr Biol: CB 11(24):1950–1957

Li J, Huang KX, Le WD (2013) Establishing a novel C. elegans model to investigate the role of autophagy in amyotrophic lateral sclerosis. Acta Pharmacol Sin 34(5):644–50

Liachko NF, Guthrie CR, Kraemer BC (2010) Phosphorylation promotes neurotoxicity in a Caenorhabditis elegans model of TDP-43 proteinopathy. J Neurosci 30(48):16208–16219. doi:10.1523/JNEUROSCI.2911-10

Link CD (1995) Expression of human beta-amyloid peptide in transgenic Caenorhabditis elegans. Proc Natl Acad Sci U S A 92(20):9368–9372

Link CD (2001) Transgenic invertebrate models of age-associated neurodegenerative diseases. Mech Ageing Dev 122(14):1639–1649

Maricq AV, Peckol E, Driscoll M, Bargmann CI (1995) Mechanosensory signalling in C. elegans mediated by the GLR-1 glutamate receptor. Nature 378(6552):78–81. doi:10.1038/378078a0

McCay CM, Crowell MF, Maynard LA (1989) The effect of retarded growth upon the length of life span and upon the ultimate body size. 1935. Nutrition 5(3):155–171; discussion 172

McIntire SL, Jorgensen E, Kaplan J, Horvitz HR (1993) The GABAergic nervous system of Caenorhabditis elegans. Nature 364(6435):337–341. doi:10.1038/364337a0

Mello CC, Kramer JM, Stinchcomb D, Ambros V (1991) Efficient gene transfer in C. elegans: extrachromosomal maintenance and integration of transforming sequences. EMBO J 10 (12):3959–3970

Morley JF, Brignull HR, Weyers JJ, Morimoto RI (2002) The threshold for polyglutamine-expansion protein aggregation and cellular toxicity is dynamic and influenced by aging in Caenorhabditis elegans. Proc Natl Acad Sci U S A 99(16):10417–10422. doi:10.1073/pnas.152161099

Nonet ML, Saifee O, Zhao H, Rand JB, Wei L (1998) Synaptic transmission deficits in Caenorhabditis elegans synaptobrevin mutants. J Neurosci 18(1):70–80

Oeda T, Shimohama S, Kitagawa N, Kohno R, Imura T, Shibasaki H, Ishii N (2001) Oxidative stress causes abnormal accumulation of familial amyotrophic lateral sclerosis-related mutant SOD1 in transgenic Caenorhabditis elegans. Hum Mol Genet 10(19):2013–2023

Riddle DL, Blumenthal T, Meyer BJ, Priess JR (1997) Introduction to C. elegans. In: Riddle DL, Blumenthal T, Meyer BJ, Priess JR (eds) C. elegans II, 2nd edn. Cold Spring Harbor Laboratory Press, Cold Spring Harbor

Saha S, Guillily MD, Ferree A, Lanceta J, Chan D, Ghosh J, Hsu CH, Segal L, Raghavan K, Matsumoto K, Hisamoto N, Kuwahara T, Iwatsubo T, Moore L, Goldstein L, Cookson M, Wolozin B (2009) LRRK2 modulates vulnerability to mitochondrial dysfunction in Caenorhabditis elegans. J Neurosci 29(29):9210–9218. doi:10.1523/JNEUROSCI.2281-09. 2009

Sakaguchi-Nakashima A, Meir JY, Jin Y, Matsumoto K, Hisamoto N (2007) LRK-1, a C. elegans PARK8-related kinase, regulatesaxonal-dendritic polarity of SV proteins. Curr Biol 17 (7):592–598

Satyal SH, Schmidt E, Kitagawa K, Sondheimer N, Lindquist S, Kramer JM, Morimoto RI (2000) Polyglutamine aggregates alter protein folding homeostasis in Caenorhabditis elegans. Proc Natl Acad Sci U S A 97(11):5750–5755. doi:10.1073/pnas.100107297

Springer W, Hoppe T, Schmidt E, Baumeister R (2005) A Caenorhabditis elegans Parkin mutant with altered solubility couples alpha-synuclein aggregation to proteotoxic stress. Hum Mol Genet 14(22):3407–3423. doi:10.1093/hmg/ddi371

Sulston JE (1983) Neuronal cell lineages in the nematode Caenorhabditis elegans. Cold Spring Harb Symp Quant Biol 48(Pt 2):443–452

Sulston J, Dew M, Brenner S (1975) Dopaminergic neurons in the nematode Caenorhabditis elegans. J Comp Neurol 163(2):215–226. doi:10.1002/cne.901630207

Sulston JE, Schierenberg E, White JG, Thomson JN (1983) The embryonic cell lineage of the nematode Caenorhabditis elegans. Dev Biol 100(1):64–119

Tabara H, Grishok A, Mello CC (1998) RNAi in C. elegans: soaking in the genome sequence. Science 282(5388):430–431

Timmons L, Fire A (1998) Specific interference by ingested dsRNA. Nature 395(6705):854. doi:10.1038/27579

To'th ML, Sigmond T, Borsos E, Barna J, Erde'lyi P, Taka'cs-Vellai K, Orosz L, Kova'cs AL, Csiko's G, Sass M, Vellai T (2008) Longevity pathways converge on autophagy genes to regulate life span in Caenorhabditis elegans. Autophagy 4(3):330–338

van Ham TJ, Thijssen KL, Breitling R, Hofstra RM, Plasterk RH, Nollen EA (2008) C. elegans model identifies genetic modifiers of alpha-synuclein inclusion formation during aging. PLoS Genet 4(3):e1000027. doi:10.1371/journal.pgen.1000027

Waggoner LE, Zhou GT, Schafer RW, Schafer WR (1998) Control of alternative behavioral states by serotonin in Caenorhabditis elegans. Neuron 21(1):203–214

Wang J, Farr GW, Hall DH, Li F, Furtak K, Dreier L, Horwich AL (2009) An ALS-linked mutant SOD1 produces a locomotor defect associated with aggregation and synaptic dysfunction when expressed in neurons of Caenorhabditis elegans. PLoS Genet 5(1):e1000350

White JG, Southgate E, Thomson JN, Brenner S (1986) The structure of the nervous system of the nematode Caenorhabditis elegans. Philos Trans R Soc Lond B Biol Sci 314(1165):1–340

Yao C, El Khoury R, Wang W, Byrd TA, Pehek EA, Thacker C, Zhu X, Smith MA, Wilson-Delfosse AL, Chen SG (2010) LRRK2-mediated neurodegeneration and dysfunction of dopaminergic neurons in a Caenorhabditis elegans model of Parkinson's disease. Neurobiol Dis 40 (1):73–81

Zhang T, Mullane PC, Periz G, Wang J (2011) TDP-43 neurotoxicity and protein aggregation modulated by heat shock factor and insulin/IGF-1 signaling. Hum Mol Genet 20(10):1952–1965

Part IV
Metabolism: Factors Affecting Tissue Aging

Chapter 10
α-Klotho in Health and Diseases

Yo-ichi Nabeshima

Abstract *α-klotho*ˉ(*α-kl*) was first identified as an aging gene and later shown to be a regulator of calcium and phosphate homeostasis. *α-kl* is predominantly expressed in tissues that are involved in mineral homeostasis, and it encodes a 130-kDa type I glycoprotein. α-Kl was first predicted to localize to the cell surface. However, large amounts of α-Kl proteins have been detected in the intra-cellular space. In addition, the extra-cellular domain is cleaved, and secreted forms have been identified in the blood, CSF and urine. These findings suggest that α-Kl has several functions that depend on its intracellular, membrane, and extra-cellular secreted forms. In fact, the intra-cellular form of α-Kl activates Ca^{2+} transport from the blood to the CSF in the choroid plexus and Ca^{2+} re-absorption in the kidney and regulates PTH secretion in parathyroid glands by controlling the trafficking of the Na^{+}-K^{+}-ATPase complex to plasma membrane. On the membrane, α-Kl forms a ternary complex with FGF23 and FGFR1 and negatively regulates 1, $25(OH)_2D$ synthesis and phosphate re-absorption in the kidney. As a down-steam event of hypervitaminosis D and hyperphosphatemia, Calpain-1 is greatly activated and is responsible for many phenotypes. Although a growing number of papers have reported the biological and clinical roles of the secreted form of α-Kl, the functions of the secreted form of α-Kl are poorly understood.

The extracellular domain of α-Kl contains two internal repeats that are homologous to family 1 β-glycosidase. However, critical amino acid residues that are essential for enzyme action are replaced. Nonetheless, α-Kl was found to exhibit a subtle but specific β-glucuronidase activity. This finding suggests that the function of α-Kl may be twofold; it may act as an enzyme or as a glycoside-binding protein. The analyses of the sugar chains of α-Kl binding proteins and revealed that α-Kl functions as a glycoside-binding protein.

Keywords Klotho • FGF23 • Na^{+},K^{+}-ATPase • Family 1 β-glycosidase • Mineral homeostasis • Aging syndromes

Y.-i. Nabeshima (✉)
Laboratory of Molecular Life Science Foundation for Biomedical Research and Innovation, Kobe, Japan
e-mail: nabemr@lmls-kobe.org

© Springer Japan 2015

N. Mori, I. Mook-Jung (eds.), *Aging Mechanisms*,
DOI 10.1007/978-4-431-55763-0_10

10.1 Introduction: Discovery of Klotho

In 1997, we identified a short-lived model mouse that displays a variety of premature aging-related syndromes (Kuro-o et al. 1997). The gene responsible for these phenotypes was named *klotho*, after the name of one of the Three Fates in Greek Mythology, who spins the thread of life. This gene was later renamed *α-klotho* (*α-kl*) because a *klotho* homologue, *β-klotho* was identified (Ito et al. 2005). $\alpha\text{-}kl^{-/-}$ mice display multiple phenotypes that resemble human aging-related syndromes, including atherosclerosis, ectopic calcifications in arteries and various soft tissues, bone malformations with abnormal calcification at the epiphysis and reduced bone mineral density in the diaphysis, kyphosis, pulmonary emphysema, senile atrophy of the skin, generalized tissue atrophy, a reduction in the white fat deposits in all examined organs, infertility, and short lifespan (Fig. 10.1) Specifically, the serum levels of 1,25-$(OH)_2D$ in $\alpha\text{-}kl^{-/-}$ mice were much higher than those of wild type mice at 2 weeks of age, which preceded multiple histological phenotypes; this level remained significantly higher throughout the lifespan of the mice

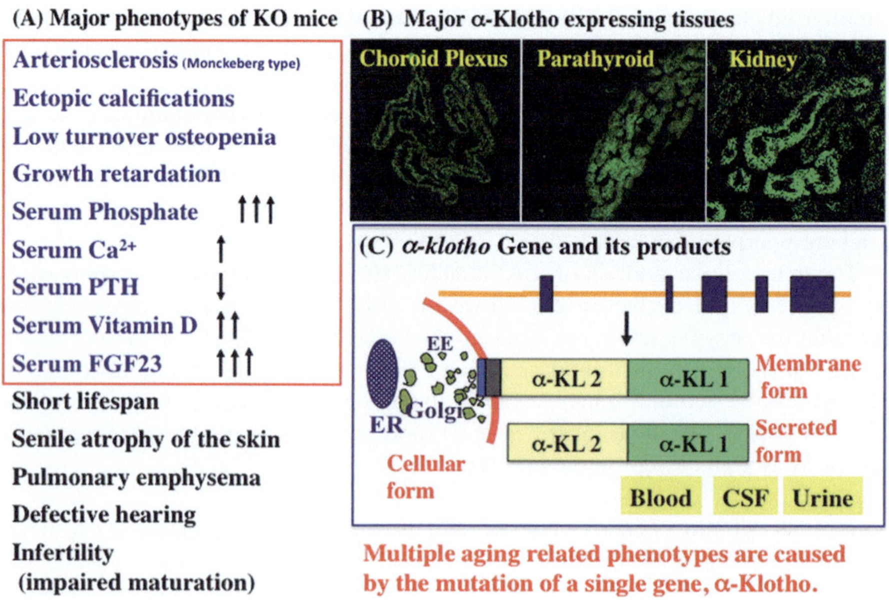

Fig. 10.1 α-Klotho: Phenotypes of KO mice, expression in tissue, and molecular characteristics of α-Klotho. (**a**) Major phenotypes of *α-kl* knockout mice. The phenotypes surrounded by *red squares* are all related to the abnormalities of mineral homeostasis. (**b**) Immunohistological expression of α-Klotho in tissues related to mineral homeostasis. (**c**) The *α-kl* gene encodes a type-1 membrane protein and thus is expressed on the cell surface. α-Kl is also abundantly identified in the cytoplasm of α-Kl-expressing cells (**b**) and the truncated forms of α-Kl were detected in the blood, CSF and urine. Importantly, all phenotypes are the result of the mutation of a single gene, *α-kl*

(Yoshida et al. 2002; Tsujikawa et al. 2003). These high serum levels of 1,25-$(OH)_2D$ in α-$kl^{-/-}$ mice are also associated with the dysregulation of serum calcium, phosphate and fibroblast growth factor 23 (FGF23), which characterize the serological phenotypes of α-$kl^{-/-}$ mice. Taken together, we concluded that the mutation of a single gene, α-Klotho, resulted in multiple aging-related phenotypes (Kuro-o et al. 1997).

10.2 Characterization of α-Klotho

α-kl encodes a 130-kDa type I membrane glycoprotein (Kuro-o et al. 1997). The extracellular domain of α-Kl contains two internal repeats (termed α-KL1 and α-KL2) that are similar to the family 1 β-glycosidase (Henrissat 1991; Henrissat and Bairoch 1993, 1996), which hydrolyze the glycosidic bond between two carbohydrates or between carbohydrate and non-carbohydrate moieties (Davies and Henrissat 1995). The glycosidase family has been classified into approximately 90 families based on amino acid similarities, and they participate in the essential steps of the synthesis and degradation of oligosaccharides and polysaccharides (Herscovics 1999); these saccharides are involved in pathogen defense systems, detoxification, the control of signal transduction, and the modification of hormones (Wittstock and Halkier 2002; Rask et al. 2000; Gopalan et al. 1992; LaMarco and Glew 1986; Wells et al. 2001; Brzobohaty et al. 1993). Members of the family 1 β-glycosidase group (LaMarco and Glew 1986; Wells et al. 2001; Brzobohaty et al. 1993), such as β-glucosidase, β-galactosidase, 6-phospho-β-glucosidase, 6-phospho-β-galactosidase, myrosinase, and lactase-phlorizin hydrolase (LPH), are widely expressed in bacteria, prokaryotes and eukaryotes.

The α-*kl* gene is predominantly expressed in tissues involved in mineral homeostasis; namely, the parathyroid glands, kidney DCT nephrons, and choroid plexus of the brain (Kuro-o et al. 1997; Tsujikawa et al. 2003). The α-*kl* gene encodes a type I membrane protein and is thus predicted to localize on the cell surface. However, large amounts of α-Kl proteins are detectable in the intra-cellular space (Imura et al. 2007). In addition, the extra-cellular domain is cleaved at the N-terminal side of the trans-membrane sequence, and secreted forms have been identified in the blood, CSF and urine (Imura et al. 2004). These findings suggest that α-Kl has several functions depending on its intracellular, membrane, and extra-cellular secreted forms (Fig. 10.1).

10.3 Biological Functions of α-Klotho

Intra-cellular α-Kl binds the Na^+-,K^+-ATPase complex in the choroid plexus, kidney and parathyroid glands. Na^+K^+-ATPase/α-Kl complexes have been detected in trans-Golgi networks and are largely accumulated in endosome fractions, and they are quickly recruited to the cell surface in response to decreasing extra-cellular calcium levels. Simultaneously α-Kl is secreted into the extra-cellular spaces (Imura et al. 2007). The Na^+-,K^+-ATPase complex has been identified as a membrane-bound pump that transports two K^+ ions into cells in exchange for the export of three Na^+ ions (Skou 1988). In addition, the Na^+ gradient created by the Na^+-K^+-ATPase complex drives the actions of the other ion exchangers and channels, such as the Na^+/Ca^{2+} exchanger (Blaustein and Lederer 1999; Lytton 2007). Furthermore, the increased electrochemical gradient drives the docking of secretion granules to cell membrane. In fact, the increase in the plasma-membrane recruitment of the Na^+-K^+-ATPase complex drives the trans-epithelial transport of Ca^{2+} from the bloodstream to the CSF in the choroid plexus, controls calcium re-absorption in the kidney DCT nephrons, and regulates the secretion of PTH in parathyroid glands (Imura et al. 2007) (Fig. 10.2). All of these systems are impaired in α-$kl^{-/-}$ mice; thus, the calcium concentration in the CSF is lower in α-$kl^{-/-}$ mice than in wild type mice; regulated PTH secretion is not induced in the knockout (Imura et al. 2007), and the excess calcium is excreted into the urine (Tsuruoka et al. 2006).

The function of the membrane form of α-Kl was discovered due to the similarity of the phenotypes of α-$kl^{-/-}$ mice and *Fgf23* knockout mice (Urakawa et al. 2006).

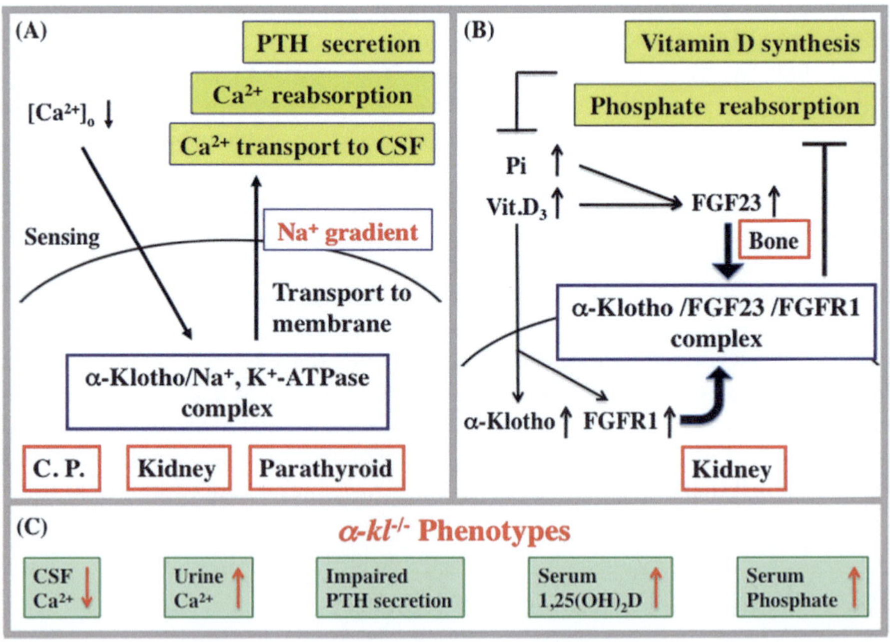

Fig. 10.2 Two α-Klotho functions in mineral homeostasis α-Kl, together with its binding proteins, positively regulates extra-cellular Ca^{2+} decline-induced PTH secretion, Ca^{2+} re-absorption from the urine to the blood and Ca^{2+} transport from the blood to the CSF (**a**). α-Kl negatively controls vitamin D synthesis and phosphate re-absorption in PCT nephrons (**b**). We can detect phenotypes that are consistent with these abnormalities in *α-kl* knockout mice (**c**)

FGF23 is a member of the metabolic/hormonal FGF subfamily, which also includes FGF15 and FGF21 (Itoh and Ornitz 2004). FGF23 is mainly synthesized in osteocytes, secreted into the bloodstream and transported to the kidney. FGF23 directly binds to α-Kl, and the interaction between α-Kl and FGFR1 constructs FGF23/α-Kl/FGFR1 complexes (Kurosu et al. 2006; Tomiyama et al. 2010) (Fig. 10.2). In the kidney, FGF23 suppresses Cyp27b1 gene expression and up-regulates the expression of Cyp24, resulting in the negative feedback regulation of vitamin D synthesis. Because the FGF23 signal is reportedly critical for the suppression of phosphate re-absorption in the kidney (Nakatani et al. 2009), the vitamin D activity and serum phosphate levels are greatly increased in $\alpha\text{-}kl^{-/-}$ mice and *Fgf23* knockout mice (Bhattacharyya et al. 2012; Kovesdy and Quarles 2013) (Fig. 10.2).

The secreted form of α-Kl reportedly acts as a humoral factor and regulates glycoproteins on the cell surface, including ion channels and growth factors (Cheng et al. 2011; Drüeke and Prié 2007; Razzaque 2009). Secreted α-Kl has also been suggested to exert phosphaturic effects independently of FGF-23. Although a growing number of papers have reported the biological and clinical roles of the secreted form of α-Kl, the molecular actions of its secreted form have remained poorly understood and in some cases, controversial. The exact functions of soluble α-Kl are yet to be defined. Most importantly, target proteins with which secreted form of α-Kl interacts need to be identified. Particularly, the proposed/putative α-Kl receptor that attenuates the actions of insulin and/or insulin-like growth factor signaling has not yet been successfully identified, even though the secreted form of α-Kl has been found to extend lifespan (Kurosu et al. 2005). Therefore, this study needs to be carefully followed to conclusively determine the role of secreted α-Kl in life-span elongation.

An ELISA system to measure the serum levels of α-Kl has been established (Yamazaki et al. 2010). A multivariate linear regression analysis that included healthy children and adults demonstrated that the secreted α-Kl levels are significantly higher in children than in adults, and these levels correlated negatively with age and Ca and positively with Pi. Data on the levels of soluble α-Kl in patients with CKD and/or those who are undergoing dialysis are scarce. Kim et al. reported that low circulating α-kl levels were associated with adverse kidney disease outcome, suggesting that α-Kl is a novel biomarker for CKD progression (Kim et al. 2013). In contrast, Seiler et al. reported that the plasma levels of secreted α-Kl were not associated with kidney function or the parameters of calcium-phosphate metabolism, suggesting that the soluble α-Kl levels are not prognostic (Seiler et al. 2013). More data from larger prospective longitudinal studies are required to validate the importance of the secreted form of α-Kl.

10.4 α-Klotho Is a Regulator of Calcium Homeostasis

Calcium homeostasis regulation has been intensively studied in recent decades and was thought to be fully understood (Parfitt and Kleerekoper 1980; Brown et al. 1993; Bringhurst et al. 2011). However, the study of α-Kl is significantly re-shaping our knowledge of calcium homeostasis. Thus, an overview of α-Kl function is summarized in Fig. 10.3 to clarify the importance of α-Kl in the regulation of calcium homeostasis. In response to hypocalcemic stimuli, transepithelial Ca^{2+} transport into the choroid plexus and kidney is triggered by a Na^{+} gradient, which results from the increased plasma-membrane recruitment of the Na^{+}-K^{+}-ATPase complex and quickly adjusts the Ca^{2+} concentrations of the CSF and blood/body fluid. Decreased Ca^{2+} levels also trigger PTH secretion. Secreted PTH increases the serum Ca^{2+} levels and induces vitamin D synthesis, which enhances Ca^{2+} resorption in the bone, Ca^{2+} re-absorption in kidney and Ca^{2+} and phosphate incorporation in the intestine. Increased Ca^{2+} levels suppress trans-epithelial Ca^{2+} transport and PTH secretion. In addition, $1,25(OH)_2D$ inhibits the production of PTH and suppresses Cyp27b1 gene expression in conjunction with the negative feedback action of FGF23. As shown in Fig. 10.3, three systems up-regulate the serum Ca^{2+}. Three types of responses that are triggered by the

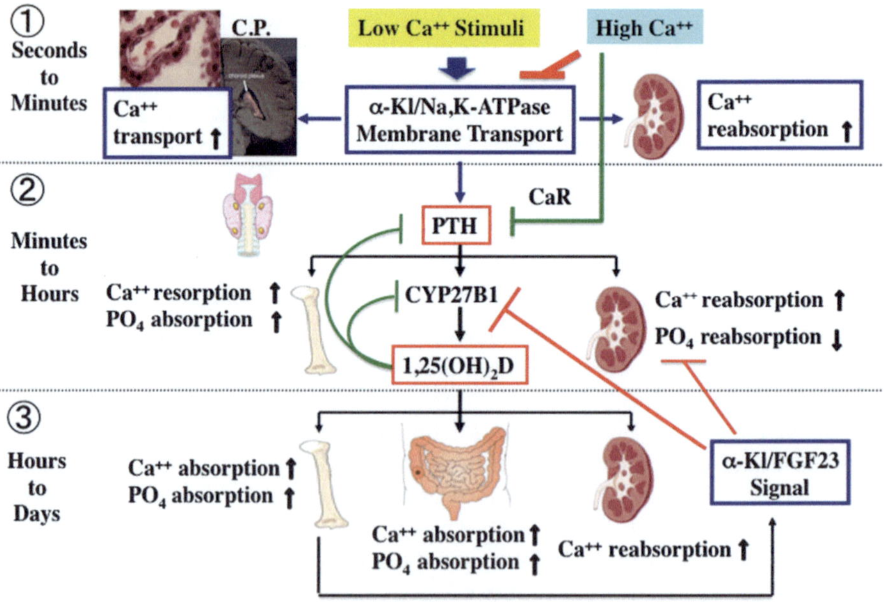

Fig. 10.3 The roles of α-Klotho in the regulation of calcium homeostasis α-Kl binds to the Na^{+}-K^{+}-ATPase complex, and the cell surface recruitment of the Na^{+}-K^{+}-ATPase complex is triggered in response to lowered $[Ca^{2+}]_o$. Increased Na^{+} gradient/electrochemical gradient created by the membrane-trafficked Na^{+}-K^{+}-ATPase complex controls the calcium re-absorption in kidney DCT cells, the calcium transport across the choroid plexus into the CSF, and the PTH secretion in the parathyroid glands. α-Kl, as a co-rector for FGF23/FGFR1 system, is also involved in the signal transduction of Fgf23 that suppresses Cyp27b1 expression in PCT nephrons, leading to the suppression of $1,25(OH)_2D_3$ synthesis. This figure was partly quoted from Fig. 3 published in Nabeshima (2009)

increased membrane trafficking of the Na^+-K^+-ATPase complex (upper panel) are immediately activated after the decline of the extra-cellular Ca^{2+} concentrations and persist for a short time. Thus, they can be placed into the "seconds to minutes order regulation". The PTH-mediated increase in Ca^{2+}, such as Ca^{2+} reabsorption in the kidney and Ca^{2+} resorption in the bone (middle panel), continues for hours and thus constitutes "minutes to hours order regulation". The production of 1,25 $(OH)_2D_3$ and subsequent 1,25 $(OH)_2D_3$-mediated up-regulation of Ca^{2+} (lower panel) continues for hours to days and thus constitutes "hours to days order regulation". Taken together, these findings indicate that complex and reciprocal actions along with feedback mechanisms that require between seconds and day (s) to complete control calcium metabolism. Thus, the extra-cellular calcium concentrations are rapidly adjusted and continuously maintained within strictly narrow ranges. These findings indicate that α-Kl plays critical roles in calcium homeostasis. Thus, we proposed that α-Kl is a regulator of calcium homeostasis.

10.4.1 Major Causes of Multiple Aging Related Phenotypes

Because the phenotypes of *α-kl*-deficient mice largely overlap with those of *Fgf23*-null mice (Urakawa et al. 2006) and the phenotypes of *α-kl* and *Fgf23* double knockout mice are identical to those of their single knockout counterparts (Nakatani et al. 2009), we predicted that the phenotypes detected in $\alpha\text{-}kl^{-/-}$ and $Fgf23^{-/-}$ are caused by abnormalities in the pathway(s) that include α-Kl and FGF23. In fact, the FGF23 secreted from osteocytes is transported to the kidney, where FGF23 binds to α-Kl and leads to the conversion of the canonical FGF receptor 1 (FGFR1) to a receptor that is specific for FGF23 (Urakawa et al. 2006). FGF23 suppresses Cyp27b1 expression and phosphate re-absorption in the kidney. These two negative signals are similarly impaired in $\alpha\text{-}kl^{-/-}$ and *Fgf23*-null mice and lead to increases in the serum levels of $1{,}25(OH)_2D$ and phosphate. This interaction suggests that 1,25(OH)2D overproduction and severe hyperphosphatemia are the common major causes of the tissue-damage phenotypes seen in both $\alpha\text{-}kl^{-/-}$ and $Fgf23^{-/-}$ mice (Bhattacharyya et al. 2012; Kovesdy and Quarles 2013). Consistently, many phenotypes could be prevented by decreasing the 1,25(OH)2D activity via (i) dietary restriction (a regimen in which $\alpha\text{-}kl^{-/-}$ mice are fed a vitamin D-deficient diet) (Tsujikawa et al. 2003) or (ii) the genetic ablation of *Cyp27b1* in $\alpha\text{-}kl^{-/-}$ mice or in $Fgf23^{-/-}$ mice (Ohnishi et al. 2006; Razzaque et al. 2006) and by normalizing the phosphate levels via (iii) the genetic ablation of NaPi-IIa gene in $\alpha\text{-}kl^{-/-}$ mice (Ohnishi and Razzaque 2010) (Fig. 10.4).

Next, we discuss the direct cause of phenotypes and the mechanisms that result in tissue-damage phenotypes. In $\alpha\text{-}kl^{-/-}$ mice, calpain-1, a calcium-dependent cytosolic cysteine protease activated by micromolar concentrations of calcium, is first activated at 2 weeks of age. However, at this stage, α-II-spectrin, a critical cytoskeletal target of calpain-1, remains almost intact. At 3 weeks of age, the abnormally activated calpain-1 has largely digested calpastatin, an endogenous

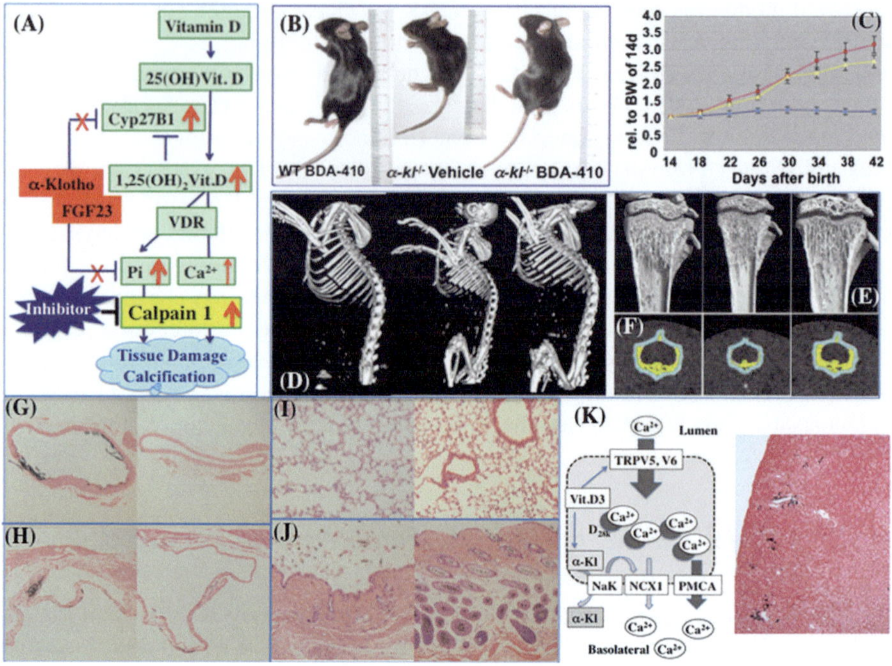

Fig. 10.4 α-Klotho-deficient morphological phenotypes are primarily due to Calpain 1 activation. (**a**) A common pathway causes similar phenotypes in α-$kl^{-/-}$ mice and $Fgf^{-/-}$ mice. The administration of Calpain-1 inhibitor ameliorates α-kl-deficient phenotypes. (**b**) The growth features of WT, KO and rescued KO mice. (**c**) Body weight gains of WT (*Red*), KO (*Blue*) and rescued KO (*Yellow*) mice. (**d–f**) μCT imaging pictures of WT (*left*), KO (*center*) and rescued KO (*right*) mice. (**g–j**)are histological examinations of aorta (Kossa stain), heart valve (Kossa stain), lung, and skin from KO mice (*left*) and rescued KO mice (*right*), respectively. (**k**) Ca^{2+} re-absorption mechanism (*left*) in DCT nephrons and very small calcium depositions in the cortical layers of kidneys from KO mice (*right*). In this figure, I re-arranged figures that were published in Nabeshima et al. (2014)

calpain-1 inhibitor, and begins to cleave α-II-spectrin, which results in cellular and tissue damage (Manya et al. 2002). This mechanism may be responsible for the manifestation of tissue abnormalities in α-$kl^{-/-}$ mice starting at 3 weeks of age (Kuro-o et al. 1997; Imura et al. 2007; Tsujikawa et al. 2003) and is consistent with the idea that calpain-1 activation is a major and direct cause of tissue damage in α-$kl^{-/-}$ mice. This idea was confirmed by analyzing the effects of a calpain-1 inhibitor on the phenotypes of α-$kl^{-/-}$ mice (Nabeshima et al. 2014).

As shown in Fig. 10.4, the administration of a Calpain-1 inhibitor clearly ameliorated the abnormalities in body size, body weight gains and the growth structures of α-$kl^{-/-}$ mice. Concomitantly, atrophic changes in these organs in α-$kl^{-/-}$ mice were all ameliorated in proportion to the recovery of whole-body size and weight in inhibitor-treated animals. The administrations of inhibitor also rescued the histological abnormalities of $\alpha\text{-}kl^{-/-}$ mice, such as the senile atrophy of the skin, pulmonary emphysema, ectopic calcification and arteriosclerosis. Specifically, cardiovascular calcification in the aorta and the calcification-associated thickening of vessel walls were no longer detectable after calpain-1 inhibitor administration, and heart valve calcifications were vastly improved (Nabeshima et al. 2014). Because $\alpha\text{-}kl^{-/-}$ mice are a good model of the emphysematous changes often seen in aging populations (Sato et al. 2007), the capain-1 activity may constitute a novel therapeutic target for the treatment of pulmonary emphysema and chronic obstructive pulmonary disease (COPD). The characteristic features of skin aging in $\alpha\text{-}kl^{-/-}$ mice mimic senile atrophoderma in aging humans (Calleja-Agius et al. 2007) and were markedly improved by calpain-1 inhibitor treatment (Fig. 10.4). μCT imaging clearly showed the characteristic bone phenotypes of $\alpha\text{-}kl^{-/-}$ mice, such as a generalized decrease in the bone mineral density, significantly reduced trabecular and cortical thickness, lower trabecular and cortical bone volume, a severely reduced trabecular bone fraction, malformation of the spongy bone, the structural deformities of bones and the irregular/rough appearances of bones (Nabeshima et al. 2014) (Fig. 10.4). Calpain-1 inhibitor administration restored these abnormalities to the wild-type levels. However, the administration of inhibitor did not improve the increases in the serum calcium, phosphate and 1,25 $(OH)_2D$ levels, suggesting that the abnormal activation of calpain-1 and consequent tissue damages are downstream of the onset of hyper-vitaminosis D and hyper-phosphatemia (Nabeshima et al. 2014). Because $1{,}25(OH)_2D_3$ administration reportedly induces Ca^{2+}-mediated apoptosis in adipocytes and breast cancer cells via increased calcium influx and the subsequent activation of calpain-1 and caspase-12 (Sergeev 2005, 2009), the overproduction of $1{,}25(OH)_2D$ can reasonably be speculated to disrupt physiological processes that regulate calpain-1 activity, which results in failure to maintain cellular and tissue integrity. Therefore, the modulation of calpain-1 activity may prove useful in the alleviation of aging-related syndromes. However, the therapeutic use of calpain inhibitors is currently associated with limitations (Goll et al. 2003; Campbell and Davies 2012). Nevertheless, calpains are likely to cleave polypeptides at limited sites, which may limit the undesired side effects of calpain inhibitors (Wang and Yuen 1994; Lee et al. 2000). However, tissue-, cell- and/or target-calpain type-specific inhibition remains desirable to minimize unexpected side effects in the treatment of aging-related diseases.

Tiny calcium precipitates were detected in the cortical layer of nephron segments in $\alpha\text{-}kl^{-/-}$ mice, even after the administration of Calpain-1 inhibitor (Fig. 10.4). The small calcium precipitates were detected at the nephron segments that expressed the LacZ-neomycin fusion protein gene (β-geo gene) which was integrated directly downstream of the translational initiation codon of the α-kl gene (Takeshita et al. 2004). Similar patterns of calcium precipitates have been reported within calbindin-D28K-positive nephrons in $\alpha\text{-}kl^{-/-}$ mice (Alexander et al. 2009). Because *calbindin-*D_{28K} and *α-kl* are reportedly co-expressed with *TRPV5 and NCX-1* in the nephron segments responsible for regulated Ca^{2+} re-absorption in the kidney (Imura et al 2007; Hoenderop et al. 2005), small calcium precipitates are expected in the nephrons responsible for regulated Ca^{2+} re-absorption. In these cells, Ca^{2+} in the urine enters cells at the luminal membrane (apical) via TRPV5 or TRPV6 and is sequestered by calbindin-D_{28K} or -D_{9K}. Bound Ca^{2+} then diffuses to the basolateral cell surface, where Ca^{2+} ions are extruded into the blood compartment via NCX-1 and plasma membrane calcium ATPase 1b (PMCA1b). 1,25 $(OH)_2D$ enhances the expression and function of TRPV5 and thereby up-regulates Ca^{2+} re-absorption/influx (Hoenderop et al. 2005). At the basolateral side, α-Kl has been proposed to play a critical role in the efflux of Ca^{2+} by regulating the plasma-membrane recruitment of the Na^+-K^+-ATPase complex (Imura et al. 2007). Therefore, the overproduction of 1,25$(OH)_2D$ and α-Kl deficiency likely lead to the excessive influx of Ca^{2+} and down-regulated efflux of Ca^{2+}, respectively, resulting in the overflow of Ca^{2+} in the cytoplasm. This action may be responsible for the calcium precipitates that specifically accumulated in the distal nephron segments in α-$kl^{-/-}$ mice but not in *cyp27b1*$^{-/-}$/α-$kl^{-/-}$ mice, in which 1,25$(OH)_2D$ production is down-regulated (Ohnishi et al. 2006).

10.5 α-Klotho and FGF23 in the Pathogeneses of CKD

Notably, the aging-related phenotypes seen in α-$kl^{-/-}$ mice are very similar to many of the complications that develop in patients suffering from advanced chronic kidney disease (CKD) (London and Drueke 1997; Rostand and Drueke 1999; Milliner et al. 1990; Urena and De Vernejoul 1999; Kuizon and Salusky 1999). This similarity is further supported by the following: (i) the expression of *α-kl* mRNA and α-Kl protein are severely reduced in these patients (Koh et al. 2001); (ii) high serum phosphate, the major cause of abnormalities of α-$kl^{-/-}$ mice, is reportedly closely associated with high levels of cardiovascular disease morbidity and mortality in patients with CKD, particularly in patients with end-stage renal disease (Mathew et al. 2008; Komaba and Fukagawa 2012; Quarles 2012); and (iii) defects in FGF23 (Stubbs et al. 2007) and α-Kl (Kuro-o et al. 1997), together with the dysregulation of endogenous anti-calcification factors (Wallin et al. 2001; Proudfoot and Shanahan 2006; Moe et al. 2005; Mellgren and Huang 2007), are considered to play an important role in cardiovascular calcification, a dire complication of CKD. These observations suggest that α-Kl and FGF23 are involved in the pathogeneses of not only aging-related syndromes but also the complications of CKD. Thus, α-Kl, FGF23 and their downstream molecules are candidate targets for therapeutic approaches that intend to ameliorate or delay age-related syndromes and CKD complications.

As predicted from the report that cardiovascular calcification is an active process that recapitulates regulated osteogenesis (Sage et al. 2010; Shao et al. 2010), the induced expressions of *Fgf23* and osteogenesis-related genes, Runx2, Osteopontin, and RANKL were identified in the calcified arteries of α-$kl^{-/-}$ mice (Fig. 10.5). However, they were not detected when ectopic calcification was blocked with calpain-1 inhibitor treatment and in the arteries of young α-$kl^{-/-}$ mice (2 weeks of age) prior to the development of cardio-vascular calcification, which suggests

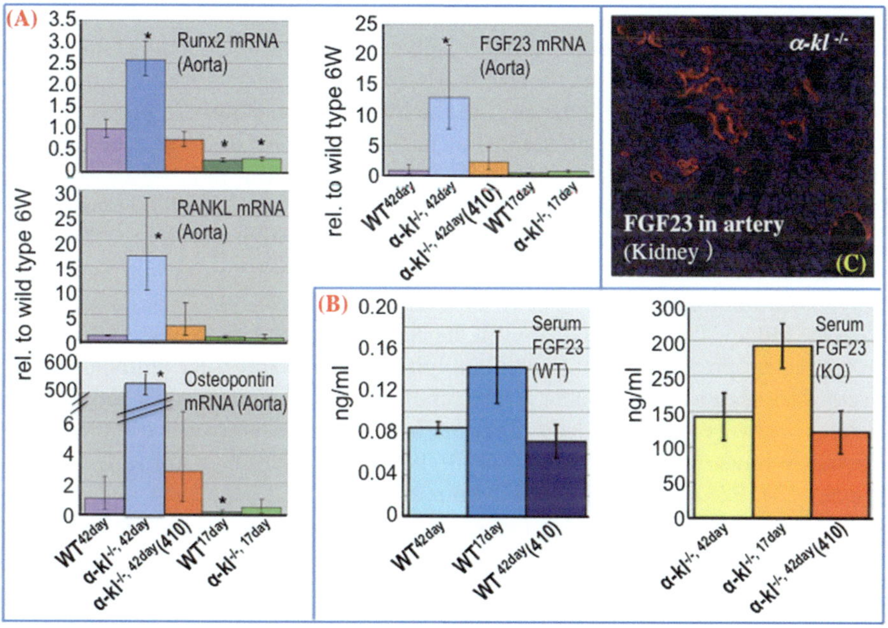

Fig. 10.5 Expression of Runx2, RANKL, Osteopontin and FGF23 in calcified arteries. (**a**) The expressions of osteogenesis genes and FGF23 in the calcified arteries of α-kl$^{-/-}$ mice. (**b**) Serum levels of FGF23 in wild type, α-kl$^{-/-}$ mice and rescued α-kl$^{-/-}$ mice. (**c**) FGF23 expression in the kidney of α-kl$^{-/-}$ mice. The figures shown are quoted from Nabeshima et al. (2014)

that the ectopic expression of FGF23 is induced in calcified arteries in parallel with the progression of cardio-vascular calcification. The ectopic expression of FGF23 at least partly mirrors the progression of cardiovascular calcification, a major risk factor for cardiovascular mortality and thus may be responsible for the close association between the increased plasma levels of FGF23 and high cardiovascular mortality, particularly in late- to end-stage CKD patients (Isakova et al. 2011; Kendrick and Chonchol 2011).

10.6 Molecular Functions of α-Klotho

The extracellular domain of α-Kl contains two internal repeats (α-KL1 and α-KL2) that are very similar to the family 1 β-glycosidase. Members of the family 1 - β-glycosidase operate according to a molecular mechanism that leads to the overall retention of the anomeric configuration and involves the formation and breakdown of a covalent glycosyl enzyme intermediate. Two conserved active center sequences are involved in this catalytic mechanism. Particularly, two glutamic acid (Glu) residues at the putative active centers are highly conserved and are believed to be critical for catalytic action; one acts as a nucleophile and the other acts as an acid/base (Rye and Withers 2000; McCarter and Withers 1994). Intriguingly, unlike other members of family 1 β-glycosidase, the nucleophilic Glu residue is present in the α-KL1 region (N-terminal domain), but the acid/base Glu residue is replaced by Asn residue. Inversely, the acid/base Glu residue is present in the α-KL2 region (C-terminal domain), but the nucleophile Glu is replaced by Ala and Ser residues in mouse and human α-Kl, respectively (Kuro-o et al. 1997) (Fig. 10.6). Thus, the enzymatic action of α-Kl has been questioned. Nonetheless, α-Kl was found to exhibit subtle but specific β-glucuronidase activity (Tohyama et al. 2004). This activity suggests α-Kl may have dual functions as an enzyme or as a glycoside-binding protein.

In the kidney, certain proportions of the Na^{+}-K^{+}-ATPase β-subunit, FGF-R1 and several other proteins are immuno-precipitated using anti-α-Kl antibodies. Our recent data suggest that specific N-glycan chains are commonly attached to these proteins. However, FGF23 does not contain this type of N-glycan, even though it efficiently binds to α-Kl. Instead, an O-glycan was identified in the C-terminal region of FGF23. The O-glycan is required for the α-Kl-dependent kidney accumulation of FGF23, critical for the stabilized/facilitated interaction of α-Kl and FGF23 (α-Kl efficiently binds to the native form of FGF23 that contains this O-glycan but weakly to non-glycosylated FGF23 synthesized in E. coli), necessary for FGF23 signal transduction at physiological doses and directly bound to α-Kl. When α-Kl was co-incubated with the O-glycan, the binding affinities of

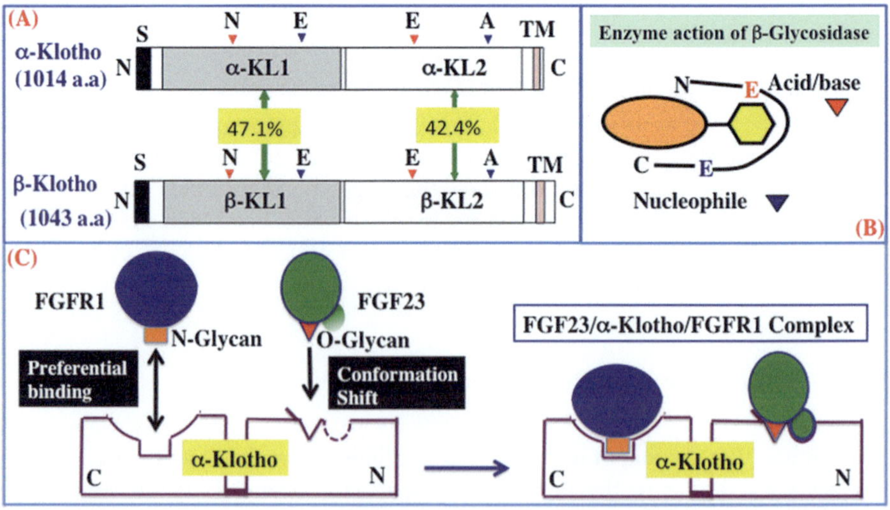

Fig. 10.6 The molecular roles of O-glycan and N-glycan in protein-protein interaction. (**a**) Schematic structures of α-Klotho and β-Klotho. "E" shows the glutamate residue of the putative active center. N and A are asparagine and alanine residues, respectively. (**b**) Schematic representation of enzyme action mechanism of β-glycosidase. (**c**) Predicted model of FGF23/α-Kl/FGFR1 complex formation. The distinct roles of N-glycan and O-glycan are proposed in this model

non-glycosylated FGF23$^{E.coli}$ to O-glycan-docked-α-Kl were dose-dependently strengthened and reached the affinity levels between α-Kl and O-glycosylated FGF23. These findings suggest an unexpected role for O-glycan in the interaction between α-Kl and FGF23. Because co-incubated glycan does not covalently bind to FGF23, glycan can be reasonably assumed to induce conformational changes at the binding site and affect the conformation of α-Kl at wider regions, sifting α-Kl toward high-affinity states for FGF23$^{E.coli}$. This prediction was further supported by calorimetric analyses. O-glycan, by binding to α-Kl, increases the thermodynamic stability of α-Kl and induces the conformational change of α-Kl, which may convert the surface structure of α-Kl from a hydrophobic state to a hydrophilic state. Figure 10.6 summarizes the roles of N-glycan and O-glycan in the ternary complex formation of FGF23, α-Kl and FGFR1. FGFR1, which contains a specific N-glycan, preferentially and selectively binds to α-Kl. In contrast, the O-glycan of FGF23 induces a conformational shift of α-Kl that leads to the stable binding of FGF23 and α-Kl. Taken together, these results support our proposal that α-Kl functions as a glycoside-binding protein.

References

Alexander RT, Woudenberg-Vrenken TE, Buurman J, Dijkman H, van der Eerden BC, van Leeuwen JP, Bindels RJ, Hoenderop JG (2009) Klotho prevents renal calcium loss. J Am Soc Nephrol 20:2371–2379

Bhattacharyya N, Chong WH, Gafni RI, Collins MT (2012) Fibroblast growth factor 23: state of the field and future directions. Trends Endocrinol Metab 23:610–618

Blaustein MP, Lederer WJ (1999) Sodium/calcium exchange: its physiological implications. Physiol Rev 79:763–854

Bringhurst FR, Demay MB, Kronenberg H (2011) Hormones and disorders of mineral metabolism. In: Bringhurst FR, Demay MB, Kronenberg HM (eds) William's textbook of endocrinology, 12th edn. Elsevier Saunders, Philadelphia, pp 1237–1304

Brown EM, Gamba G, Riccardi D, Lombardi M, Butters R, Kifor O et al (1993) Cloning and characterization of an extracellular Ca^{2+} sensing receptor from bovine parathyroid. Nature 366:575–580

Brzobohaty B, Moore I, Kristoffersen P, Bako L, Campos N, Schell J, Palme K (1993) Release of active cytokinin by a beta-glucosidase localized to the maize root meristem. Science 262:1051–1054

Calleja-Agius J, Muscat-Baron Y, Brincat MP (2007) Skin ageing. Menopause Int 13:60–64

Campbell RL, Davies PL (2012) Structure-function relationships in calpains. Biochem J 447:335–351

Cheng CY, Kuro-o M, Razzaque MS (2011) Molecular regulation of phosphate metabolism by fibroblast growth factor-23-klotho system. Adv Chronic Kidney Dis 18:91–97

Davies G, Henrissat B (1995) Structures and mechanisms of glycosyl hydrolases. Structure 3:853–859

Drüeke TB, Prié D (2007) Klotho spins the thread of life – what does klotho do to the receptors of fibroblast growth factor-23 (FGF23)? Nephrol Dial Transplant 22:1524–1526

Goll DE, Thompson VF, Li H, Wei W, Cong J (2003) The calpain system. Physiol Rev 83:731–801

Gopalan V, Pastuszyn A, Galey WR Jr, Glew RH (1992) Exolytic hydrolysis of toxic plant glucosides by guinea pig liver cytosolic beta-glucosidase. J Biol Chem 267:14027–14032
Henrissat B (1991) A classification of glycosyl hydrolases based on amino acid sequence similarities. Biochem J 280:309–316
Henrissat B, Bairoch A (1993) New families in the classification of glycosyl hydrolases based on amino acid sequence similarities. Biochem J 293:781–788
Henrissat B, Bairoch A (1996) Updating the sequence-based classification of glycosyl hydrolases. Biochem J 316:695–696
Herscovics A (1999) Importance of glycosidases in mammalian glycoprotein biosynthesis. Biochim Biophys Acta 1473:96–107
Hoenderop JGJ, Nilius B, Bindels RJM (2005) Calcium absorption across epithelia. Physiol Rev 85:373–422
Imura A, Iwano A, Kita N, Thoyama O, Fujimori T, Nabeshima Y (2004) Secreted klotho protein in sera and CSF: implication for post-translational cleavage in release of klotho protein from cell membrane. FEBS Lett 565:143–147
Imura A, Tsuji Y, Murata M, Maeda R, Kubota K, Iwano A, Obuse C, Togashi K, Tominaga M, Kita N, Tomiyama K, Iijima J, Nabehsima Y, Fujioka M, Asato R, Tanaka S, Kojima K, Ito J, Nozaki K, Hashimoto N, Ito T, Nishio T, Uchiyama T, Fujimori T, Nabehsima Y (2007) Alpha-klotho as a regulator of calcium homeostasis. Science 316:1615–1618
Isakova T, Wahl P, Vargas GS, Gutiérrez OM, Scialla J, Xie H, Appleby D, Nessel L, Bellovich K, Chen J, Hamm L, Gadegbeku C, Horwitz E, Townsend RR, Anderson CA, Lash JP, Hsu CY, Leonard MB, Wolf M (2011) Fibroblast growth factor 23 is elevated before parathyroid hormone and phosphate in chronic kidney disease. Kidney Int 79:1370–1378
Ito S, Fujimori T, Furuya A, Satoh J, Nabeshima Y, Nabeshima Y (2005) Impaired negative feedback suppression of bile acid synthesis in mice lacking betaklotho. J Clin Invest 115:2202–2208
Itoh N, Ornitz DM (2004) Evolution of the Fgf and Fgfr gene families. Trends Genet 20:563–569
Kendrick J, Chonchol M (2011) The role of phosphorus in the development and progression of vascular calcification. Am J Kidney Dis 58:826–834
Kim HR, Nam BY, Kim DW, Kang MW, Han JH, Lee MJ, Shin DH, Doh FM, Koo HM, Ko KI, Kim CH, Oh HJ, Yoo TH, Kang SW, Han DS, Han SH (2013) Circulating α-klotho levels in CKD and relationship to progression. Am J Kidney Dis 61:899–909
Koh N, Fujimori T, Tamori A, Nishiguchi S, Shiomi S, Nakatani T, Sugimura K, Kishimoto T, Kuroki T, Nabeshima Y (2001) Severely reduced expression of klotho gene in human chronic renal failure kidney. Biochem Biophys Res Commun 280:1015–1020
Komaba H, Fukagawa M (2012) The role of FGF23 in CKD – with or without klotho. Nat Rev Nephrol 8:484–490
Kovesdy CP, Quarles LD (2013) Fibroblast growth factor-23: what we know, what we don't know, and what we need to know. Nephrol Dial Transplant 28(9):2228–2236
Kuizon BD, Salusky IB (1999) Growth retardation in children with chronic renal failure. J Bone Miner Res 14:1680–1690
Kuro-o M, Matsumura Y, Aizawa H, Kawaguchi H, Suga T, Utsugi T, Ohyama Y, Kaname T, Kume E, Iwasaki H, Iida A, Shiraki-Iida T, Nishikawa S, Ryozo N, Nabeshima Y (1997) Mutation of the mouse *klotho* gene leads to a syndrome resembling ageing. Nature 390:45–51
Kurosu H, Yamamoto M, Clark AJD, Pastor JV, Nandi A, Grunani P et al (2005) Suppression of aging in mice by the hormone klotho. Science 309:1829–1833
Kurosu H, Ogawa Y, Miyoshi M, Yamamoto M, Nandi A, Rosenblatt KP, Baum MG, Schiavi S, Hu MC, Moe OW, Kuro-o M (2006) Regulation of fibroblast growth factor-23 signaling by klotho. J Biol Chem 281:6120–6123
LaMarco KL, Glew RH (1986) Hydrolysis of a naturally occurring beta-glucoside by a broad-specificity beta-glucosidase from liver. Biochem J 237:469–476
Lee MS, Kwon YT, Li M, Peng J, Friedlander RM, Tsai LH (2000) Neurotoxicity induces cleavage of p35 to p25 by calpain. Nature 405:360–364

London GM, Drueke TB (1997) Atherosclerosis and arteriosclerosis in chronic renal failure. Kidney Int 51:1678–1695
Lytton J (2007) Na./Ca. exchangers: three mammalian gene families control C^{2+} transport. Biochem J 406:365–382
Manya H, Fujimori T, Nabeshima Y, Endo T (2002) Klotho protein deficiency leads to overactivation of (Mu)-calpain. J Biol Chem 277:35503–35508
Mathew S, Tustison KS, Sugatani T, Chaudhary LR, Rifas L, Hruska KA (2008) The mechanism of phosphorus as a cardiovascular risk factor in CKD. J Am Soc Nephrol 19:1092–1105
McCarter JD, Withers SG (1994) Curr Opin Struct Biol 4:885–892
Mellgren RL, Huang X (2007) Fetuin a stabilizes μ-calpain and facilitates plasma membrane repair. J Biol Chem 282:35868–35877
Milliner DS, Zinsmeister AR, Lieberman E, Landing B (1990) Soft tissue calcification in pediatric patients with end-stage renal disease. Kidney Int 38:931–936
Moe SM, Reslerova M, Ketteler M, O'neill K, Duan D, Koczman J, Westenfeld R, Jahnen-Dechent W, Chen NX (2005) Role of calcification inhibitors in the pathogenesis of vascular calcification in chronic kidney disease (CKD). Kidney Int 67:2295–2304
Nabeshima Y (2009) Discovery of α-Klotho unveiled new insights into calcium and phosphate homeostasis. Proc Jpn Acad Ser B 85:125–141
Nabeshima Y, Washida M, Tamura M, Maeno A, Ohnishi M, Shiroishi T, Imura A, Razzaque MS, Nabeshima Y (2014) Calpain 1 inhibitor BDA-410 ameliorates α-klotho-deficiency phenotypes resembling human aging-related syndromes. Sci Rep 4:5847
Nakatani T, Sarraj B, Ohnishi M, Densmore MJ, Taguchi T, Goetz R, Mohammadi M, Lanske B, Razzaque MS (2009) In vivo genetic evidence for klotho-dependent, fibroblast growth factor 23 (Fgf23) -mediated regulation of systemic phosphate homeostasis. FASEB J 23:433–441
Ohnishi M, Razzaque MS (2010) Dietary and genetic evidence for phosphate toxicity accelerating mammalian aging. FASEB J 24:3562–3571
Ohnishi M, Nakatani T, Lanske B, Razzaque MS (2006) Reversal of mineral ion homeostasis and soft-tissue calcification of klotho knockout mice by deletion of vitamin D 1α-hydroxylase. Kidney Int 75:1166–1172
Parfitt AM, Kleerekoper M (1980) The divalent ion homeostatic system: physiology and metabolism of calcium, phosphorus, magnesium, and bone. In: Maxwell MH, Kleeman CR (eds) Clinical disorders of fluid and electrolyte metabolism, 3rd edn. McGraw-Hill, New York, pp 269–398
Proudfoot D, Shanahan CM (2006) Molecular mechanisms mediating vascular calcification: role of matrix Gla protein. Nephrology 11:455–461
Quarles LD (2012) Role of FGF23 in vitamin D and phosphate metabolism: implications in chronic kidney disease. Exp Cell Res 318:1040–1048
Rask L, Andreasson E, Ekbom B, Eriksson S, Pontoppidan B, Meijer J (2000) Myrosinase: gene family evolution and herbivore defense in brassicaceae. Plant Mol Biol 42:93–113
Razzaque MS (2009) The FGF23-Klotho axis: endocrine regulation of phosphate homeostasis. Nat Rev Endocrinol 5:611–619
Razzaque MS, Sitara D, Taguchi T, St-Arnaud R, Lanske B (2006) Premature aging-like phenotype in fibroblast growth factor 23 null mice is a vitamin D mediated process. FASEB J 20:720–722
Rostand SG, Drueke TB (1999) Parathyroid hormone, vitamin D, and cardiovascular disease in chronic renal failure. Kidney Int 56:383–392
Rye CS, Withers SG (2000) Curr Opin Chem Biol 4:573–580
Sage AP, Tintut Y, Demer LL (2010) Regulatory mechanisms in atherosclerotic calcification. Nat Rev Cardiol 7:528–536
Sato A, Hirai T, Imura A, Kita A, Iwano A, Muro S, Nabeshima Y, Suki B, Mishima M (2007) Morphological mechanism of the development of pulmonary emphysema in klotho mice. Proc Natl Acad Sci U S A 104:2361–2365
Seiler S, Wen M, Roth HJ, Fehrenz M, Flügge F, Herath E, Weihrauch A, Fliser D, Heine GH (2013) Plasma klotho is not related to kidney function and does not predict adverse outcome in patients with chronic kidney disease. Kidney Int 83:121–128

Sergeev IN (2005) Calcium signaling in cancer and vitamin D. J Steroid Biochem Mol Biol 97:145–151

Sergeev IN (2009) 1,25-Dihydroxyvitamin D3 induces Ca^{2+}-mediated apoptosis in adipocytes via activation of calpain and caspase-12. Biochem Biophys Res Commun 384:18–21

Shao JS, Cheng SL, Sadhu J, Towler DA (2010) Inflammation and the osteogenic regulation of vascular calcification: a review and perspective. Hypertension 55:579–592

Skou JC (1988) The Na., K.-pump. Methods Enzymol 156:1–25

Stubbs JR, Liu S, Tang W, Zhou J, Wang Y, Yao X, Quarles LD (2007) Role of hyperphosphatemia and 1,25-dihydroxyvitamin D in vascular calcification and mortality in fibroblastic growth factor 23 null mice. J Am Soc Nephrol 18:2116–2124

Takeshita K, Fujimori T, Kurotaki Y, Honjo H, Tsujikawa H, Yasui K, Lee J-K, Kamiya K, Kitaichi K, Yamamoto K, Ito M, Kondo T, Iino S, Inden Y, Hirai M, Murohara T, Kodama I, Nabeshima Y (2004) Sinoatrial node dysfunction and early unexpected death of mice with a defect of *klotho* gene expression. Circulation 109:1776–1782

Tohyama O, Imura A, Iwano A, Freund JN, Henrissat B, Fujimori T, Nabeshima Y (2004) Klotho is a novel β-glucuronidase capable of hydrolyzing steroid β-glucuronides. J Biol Chem 273:9777–9784

Tomiyama K, Maeda R, Urakawa I, Yamazaki Y, Tanaka T, Ito S, Nabeshima Y, Tomita T, Odori S, Hosoda K, Nakao K, Imura A, Nabeshima Y (2010) Relevant use of Klotho in FGF19 subfamily signaling system *in vivo*. Proc Natl Acad Sci U S A 107:1666–1671

Tsujikawa H, Kurotaki Y, Fujimori T, Fukuda K, Nabeshima Y (2003) Klotho, a gene related to a syndrome resembling human premature aging, functions in a negative regulatory circuit of vitamin D endocrine system. Mol Endocrinol 17:2393–2403

Tsuruoka S, Nishiki K, Ioka T, Ando H, Saito Y, Kurabayashi M, Nagai R, Fujimura A (2006) Defect in parathyroid-hormone-induced luminal calcium absorption in connecting tubules of klotho mice. Nephrol Dial Transplant 21:2762–2767

Urakawa I, Yamazaki Y, Shimada T, Iijima K, Hasegawa H, Okawa K, Fujita T, Fukumoto S, Yamashita T (2006) Klotho converts canonical FGF receptor into a specific receptor for FGF23. Nature 444:770–774

Urena P, De Vernejoul MC (1999) Circulating biochemical markers of bone remodeling in uremic patients. Kidney Int 55:2141–2156

Wallin R, Wajih N, Greenwood GT, Sane DC (2001) Arterial calcification: a review of mechanisms, animal models, and the prospects for therapy. Med Res Rev 21:274–301

Wang KK, Yuen PW (1994) Calpain inhibition: an overview of its therapeutic potential. Trends Pharmacol Sci 15:412–419

Wells L, Vosseller K, Hart GW (2001) Glycosylation of nucleocytoplasmic proteins: signal transduction and O-GlcNAc

Wittstock U, Halkier BA (2002) Glucosinolate research in the arabidopsis era. Trends Plant Sci 7:263–270

Yamazaki Y, Imura A, Urakawa I, Shimada T, Murakami J, Aono Y, Hasegawa H, Yamashita T, Nakatani K, Saito Y, Okamoto N, Kurumatani N, Namba N, Kitaoka T, Ozono K, Sakai T, Hataya H, Ichikawa S, Imel EA, Econs MJ, Nabeshima Y (2010) Establishment of sandwich ELISA for soluble alpha-klotho measurement: age-dependent change of soluble alpha-klotho levels in healthy subjects. Biochem Biophys Res Commun 398:513–518

Yoshida T, Fujimori T, Nabeshima Y (2002) Mediation of unusually high concentrations of 1,25-dihydroxyvitamin D3 in homozygous klotho mutant mice by increased expression of renal 1alpha-hydroxylase gene. Endocrinology 143:683–689

Chapter 11
Role of the Forkhead Box O Family and Neuropeptide Y in Calorie Restriction

Ryoichi Mori, Seongjoon Park, and Isao Shimokawa

Abstract Calorie restriction (CR) is known to extend lifespan and reduce the risk of various morbidities in lower animals to primates. The effects of CR are regulated by the expression of several age-related genes and a signaling pathway, including forkhead box O (FOXO) and neuropeptide Y (NPY). Heterozygous Foxo1-deficient mice under CR exhibit alterations in cancer development but not lifespan, indicating that FOXO1 contributes to anti-cancer effects in aging. NPY deficiency in mice under CR attenuates lifespan and adiposity but does not affect anticancer efficacy and oxidant stress tolerance, suggesting that NPY might play a role in the regulation of age-related metabolic diseases. Furthermore, intracellular organelles including mitochondria contribute to CR benefits. Recently, CR effects have been examined in individuals, leading to development of CR mimetic compounds. This review summarizes the current insights into the molecular mechanisms linking regulation of age-related genes and CR benefits in mammals.

Keywords Calorie restriction • Calorie restriction mimetics • Longevity • FOXO • NPY

11.1 Introduction

The most effective method to delay aging and extend lifespan is CR. Conversely, it is known that overeating accelerates the onset of lifestyle-related diseases and aging. The reason for these observations is that regulation of aging and longevity are maintained by a cooperative metabolic control network between various distant organs. These energy and metabolism regulation mechanisms (e.g., feeding regulation and fat accumulation) involve various endocrine, autocrine, and paracrine systems in each internal organ, which are systematized in individuals.

On the other hand, cells stimulated by nutrients such as glucose and endocrine factors actively perform bioactive phenomenon including metabolism, stress

R. Mori • S. Park • I. Shimokawa (✉)
Department of Pathology, Graduate School of Biomedical Sciences, Nagasaki University, 1-12-4 Sakamoto, Nagasaki 852-8523, Japan
e-mail: shimo@nagasaki-u.ac.jp

© Springer Japan 2015

N. Mori, I. Mook-Jung (eds.), *Aging Mechanisms*,
DOI 10.1007/978-4-431-55763-0_11

responses, DNA repair, and epigenetic control to regulate homeostasis. Therefore, it is thought that the life of an individual is maintained by relationships between intracellular metabolism and the metabolism of internal organs, and their breakdown induces aging and morbidity. Taken together, the balance of energy consumption as an individual, such as CR, maintains low stress in internal organs. Furthermore, it is thought that CR leads to longevity and disease onset prevention.

McCay and colleagues reported that CR led to lifespan extension, although the molecule mechanism was unknown (McCay et al. 1935). In studies linking molecular mechanisms with longevity and gene regulation, Friedman and Johnson identified *age-1* as an age-related gene [phosphoinositide 3-kinase (PI3K) homolog in mammals] (Friedman and Johnson 1988) because *Caenorhabditis elegans* (*C. elegans*) with mutant *age-1* lived longer than controls, suggesting that a single gene might regulate lifespan. Analyses of *age-1* and insulin receptor homolog *daf-2* showed that the insulin pathway, which is involved in glucose metabolism, might regulate aging and longevity (Kenyon et al. 1993; Kimura et al. 1997). Subsequently, it was shown that the insulin-like growth factor (IGF)/insulin pathway participated in the lifespan of *Drosophila melanogaster* (*D. melanogaster*) and *Mus musculus* (*M. musculus*). These findings indicate that (1) a single gene influences lifespan and (2) lifespan control is evolutionarily conserved from yeast and lower organisms such as *C. elegans* to mammals such as mice and humans.

It is clear that the IGF/insulin pathway related to longevity has been conserved during evolution. Activation of the IGF/insulin pathway can be controlled by CR in mammals. Molecules in the IGF/insulin pathway have been studied extensively. Of the various molecular groups identified thus far, the FOXO family and NPY are particularly interesting. Therefore, in this review, we will provide an outline of the molecular mechanisms of the FOXO family and NPY that are at the core of present research into aging.

11.2 Role of the Foxo Family in Aging and Homeostasis

A dwarf mouse, which exhibits a low serum concentration of IGF-1 because of inadequate secretion of growth hormone (GH), shows a phenotype of longevity (Bartke and Brown-Borg 2004). We reported that overexpression of GH antisense oligodeoxynucleotides in rats is longevity concomitant with attenuation of the concentrations of IGF-1 and insulin (Shimokawa et al. 2002). The IGF/insulin pathway participates in activation of the FOXO transcription factor family (FOXO1, FOXO3, FOXO4, and FOXO6 in mammals) through the PI3K-protein kinase B/thymoma viral proto-oncogene (AKT) pathway. FOXO family members control glucagon production and glucose metabolism. Therefore, it has been thought that the FOXO family participates in aging under CR, and its molecular mechanisms have been clarified (Nakae et al. 2008).

Mutation of the *daf-2* gene encoding the IGF receptor in *C. elegans* extends lifespan (Kenyon et al. 1993; Kimura et al. 1997). Lifespan extension through an

IGF pathway is indispensable to increase activation of *daf-16* that is known as Foxo in *C. elegans* (Ogg et al. 1997; Lin et al. 1997). Heterozygous type 1 IGF receptor-deficient mice and adipose tissue-specific insulin receptor-deficient mice have extended lifespans (Holzenberger et al. 2003; Bluher et al. 2003), suggesting that Foxo, which is downstream in the IGF pathway, participates in lifespan extension.

The lifespan of Foxo1/Foxo3a/Foxo4 triple-conditional deficient mice is shorter, but each Foxo conditional deficient mouse shows no alteration in lifespan (Paik et al. 2007), suggesting compensation by other FOXO proteins (only one kind of FOXO protein exists in *C. elegans*). Taken together, a single FOXO protein is involved in lifespan extension, although a FOXO isoform is considered to synergistically participate in lifespan extension because each Foxo conditional deficient mouse shows no alteration in lifespan.

Foxo1 deficiency ($Foxo1^{-/-}$) in embryonic mice has been shown to be lethal, causing abnormal vascular development (Furuyama et al. 2004; Hosaka et al. 2004). Therefore, we investigated the function of Foxo1 in aging processes under CR using heterozygous Foxo1-deficient ($Foxo1^{+/-}$) mice (Yamaza et al. 2010). $Foxo1^{+/-}$ mice are viable despite expressing <50 % of both FOXO1 in intact skin and Foxo1 mRNA in liver, spleen, muscle, adipose, and hippocampal tissues. The rate of lifespan extension in CR-$Foxo1^{+/-}$ mice is similar to that in CR-WT mice. Interestingly, compared with CR-WT mice, there is a significant increase in the incidence rate of malignant lymphoma and liver cancer in CR-$Foxo1^{+/-}$ mice. One of the features of CR is an anti-cancer effect. These results suggest that FOXO1 might contribute to protection against tumor growth under CR.

11.3 Role of NPY in Aging and Energy Metabolism

Animals have a system that responds to starvation. When food is plentiful, they readily consume food and reserve energy as fat. Under starvation or CR, the fat is metabolized to produce energy. Furthermore, to exclude unnecessary energy consumption, there is attenuation of fundamental physiological functions such as reproduction, growth, and heat generation.

In mammals, neurons of the hypothalamus arcuate nucleus control secretion of neuroendocrine factors that increase food consumption and body weight. They secrete the orexigenic neuropeptide NPY and agouti-related protein (Agrp) to increase blood insulin and leptin, and decrease ghrelin (Schwartz et al. 2000). Conversely, other kinds of neurons express anorexigen neuropeptides such as proopiomelanocortin (Pomc) (Sainsbury and Zhang 2012). Two different kinds of neurons control the balance of appetite and fundamental physiological phenomena (the so-called neuroendocrine hypothesis) (Stanley et al. 2005; Schwartz et al. 2000; Sainsbury and Zhang 2012). Several studies have reported that NPY is regulated by mammalian target of rapamycin (mTOR) and SIRT1-FOXO1 signaling pathways (Hong et al. 2012; Martins et al. 2012). For example, an mTOR inhibitor, rapamycin, inhibits the orexigenic action of ghrelin through suppression of NPY and Agrp, as well as phosphorylated (p)-cAMP response element-binding protein (CREB) and p-FOXO1 in rats (Martins et al. 2012). Furthermore, mRNA expression of NPY is regulated by SIRT1 and FOXO1 in mouse hypothalamic cells (Hong et al. 2012). In the next section, we will discuss molecular mechanisms linking CR and the neuroendocrine hypothesis.

NPY plays roles in energy consumption control, stress responses, and protection of neurons. Overexpression of NPY in rats leads to lifespan extension in the absence of CR (Michalkiewicz et al. 2003). We have reported that $NPY^{-/-}$ mice show attenuation of lifespan under CR (20 % prolonged lifespan in WT-CR mice vs 7 % prolonged lifespan in $NPY^{-/-}$-CR mice compared with AL-WT mice) (Chiba et al. 2014). The anticancer efficacy and oxidant stress tolerance in CR-$NPY^{-/-}$ mice are unaltered compared with CR-WT mice, suggesting that NPY has an important role in anti-aging under CR.

NPY is not only involved in hypothalamic regulation of appetite but also peripheral regulation of lipid metabolism. Administration or overexpression of NPY promotes adiposity (Ruohonen et al. 2012; Zarjevski et al. 1993), whereas inhibition of NPY or its receptor reduces adiposity in diet-induced obese mice (Chao et al. 2011; Ishihara et al. 2006). These results indicate that antagonism of NPY may prevent diabetes and obesity. Some studies have shown that NPY regulates adipogenesis and lipolysis in adipocytes, but the precise mechanism by which NPY regulates adipocyte metabolism is poorly understood. We recently showed that deficiency of NPY activates lipolytic actions through augmentation of SIRT1, p-hormone sensitive lipase (HSL) (ser563), and CGI-58 in white adipose tissue. These metabolic changes due to NPY antagonism in adipose tissue lead to amelioration of factors contributing to the age-related imbalance of adipose tissue metabolism, such as increased inflammatory cytokines, decreased de novo lipogenesis, and reduced thermogenic activity in white adipose tissue (Park et al. 2014). However, the functions of NPY in regulation of fat metabolism appear to be gender specific. For example, NPY knockout mice show reduced adiposity in female mice but not in male mice (Park et al. 2014). Further studies should be performed to verify gender differences in the role of NPY in regulation of adipose tissue metabolism.

11.4 CR Applications in Humans

The aim of aging and CR research is establishment of a method to control human aging and its associated diseases. There are many similarities in CR effects between primates and rodents (Table 11.1) (Lane et al. 2001). Two institutes, the National Institute on Aging (NIA) and Wisconsin National Primate Research Center (WNPRC), currently perform CR studies using primates (Colman et al. 2009; Mattison et al. 2012). Indicators of CR such as body temperature, attenuation of the GH/IGF-1 signaling pathway, and the serum concentration of insulin in

Table 11.1 Similarity of CR effects in primates and rodents

Item	Sign	Similarity
Weight	Down	Yes
Amount of fat and muscle	Down	Yes
Age of sexual maturation	Delay	Yes
Glucose and insulin (fasted)	Down	Yes
Insulin sensitivity	Up	Yes
Metabolic activity (short span)	Down	Yes
Metabolic activity (long span)	No change	Yes
Body temperature	Down	Yes
Locomotion	No change or up	Yes
Serum triglyceride	Down	Yes
Serum HDL2B	Up	Yes
GH/IGF-1	Down	Yes
Interleukin-6	Down	Yes
Testosterone	No change	Yes
Estradiol	No change	Yes
Luteinizing hormone	No change	Yes
Follicle-stimulating hormone	No change	Yes
Progesterone	No change	Yes
Wound healing	No change	Yes
Fibroblast clonal proliferation	No change	No report
β-gal positive senescence cell	No change	No report
Rate of decline in DHEAS	Inhibition	No report
Lymphocyte number	Down	Yes
Lymphocyte calcium response	No change	No

Adapted from Lane et al. (2001)
Abbreviations: *HDL* high density lipoprotein, *DHEAS* dehydroepiandrosterone sulfate

primates are also present in rodents. These observations indicate that CR studies using primates and rodents apply to humans.

In recent years, there has been development of CR mimetics (CRM) that act on molecules relevant to the CR effect (Chiba et al. 2010). Criteria for a CRM are as follows (Ingram et al. 2006): (i) similarity to the metabolic, hormonal, and physiological effects of CR; (ii) it does not significantly reduce long-term food intake; (iii) it activates stress response pathways observed in CR and provides protection against a variety of stressors; (iv) it produces CR-like effects on longevity, reduction of age-related disease, and maintenance of functions. Phytochemicals in plants have recently attracted considerable attention (Son et al. 2008). CRMs are involved in hormesis, inhibition of the insulin/IGF-1 pathway, metabolism, and activation of the sirtuin pathway (Table 11.2). Ultimately, it is expected that life expectancy will be extended and age-related disease will be overcome by CRMs in the future.

Table 11.2 CRM candidates

Name	Possibility for effect	Model	Target	References
Juglone	Stress resistance	C	Foxo1, Nrf2	Przybysz et al. (2009)
Sulforaphane	Stress resistance	CC, D, M	Nrf2/ARE pathway	Son et al. (2008)
Curcumin	Stress resistance	CC, M	Induction of HO-1	Son et al. (2008)
Resveratrol	Extend life span	Obesity M	SIRT	Baur et al. (2006)
Rapamycin	Extend life span	M	mTOR pathway	Miller et al. (2011)
2DG	Extend life span	C	Glucose metabolism	Schulz et al. (2007)
Metformin	Extend life span	C, M	AMPK	Onken and Driscoll (2010) and Martin-Montalvo et al. (2013)
Statin	Extend life span	D	Oxidative stress	Spindler et al. (2012)
Metoprolol	Extend life span	D, M	β1-Adrenergic receptor	Spindler et al. (2013)
Nebivolol	Extend life span	D, M	β1-Adrenergic receptor	Spindler et al. (2013)
STAC	Possibility for improving metabolic/age-related diseases	CC	SIRT1	Hubbard et al. (2013)

Abbreviations: *C C. elegans*, *CC* cultured cells, *D D. melanogaster*, *M M. musculus*, *Nrf2* NF-E2-related factor 2, *ARE* antioxidant response element, *HO-1* heme-oxygenase 1, *mTOR* mechanistic target of rapamycin, *2DG* 2-deoxy-D-glucose, *AMPK* AMP-activated protein kinase, *STAC* sirtuin-activating compounds

11.5 Summary and Conclusion

Longevity-related genes have been identified in several organisms including *C. elegans*, *D. melanogaster*, and *M. musculus*. These molecules play major roles in homeostasis and age-related diseases. We believe that the mechanisms of CR include hierarchical clustering in which the hypothalamus controls neuroendocrine factors, blood hormone secretion, and intracellular organelle functions (Fig. 11.1). In mammals, redundancy might improve homeostasis upon deletion of one gene, making analysis difficult. However, transgenic mice and CRMs are useful tools for CR studies.

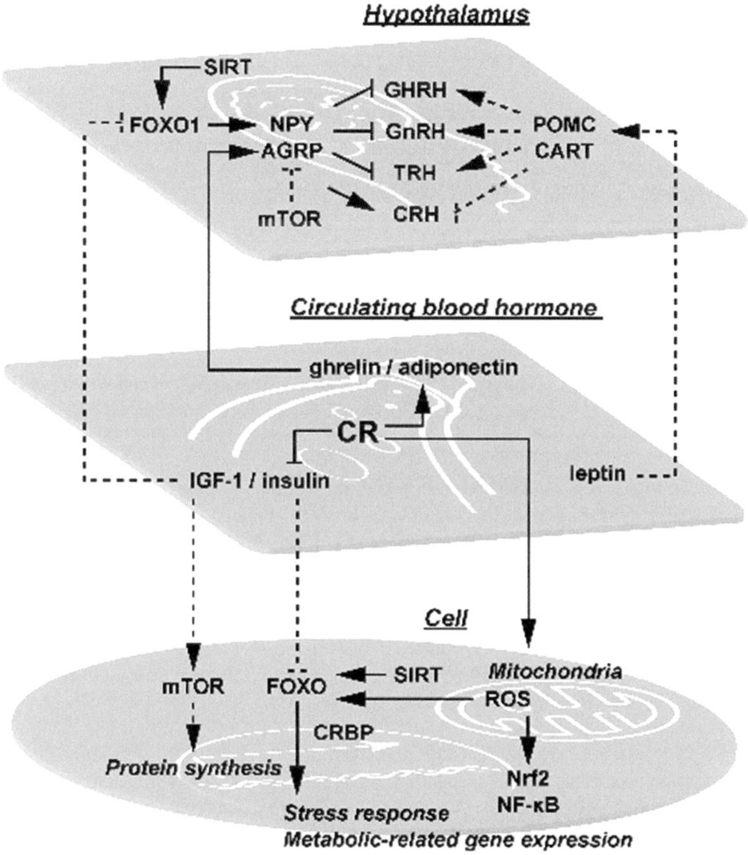

Fig. 11.1 Hierarchical clustering mechanisms of anti-aging effects and longevity in CR. Individuals adapt to CR because of a decrease in calorie intake and alterations in the hypothalamus and hormone balance (neuroendocrine hypothesis). mTOR is inhibited because of a decrease in blood insulin and IGF-1. Inhibition of FOXO is reduced and susceptible to redox. Changes in the constitution of mitochondrial respiratory chains lead to increases in the efficiency of respiratory chains and variations in reactive oxygen species (ROS). There are alterations in the expression of several transcription factors (i.e. FOXO, Nrf2, CREB, and NF-κB) and regulation of the expression of redox-activated target genes with roles in anti-aging and anti-cancer effects (mitohormesis hypothesis). *Arrows*: potentiation of signaling; *dashed arrows*; attenuation of signaling; *lines*: potentiation of inhibition signaling; *dashed lines*; attenuation of inhibitive signaling

References

Bartke A, Brown-Borg H (2004) Life extension in the dwarf mouse. Curr Top Dev Biol 63:189–225. doi:10.1016/s0070-2153(04)63006-7

Baur JA, Pearson KJ, Price NL, Jamieson HA, Lerin C, Kalra A, Prabhu VV, Allard JS, Lopez-Lluch G, Lewis K, Pistell PJ, Poosala S, Becker KG, Boss O, Gwinn D, Wang M, Ramaswamy S, Fishbein KW, Spencer RG, Lakatta EG, Le Couteur D, Shaw RJ, Navas P,

Puigserver P, Ingram DK, de Cabo R, Sinclair DA (2006) Resveratrol improves health and survival of mice on a high-calorie diet. Nature 444(7117):337–342. doi:10.1038/nature05354

Bluher M, Kahn BB, Kahn CR (2003) Extended longevity in mice lacking the insulin receptor in adipose tissue. Science 299(5606):572–574. doi:10.1126/science.1078223

Chao PT, Yang L, Aja S, Moran TH, Bi S (2011) Knockdown of NPY expression in the dorsomedial hypothalamus promotes development of brown adipocytes and prevents diet-induced obesity. Cell Metab 13(5):573–583. doi:10.1016/j.cmet.2011.02.019

Chiba T, Tsuchiya T, Komatsu T, Mori R, Hayashi H, Shimokawa I (2010) Development of calorie restriction mimetics as therapeutics for obesity, diabetes, inflammatory and neurodegenerative diseases. Curr Genomics 11(8):562–567. doi:10.2174/138920210793360934

Chiba T, Tamashiro Y, Park D, Kusudo T, Fujie R, Komatsu T, Kim SE, Park S, Hayashi H, Mori R, Yamashita H, Chung HY, Shimokawa I (2014) A key role for neuropeptide Y in lifespan extension and cancer suppression via dietary restriction. Sci Rep 4:4517. doi:10.1038/srep04517

Colman RJ, Anderson RM, Johnson SC, Kastman EK, Kosmatka KJ, Beasley TM, Allison DB, Cruzen C, Simmons HA, Kemnitz JW, Weindruch R (2009) Caloric restriction delays disease onset and mortality in rhesus monkeys. Science 325(5937):201–204. doi:10.1126/science.1173635

Friedman DB, Johnson TE (1988) Three mutants that extend both mean and maximum life span of the nematode, Caenorhabditis elegans, define the age-1 gene. J Gerontol 43(4):B102–B109

Furuyama T, Kitayama K, Shimoda Y, Ogawa M, Sone K, Yoshida-Araki K, Hisatsune H, Nishikawa S, Nakayama K, Ikeda K, Motoyama N, Mori N (2004) Abnormal angiogenesis in Foxo1 (Fkhr)-deficient mice. J Biol Chem 279(33):34741–34749. doi:10.1074/jbc.M314214200

Holzenberger M, Dupont J, Ducos B, Leneuve P, Geloen A, Even PC, Cervera P, Le Bouc Y (2003) IGF-1 receptor regulates lifespan and resistance to oxidative stress in mice. Nature 421(6919):182–187. doi:10.1038/nature01298

Hong SH, Lee KS, Kwak SJ, Kim AK, Bai H, Jung MS, Kwon OY, Song WJ, Tatar M, Yu K (2012) Minibrain/Dyrk1a regulates food intake through the Sir2-FOXO-sNPF/NPY pathway in Drosophila and mammals. PLoS Genet 8(8):e1002857. doi:10.1371/journal.pgen.1002857

Hosaka T, Biggs WH 3rd, Tieu D, Boyer AD, Varki NM, Cavenee WK, Arden KC (2004) Disruption of forkhead transcription factor (FOXO) family members in mice reveals their functional diversification. Proc Natl Acad Sci U S A 101(9):2975–2980. doi:10.1073/pnas.0400093101

Hubbard BP, Gomes AP, Dai H, Li J, Case AW, Considine T, Riera TV, Lee JE ESY, Lamming DW, Pentelute BL, Schuman ER, Stevens LA, Ling AJ, Armour SM, Michan S, Zhao H, Jiang Y, Sweitzer SM, Blum CA, Disch JS, Ng PY, Howitz KT, Rolo AP, Hamuro Y, Moss J, Perni RB, Ellis JL, Vlasuk GP, Sinclair DA (2013) Evidence for a common mechanism of SIRT1 regulation by allosteric activators. Science 339(6124):1216–1219. doi:10.1126/science.1231097

Ingram DK, Zhu M, Mamczarz J, Zou S, Lane MA, Roth GS, deCabo R (2006) Calorie restriction mimetics: an emerging research field. Aging Cell 5(2):97–108. doi:10.1111/j.1474-9726.2006.00202.x

Ishihara A, Kanatani A, Mashiko S, Tanaka T, Hidaka M, Gomori A, Iwaasa H, Murai N, Egashira S, Murai T, Mitobe Y, Matsushita H, Okamoto O, Sato N, Jitsuoka M, Fukuroda T, Ohe T, Guan X, MacNeil DJ, Van der Ploeg LH, Nishikibe M, Ishii Y, Ihara M, Fukami T (2006) A neuropeptide Y Y5 antagonist selectively ameliorates body weight gain and associated parameters in diet-induced obese mice. Proc Natl Acad Sci U S A 103(18):7154–7158. doi:10.1073/pnas.0510320103

Kenyon C, Chang J, Gensch E, Rudner A, Tabtiang R (1993) A C. elegans mutant that lives twice as long as wild type. Nature 366(6454):461–464. doi:10.1038/366461a0

Kimura KD, Tissenbaum HA, Liu Y, Ruvkun G (1997) daf-2, an insulin receptor-like gene that regulates longevity and diapause in Caenorhabditis elegans. Science 277(5328):942–946

Lane MA, Black A, Handy A, Tilmont EM, Ingram DK, Roth GS (2001) Caloric restriction in primates. Ann N Y Acad Sci 928:287–295

Lin K, Dorman JB, Rodan A, Kenyon C (1997) daf-16: an HNF-3/forkhead family member that can function to double the life-span of Caenorhabditis elegans. Science 278(5341):1319–1322

Martin-Montalvo A, Mercken EM, Mitchell SJ, Palacios HH, Mote PL, Scheibye-Knudsen M, Gomes AP, Ward TM, Minor RK, Blouin MJ, Schwab M, Pollak M, Zhang Y, Yu Y, Becker KG, Bohr VA, Ingram DK, Sinclair DA, Wolf NS, Spindler SR, Bernier M, de Cabo R (2013) Metformin improves healthspan and lifespan in mice. Nat Commun 4:2192. doi:10.1038/ncomms3192

Martins L, Fernandez-Mallo D, Novelle MG, Vazquez MJ, Tena-Sempere M, Nogueiras R, Lopez M, Dieguez C (2012) Hypothalamic mTOR signaling mediates the orexigenic action of ghrelin. PLoS One 7(10):e46923. doi:10.1371/journal.pone.0046923

Mattison JA, Roth GS, Beasley TM, Tilmont EM, Handy AM, Herbert RL, Longo DL, Allison DB, Young JE, Bryant M, Barnard D, Ward WF, Qi W, Ingram DK, de Cabo R (2012) Impact of caloric restriction on health and survival in rhesus monkeys from the NIA study. Nature 489 (7415):318–321. doi:10.1038/nature11432

McCay CM, Crowell MF, Maynard LA (1935) The effect of retarded growth upon the length of life span and upon the ultimate body size. J Nutr 10(1):63–79

Michalkiewicz M, Knestaut KM, Bytchkova EY, Michalkiewicz T (2003) Hypotension and reduced catecholamines in neuropeptide Y transgenic rats. Hypertension 41(5):1056–1062. doi:10.1161/01.hyp.0000066623.64368.4e

Miller RA, Harrison DE, Astle CM, Baur JA, Boyd AR, de Cabo R, Fernandez E, Flurkey K, Javors MA, Nelson JF, Orihuela CJ, Pletcher S, Sharp ZD, Sinclair D, Starnes JW, Wilkinson JE, Nadon NL, Strong R (2011) Rapamycin, but not resveratrol or simvastatin, extends life span of genetically heterogeneous mice. J Gerontol A Biol Sci Med Sci 66(2):191–201. doi:10.1093/gerona/glq178

Nakae J, Oki M, Cao Y (2008) The FoxO transcription factors and metabolic regulation. FEBS Lett 582(1):54–67. doi:10.1016/j.febslet.2007.11.025

Ogg S, Paradis S, Gottlieb S, Patterson GI, Lee L, Tissenbaum HA, Ruvkun G (1997) The Fork head transcription factor DAF-16 transduces insulin-like metabolic and longevity signals in C. elegans. Nature 389(6654):994–999. doi:10.1038/40194

Onken B, Driscoll M (2010) Metformin induces a dietary restriction-like state and the oxidative stress response to extend C. elegans healthspan via AMPK, LKB1, and SKN-1. PLoS One 5(1): e8758. doi:10.1371/journal.pone.0008758

Paik JH, Kollipara R, Chu G, Ji H, Xiao Y, Ding Z, Miao L, Tothova Z, Horner JW, Carrasco DR, Jiang S, Gilliland DG, Chin L, Wong WH, Castrillon DH, DePinho RA (2007) FoxOs are lineage-restricted redundant tumor suppressors and regulate endothelial cell homeostasis. Cell 128(2):309–323. doi:10.1016/j.cell.2006.12.029

Park S, Fujishita C, Komatsu T, Kim SE, Chiba T, Mori R, Shimokawa I (2014) NPY antagonism reduces adiposity and attenuates age-related imbalance of adipose tissue metabolism. FASEB J. doi:10.1096/fj.14-258384

Przybysz AJ, Choe KP, Roberts LJ, Strange K (2009) Increased age reduces DAF-16 and SKN-1 signaling and the hormetic response of Caenorhabditis elegans to the xenobiotic juglone. Mech Ageing Dev 130(6):357–369. doi:10.1016/j.mad.2009.02.004

Ruohonen ST, Vahatalo LH, Savontaus E (2012) Diet-induced obesity in mice overexpressing neuropeptide y in noradrenergic neurons. Int J Pept 2012:452524. doi:10.1155/2012/452524

Sainsbury A, Zhang L (2012) Role of the hypothalamus in the neuroendocrine regulation of body weight and composition during energy deficit. Obes Rev 13(3):234–257. doi:10.1111/j.1467-789X.2011.00948.x

Schulz TJ, Zarse K, Voigt A, Urban N, Birringer M, Ristow M (2007) Glucose restriction extends Caenorhabditis elegans life span by inducing mitochondrial respiration and increasing oxidative stress. Cell Metab 6(4):280–293. doi:10.1016/j.cmet.2007.08.011

Schwartz MW, Woods SC, Porte D Jr, Seeley RJ, Baskin DG (2000) Central nervous system control of food intake. Nature 404(6778):661–671. doi:10.1038/35007534

Shimokawa I, Higami Y, Utsuyama M, Tuchiya T, Komatsu T, Chiba T, Yamaza H (2002) Life span extension by reduction in growth hormone-insulin-like growth factor-1 axis in a transgenic rat model. Am J Pathol 160(6):2259–2265. doi:10.1016/s0002-9440(10)61173-x

Son TG, Camandola S, Mattson MP (2008) Hormetic dietary phytochemicals. Neuromolecular Med 10(4):236–246. doi:10.1007/s12017-008-8037-y

Spindler SR, Li R, Dhahbi JM, Yamakawa A, Mote P, Bodmer R, Ocorr K, Williams RT, Wang Y, Ablao KP (2012) Statin treatment increases lifespan and improves cardiac health in Drosophila by decreasing specific protein prenylation. PLoS One 7(6):e39581. doi:10.1371/journal.pone.0039581

Spindler SR, Mote PL, Li R, Dhahbi JM, Yamakawa A, Flegal JM, Jeske DR, Lublin AL (2013) beta1-Adrenergic receptor blockade extends the life span of Drosophila and long-lived mice. Age (Dordr) 35(6):2099–2109. doi:10.1007/s11357-012-9498-3

Stanley S, Wynne K, McGowan B, Bloom S (2005) Hormonal regulation of food intake. Physiol Rev 85(4):1131–1158. doi:10.1152/physrev.00015.2004

Yamaza H, Komatsu T, Wakita S, Kijogi C, Park S, Hayashi H, Chiba T, Mori R, Furuyama T, Mori N, Shimokawa I (2010) FoxO1 is involved in the antineoplastic effect of calorie restriction. Aging Cell 9(3):372–382. doi:10.1111/j.1474-9726.2010.00563.x

Zarjevski N, Cusin I, Vettor R, Rohner-Jeanrenaud F, Jeanrenaud B (1993) Chronic intracerebroventricular neuropeptide-Y administration to normal rats mimics hormonal and metabolic changes of obesity. Endocrinology 133(4):1753–1758. doi:10.1210/endo.133.4.8404618

Chapter 12
Oxidative Stress and Endoplasmic Reticulum Stress in Kidney Aging: Impact of Uremic Toxins as the Cause of Stress

Reiko Inagi

Abstract Population aging continues all over the world. Kidney aging is a major factor in the development and progression of chronic kidney disease (CKD), which shows premature phenotypic changes. Kidney aging associated with a decline in kidney function affects the functions of remote organs (heart, blood vessels, and brain), and CKD is a significant risk factor in cardiovascular disease (CVD) and stroke, both of which are age-related diseases. This association strongly indicates that kidney aging accelerates the aging of remote organs and that the kidney is a central organ in the maintenance of life.

Recent studies of the pathophysiology of CKD emphasize the contribution of endoplasmic reticulum (ER) stress and the link between ER stress and other pathogenic stresses such as oxidative stress and glycative stress, both of which are known to lead to premature aging. In particular, uremic toxins have received attention as causal factors of a vicious cycle of these stress signals, resulting in progression of CKD, rather than being a consequence of CKD. This article summarizes the impact and pathophysiological significance of oxidative stress and glycative stress, and ER stress. It also discusses the crosstalk of these stresses, novel evidence regarding uremic toxins as inducers of these stresses, and finally the contribution of these stresses to kidney premature aging.

Keywords Endoplasmic reticulum (ER) stress • Unfolded protein response (UPR) • Chronic kidney disease (CKD) • Uremic toxins • Glycative stress • Oxidative stress • Indoxyl sulfate • Methylglyoxal • Glyoxalase 1 • Advanced glycation endproducts (AGEs)

R. Inagi, Ph.D. (✉)
Division of Chronic Kidney Disease Pathophysiology, The University of Tokyo, Graduate School of Medicine, 7-3-1, Hongo, Bunkyo-ku, Tokyo 113-8655, Japan
e-mail: inagi-npr@umin.ac.jp

© Springer Japan 2015

N. Mori, I. Mook-Jung (eds.), *Aging Mechanisms*,
DOI 10.1007/978-4-431-55763-0_12

12.1 Kidney Aging and Chronic Kidney Disease (CKD)

CKD is a serious public health problem. In Japan, more than 13.3 million people, or roughly one in every eight adults, suffer from CKD, and the number continues to increase year by year. CKD is a major cause of the increase in number of hemodialysis patients who suffer from end-stage kidney failure. The cost and resource utilization of renal replacement therapy for these patients continues to grow.

Although early detection and treatment can often halt the progression of CKD, the kidney is called a silent organ, and CKD progresses without clinical symptoms. Progression can eventually lead to kidney failure, which requires dialysis or a kidney transplant to maintain life. Even before the initiation of renal replacement therapy, asymptomatic CKD, which progresses over time, carries a number of risks. In particular, CKD leads to damage in remote organs, including the heart, blood vessels, and brain, resulting in a poor quality of life (Fig. 12.1). The high susceptibility of CKD patients to cardiovascular disease is now recognized as the "cardiorenal connection". In sum, CKD has high morbidity and mortality, and treatment of CKD imposes a heavy economic burden.

Previously, the main cause of CKD was immune-mediated glomerulonephritis. Importantly, the recent increase in the number of patients with CKD is closely linked with the aging of the population and an increase in the prevalence of life-

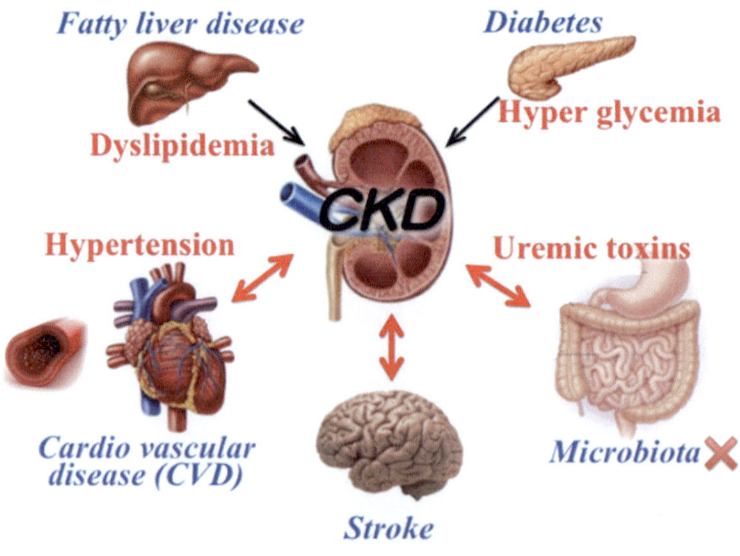

Fig. 12.1 Impact of CKD on the crosstalk between remote organs. The various factors, including hyper glycemia, dyslipidemia, hypertension, and uremic toxins (derived from gut microbiota), significantly accelerate kidney damage, leading to the development and progression of chronic kidney disease (CKD). Importantly, CKD increases the risk of cardiovascular events, stroke, and alteration of gut microbiota, indicating that the kidney is an important organ for the maintainance of the remote organs

style diseases such as diabetes, hypertension. In fact, CKD becomes more common with aging: while one in three or four people between 65 and 74 years suffer from CKD, this increases to half in people aged 75 or more. Many pathological conditions related with CKD progression are more common in elderly people, including diabetes, high blood pressure and heart disease, indicating that simple measures to control these conditions can slow the progression of CKD regardless of age and prevent damage to remote organs, such as cardiovascular events and stroke. In an aging society, CKD is thus important in improving the quality of life in both medical and economic terms. Reflecting this, the theme of the International Society of Nephrology's 2014 World Kidney Day was "*Chronic Kidney Disease and Aging*" (www.worldkidneyday.org).

Patients with CKD suffer from oxidative stress and endoplasmic reticulum (ER) stress, exhibiting a 'premature aging' phenotype (Kooman et al. 2014). Moreover, the premature senescence, or stress-induced senescence, which occurs in CKD contributes to the progression of CKD (Table 12.1). Premature aging is associated with the progressive accumulation of harmful changes and an increase in the risk of various diseases and death. Why is the premature aging phenotype, particular cellular senescence, so predominant in CKD? Many toxic factors that induce cellular stress and premature senescence are excreted by the kidneys, and any loss of kidney function with aging accelerates their accumulation in the body. This acceleration of accumulation in turn accelerates the progression of CKD and the synergistic derangement of physiological kidney aging, including DNA and mitochondrial alterations, chronic inflammation, phosphate toxicity, and dysregulation of the klotho-FGF-23 axis (signal transduction pathway related to aging phenotypes), among others (Kuro-O 2011; Shanahan 2013; Stenvinkel and Larsson 2013).

Recent evidence highlights the molecular mechanism by which the accumulation of uremic toxins due to loss of kidney function contributes to premature aging, or cellular senescence, in the kidney (Adijiang et al. 2011; Shimizu et al. 2012). In this article, we focus on uremic toxins as key players in premature aging through oxidative stress and ER stress.

Table 12.1 Premature aging in CKD

Common evidence between kidney aging and CKD
Decline of kidney function (eGFR ↓)
Accumulation of uremic toxins
Increase in ROS production
Oxidative stress ↑, glycative stress ↑
Decline of defense system against glycation (glyoxalase 1)
Glycative stress ↑
Decrease in quality of proteins due to oxidative or glycative posttranslational modification
ER stress ↑

CKD chronic kidney disease, *eGFR* estimated glomerular filtration, *ER* endoplasmic reticulum, *ROS* reactive oxygen species

12.2 Impact of Oxidative Stress in Kidney Aging

Oxidative stress is one of the major pathogenic stresses in the kidney. It is caused by various renal pathogens, including hypertension, metabolic disorders (hyperglycemia, dyslipidemia), ischemia, and nephrotoxic drugs. Oxidative stress activates numerous intracellular signaling pathways via reactive oxygen species (ROS)-mediated modulation of various enzymes and critical transcription factors. Oxidative free radicals can induce the age-related deterioration of organs through increased apoptosis or DNA mutations. Thus, increased production of ROS is a key player in oxidative stress-related tissue/organ damage and kidney aging. This "oxidative stress theory" holds that a progressive and irreversible accumulation of oxidative damage caused by ROS impacts critical aspects of the aging process and contributes to impaired physiological function in the aged population, an increased incidence of age-associated disease, and a reduction in life span.

While the free radical theory of aging remains valid, attempts to induce longevity by eliminating oxidative stress in genetically engineered mice have not necessarily been successful (Pérez et al. 2009a, b). Contrary to expectations, for example, mice overexpressing major antioxidant enzymes do not have an extended lifespan. This evidence strongly suggests the involvement of factors other than free radicals in aging, and that various factors induce senescence phenotypes in a context-dependent manner.

Recent evidence emphasizes that uremic toxins induce oxidative stress and may accelerate kidney aging. In the following sections, I summarize the impact of uremic toxins as premature aging factors which induce oxidative stress, and the crosstalk between oxidative stress and other stresses.

12.3 The Link Between Oxidative Stress and Hypoxia by Uremic Toxins

Superoxide, a representative ROS, is produced by an imbalance in oxygen metabolism. In addition to causing direct damage to the kidney, superoxide radicals decrease the bioavailability of nitric oxide through the formation of peroxynitrite. Reduced levels of nitric oxide lead to a decrease in regional blood flow. Furthermore, nitric oxide suppresses mitochondrial respiration, and depletion of nitric oxide by oxidative stress stimulates mitochondrial respiration and uncouples it from chemical energy consumption, resulting in excessive oxygen consumption and tissue hypoxia. Thus, oxidative stress and hypoxia are intricately linked.

This pathogenic linkage of oxidative stress and hypoxia is also induced by uremic toxins. Indoxyl sulfate, a representative uremic toxin, increases oxygen consumption in freshly isolated rat and human proximal tubules in an oxidative stress-dependent manner (Palm et al. 2010). Furthermore, an increase in the serum level of indoxyl sulfate is associated with a decline in renal function via the dysregulation of oxygen metabolism in tubular cells in experimental animals (Palm et al. 2010).

To make matters worse, uremia suppresses the activation of hypoxia-inducible factor (HIF), a master transcriptional regulator of adaptive responses against hypoxia. The expression level of HIF is regulated by degradation of the a subunit via hydroxylation of the proline residues. This proline hydroxylation is controlled by the oxygen-dependent activity of prolyl hydroxylases (PHD). Hypoxia inactivates PHD, leading to accumulation of the α subunit, which forms a heterodimer with a constitutively expressed β subunit, and transactivates 100–200 target genes. Representative HIF target genes include erythropoietin, vascular endothelial growth factor, glucose transporters, and glycolytic enzymes. These target genes all evoke powerful responses in adaptation against hypoxia. However, indoxyl sulfate induces the expression of HIF-inhibitory CITED2 (Tanaka et al. 2013), making uremic patients susceptible to ischemic diseases such as cardiovascular disease and CKD per se.

12.4 Impact of Glycative Stress Associated with Oxidative Stress Due to Uremic Toxins in Kidney Aging

Glycation is a non-enzymatic reaction of proteins or DNA with carbohydrate, which generates advanced glycation endproducts (AGEs). Carbonyl compound derivatives, which are mainly generated by carbohydrate oxidation and glycolysis, react with lysine or arginine residues of proteins and cause glycative modification, which is a type of posttranslational modification. Among the many carbonyl compounds, glyoxal and methylglyoxal are highly reactive compounds, and are well known as glycation precursors that contribute to AGE formation. Importantly, these representative glycation precursors are also representative uremic toxins.

The state of physiological glycation is regulated by a balance between the formation and clearance of glycated proteins, or AGEs. Diabetic conditions and oxidative stress accelerate AGE formation, in turn leading to glycative stress. In addition, uremic toxins classified as glycation precursors are excluded by the kidney, demonstrating that kidney dysfunction also induces glycative stress through the defective clearance of AGEs. These findings strongly support the idea that aging that is associated with metabolic disorders, and that the decline in kidney function is a critical trigger of glycative stress.

Serum AGE levels increase with age in a large cohort of normal subjects, correlating well with levels of established markers of oxidative stress and inflammation (Uribarri et al. 2007). These results imply a link between oxidative stress and glycative stress. Importantly, AGE accumulation associated with oxidative stress is closely linked with the development or progression of age-related diseases (cardiovascular disease and stroke, as well as CKD) and lifespan, suggesting that glycative stress is likely to be a cause rather than a consequence of the aging process, at least in part (Rabbani and Thornalley 2011; Kizer et al. 2014).

12.5 Anti-glycation Enzyme, Glyoxalase 1

AGE precursors, some of which are uremic toxins, particular the highly reactive carbonyl compound methyglyoxal, are detoxified by anti-glycation enzymes, such as glyoxalase 1 (GLO1). GLO1 contributes to the regulation of physiological glycation state (Rabbani and Thornalley 2011). When the level of AGE precursor formation is increased or GLO1 activity is decreased under pathogenic microenvironments, such as hyperglycemia, hypoxia, or aging, AGE formation is accelerated. AGEs then accumulate in the body, leading to glycative stress.

We previously reported that glycative stress significantly accelerates both kidney aging and kidney disease and that GLO1 is a beneficial renoprotective molecule (Ikeda et al. 2011). We also showed that the overexpression of GLO1 ameliorates senescence phenotypes of the kidney via the detoxification of AGE precursors (Ikeda et al. 2011). For example, compared with young rat kidneys (10 weeks), aged rat kidneys (14 months) showed increased renal accumulation of AGEs: the methylglyoxal adduct CEL (N(e)-carboxyethyllysine), a representative marker of glycative stress, was mainly accumulated in age-related damaged tubules associated with tubulointerstitial thickening. We elucidated that age-induced glycative stress was significantly correlated with a decrease in renal GLO1 activity. To evaluate the role of GLO1 in kidney aging, we established transgenic (Tg) rats which overexpress human GLO1 systemically, and showed that GLO1 overexpression significantly ameliorated age-induced glycative stress in the kidney. These GLO1 Tg rats also showed the amelioration of not only glycative stress but also oxidative stress: while levels of renal 4-HNE and urinary 8OH-dG, two representative oxidative stress markers, increased with age, GLO1 overexpression significantly decreased them. Importantly, an age-related increase in both renal oxidative stress and glycative stress via a decline in renal GLO1 activity was correlated with an increase in expression of cellular senescence markers in tubules, such as P53, P21, and P16, and improvement of these stresses decreased the expression of cellular senescence markers at the protein and mRNA levels.

As one age-related morphological change in the kidney of elderly humans, we detected interstitial thickening, estimated by Masson's trichrome staining, in association with a decline in estimated glomerular filtration rate (eGFR). This thickening was also observed in the aged rat kidney and was ameliorated by GLO1 overexpression (Ikeda et al. 2011). Interestingly, this phenomenon was closely associated with an alteration in renal senescence state: an increase in the number of senescence-associated b-galactosidase (SABG)-positive senescent tubules was observed in elderly rats, and was significantly reduced by GLO1 overexpression. These improvements indeed ameliorated age-related renal dysfunction; namely, they suppressed the decline in eGFR with age. Together, our results demonstrated that kidney aging is associated with both oxidative stress and glycative stress, and that GLOI delayes these renal senescence phenotypes via the alteration of these stresses. In other words, GLO1 may be a key player in reducing AGE precursors of uremic toxins that accumulate with age, and thereby retard renal senescence.

Furthermore, we also observed that glycative stress contributes to vascular aging (Jo-Watanabe et al. 2014). Our GLO1 Tg rats showed a threefold increase in vascular GLO1 activity in association with a reduction in vascular glycative stress with age. Vascular GLO1 overexpression partly ameliorated age-related vascular dysfunction, in association with enhancement of NO production. Further, we found that GLO1 overexpression suppressed the inhibitory phosphorylation of eNOS (Thr495) and showed a tendency to increase the active phosphorylation (Ser1177) of eNOS. These in vivo data are consistent with an in vitro study utilizing cultured human endothelial cells which showed that methylglyoxal increased the inhibitory phosphorylation level of Thr495 in cultured human endothelial cells in a dose- and time-dependent manner, but that this effect was markedly reduced in cells overexpressing human GLO1 cDNA. These data demonstrate that the age-related increase in methylglyoxal in vessels, which is mainly caused by the age-dependent loss of kidney function, increases the inhibitory phosphorylation level of eNOS, resulting in the suppression of eNOS activity and subsequent endothelial damage. Importantly, this pathway was significantly inhibited by GLO1 to maintain vascular homeostasis.

These results emphasize the pathophysiological role of glycative stress and oxidative stress in aging of the kidney and vessels. Uremic toxins, namely glycation precursors, accumulate with age and induce oxidative or glycative stress, which leads to kidney and vascular dysfunction, as well as the aging of these organs (Ikeda et al. 2011; Jo-Watanabe et al. 2014). This kidney dysfunction then further exacerbates the accumulation of these uremic toxins. To prevent these effects, we identified the effectiveness of renal GLO1 as an anti-glycation enzyme, or an anti-kidney aging enzyme.

12.6 Endoplasmic Reticulum (ER) Stress as a Novel Pathogenic Stress in CKD in Kidney Aging

Protein homeostasis, or proteostasis, is achieved via the operation of sophisticated networks of mechanisms which act to maintain the quality of proteins. The quality control of proteins is regulated by the balance of protein synthesis, folding, and degradation. The most important point is that because proteostasis significantly influences the structure and function of cells, it contributes to cell fate determination. Defective proteostasis occurring in kidney cells under pathogenic conditions would lead to kidney disease as well as kidney aging (Inagi 2010, 2014) (Fig. 12.2).

The major site of proteostasis networks is the ER. The ER plays an important role in the quality control of proteins via its regulation of synthesis, folding and

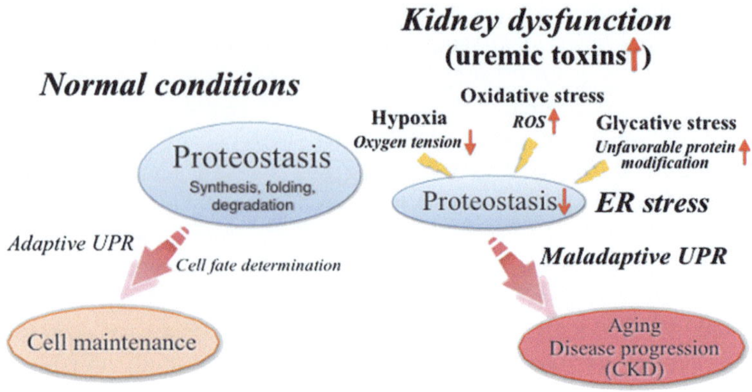

Fig. 12.2 Pathophysiological link of ER stress with common pathogenic stresses. Protein homeostasis, or proteostasis, in the ER is regulated by protein synthesis, folding, and degradation through the UPR pathway. Under normal conditions, the adaptive UPR pathway determines cell fate by maintaining cell development, structure and function (*left side*). In contrast, when kidney function declines, resulting in the accumulation of uremic toxins (indoxyl sulfate, glycation precursors), low oxygen tension (hypoxia), increased production of ROS by oxidative stress and loss of protein function by glycative stress occur in kidney cells. These conditions induce overwhelming activation of the adaptive and/or apoptotic UPR pathways (referred to as ER stress). This maladaptive UPR system leads to a defective proteostasis, and subsequently to kidney ageing as well as kidney disease. Abbreviations: *CKD* chronic kidney disease, *ER* endoplasmic reticulum, *ROS* reactive oxygen species, *UPR* unfolded protein response

trafficking (Wang and Kaufman 2014). The balance of capacity between protein synthesis and folding is important for ER homeostasis, or in other words, ER proteostasis. Various disturbances decrease folding capacity, however, including folding mutation, hypoxia, or oxidative stress. This decrease in folding capacity induces the accumulation of malfolded proteins in the ER, which leads to ER stress. This ER stress in turn triggers the unfolded protein response (UPR) as an adaptive response to ER stress. The ER stress-induced UPR pathway is known as a stress signal which normalizes ER function and ER proteostasis. In contrast, maladaptation due to excessive or prolonged UPR activation induces the apoptotic UPR and contributes to the progression of kidney damage.

12.6.1 Adaptive UPR Pathway

The UPR pathway is an adaptive signal transduction induced by ER stress which acts to maintain ER function (Fig. 12.3). The primary purpose of the UPR pathway is to facilitate adaptation to the changing environment which evokes it and to maintain ER function. Activation of the adaptive UPR pathway enhances protein-folding capacity by activating the transcription of UPR target genes such as ER chaperones, including glucose-regulated protein 78 (GRP78), GRP94, and

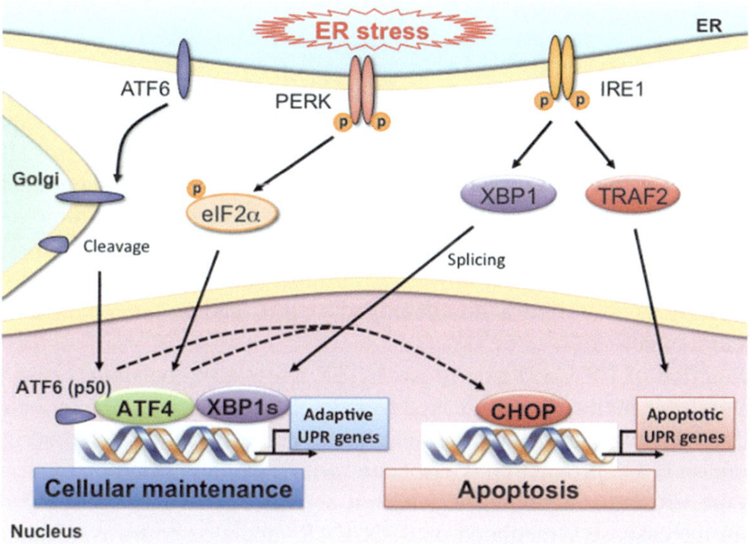

Fig. 12.3 Adaptive and apoptotic UPR pathways as ER stress signals. Under ER stress, the UPR modulators PERK and IRE1 are activated by dimerization followed by phosphorylation, and ATF6 is cleaved in the Golgi apparatus, inducing the activation of UPR transcription factors ATF6 (p50), ATF4 and XBP1s. These transcription factors mainly upregulate the adaptive UPR pathway and normalize ER function to maintain the cells via the ATF6, PERK–eIF2a–ATF4 or IRE1-XBP1 pathways. Under long-term ER stress, the adaptive UPR pathway fails to rescue the cells, and the apoptotic UPR pathway, namely the PERK-eIF2a-ATF4-CHOP or IRE1-TRAF2 pathway, is induced to eliminate the defective cells. Abbreviations: *ATF* activating transcription factor, *CHOP* C/EBP-homologous protein, *eIF2a* eukaryotic translation initiation factor 2a, *ER* endoplasmic reticulum, *IRE1* inositol-requiring protein 1, *P* phosphate, *PERK* double-stranded RNA-activated protein kinase (PKR)-like endoplasmic reticulum kinase, *TRAF2* tumour necrosis factor receptor-associated factor 2, *UPR* unfolded protein response, *XBP1* X-box binding protein 1, *XBP1s* spliced form of XBP1

calreticulin. GRP78, also referred to as BiP, is a central regulator of ER function under pathogenic conditions. The N-termini of transmembrane ER proteins, such as inositol-requiring enzyme 1 (IRE1), double-stranded RNA-activated protein kinase-like ER kinase (PERK), and activating transcription factor 6 (ATF6), are normally held in an inactive (or physiological) state in the ER membrane by binding to GRP78. When unfolded proteins accumulate in the ER lumen, GRP78 releases the transmembrane ER proteins, allowing them to either dimerize (or polymerize) in the case of PERK or IRE1, or move to the Golgi in the case of ATF6 within the cell. PERK and IRE1 activate the downstream transcription factors ATF4 and XBP1, respectively. ATF6 is activated by cleavage in the Golgi, and the truncated ATF6 acts as a transcription factor. The PERK-ATF4, IRE-XBP1, and ATF6 axes initiate expression of the downstream target genes (ER chaperones and the molecules which act to degrade malfolded proteins) to maintain ER homeostasis, or proteostasis.

12.6.2 Apoptotic UPR Pathway

When cells exposed to excessive or prolonged ER stress fail to resolve the protein-folding defect or rescue ER homeostasis via the adaptive UPR pathway, apoptosis mediated by the maladaptive UPR is induced (Fig. 12.3). This is called the apoptotic UPR pathway. To induce apoptosis, the PERK-ATF4 axis or ATF6 axis activates proapoptotic CHOP (CCAAT/enhance-binding protein homologous protein) expression, and the IRE1 axis activates TRAF2 (tumor necrosis factor receptor–associated factor 2) followed by caspase 12 or caspase 4 activation. The UPR pathway is thus like a double-edged sword, and is altered in a context-dependent manner.

Dysfunction of ER Ca^+ homeostasis by ER stress also contributes to apoptosis through multiple pathways: Ca^+ leaked from the ER lumen enters the mitochondria, depolarizes the inner mitochondrial membrane, and generates mitochondrial ROS. This increased ROS production is associated with a vicious cycle of oxidative stress in both the ER and mitochondria, where it activates several apoptotic pathways, including the caspase 9-mediated or BAX/BAK-mediated pathways. This dispensation with dysfunctional cells represents a last resort on the part of multicellular organisms.

12.6.3 Link Between ER Stress, Oxidative Stress, and Glycative Stress

Persistent oxidative stress induces ER dysfunction in association with prolonged or severe UPR activation, and thereby initiates ER stress-associated apoptotic cascades. This ultimately represents a vicious cycle of ER stress and oxidative stress. The cycle might be mediated by ROS produced by the dysfunction of ER Ca^+ homeostasis, as described above.

As another way to link oxidative stress to ER stress, Hasnain et al. (2014) demonstrated the contribution of inflammation. Oxidative stress and ER stress are often accompanied by hyperglycemia-induced pancreatic beta cell damage (impaired insulin biosynthesis and secretion). These authors showed that multiple inflammatory cytokines elevated in diabetic pancreatic islets induced beta cell oxidative and ER stress, with interleukin-23 (IL-23), IL-24 and IL-33 being the most potent. In contrast, islet-endogenous and -exogenous IL-22 suppressed both the oxidative stress and ER stress caused by cytokines or glucolipotoxicity. These results were confirmed by in vivo studies: in obese mice, antibody neutralizing IL-23 or IL-24 partially reduced ER stress in beta cells and improved glucose tolerance, whereas IL-22 administration modulated oxidative stress regulatory genes in islets, resulting in the suppression of ER stress and the restoration of glucose homeostasis. Given that chronic inflammation is one of the key factors of the aging process in various organs (Jurk et al. 2014; Walke et al. 2014) and that diabetes is a major cause of CKD, we suggest that oxidative stress and ER stress induced by hyperglycemia-induced inflammation in diabetes may enhance exacerbate premature aging of the pancreas and kidney, resulting in the development of CKD with diabetes.

Recent evidence also emphasizes that ER stress links to not only oxidative stress, but also glycative stress (Inagi 2011; Zhuang and Forbes 2014). It has been reported that AGEs accumulated under glycative stress conditions induce ER stress directly. For example, glycated serum albumin (AGE-bovine serum albumin, AGE-BSA) induced ER stress, as estimated by GRP78 expression, and the apoptotic UPR induced apoptosis in a dose- and time-dependent manner in mouse podocytes. This phenomenon was caused by an increase in intracellular Ca^{2+} concentration. The ER stress inhibitor taurine-conjugated ursodeoxycholic acid (TUDCA), which acts as a chemical chaperone that enhances the folding and trafficking of proteins, prevents AGE-induced podocyte apoptosis, suggesting that this apoptosis is mediated by ER stress, namely the apoptotic UPR pathway, rather than by signaling through the receptor for AGE (RAGE) (Chen et al. 2008).

Further, Liu et al. (2014) showed that RAGE promotes senescence of proximal tubular epithelial cells via activation of ER stress-dependent p21 signaling. The elevated expressions of RAGE, ER chaperone GRP78 as a marker of UPR pathway activation, and cellular senescence marker p21 were correlated with an accumulation of senescence-associated β-galactosidase in diabetic nephropathy patients. In vitro studies revealed that AGE-BSA induced the expression of RAGE, GRP78, and p21 in cultured mouse proximal tubular cells. These data strongly indicate that premature senescence in tubules is induced by glycative stress via the induction of ER stress, or the maladaptive UPR pathway.

Glycated LDL (low-density lipoprotein) also causes aberrant ER stress, as estimated by increased GRP78, PERK phosphorylation, and ATF6 activation, in cultured bovine aortic endothelial cells (Dong et al. 2010). This is followed by endothelial dysfunction, such as impaired endothelium-dependent vasorelaxation. Similar results were observed in in vivo studies utilizing isolated aorta from mice fed an atherogenic diet. Importantly, chronic administration of Tempol, a potent antioxidant, attenuated the endothelial dysfunction induced by glycated LDL in association with the amelioration of ER stress (Dong et al. 2010). These findings also suggest that oxidative stress, glycative stress, and ER stress crosstalk during the progression of atherosclerosis, and that this stress network contributes to the phenotypic changes of endothelial cells.

Like glycated albumin and LDL, AGE precursors also lead to ER stress-induced apoptosis. Extracellular matrix is frequently modified by AGE precursors in the skin of diabetic patients. Type I collagen modified by an AGE precursor such as methylglyoxal or 3-deoxyglucosone, both of which are uremic toxins, causes ER stress-mediated apoptosis via ROS-mediated CHOP activation in dermal fibroblasts, suggesting a pathophysiological role for the link between glycative stress and ER stress in diabetic wounds (Loughlin and Artlett 2011).

Another study utilizing diabetic mice also showed that ER stress in 22-month-old diabetic kidneys resulted in the up-regulation of CHOP (Wu et al. 2010). CHOP might play a role in exacerbating kidney lesions given that CHOP-deficient proximal tubular cells were resistant to ER stress-induced cell death, and that CHOP-deficient mice were protected from diabetic nephropathy. CHOP deficiency also protected the mouse kidney from tubulointerstitial fibrosis, a major renal aging phenotype which is accelerated with age. Taken together with the finding that diabetic nephropathy is also exacerbated with aging, these findings suggest that ER stress associated with unfavorable CHOP activation might contribute to the progression of premature kidney aging, in turn leading to age-related kidney damage.

12.7 Conclusion

Kidney aging is a multifactorial process in which several key-mediators play a significant role, including oxidative stress, ER stress, and glycative stress. The defective crosstalk of these stresses under pathogenic conditions often leads to a vicious cycle and contributes to kidney premature aging phenotypes. Further investigation of these stress pathways is needed to develop pharmacological approaches to retarding these aging processes or the decline in renal function in both the elderly and the general population.

References

Adijiang A, Shimizu H, Higuchi Y, Nishijima F, Niwa T (2011) Indoxyl sulfate reduces klotho expression and promotes senescence in the kidneys of hypertensive rats. J Ren Nutr 21:105–109

Chen Y, Liu CP, Xu KF, Mao XD, Lu YB, Fang L, Yang JW, Liu C (2008) Effect of taurine-conjugated ursodeoxycholic acid on endoplasmic reticulum stress and apoptosis induced by advanced glycation end products in cultured mouse podocytes. Am J Nephrol 28:1014–1022

Dong Y, Zhang M, Wang S, Liang B, Zhao Z, Liu C, Wu M, Choi HC, Lyons TJ, Zou MH (2010) Activation of AMP-activated protein kinase inhibits oxidized LDL-triggered endoplasmic reticulum stress in vivo. Diabetes 59:1386–1396

Hasnain SZ, Borg DJ, Harcourt BE, Tong H, Sheng YH, Ng CP, Das I, Wang R, Chen AC, Loudovaris T, Kay TW, Thomas HE, Whitehead JP, Forbes JM, Prins JB, McGuckin MA (2014) Glycemic control in diabetes is restored by therapeutic manipulation of cytokines that regulate beta cell stress. Nat Med 20:1417–1426

Ikeda Y, Inagi R, Miyata T, Nagai R, Arai M, Miyashita M, Itokawa M, Fujita T, Nangaku M (2011) Glyoxalase I retards renal senescence. Am J Pathol 179:2810–2821

Inagi R (2010) Endoplasmic reticulum stress as a progression factor for kidney injury. Curr Opin Pharmacol 10:156–165

Inagi R (2011) Inhibitors of advanced glycation and endoplasmic reticulum stress. Methods Enzymol 491:361–380

Inagi R, Ishimoto Y, Nangaku M (2014) Proteostasis in endoplasmic reticulum – new mechanisms in kidney disease. Nat Rev Nephrol 10:369–378

Jo-Watanabe A, Ohse T, Nishimatsu H, Takahashi M, Ikeda Y, Wada T, Shirakawa J, Nagai R, Miyata T, Nagano T, Hirata Y, Inagi R, Nangaku M (2014) Glyoxalase I reduces glycative and oxidative stress and prevents age-related endothelial dysfunction through modulation of endothelial nitric oxide synthase phosphorylation. Aging Cell 13:519–528
Jurk D, Wilson C, Passos JF, Oakley F, Correia-Melo C, Greaves L, Saretzki G, Fox C, Lawless C, Anderson R, Hewitt G, Pender SL, Fullard N, Nelson G, Mann J, van de Sluis B, Mann DA, von Zglinicki T (2014) Chronic inflammation induces telomere dysfunction and accelerates ageing in mice. Nat Commun 2:4172
Kizer JR, Benkeser D, Arnold AM, Ix JH, Mukamal KJ, Djousse L, Tracy RP, Siscovick DS, Psaty BM, Zieman SJ (2014) Advanced glycation/glycoxidation endproduct carboxymethyl-lysine and incidence of coronary heart disease and stroke in older adults. Atherosclerosis 235:116–121
Kooman JP, Kotanko P, Schols AM, Shiels PG, Stenvinkel P (2014) Chronic kidney disease and premature ageing. Nat Rev Nephrol 10:732–742
Kuro-O M (2011) A phosphate-centric paradigm for pathophysiology and therapy of chronic kidney disease. Kidney Int Suppl 3:420–426
Liu J, Huang K, Cai GY, Chen XM, Yang JR, Lin LR, Yang J, Huo BG, Zhan J, He YN (2014) Receptor for advanced glycation end-products promotes premature senescence of proximal tubular epithelial cells via activation of endoplasmic reticulum stress-dependent p21 signaling. Cell Signal 26:110–121
Loughlin DT, Artlett CM (2011) Modification of collagen by 3-deoxyglucosone alters wound healing through differential regulation of p38 MAP kinase. PLoS One 6:e18676
Palm F, Nangaku M, Fasching A, Tanaka T, Nordquist L, Hansell P, Kawakami T, Nishijima F, Fujita T (2010) Uremia induces abnormal oxygen consumption in tubules and aggravates chronic hypoxia of the kidney via oxidative stress. Am J Physiol Renal Physiol 299:F380–F386
Pérez VI, Bokov A, Remmen HV, Mele J, Ran Q, Ikeno Y, Richardson A (2009a) Is the oxidative stress theory of aging dead? Biochim Biophys Acta 1790:1005–1014
Pérez VI, Van Remmen H, Bokov A, Epstein CJ, Vijg J, Richardson A (2009b) The overexpression of major antioxidant enzymes does not extend the lifespan of mice. Aging Cell 8:73–75
Rabbani N, Thornalley PJ (2011) Glyoxalase in diabetes, obesity and related disorders. Semin Cell Dev Biol 22:309–317
Shanahan CM (2013) Mechanisms of vascular calcification in CKD-evidence for premature ageing? Nat Rev Nephrol 9:661–670
Shimizu H, Yisireyili M, Nishijima F, Niwa T (2012) Stat3 contributes to indoxyl sulfate-induced inflammatory and fibrotic gene expression and cellular senescence. Am J Nephrol 36:184–189
Stenvinkel P, Larsson TE (2013) Chronic kidney disease: a clinical model of premature aging. Am J Kidney Dis 62:339–351
Tanaka T, Yamaguchi J, Higashijima Y, Nangaku M (2013) Indoxyl sulfate signals for rapid mRNA stabilization of Cbp/p300-interacting transactivator with Glu/Asp-rich carboxy-terminal domain 2 (CITED2) and suppresses the expression of hypoxia-inducible genes in experimental CKD and uremia. FASEB J 27:4059–4075
Uribarri J, Cai W, Peppa M, Goodman S, Ferrucci L, Striker G, Vlassara H (2007) Circulating glycotoxins and dietary advanced glycation endproducts: two links to inflammatory response, oxidative stress, and aging. J Gerontol A Biol Sci Med Sci 62:427–433
Walker SR, Wagner M, Tangri N (2014) Chronic kidney disease, frailty, and unsuccessful aging. J Ren Nutr 24:364–370
Wang M, Kaufman RJ (2014) The impact of the endoplasmic reticulum protein-folding environment on cancer development. Nat Rev Cancer 14:581–597
Wu J, Zhang R, Torreggiani M, Ting A, Xiong H, Striker GE, Vlassara H, Zheng F (2010) Induction of diabetes in aged C57B6 mice results in severe nephropathy: an association with oxidative stress, endoplasmic reticulum stress, and inflammation. Am J Pathol 176:2163–2176
Zhuang A, Forbes JM (2014) Stress in the kidney is the road to pERdition: is endoplasmic reticulum stress a pathogenic mediator of diabetic nephropathy? J Endocrinol 222:R97–R111

Chapter 13
Vitamin K Benefits in Aging and Cancer

Kotaro Azuma and Satoshi Inoue

Abstract Vitamin K is well known as a critical blood coagulation factor. In addition, epidemiological studies suggest that reduced vitamin K intake is associated with several geriatric diseases, such as osteoporosis, osteoarthritis, dementia, and arteriosclerosis. Indeed, the therapeutic benefits of vitamin K have been demonstrated in osteoporosis patients in several clinical studies. It has also been proposed that vitamin K may contribute to the prevention and treatment of some types of malignancies. Vitamin K functions as a co-factor of γ-glutamyl carboxylase (GGCX) and regulates the activity of vitamin K-dependent proteins expressed in various tissues via posttranscriptional modifications. As an additional mechanism of vitamin K action, we revealed that vitamin K activates the Steroid and Xenobiotic receptor (SXR), a member of nuclear receptor superfamily, leading to changes in gene transcription. Analysis of the bones of pregnane X receptor (PXR)-deficient mice demonstrated that the bone protective effects of vitamin K are partially mediated by SXR/PXR-dependent signaling. In certain malignancies, vitamin K exerts therapeutic activity in an SXR/PXR-dependent manner. Another distinct function of vitamin K involves apoptotic induction of leukemia cells, which may be attributed to the covalent association of vitamin K2 epoxide with pro-apoptotic protein Bak. The progress of vitamin K research has opened up new possibilities that vitamin K can be useful for prevention and treatment of a variety of diseases.

K. Azuma
Department of Geriatric Medicine, Graduate School of Medicine, The University of Tokyo, 7-3-1 Hongo, Bunkyo-ku, Tokyo 113-8655, Japan

S. Inoue, M.D., Ph.D. (✉)
Department of Geriatric Medicine, Graduate School of Medicine, The University of Tokyo, 7-3-1 Hongo, Bunkyo-ku, Tokyo 113-8655, Japan

Department of Anti-Aging Medicine, Graduate School of Medicine, The University of Tokyo, 7-3-1 Hongo, Bunkyo-ku, Tokyo 113-8655, Japan

Division of Gene Regulation and Signal Transduction, Research Center for Genomic Medicine, Saitama Medical School, 1397-1 Yamane, Hidaka, Saitama 350-1241, Japan
e-mail: INOUE-GER@h.u-tokyo.ac.jp

© Springer Japan 2015

N. Mori, I. Mook-Jung (eds.), *Aging Mechanisms*,
DOI 10.1007/978-4-431-55763-0_13

Keywords Vitamin K • Steroid and xenobiotic receptor (SXR) • Pregnane X receptor (PXR) • γ-glutamyl carboxylase (GGCX) • Aging • Osteoporosis • Osteoarthritis • Alzheimer's disease • Cancer

13.1 Introduction

Vitamin K is a fat-soluble vitamin that was discovered around 1930 as an essential factor for blood coagulation (Dam 1935). The mechanism by which vitamin K promotes blood coagulation was clarified in the 1970s, reporting that vitamin K is involved in the posttranscriptional modification of coagulation factors (Nelsestuen et al. 1974; Stenflo et al. 1974). Vitamin K2 is administered to newly born babies in order to prevent intracranial hemorrhage. On the other hand, warfarin, which inhibits vitamin K function, is widely used as an anticoagulant. More recently, vitamin K has been examined in connection with various physiological and pathological conditions related to aging and cancer. In this chapter, we will discuss the recently discovered functional mechanisms and roles of vitamin K in aging and cancer.

13.2 Structure of Vitamin K

Vitamin K is actually a group of compounds, existing in any of seven possible forms (Fig. 13.1). Among these, only vitamin K1 (phylloquinone) and K2 (menaquinone) are naturally occurring products. Vitamin K1 is present in vegetables while vitamin K2 is synthesized by microorganisms (Booth and Suttie 1998). The traditional Japanese food "Natto," which consists of fermented soybeans, contains high concentrations of vitamin K2. Vitamins K1 and K2 differ only in their side chains, and it was recently observed that vitamin K1 is actually a precursor of vitamin K2, as K1 is converted to K2 to become active in the body (Okano et al. 2008; Nakagawa et al. 2010). Vitamin K2 is abbreviated as 'MK-n' according to the number of isoprene units in its side chain. MK-4 has been shown to be the most abundant form of vitamin K in the human body (Thijssen and Drittij-Reijnders 1996), while 'Natto' contains MK-7. Two forms of synthetic vitamin K, vitamin K3 (menadione) and vitamin K4 (menadiol), were clinically used in the past but have since been discontinued due to toxicity from the production of reactive oxygen species and suppression of bilirubin metabolizing enzymes.

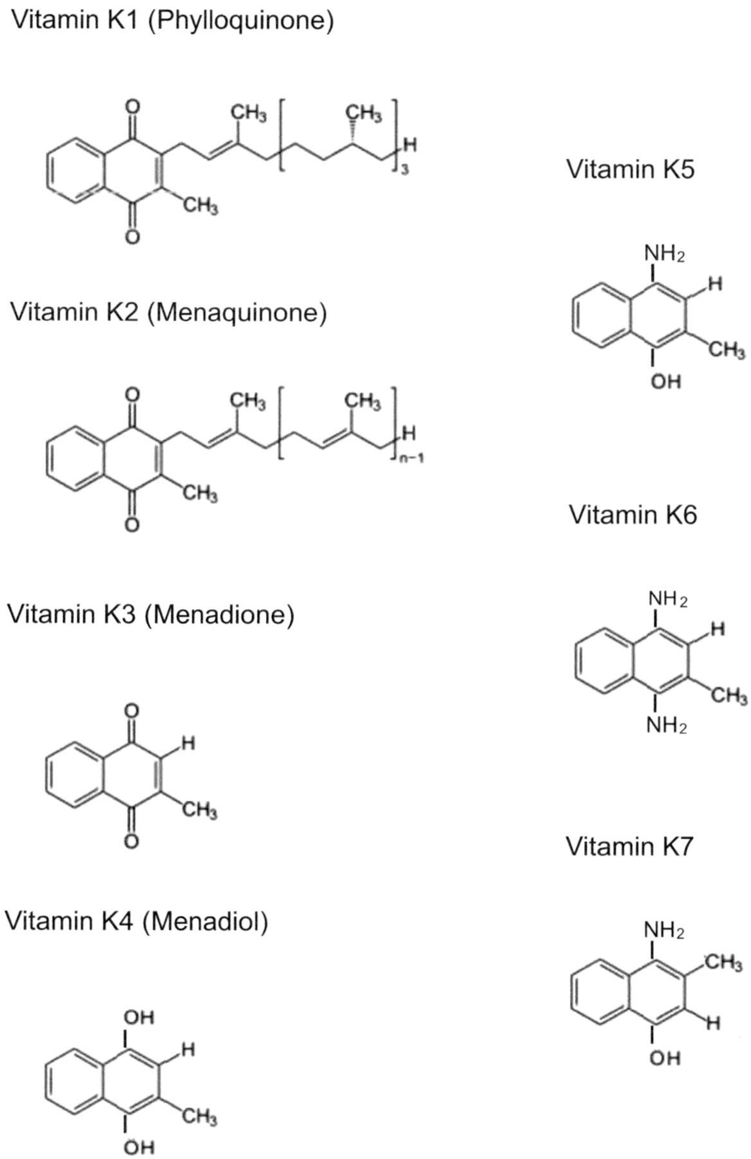

Fig. 13.1 Molecular structures of the seven forms of vitamin K. Vitamin K1, K2, and K3 that have a common naphthoquinone ring but differ in their side chains. Vitamin K1 has a phytyl side chain while the side chain of vitamin K2 is composed of a varying number of isoprenoid residues. In contrast, vitamin K3 lacks a side chain. Vitamin K1 and K2 are naturally occurring products, whereas the rest are synthetically produced

13.3 Vitamin K Function Mediated by Protein Modification

To date, several vitamin K mechanisms have been elucidated or proposed (Fig. 13.2). It can act as a co-factor for γ-glutamyl carboxylase (GGCX), and it also serves as a ligand for the nuclear steroid and xenobiotic receptor (SXR), as well as its murine homolog, pregnane X receptor (PXR).

The role of vitamin K as a co-factor of GGCX was first elucidated in the 1970s (Nelsestuen et al. 1974; Stenflo et al. 1974). GGCX mediates the addition of a carboxyl group to glutamate residues in substrate proteins. The most well-known substrates for this reaction are the vitamin K-dependent coagulation factors II, VII, IX, and X. These coagulation factors become active only when several glutamate residues are γ-carboxylated (known as Gla residues). Proteins involved in the fibrinolysis cascade, including proteins C, S, and Z (Rezaie et al. 2008), are also known to be substrates for GGCX. To date, 18 human proteins are known to be γ-carboxylated, including growth arrest specific-6 (Gas6) (Varnum et al. 1995), which regulates endothelial function; osteocalcin (also called bone Gla protein;

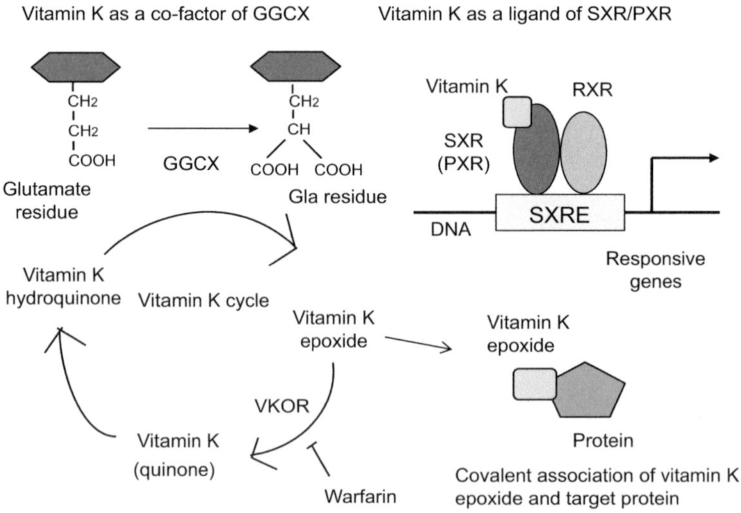

Fig. 13.2 Model of mechanisms of vitamin K actions. Vitamin K functions as a co-factor of γ-glutamyl carboxylase (GGCX). GGCX catalyzes conversion of glutamate residues into Gla residues by incorporating an additional carboxyl group onto the glutamate. This reaction requires the cyclic use of vitamin K. Vitamin K epoxide reductase (VKOR) is required for recycling vitamin K, which is oxidized during γ-glutamyl carboxylation. Warfarin inhibits VKOR and vitamin K recycling, and as a result, suppresses GGCX activity. Vitamin K also functions as a ligand of steroid and xenobiotic receptor (SXR), and its murine homolog, pregnane X receptor (PXR). SXR/PXR form heterodimers with 9-cis-retinoid acid receptor (RXR), and this complex binds to SXR-responsive elements (SXRE) within the promoter or enhancer regions of target genes. Covalent binding of vitamin K epoxide and its target proteins has been also proposed as a novel mechanism of vitamin K action

Table 13.1 The expression sites and known functions of γ-carboxylated proteins. Including transthyretin, which only contains one Gla residue, and the conantokins, which are not expressed in vertebrates, the total of γ-carboxylated proteins are known in total comes to around 20

	Expression	Function
Coagulation factor II (prothrombin)	Liver, kidney, adrenal gland, heart, parathyroid gland, etc.	Blood coagulation
Coagulation factor VII	Liver, kidney, intestine, testis, adrenal gland, etc.	Blood coagulation
Coagulation factor IX	Liver	Blood coagulation
Coagulation factor X	Liver, etc.	Blood coagulation
Protein C	Liver, kidney, testis, etc.	Anti-coagulation
Protein S	Liver, endothelium, monocyte, etc.	Anti-coagulation
Protein Z	Liver	Anti-coagulation
Osteocalcin (bone Gla protein, BGP)	Osteoblast	Regulation of bone calcification Regulation of glucose metabolism? Enhancing male fertility?
Matrix Gla protein (MGP)	Chondrocyte, vascular smooth muscle cell, etc.	Regulation of calcification
Growth arrest specific-6 (GAS6)	Lung, heart, kidney, intestine, endothelium, vascular smooth muscle cell, bone marrow, oseteoblast, osteoclast, monocyte, etc.	Thrombus formation Inflammation Cell proliferation
Proline-rich Gla protein 1 (PRGP1)	Spinal cord, etc.	Unknown
Proline-rich Gla protein 2 (PRGP2)	Thyroid, trachea, prostate, intestine, kidney, etc.	Unknown
Transmembrane Gla protein 3 (TMG3)	Brain, heart, lung, kidney, etc.	Unknown
Transmembrane Gla protein 4 (TMG4)	Kidney, pancreas, etc.	Unknown
Periostin	Periosteum, periodontal ligament, heart valve, adrenal gland, lung, thyroid, intestine, ovary, testis, prostate	Maintenance of periodontal ligament Bone development Maintenance of bone mineral density Tissue repair Enhanced fibrosis
TGFβ induced (TGFBI)	Bone, joint, skin, cornea, kidney	Bone development Maintenance of cornea

(continued)

Table 13.1 (continued)

	Expression	Function
Gla-rich protein (GRP)/Upper zone of growth plate and cartilage matrix associated protein (Ucma)	Chondrocyte, osteoblast, osteocyte, vascular smooth muscle cell, skin	Unknown
Transthyretin (Prealbumin)	Liver, kidney, pancreas, choroid plexus, retina	Transfer of thyroid hormone and retinoid
γ-glutamyl carboxylase (GGCX)	Systemic (High expression in liver)	γ-carboxylation of vitamin K-dependent proteins
Conantokin G	Venom of the marine cone snail	Inhibition of NMDA receptor
Conantokin T	Venom of the marine cone snail	Inhibition of NMDA receptor

BGP) (Price et al. 1976), which is expressed in osteoblasts; matrix Gla protein (MGP) (Luo et al. 1997), an inhibitor of calcification in cartilage; periostin (Coutu et al. 2008), which is expressed in periosteum, periodontal ligament, and heart; TGFβ induced (TGFBI) (Coutu et al. 2008), which is involved in corneal dystrophy; proline-rich Gla protein 1 and 2 (Kulman et al. 1997); transmembrane Gla protein 3 and 4 (Kulman et al. 2001); Gla-rich protein (Viegas et al. 2008); and transthyretin (Rüggeberg et al. 2008) (Table 13.1). Even GGCX itself has been shown to be γ-carboxylated (Berkner and Pudota 1984). In non-vertebrates, conantokin G (McIntosh et al. 1984) and conantokin T (Warder et al. 1998), toxic compounds produced by a marine cone snail, have also been reported to be γ-carboxylated. Altogether, these γ-carboxylated proteins are known as Gla proteins, or vitamin K-dependent proteins, as vitamin K-dependent γ-carboxylation is assumed to be required for their activity.

GGCX localizes to the membrane of the endoplasmic reticulum (ER), and most of its substrates have a highly conserved pro-peptide at their N-terminus containing a carboxylase recognition site (Coutu et al. 2008). GGCX substrates are first tagged on their pro-peptide with GGCX, and then multiple glutamate residues are sequentially γ-carboxylated (Fig. 13.3) (Morris et al. 1995). The glutamate residues to be γ-carboxylated form clusters, known as the “Gla domain,” near the N-terminus of GGCX substrates. The typical sequence of the Gla domain is Gla-x-x-x-Gla-x-Cys, and this sequence is highly conserved among vitamin K-dependent proteins. After all glutamate residues in the Gla domain have been carboxylated, the pro-peptide at the N-terminus is cleaved off and the rest of the protein is transmitted to the Golgi body. Among γ-carboxylated proteins, transthyretin is exceptional in that it lacks pro-peptide and only has one reported Gla residue. The γ-carboxylated transthyretin is only detected in the cerebrospinal fluid of subjects with moyamoya disease, suggesting γ-carboxylation of transthyretin may contribute to pathophysiological function.

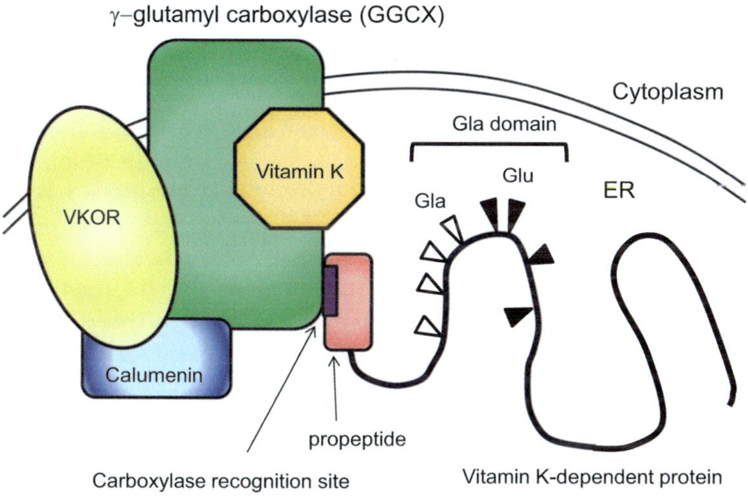

Fig. 13.3 γ-Carboxylation of vitamin K-dependent proteins. In the endoplasmic reticulum (ER), vitamin K-dependent proteins associate with GGCX via their N-terminal pro-peptides, which contain a carboxylase recognition site. Multiple glutamate residues (Glu) are then sequentially converted into Gla residues in this complex. The glutamate residues to be γ-carboxylated form a cluster known as the "Gla domain," which consists of the highly conserved "Gla-x-x-x-Gla-x-Cys" sequence near the N-terminus of GGCX substrates. On the ER lumen, GGCX is co-localized with VKOR, which has an important role in the vitamin K cycle. A protein called calumenin, which negatively regulates GGCX activity, is also found in this complex

In serving as a co-factor for GGCX, cyclic use of vitamin K is essential (Stafford 2005). When vitamin K-dependent proteins are carboxylated, vitamin K is oxidated into vitamin K epoxide. They must then be recycled before functioning as a GGCX co-factor again. Specifically, vitamin K epoxide is reduced by an enzyme known as vitamin K epoxide reductase (VKOR). Warfarin, which is clinically used as an anti-coagulant, is an inhibitor of VKOR and leads to a subsequent decrease in GGCX activity (Fig. 13.2). VKOR co-localizes with GGCX and the regulatory protein calumenin, which functions as a GGCX inhibitor (Fig. 13.3).

Recently, a novel vitamin K action, mediated by protein modification, was reported in a study of pro-apoptotic effects of vitamin K on leukemia cells. Karasawa et al. found that vitamin K promoted modification of the pro-apoptotic protein Bak (Karasawa et al. 2013). Based on mass spectrometric analysis, they proposed Bak is covalently modified by the epoxide form of vitamin K2 (Fig. 13.2). Although GGCX activity is required to generate this epoxide form, this mechanism of action appears to be distinct and further validation, as well as a search for potential binding partners, is required.

13.4 Vitamin K Function Mediated by Transcriptional Regulation

In addition to the effects of vitamin K that are mediated by posttranscriptional modifications, we recently identified secondary mechanism of its action, occurring at the level of transcriptional regulation (Tabb et al. 2003). Vitamin K functions as a ligand of the nuclear receptor SXR and its murine ortholog PXR (Tabb et al. 2003). This receptor is also known as NR1I2, according to standardized nomenclature designated by the nuclear receptor committee. Blumberg et al. cloned and characterized SXR/PXR as a novel nuclear receptor that is mainly expressed in the liver and intestine (Blumberg et al. 1998). SXR/PXR functions as a ligand-dependent transcription factor and is activated by various pharmaceutical agents and xenobiotic compounds (Zhou et al. 2009). Since endogenous ligands of SXR/PXR were unknown at the time it was cloned, SXR/PXR was originally classified as an orphan receptor. However, it is now reported that secondary bile acids are endogenous ligands for this receptor (Staudinger et al. 2001; Xie et al. 2001). When a ligand binds to SXR/PXR, it forms a heterodimer with 9-cis retinoid acid receptor (RXR). This complex then binds to SXR-responsive elements (SXRE) in the promoter or enhancer regions of target genes (Fig. 13.2). Typical target genes include the drug metabolizing enzyme CYP3A4, and the ABC family transporter MDR1, indicating that SXR/PXR functions as a xenobiotic sensor by inducing genes involved in detoxification and drug excretion (Synold et al. 2001). SXR/PXR expression has also been detected in the kidney, lung (Miki et al. 2005), and peripheral mononuclear cells (Albermann et al. 2005).

We have previously shown that SXR is expressed in osteoblastic cell lines and is activated by vitamin K2 (Tabb et al. 2003). We also demonstrated that vitamin K2 can serve as a ligand for PXR in osteoblasts using PXR-deficient and wild typemice (Tabb et al. 2003). The fact that it can function as a ligand for SXR/PXR has suggested the possibility that vitamin K may influence many physiological and pathological processes through regulation of PXR/SXR target genes.

13.5 Roles of Vitamin K in Age-Related Diseases

In Japan and several other Asian countries, vitamin K is used in the treatment of osteoporosis. Interestingly, epidemiological studies have revealed that intake of Natto is negatively correlated with incidence of hip fracture (Kaneki et al. 2001). In fact, vitamin K is the only nutrient that is significantly correlated with hip fracture incidence (Yaegashi et al. 2008). Clinical studies, which included meta-analysis, have confirmed the ability of vitamin K to prevent hip fracture (Cockayne et al. 2006).

We previously demonstrated a functional single nucleotide polymorphism in GGCX that is associated with bone mineral density in elderly Japanese women

(Kinoshita et al. 2007), suggesting the bone protective function of vitamin K may occur at the level of GGCX activity. However, effects of vitamin K on bone tissue mediated by γ-carboxylation are difficult to evaluate, as its hepatic functions hamper the analysis of extra hepatic roles of GGCX. GGCX-deficient mice die before birth or on the day of birth due to peritoneal hemorrhage (Zhu et al. 2007). Recently, we developed mouse models with tissue-specific ablation of GGCX in order to overcome these issues (Azuma et al. 2014). In an epidemiological study, high concentrations of non-carboxylated osteocalcin, designated as undercarboxylated-osteocalcin, are shown to be associated with a high risk of hip fracture (Vergnaud et al. 1997). Considering osteocalcin is a major substrate of GGCX in bone tissue, these findings may indicate that reduced activity of GGCX promotes bone fragility. However, osteocalcin-deficient mice have been shown to have mechanically stronger bones than wild type mice (Ducy et al. 1996), demanding a reconsideration of its actual function in bone tissue.

According to results from recent investigations, including our own, activation of SXR/PXR in bone tissue could mediate the beneficial effects of vitamin K. We established cells stably overexpressing SXR using the human osteoblastic cell line MG63. By comprehensive transcriptome analysis using microarrays, we identified genes induced by vitamin K in an SXR-dependent manner (Ichikawa et al. 2006), including tsukushi (TSK), matrilin-2 (MATN2), and CD14. Among these, TSK encodes a protein that has a collagen-accumulating effect (Ohta et al. 2004), MATN2 is a protein comprising extracellular matrix as like collagen (Wagener et al. 2005), while CD14 regulates osteoblastogenesis (Roman-Roman et al. 2003) and osteoclastogenesis by inducing differentiation of B cells (Filipp et al. 2001; Manabe et al. 2001). Induction of these genes is not inhibited by warfarin treatment, indicating a GGCX-dependent mechanism is not involved.

We also confirmed the involvement of SXR/PXR signaling in formation of bone tissue by analyzing the bone phenotype of PXR-deficient mice. Contrary to GGCX-deficient mice, PXR-deficient mice are viable after birth and are fertile, which enabled us to investigate in vivo effects of SXR/PXR signaling on bone tissue. We found that 4-month-old female PXR-deficient mice displayed lower femoral bone mineral density. Micro CT images of the femoral trabecular bone revealed a fragile structure in these mice. Histomorphometric analysis revealed enhanced bone resorption and reduced bone formation in PXR-deficient mice. The mechanical weakness of bone from PXR-deficient mice was confirmed using the 3-point bending test (Azuma et al. 2010). Considering the bone turnover status of PXR-deficient mice, there may be other mechanisms involved in bone tissue formation, since enhanced bone resorption may not be explained by known target genes in osteoblasts. One candidate mechanism involves the suppression of NF-κB by SXR/PXR signaling. SXR activation has been shown to repress NF-κB activity, and PXR-deficient mice display small bowel inflammation (Zhou et al. 2006), suggesting a proinflammatory diathesis was probably caused by lack of NF-κB suppression. This pro-inflammatory response might contribute to enhanced bone resorption.

Epidemiological studies have suggested that vitamin K deficiencies may be related to other skeletal diseases, such as osteoarthritis. In both North America and Japan, low vitamin K intake has been correlated to the prevalence of osteoarthritis (Neogi et al. 2006; Oka et al. 2009; Misra et al. 2013). However, a randomized controlled study found vitamin K to have no therapeutic benefit in patients with osteoarthritis (Neogi et al. 2008). To clarify whether the study length was insufficient to capture the positive effect, or if vitamin K's effects are only preemptive, further studies are required.

Vitamin K deficiency has also been reported in dementia, another aging-related disease. Epidemiological study of patients showing early-stage Alzheimer's disease found they consume less vitamin K when compared with cognitively intact controls (Shatenstein et al. 2007). In vitro analysis demonstrated the protective effect of Gas6, a vitamin K-dependent protein, in amyloid beta-induced apoptosis of cortical neurons (Yagami et al. 2002). These findings may provide a mechanistic explanation for the benefits of vitamin K in Alzheimer's disease. Vitamin K has also been linked to neurological function through the regulation of key enzymatic activities involved in sphingolipid metabolism (Sundaram and Lev 1988), a process whose dysfunction has been identified in neuro degenerative disorders. However, to the best of our knowledge, there is no report of an intervention study to evaluate the therapeutic effect of vitamin K in such a setting.

Finally, in this section we would like to describe the relationship of vitamin K and atherosclerosis. There are independent epidemiological studies reporting the association of vitamin K intake and lower incidence of coronary heart disease (Beulens et al. 2009; Gast et al. 2009). Additionally, there is an interventional trial showing administration of vitamin K1 suppressed the progression of coronary artery calcification (Shea et al. 2009). Two vitamin K-dependent proteins, Gas6 and MGP, are involved in atherosclerosis, providing a possible explanation of how vitamin K functions in this disease. Gas6 is known to suppress atherosclerosis through anti-apoptotic effects in vascular smooth muscle cells (Son et al. 2006). The anti-calcification function of MGP was demonstrated by re-expressing MGP in vascular smooth muscle cells of MGP knockout mice, which rescued the arterial calcification phenotype (Murshed et al. 2004).

13.6 Roles of Vitamin K in Malignancies

Vitamin K may also contribute to the prevention and treatment of malignant diseases, which increase in incidence according to age. Although many studies have focused on the cytotoxic activity of synthetic vitamin K3 (Lamson and Plaza 2003), the vitamins K1 and K2 have also been shown to have antitumorigenic activities.

An epidemiological study in North America revealed that postmenopausal women treated with vitamin K1 had a lower risk of cancer than those who received placebo (Cheung et al. 2008). In this study, almost half of the detected malignancies were breast cancers. Interestingly, an epidemiological study carried out in Europe found dietary intake of vitamin K2, rather than vitamin K1, led to reduced risk of malignancies. The reduction in risk was statistically significant in male population, and was driven by a pronounced reduction in prostate and lung cancer (Nimptsch et al. 2010).

Clinical trials carried out in Japan found that the administration of vitamin K2 reduced the *de novo* occurrence of hepatocellular carcinoma, a condition associated with liver cirrhosis (Habu et al. 2004). It has also been reported by two different groups in Japan that administration of vitamin K2 prolongs incident free time in hepatocellular carcinoma patients following treatment initial (Mizuta et al. 2006; Kakizaki et al. 2007). However, one of these groups reported no beneficial effect of vitamin K in a subsequent, larger, randomized controlled study (Yoshida et al. 2011). As for hematopoetic diseases, vitamin K2 may alleviate cytopenia in some patients with myelodysplastic syndrome (MDS) and post-MDS acute myeloid leukemia (Miyazawa et al. 2000).

The antitumor activity of vitamin K could be explained both by changes in the activity of GGCX, as well as the transcriptional activation of SXR/PXR. The growth of hepatocellular carcinoma is stimulated by des-γ-carboxyprothrombin (DCP; also known as PIVKA II), which is produced in response to vitamin K-deficiencies (Ma et al. 2009). Administration of vitamin K prevents DCP production and leads to reduced proliferation. It has been reported that an abnormality in GGCX mRNA splicing deregulates carboxylase activity in hepatocellular carcinoma cell lines (Ueda et al. 2008). Meanwhile, we hypothesized that the tumor-suppressing effect of vitamin K could be attributed to its SXR/PXR-dependent functions, as the receptor is abundantly expressed in the liver. We confirmed endogenous SXR expression in the hepatocellular carcinoma cell lines, HepG2 and HuH7. Overexpression of exogenous SXR in HuH7 cells suppressed proliferation and cellular motility (Azuma et al. 2009). The effects of SXR overexpression were further enhanced when cells were stimulated with vitamin K, again supporting the idea that these are SXR-dependent vitamin K effects (Fig. 13.4).

Verma et al. demonstrated that several SXR activators inhibit proliferation of the breast cancer cell lines, MCF-7 and ZR-75-1 (Verma et al. 2009). This may explain the prophylactic effects of vitamin K in the treatment of breast cancer. Conversely, a clinicopathological study reported that SXR expression was positively correlated with histological grade and lymph node status in estrogen receptor-positive breast cancer (Miki et al. 2006).

Regarding hematopoietic dysfunction, it was recently reported that expression of SXR in myeloid progenitor cells and that treatment with vitamin K2 or rifampicin, promoted CD11b and CD14 expression in these cells (Sada et al. 2010). Notably CD14, which triggers differentiation of B cells, was also induced in an osteoblastic cell line in an SXR-dependent manner (Ichikawa et al. 2006). This indicates that SXR-dependent vitamin K actions promote differentiation of myeloid progenitor cells, preventing their transition to leukemia. Recently, it was reported that PXR-deficient mice develop lymphoma at older ages, implying that SXR/PXR signaling would also be involved in normal differentiation and proliferation of lymphocytic

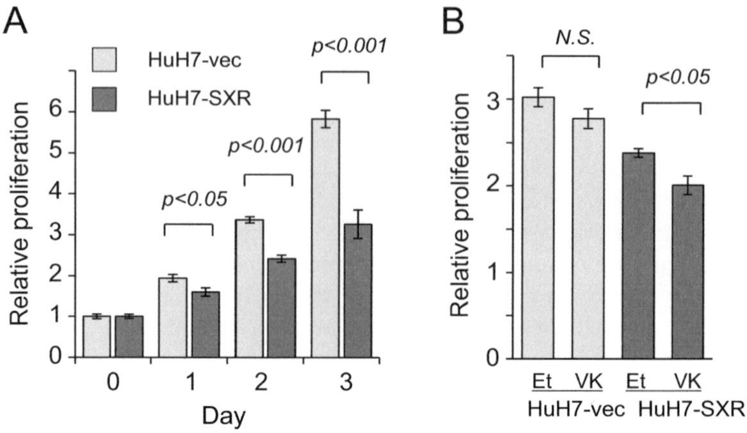

Fig. 13.4 Vitamin K-dependent growth suppression in the hepatocellular carcinoma cells. SXR was stably expressed in the hepatocellular carcinoma cell line, HuH7 (HuH7-SXR cells). (**a**) HuH7-SXR cells and negative control HuH-vec cells were seeded at a density of 1,000 cells/well and cultured in DMEM with 10 % FCS. Cell growth was assayed using WST-8 tetrazolium salt. Mean ± SEM of relative absorbance at 450 nm for each clone on each day, normalized to values at day 0, is shown (n = 3). (**b**) HuH7-SXR and HuH7-vec cells were seeded at a density of 1,000 cells/well and cultured in phenol red-free DMEM with charcoal/dextran-treated FCS (5 %) and MK-4 (VK; 10 μM) or ethanol (Et). Cell growth was assayed using WST-8 tetrazolium salt. *Bars* represent the mean ± SEM of relative absorbance at 450 nm for each clone on day 4 normalized to values at day 0 (n = 4). *N.S.* not significant

lineages (Casey et al. 2011). Other than promoting differentiation, vitamin K has also been reported to negatively affect leukemic cells by inducing apoptosis (Yaguchi et al. 1997). This is proposed to occur through covalent modification of the pro-apoptotic protein Bak in response to reaction with the epoxide form of vitamin K2 (Karasawa et al. 2013).

In cholangiocellular carcinoma and colon cancer (Enomono et al. 2007; Kawakita et al. 2009), vitamin K2 is reported to induce autophagic cell death; a type of programmed cell death that is distinct from apoptosis. However, the molecular mechanism of this phenomenon remains unclear.

Lastly, it should be noted that expression of SXR correlates with clinical prognosis in several types of cancer. In esophageal cancer, SXR is associated with a favorable prognosis (Takeyama et al. 2010), and we have demonstrated that expression of SXR, and its target gene CYP3A4, correlate with favorable prognoses in prostate cancer (Fujimura et al. 2012). Thus, esophageal and prostate cancers may be potential targets for therapy with SXR ligands. In contrast, prognoses of breast (Miki et al. 2006) and ovarian cancers (Yue et al. 2010) have been reported to be inversely correlative with SXR expression as determined by histological examination.

13.7 Conclusion

We have discussed the molecular mechanisms underlying vitamin K effects on aging-related diseases and cancer. We have also introduced a novel mechanism of vitamin K action mediated by SXR/PXR activation, and possible covalent modification with vitamin K epoxide. Theoretically, vitamin K can exert functional effects in many organs and tissues where SXR/PXR is expressed, in addition to those where Gla proteins are present. Further investigation of vitamin K will shed light on the physiological and pathological manifestations of aging and cancer. Considering it is a naturally occurring compound, and is known to be relatively safe in clinical use, vitamin K and its related compounds would be promising therapeutic and preventive agents for these diseases.

Acknowledgments This work was supported by Grants-in-Aid from the MHLW, Cell Innovation Program and Grants-in-Aid from the MEXT, the Program for Promotion of Fundamental Studies in Health Sciences from the NIBIO, a grant from the Novartis Foundation for Gerontological Research, and also, in part, by the Asian CORE program of JSPS.

References

Albermann N, Schmitz-Winnenthal FH, Z'graggen K, Volk C, Hoffmann MM, Haefeli WE, Weiss J (2005) Expression of the drug transporters MDR1/ABCB1, MRP1/ABCC1, MRP2/ABCC2, BCRP/ABCG2, and PXR in peripheral blood mononuclear cells and their relationship with the expression in intestine and liver. Biochem Pharmacol 70:949–958

Azuma K, Urano T, Ouchi Y, Inoue S (2009) Vitamin K2 suppresses proliferation and motility of hepatocellular carcinoma cells by activating steroid and xenobiotic receptor. Endocr J 56:843–849

Azuma K, Casey SC, Ito M, Urano T, Horie K, Ouchi Y, Kirchner S, Blumberg B, Inoue S (2010) Pregnane X receptor knockout mice display osteopenia with reduced bone formation and enhanced bone resorption. J Endocrinol 207:257–263

Azuma K, Tsukui T, Ikeda K, Shiba S, Nakagawa K, Okano T, Urano T, Horie-Inoue K, Ouchi Y, Ikawa M, Inoue S (2014) Liver-specific γ-glutamyl carboxylase-deficient mice display bleeding diathesis and short life span. PLoS One 9:e88643

Berkner KL, Pudota BN (1984) Vitamin K-dependent carboxylation of the carboxylase. Proc Natl Acad Sci U S A 95:466–471

Beulens JW, Bots ML, Atsma F, Bartelink ML, Prokop M, Geleijnse JM, Witteman JC, Grobbee DE, van der Schouw YT (2009) High dietary menaquinone intake is associated with reduced coronary calcification. Atherosclerosis 203:489–493

Blumberg B, Sabbagh W Jr, Juguilon H, Bolado J Jr, van Meter CM, Ong ES, Evans RM (1998) SXR, a novel steroid and xenobiotic-sensing nuclear receptor. Genes Dev 12:3195–3205

Booth SL, Suttie JW (1998) Dietary intake and adequacy of vitamin K. J Nutr 128:785–788

Casey SC, Nelson EL, Turco GM, Janes MR, Fruman DA, Blumberg B (2011) B-1 cell lymphoma in mice lacking the steroid and xenobiotic receptor, SXR. Mol Endocrinol 25:933–943

Cheung AM, Tile L, Lee Y, Tomlinson G, Hawker G, Scher J, Hu H, Vieth R (2008) Vitamin K supplementation in postmenopausal women with osteopenia (ECKO trial): a randomized controlled trial. PLoS Med 5:e196

Cockayne S, Adamson J, Lanham-New S, Shearer MJ, Gilbody S, Torgerson DJ (2006) Vitamin K and the prevention of fractures: systematic review and meta-analysis of randomized controlled trials. Arch Intern Med 166:1256–1261

Coutu DL, Wu JH, Monette A, Rivard GE, Blostein MD, Galipeau J (2008) Periostin, a member of a novel family of vitamin K-dependent proteins, is expressed by mesenchymal stromal cells. J Biol Chem 283:17991–18001

Dam H (1935) The antihemorrhagic vitamin of the chick: occurrence and chemical nature. Nature 135:652–653

Ducy P, Desbois C, Boyce B, Pinero G, Story B, Dunstan C, Smith E, Bonadio J, Goldstein S, Gundberg C, Bradley A, Karsenty G (1996) Increased bone formation in osteocalcin-deficient mice. Nature 382:448–452

Enomoto M, Tsuchida A, Miyazawa K, Yokoyama T, Kawakita H, Tokita H, Naito M, Itoh M, Ohyashiki K, Aoki T (2007) Vitamin K2-induced cell growth inhibition via autophagy formation in cholangiocellular carcinoma cell lines. Int J Mol Med 20:801–808

Filipp D, Alizadeh-Khiavi K, Richardson C, Palma A, Paredes N, Takeuchi O, Akira S, Julius M (2001) Soluble CD14 enriched in colostrum and milk induces B cell growth and differentiation. Proc Natl Acad Sci U S A 98:603–608

Fujimura T, Takahashi S, Urano T, Tanaka T, Zhang W, Azuma K, Takayama K, Obinata D, Murata T, Horie-Inoue K, Kodama T, Ouchi Y, Homma Y, Inoue S (2012) Clinical significance of steroid and xenobiotic receptor and its targeted gene CYP3A4 in human prostate cancer. Cancer Sci 103:176–180

Gast GC, de Roos NM, Sluijs I, Bots ML, Beulens JW, Geleijnse JM, Witteman JC, Grobbee DE, Peeters PH, van der Schouw YT (2009) A high menaquinone intake reduces the incidence of coronary heart disease. Nutr Metab Cardiovasc Dis 19:504–510

Habu D, Shiomi S, Tamori A, Takeda T, Tanaka T, Kubo S, Nishiguchi S (2004) Role of vitamin K2 in the development of hepatocellular carcinoma in women with viral cirrhosis of the liver. JAMA 292:358–361

Ichikawa T, Horie-Inoue K, Ikeda K, Blumberg B, Inoue S (2006) Steroid and xenobiotic receptor SXR mediates vitamin K2-activated transcription of extracellular matrix-related genes and collagen accumulation in osteoblastic cells. J Biol Chem 281:16927–16934

Kakizaki S, Sohara N, Sato K, Suzuki H, Yanagisawa M, Nakajima H, Takagi H, Naganuma A, Otsuka T, Takahashi H, Hamada T, Mori M (2007) Preventive effects of vitamin K on recurrent disease in patients with hepatocellular carcinoma arising from hepatitis C viral infection. J Gastroenterol Hepatol 22:518–522

Kaneki M, Hodges SJ, Hosoi T, Fujiwara S, Lyons A, Crean SJ, Ishida N, Nakagawa M, Takechi M, Sano Y, Mizuno Y, Hoshino S, Miyao M, Inoue S, Horiki K, Shiraki M, Ouchi Y, Orimo H (2001) Japanese fermented soybean food as the major determinant of the large geographic difference in circulating levels of vitamin K2: possible implications for hip-fracture risk. Nutrition 17:315–321

Karasawa S, Azuma M, Kasama T, Sakamoto S, Kabe Y, Imai T, Yamaguchi Y, Miyazawa K, Handa H (2013) Vitamin K2 covalently binds to Bak and induces Bak-mediated apoptosis. Mol Pharmacol 83:613–620

Kawakita H, Tsuchida A, Miyazawa K, Naito M, Shigoka M, Kyo B, Enomoto M, Wada T, Katsumata K, Ohyashiki K, Itoh M, Tomoda A, Aoki T (2009) Growth inhibitory effects of vitamin K2 on colon cancer cell lines via different types of cell death including autophagy and apoptosis. Int J Mol Med 23:709–716

Kinoshita H, Nakagawa K, Narusawa K, Goseki-Sone M, Fukushi-Irie M, Mizoi L, Yoshida H, Okano T, Nakamura T, Suzuki T, Inoue S, Orimo H, Ouchi Y, Hosoi T (2007) A functional single nucleotide polymorphism in the vitamin-K-dependent gamma-glutamyl carboxylase gene (Arg325Gln) is associated with bone mineral density in elderly Japanese women. Bone 40:451–456

Kulman JD, Harris JE, Haldeman BA, Davie EW (1997) Primary structure and tissue distribution of two novel proline-rich gamma-carboxyglutamic acid proteins. Proc Natl Acad Sci U S A 94:9058–9062

Kulman JD, Harris JE, Xie L, Davie EW (2001) Identification of two novel transmembrane gamma-carboxyglutamic acid proteins expressed broadly in fetal and adult tissues. Proc Natl Acad Sci U S A 98:1370–1375

Lamson DW, Plaza SM (2003) The anticancer effects of vitamin K. Altern Med Rev 8:303–318

Luo G, Ducy P, McKee MD, Pinero GJ, Loyer E, Behringer RR, Karsenty G (1997) Spontaneous calcification of arteries and cartilage in mice lacking matrix GLA protein. Nature 386:78–81

Ma M, Qu XJ, Mu GY, Chen MH, Cheng YN, Kokudo N, Tang W, Cui SX (2009) Vitamin K2 inhibits the growth of hepatocellular carcinoma via decrease of des-gamma-carboxy prothrombin. Chemotherapy 55:28–35

Manabe N, Kawaguchi H, Chikuda H, Miyaura C, Inada M, Nagai R, Nabeshima Y, Nakamura K (2001) Connection between B lymphocyte and osteoclast differentiation pathways. J Immunol 167:2625–2631

McIntosh JM, Olivera BM, Cruz LJ, Gray WR (1984) Gamma-carboxyglutamate in a neuroactive toxin. J Biol Chem 259:14343–14346

Miki Y, Suzuki T, Tazawa C, Blumberg B, Sasano H (2005) Steroid and xenobiotic receptor (SXR), cytochrome P450 3A4 and multidrug resistance gene 1 in human adult and fetal tissues. Mol Cell Endocrinol 231:75–85

Miki Y, Suzuki T, Kitada K, Yabuki N, Shibuya R, Moriya T, Ishida T, Ohuchi N, Blumberg B, Sasano H (2006) Expression of the steroid and xenobiotic receptor and its possible target gene, organic anion transporting polypeptide-A, in human breast carcinoma. Cancer Res 66:535–542

Misra D, Booth SL, Tolstykh I, Felson DT, Nevitt MC, Lewis CE, Torner J, Neogi T (2013) Vitamin K deficiency is associated with incident knee osteoarthritis. Am J Med 126:243–248

Miyazawa K, Nishimaki J, Ohyashiki K, Enomoto S, Kuriya S, Fukuda R, Hotta T, Teramura M, Mizoguchi H, Uchiyama T, Omine M (2000) Vitamin K2 therapy for myelodysplastic syndromes (MDS) and post-MDS acute myeloid leukemia: information through a questionnaire survey of multi-center pilot studies in Japan. Leukemia 14:1156–1157

Mizuta T, Ozaki I, Eguchi Y, Yasutake T, Kawazoe S, Fujimoto K, Yamamoto K (2006) The effect of menatetrenone, a vitamin K2 analog, on disease recurrence and survival in patients with hepatocellular carcinoma after curative treatment: a pilot study. Cancer 106:867–872

Morris DP, Stevens RD, Wright DJ, Stafford DW (1995) Processive post-translational modification. Vitamin K-dependent carboxylation of a peptide substrate. J Biol Chem 270:30491–30498

Murshed M, Schinke T, McKee MD, Karsenty G (2004) Extracellular matrix mineralization is regulated locally; different roles of two gla-containing proteins. J Cell Biol 165:625–630

Nakagawa K, Hirota Y, Sawada N, Yuge N, Watanabe M, Uchino Y, Okuda N, Shimomura Y, Suhara Y, Okano T (2010) Identification of UBIAD1 as a novel human menaquinone-4 biosynthetic enzyme. Nature 468:117–121

Nelsestuen GL, Zytkovicz TH, Howard JB (1974) The mode of action of vitamin K. Identification of gamma-carboxyglutamic acid as a component of prothrombin. J Biol Chem 249:6347–6350

Neogi T, Booth SL, Zhang YQ, Jacques PF, Terkeltaub R, Aliabadi P, Felson DT (2006) Low vitamin K status is associated with osteoarthritis in the hand and knee. Arthritis Rheum 54:1255–1261

Neogi T, Felson DT, Sarno R, Booth SL (2008) Vitamin K in hand osteoarthritis: results from a randomized clinical trial. Ann Rheum Dis 67:1570–1573

Nimptsch K, Rohrmann S, Kaaks R, Linseisen J (2010) Dietary vitamin K intake in relation to cancer incidence and mortality: results from the Heidelberg cohort of the European Prospective Investigation into Cancer and Nutrition (EPIC-Heidelberg). Am J Clin Nutr 91:1348–1358

Ohta K, Lupo G, Kuriyama S, Keynes R, Holt CE, Harris WA, Tanaka H, Ohnuma S (2004) Tsukushi functions as an organizer inducer by inhibition of BMP activity in cooperation with chordin. Dev Cell 7:347–358

Oka H, Akune T, Muraki S, En-yo Y, Yoshida M, Saika A, Sasaki S, Nakamura K, Kawaguchi H, Yoshimura N (2009) Association of low dietary vitamin K intake with radiographic knee osteoarthritis in the Japanese elderly population: dietary survey in a population-based cohort of the ROAD study. J Orthop Sci 14:687–692

Okano T, Shimomura Y, Yamane M, Suhara Y, Kamao M, Sugiura M, Nakagawa K (2008) Conversion of phylloquinone (Vitamin K1) into menaquinone-4 (Vitamin K2) in mice: two possible routes for menaquinone-4 accumulation in cerebra of mice. J Biol Chem 283:11270–11279

Price PA, Otsuka AA, Poser JW, Kristaponis J, Raman N (1976) Characterization of a gamma-carboxyglutamic acid-containing protein from bone. Proc Natl Acad Sci U S A 73:1447–1451

Rezaie AR, Bae JS, Manithody C, Qureshi SH, Yang L (2008) Protein Z-dependent protease inhibitor binds to the C-terminal domain of protein Z. J Biol Chem 283:19922–19926

Roman-Roman S, Garcia T, Jackson A, Theilhaber J, Rawadi G, Connolly T, Spinella-Jaegle S, Kawai S, Courtois B, Bushnell S, Auberval M, Call K, Baron R (2003) Identification of genes regulated during osteoblastic differentiation by genome-wide expression analysis of mouse calvaria primary osteoblasts in vitro. Bone 32:474–482

Rüggeberg S, Horn P, Li X, Vajkoczy P, Franz T (2008) Detection of a gamma-carboxy-glutamate as novel post-translational modification of human transthyretin. Protein Pept Lett 15:43–46

Sada E, Abe Y, Ohba R, Tachikawa Y, Nagasawa E, Shiratsuchi M, Takayanagi R (2010) Vitamin K2 modulates differentiation and apoptosis of both myeloid and erythroid lineages. Eur J Haematol 85:538–548

Shatenstein B, Kergoat MJ, Reid I (2007) Poor nutrient intakes during 1-year follow-up with community-dwelling older adults with early-stage Alzheimer dementia compared to cognitively intact matched controls. J Am Diet Assoc 107:2091–2099

Shea MK, O'Donnell CJ, Hoffmann U, Dallal GE, Dawson-Hughes B, Ordovas JM, Price PA, Williamson MK, Booth SL (2009) Vitamin K supplementation and progression of coronary artery calcium in older men and women. Am J Clin Nutr 89:1799–1807

Son BK, Kozaki K, Iijima K, Eto M, Kojima T, Ota H, Senda Y, Maemura K, Nakano T, Akishita M, Ouchi Y (2006) Statins protect human aortic smooth muscle cells from inorganic phosphate-induced calcification by restoring Gas6-Axl survival pathway. Circ Res 98:1024–1031

Stafford DW (2005) The vitamin K cycle. J Thromb Haemost 3:1873–1878

Staudinger JL, Goodwin B, Jones SA, Hawkins-Brown D, MacKenzie KI, LaTour A, Liu Y, Klaassen CD, Brown KK, Reinhard J, Willson TM, Koller BH, Kliewer SA (2001) The nuclear receptor PXR is a lithocholic acid sensor that protects against liver toxicity. Proc Natl Acad Sci U S A 98:3369–3374

Stenflo J, Fernlund P, Egan W, Roepstorff P (1974) Vitamin K dependent modifications of glutamic acid residues in prothrombin. Proc Natl Acad Sci U S A 71:2730–2733

Sundaram KS, Lev M (1988) Warfarin administration reduces synthesis of sulfatides and other sphingolipids in mouse brain. J Lipid Res 29:1475–1479

Synold TW, Dussault I, Forman BM (2001) The orphan nuclear receptor SXR coordinately regulates drug metabolism and efflux. Nat Med 7:584–590

Tabb MM, Sun A, Zhou C, Grün F, Errandi J, Romero K, Pham H, Inoue S, Mallick S, Lin M, Forman BM, Blumberg B (2003) Vitamin K2 regulation of bone homeostasis is mediated by the steroid and xenobiotic receptor SXR. J Biol Chem 278:43919–43927

Takeyama D, Miki Y, Fujishima F, Suzuki T, Akahira J, Hata S, Miyata G, Satomi S, Sasano H (2010) Steroid and xenobiotic receptor in human esophageal squamous cell carcinoma: a potent prognostic factor. Cancer Sci 101:543–549

Thijssen HH, Drittij-Reijnders MJ (1996) Vitamin K status in human tissues: tissue-specific accumulation of phylloquinone and menaquinone-4. Br J Nutr 75:121–127

Ueda N, Shiraha H, Fujikawa T, Takaoka N, Nakanishi Y, Suzuki M, Matsuo N, Tanaka S, Nishina S, Uemura M, Takaki A, Shiratori Y, Yamamoto K (2008) Exon 2 deletion splice variant of gamma-glutamyl carboxylase causes des-gamma-carboxy prothrombin production in hepatocellular carcinoma cell lines. Mol Oncol 2:241–249

Varnum BC, Young C, Elliott G, Garcia A, Bartley TD, Fridell YW, Hunt RW, Trail G, Clogston C, Toso RJ, Yanagihara D, Bennett L, Sylber M, Merewether LA, Tseng A, Escobar E, Liu ET, Yamane HK (1995) Axl receptor tyrosine kinase stimulated by the vitamin K-dependent protein encoded by growth-arrest-specific gene 6. Nature 373:623–626

Vergnaud P, Garnero P, Meunier PJ, Bréart G, Kamihagi K, Delmas PD (1997) Undercarboxylated osteocalcin measured with a specific immunoassay predicts hip fracture in elderly women: the EPIDOS Study. J Clin Endocrinol Metab 82:719–724

Verma S, Tabb MM, Blumberg B (2009) Activation of the steroid and xenobiotic receptor, SXR, induces apoptosis in breast cancer cells. BMC Cancer 9:3

Viegas CS, Simes DC, Laizé V, Williamson MK, Price PA, Cancela ML (2008) Gla-rich protein (GRP), a new vitamin K-dependent protein identified from sturgeon cartilage and highly conserved in vertebrates. J Biol Chem 283:36655–36664

Wagener R, Ehlen HW, Ko YP, Kobbe B, Mann HH, Sengle G, Paulsson M (2005) The matrilins – adaptor proteins in the extracellular matrix. FEBS Lett 579:3323–3329

Warder SE, Prorok M, Chen Z, Li L, Zhu Y, Pedersen LG, Ni F, Castellino FJ (1998) The roles of individual gamma-carboxyglutamate residues in the solution structure and cation-dependent properties of conantokin-T. J Biol Chem 273:7512–7522

Xie W, Radominska-Pandya A, Shi Y, Simon CM, Nelson MC, Ong ES, Waxman DJ, Evans RM (2001) An essential role for nuclear receptors SXR/PXR in detoxification of cholestatic bile acids. Proc Natl Acad Sci U S A 98:3375–3380

Yaegashi Y, Onoda T, Tanno K, Kuribayashi T, Sakata K, Orimo H (2008) Association of hip fracture incidence and intake of calcium, magnesium, vitamin D, and vitamin K. Eur J Epidemiol 23:219–225

Yagami T, Ueda K, Asakura K, Sakaeda T, Nakazato H, Kuroda T, Hata S, Sakaguchi G, Itoh N, Nakano T, Kambayashi Y, Tsuzuki H (2002) Gas6 rescues cortical neurons from amyloid beta protein-induced apoptosis. Neuropharmacology 43:1289–1296

Yaguchi M, Miyazawa K, Katagiri T, Nishimaki J, Kizaki M, Tohyama K, Toyama K (1997) Vitamin K2 and its derivatives induce apoptosis in leukemia cells and enhance the effect of all-trans retinoic acid. Leukemia 11:779–787

Yoshida H, Shiratori Y, Kudo M, Shiina S, Mizuta T, Kojiro M, Yamamoto K, Koike Y, Saito K, Koyanagi N, Kawabe T, Kawazoe S, Kobashi H, Kasugai H, Osaki Y, Araki Y, Izumi N, Oka H, Tsuji K, Toyota J, Seki T, Osawa T, Masaki N, Ichinose M, Seike M, Ishikawa A, Ueno Y, Tagawa K, Kuromatsu R, Sakisaka S, Ikeda H, Kuroda H, Kokuryu H, Yamashita T, Sakaida I, Katamoto T, Kikuchi K, Nomoto M, Omata M (2011) Effect of vitamin K2 on the recurrence of hepatocellular carcinoma. Hepatology 54:532–540

Yue X, Akahira J, Utsunomiya H, Miki Y, Takahashi N, Niikura H, Ito K, Sasano H, Okamura K, Yaegashi N (2010) Steroid and Xenobiotic Receptor (SXR) as a possible prognostic marker in epithelial ovarian cancer. Pathol Int 60:400–406

Zhou C, Tabb MM, Nelson EL, Grün F, Verma S, Sadatrafiei A, Lin M, Mallick S, Forman BM, Thummel KE, Blumberg B (2006) Mutual repression between steroid and xenobiotic receptor and NF-kappaB signaling pathways links xenobiotic metabolism and inflammation. J Clin Invest 116:2280–2289

Zhou C, Verma S, Blumberg B (2009) The steroid and xenobiotic receptor (SXR), beyond xenobiotic metabolism. Nucl Recept Signal 7:e001

Zhu A, Sun H, Raymond RM Jr, Furie BC, Furie B, Bronstein M, Kaufman RJ, Westrick R, Ginsburg D (2007) Fatal hemorrhage in mice lacking gamma-glutamyl carboxylase. Blood 109:5270–5275

Part V
Aging Brain: Adult Neurogenesis, Synaptic Plasticity, and Brain Volume

Chapter 14
Insights into Aging of the Hippocampus: A View from the Topographic Differentiation

Shozo Jinno

Abstract More than fifty years ago, a number of studies reported the reduction of neuron numbers related to brain aging. However, later studies have concluded that the neuronal loss due to aging is limited to specific regions of the nervous system and its significance is irrelevant in both humans and non-human mammals. Instead, several other mechanisms that may underlie brain aging have attracted attention in the field of neuroscience research. Namely, some papers have indicated the relationship between memory impairment and decline in adult neurogenesis in the hippocampus during aging. It has also been reported that abnormalities of oligodendrogenesis in the mature brain may be involved in age-related cognitive decline. We herein briefly review recent findings on age-related changes in adult neurogenesis and oligodendrogenesis, and discuss their functional significance from the view point of topography of the hippocampus. Namely, the hippocampus has shown to be structurally and functionally differentiated along the longitudinal and transverse axes. In the rodent brain, the dorsal (septal) hippocampus is involved in cognition, learning, and memory, while the ventral (temporal) hippocampus contributes to regulation of emotion, mood, and anxiety. Nevertheless, the question of how topographic differentiation of the hippocampus might be affected by aging still remains largely unanswered. Our latest studies have shown that the waning of adult neurogenesis and oligodendrogenesis during aging is more relevant in the ventral hippocampus than in the dorsal hippocampus. We therefore hypothesize that the ventral-dominant decline in hippocampal neurogenesis and oligodendrogenesis may partly explain why major depression frequently precedes dementia in elderly people. These findings provide new insights into aging of the hippocampus.

Keywords Hippocampus • Differentiation • Topography • Aging • Adult neurogenesis • Oligodendrogenesis

---

S. Jinno, M.D., Ph.D. (✉)
Department of Anatomy and Neuroscience, Graduate School of Medical Sciences, Kyushu University, Maidashi, Higashi-ku, Fukuoka 812-8582, Japan
e-mail: sjnno@med.kyushu-u.ac.jp

© Springer Japan 2015

N. Mori, I. Mook-Jung (eds.), *Aging Mechanisms*,
DOI 10.1007/978-4-431-55763-0_14

14.1 Introduction

More than fifty years ago, much attention was given to changes in neuron numbers related to brain aging, and many papers reported the significant neuron loss occurred in the aged brain (Brody 1955). However, later studies have concluded that neuronal loss due to aging is limited to specific regions of the nervous system, and is irrelevant (probably no more than 10 %) in both humans and non-human animals (Curcio and Coleman 1982; Peters and Sethares 1993; Pakkenberg and Gundersen 1997). Considering that the variance of neuron numbers in humans with normal cognitive function is large, it is unlikely that a 10 % loss in neurons that is not ubiquitous is the significant factor causing various symptoms that often accompany normal aging (Pannese 2011). To date, several other mechanisms causing brain aging have attracted attention in the field of neuroscience research. For instance, there is a substantial amount of data showing that the rate of adult neurogenesis in the hippocampus radically wanes during aging (Merkley et al. 2014). It has also been suggested that age-related decline in adult neurogenesis is an important factor influencing cognitive performance (Artegiani and Calegari 2012). Similar to neurogenesis, oligodendrogenesis and white matter homeostasis might also be affected by aging (Sim et al. 2002; Miyamoto et al. 2013). We herein briefly review recent findings on age-related changes in adult neurogenesis and oligodendrogenesis, and discuss their functional significance from the view point of topography of the hippocampus.

14.2 Differentiation of the Hippocampus

The rodent hippocampus has an elongated shape with its major axis extending in a C-shaped fashion from the septal nuclei of the basal forebrain to the temporal lobe. Importantly, the structure and function of the rodent hippocampus is differentiated along the longitudinal (dorsoventral) and transverse axes (Fanselow and Dong 2010). Earlier anatomical studies have reported the topographic differentiation of the hippocampus along the dorsoventral axis (Gaarskjaer 1978). The dorsal hippocampus sends massive projections to the retrosplenial cortex and mammillary complex (Ishizuka 2001; van Groen and Wyss 2003). The ventral hippocampus has intimate reciprocal connections with the amygdala and strong projections to the nucleus accumbens (Pitkänen et al. 2000). The dorsal dentate gyrus (DG) receives afferents both from the lateral and medial area of the entorhinal cortex, whereas the ventral DG receives projections from the medial area of the entorhinal cortex (Witter et al. 1989). In a series of our studies, we have demonstrated the existence of dorsoventral differences in the cellular architecture of the mouse hippocampus (Jinno and Kosaka 2006, 2010). Functional differentiation along the dorsoventral axis has also been reported in the past two decades. Injuries to the dorsal hippocampus impair spatial learning (Moser et al. 1993), but lesions of the ventral hippocampus affect anxiety-related behavior, and have no effect on spatial learning (Bannerman et al. 2003). Genes expressed in the dorsal hippocampus are associated with the brain regions related to cognitive information-processing, while genes expressed in the ventral hippocampus are associated with regions involved in emotional behaviors (Dong et al. 2009). To date, it has been widely accepted that the dorsal hippocampus plays a preferential role in cognition, learning and memory, while the ventral hippocampus is involved in regulation of emotion, mood, and anxiety. Interestingly, the differentiation of the hippocampus is evolutionarily conserved in rodents, monkeys (Colombo et al. 1998) and humans (Small et al. 2011).

14.3 Topography of Adult Hippocampal Neurogenesis

Throughout adulthood, new granule cells are continuously generated in the subgranular zone of the DG (Altman and Das 1967; Kaplan and Hinds 1977; Cameron et al. 1993). Recent studies indicated that adult neurogenesis is involved in various hippocampal functions (Balu and Lucki 2009). Drug-induced inhibition of cell proliferation impaired learning of a hippocampus-dependent spatial memory task (Shors et al. 2001). Addition and removal of adult-born granule cells in the DG might influence spatial learning and memory (Drapeau et al. 2003; Dupret et al. 2007; Farioli-Vecchioli et al. 2008).

In our recent study (Jinno 2011a), we examined the topography of adult neurogenesis along the longitudinal (dorsal vs. ventral) and transverse (suprapyramidal vs. infrapyramidal) axes of the DG of young adult mice using endogenous neurogenesis markers (Fig. 14.1): brain lipid binding protein (BLBP), doublecortin (DCX), calretinin (CR), proliferation cell nuclear antigen (PCNA) and Ki-67. BLBP belongs to the fatty acid binding protein family (Feng et al. 1994), and is expressed in radial-glia like progenitors and intermediate progenitors, i.e., neural stem cells (NSCs). DCX is a microtubule-associated phosphoprotein (Gleeson et al. 1999; Francis et al. 1999), which labels both lineage-restricted neuronal progenitors and immature granule cells, i.e., neural lineage cells (NLCs) (Brown et al. 2003; Rao and Shetty 2004). CR belongs to the EF-hand calcium-binding protein family, and is transiently expressed in granule cells at early postmitotic stage (Liu et al. 1996). PCNA (Celis and Celis 1985) and Ki-67 (Gerdes et al. 1984) can label dividing cells. Using combinations of these markers, the cells at specific stages of adult neurogenesis can be identified (Kempermann et al. 2004).

It was reported that adult neurogenesis may be more active in the dorsal hippocampus than in the ventral hippocampus (Snyder et al. 2009). In adult male gerbils, the number of 7-day-old bromodeoxyuridine (BrdU)-labeled granule cells was larger in the dorsal DG than in the ventral DG (Dawirs et al. 1998). Similarly, the number of 14-day-old BrdU-labeled granule cells was larger in the dorsal hippocampus than in the ventral hippocampus in adult male C57BL/6J mice (Ferland et al. 2002). In agreement with these studies, our recent study has shown

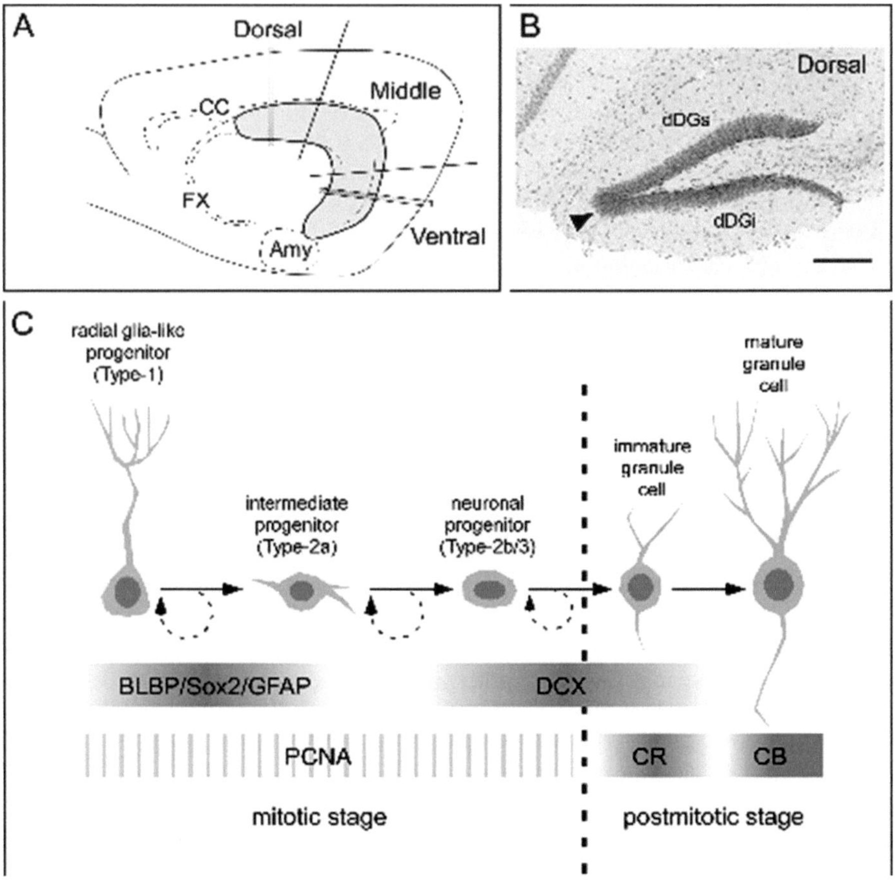

Fig. 14.1 Topography of the rodent hippocampus and endogenous markers of the adult neurogenesis. (**a**) The hippocampus (*shaded*) can be divided into the dorsal, middle and ventral regions according to the longitudinal axis. (**b**) Inverted image of 4′,6-Diamidino-2-phenylindole, dihydrochloride (DAPI)-stained transverse section of the dorsal DG. *Arrowhead* indicates the crescent. (**c**) Stages and endogenous markers of the adult neurogenesis in the hippocampus. *Scale bar* in **b** = 200 μm (applies to **b**). *Amy* amygdala, *CC* corpus callosum, *dDGi* infrapyramidal blade in dorsal DG, *dDGs* suprapyramidal blade in dorsal DG, *FX* fornix (Modified and reproduced from Jinno (2011a), with permission of the publisher)

that the numerical densities (NDs) of DCX+ NLCs and BLBP+ NSCs are significantly higher in the dorsal DG than in the ventral DG in young adult male C57BL/6J mice (Jinno 2011a). Considering the functional differentiation of the hippocampus, these results suggest that newly generated granule cells may play different roles in regulation of cognition (dorsal) and emotion (ventral).

Currently, functional dissociations along the transverse axis of the hippocampus are less intensively examined, but earlier studies reported some structural differences between two blades in the DG. For instance, the mossy fibers from the suprapyramidal blade cross directly through the stratum radiatum to reach the distal part of the stratum lucidum, whereas those arising from the infrapyramidal blade travel through the hilus and contact the proximal part of CA3 area in the rat (Claiborne et al. 1986). With regard to the inter-blade differences in the adult neurogenesis, Ambrogini et al. (2000) showed that the number of 15-day-old BrdU+ granule cells was larger in the suprapyramidal blade than in the infrapyramidal blade in rats. Similarly, we have found that the NDs of DCX+ NLCs are relatively higher in the suprapyramidal blade in the dorsal DG (Jinno 2011a). Further research is necessary to tie these findings together and elucidate the functional significance of inter-blade difference in adult neurogenesis.

14.4 Age-Related Changes in Adult Neurogenesis

There is a substantial amount of data showing that the rate of new cell production in the hippocampus radically wanes during aging (Seki and Arai 1995; Cameron and McKay 1999). However, the mechanism underlying age-related decline in adult neurogenesis is still controversial. Namely, Alonso (2001) showed that the reduction of neurogenesis in aged rats was attributable to the decline in proliferation of primary progenitors. Hattiangady and Shetty (2008) also reported that aging did not alter the number of primary progenitors, and suggested that age-related decline might be an outcome of increased quiescence of progenitors in the neurogenic region of the rat hippocampus. By contrast, Olariu et al. (2007) claimed that the decreased neurogenesis in aged rats was attributable to loss of primary progenitor cells. Aizawa et al. (2011) reported that decline in primary progenitors defined by Sox2, GFAP, and BLBP expression was specific for aged primates, and there were no alterations in the number of these cells in aged (2-year-old) ICR mice. Walter et al. (2011) reported that age-related reduction in proliferation was not only caused by a general reduction in total number of progenitor subtypes but also by a subtype-specific alteration of the proliferation rate.

In our recent study (Jinno 2011b), we examined the age-related differences in adult neurogenesis between young adult (2-month-old) and middle-aged (10-month-old) mice using endogenous markers (see, Fig. 14.2). The age-related reductions in BLBP+ NSCs were significantly larger in the ventral DG (76 % decrease) than in the dorsal DG (56 % decrease). The age-related reductions in DCX+ NLCs were more drastic than those of NSCs accompanying with a similar dorsoventral gradient: the ventral DG (95 % decrease), dorsal DG (91 % decrease). In the field of geriatric psychiatry, major depressive disorder and dementia are common conditions in old age, and frequently occur concurrently (Korczyn and Halperin 2009). These two clinical entities have a very complicated relationship, and accurate mechanisms underlying their co-occurrence are largely unclear. Some studies hypothesize that depressive disorder is a risk factor for developing dementia (Kessing and Nilsson 2003; Ownby et al. 2006), and suggest that depression is an early prodromal phase of dementia state (Schweitzer et al. 2002). Our findings

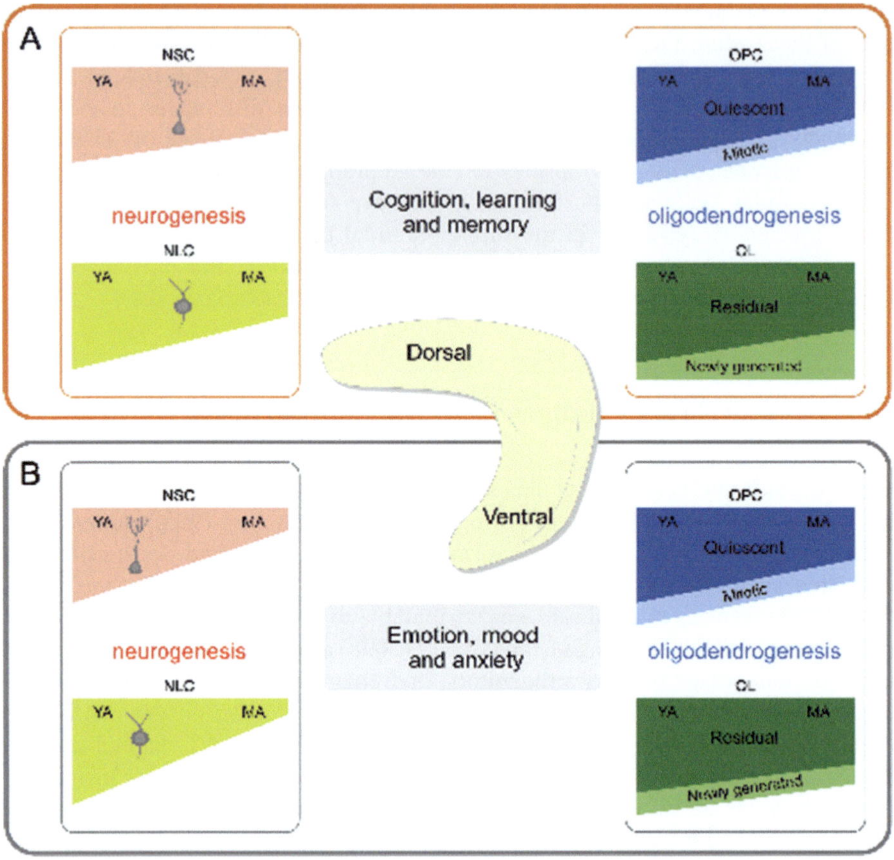

Fig. 14.2 Schematic illustration of the age-related changes in adult neurogenesis and oligodendrogenesis in the rodent hippocampus. (**a**) The dorsal hippocampus is mainly involved in cognition, learning and memory. In this region, the number of NSCs and NLCs halves from YA (young adult) to MA (middle-aged). The number of quiescent OPCs decline with age, and the fraction of mitotic OPCs remains invariant during aging. The number of residual OLs declines with age, while the fraction of newly generated OLs increases during aging. As a result, the number of OLs remain unchanged. (**b**) The hippocampus contributes to the regulation of emotion, mood and anxiety. The number of NSCs and NLCs drops sharply from YA to MA. The number of quiescent OPCs declines with age, and the fraction of mitotic OPCs remains invariant during aging. The number of residual OLs declines with age, and the fraction of newly generated OLs remains unchanged (Modified and reproduced from Jinno (2011b), with permission of the publisher)

indicate that hippocampal neurogenesis wanes faster in the ventral hippocampus than in the dorsal hippocampus during aging. Interestingly, exposure to chronic mild stress results in decreased cell proliferation in the ventral but not in the dorsal hippocampus (Jayatissa et al. 2006). Chronic treatment with agomelatin, an antidepressant, increases neurogenesis only in the ventral DG (Banasr et al. 2006). Together, these findings provide some key to understand why depression frequently precedes dementia in aged people. Future studies addressing this issue will inform us on how age-related alterations in neurogenesis are involved in late onset depression and dementia.

At this time, only a few attempts have been made at the functional significance of age-related changes in transverse differentiation of the DG. In our study, we have reported that there were no inter-blade differences in neurogenesis in middle-aged (10-month-old) mice (Jinno 2011b), while the number of BLBP+ NLCs and DCX+ NLCs were significantly higher in the suprapyramidal blade than in infrapyramidal blade in young adult (2-month-old) mice (Jinno 2011a). In this regard, there is an interesting report showing that deposition and maturation of granule cells begin near the lateral tip of the suprapyramidal blade and proceed into the infrapyramidal blade, establishing the suprapyramidal to infrapyramidal morphogenic gradient in the DG (Angevine 1965). These results indicate that dentate neurogenesis might be more active in the suprapyramidal blade than in the infrapyramidal blade only during adolescence and young adulthood, and also suggest that larger number of new granule cells in the suprapyramidal blade could be involved in the higher cognitive performance of young animals.

14.5 Oligodendrogenesis and Myelin Homeostasis

Oligodendrocytes (OLs) synthesize myelin, which is required for fast saltatory conduction of nerve impulses. The majority of myelinating OLs are born in the early postnatal period by differentiation of oligodendrocyte precursor cells (OPCs). Recent evidence shows that impaired paranode structure and function can impact neural circuitry, leading to downstream effects related to emotion and potentially to mood regulation in human psychiatric disorders (Edgar and Sibille 2012). Histological analysis using Kluver–Barrera staining method revealed that the staining intensity of deep white matter in the dorsolateral prefrontal cortex was significantly less intense in subjects who suffered from major depression (Regenold et al. 2007). A decrease in OL density has been reported in the frontopolar cortex of major depressive disorder subjects (Hayashi et al. 2011).

In the healthy adult brain, OPCs continue to divide and generate new OLs (Richardson et al. 2011). It has been shown that OPCs express NG2 proteoglycan (so they are also known as NG2 cells) and the platelet-derived growth factor receptor-alpha (PDGFαR). NG2 and PDGFαR play a critical role in proliferation, migration and survival of OPCs (Noble et al. 1988; Barres et al. 1993; Hill et al. 2013; Binamé et al. 2013). Recent studies have indicated that OPCs are not just progenitors, but are also involved in regulation of neuronal circuits, because these cells receive synaptic inputs from neurons and respond to neurotransmitters released at synapses (Bergles et al. 2000; Wigley et al. 2007). Particularly, OPCs have AMPA-type glutamate receptors, which are activated by neural activity (Lin et al. 2005). In addition, OPCs sense fine changes in extracellular K^+ concentrations during physiological neuronal activity (Maldonado et al. 2013). The authors suggest that OPCs possibly remove the excess K^+ caused by neuronal K^+ efflux at specific sites devoid of astrocytes via Kir4.1 channels. Moreover, OPCs are considered to release soluble factors, which promote neuronal survival, maintain axonal structure, and support synaptic plasticity (Wilkins et al. 2003; Sun et al. 2013). It is conceivable that OPCs sense the "state of health" of their partner neurons over the neuron-glial synapse and respond accordingly by release of neuroprotective substances (Sakry et al. 2011).

14.6 Topography of Age-Related Changes in Hippocampal Oligodendrogenesis

Until now, many papers reported the involvement of microglia and astrocytes in brain aging (Bronson et al. 1993; Sheng et al. 1996; Wu et al. 2005). Recently, age-related changes in OLs have also been well documented. For instance, several studies have shown that abnormalities of OLs and myelin may be involved in age-related cognitive decline (Peters and Kemper 2012; Kohama et al. 2012). The NDs of OLs in the rat hippocampus showed a significant aging-dependent reduction (Tanaka et al. 2005). The intensity of immunostaining for 2′,3′-cyclic nucleotide 3′-phosphodiesterase (a marker of OLs) in the rat hippocampus declined with age (Hayakawa et al. 2007). The number of OLs in the rat subcortical white matter also showed an aging-dependent reduction (Chen et al. 2011). Interestingly, the number of OLs in the mouse anterior commissure began to decline between 9 and 12 months and remained fairly low between 15 and 22 months, before rising sharply to above the 9 month level between 22 and 25 months and thereafter remaining constant (Sturrock 1987). In rhesus monkeys, the number of OLs in the optical nerve showed an age-dependent increase (Sandell and Peters 2002), while there were no significant age-related changes in the number of OLs in the occipital and prefrontal cortices (Peters and Sethares 2002).

In our recent study (Yamada and Jinno 2014), we estimated the age-related changes in oligodendrogenesis of young adult (2-month-old) and middle-aged (10-month-old) mouse hippocampi. To identify OPCs and OLs, we used a set of molecular markers, oligodendrocyte lineage transcription factor (Olig2) and PDGFαR. Intracellular dye injection shows that PDGFαR+/Olig2+ cells and PDGFαR-/Olig2+ cells can be defined as OPCs and OLs, respectively (Fig. 14.3). The quantitative analysis showed that the number of OLs declined with age in the ventral hippocampus, but they were not compromised in the dorsal hippocampus (Fig. 14.2). In this regard, it is necessary to consider two possibilities here. The first is the dorsoventral difference in death of OLs. The vulnerability of OLs increases with increasing age at differentiation as later-differentiating cells myelinate increasing the number of axonal segments (Bartzokis 2004). We thus hypothesized that mature OLs in the hippocampus might be more susceptible to aging in the ventral than the dorsal region. The second is the dorsoventral differences in production of OLs. Differently from neurogenesis, previous studies have shown

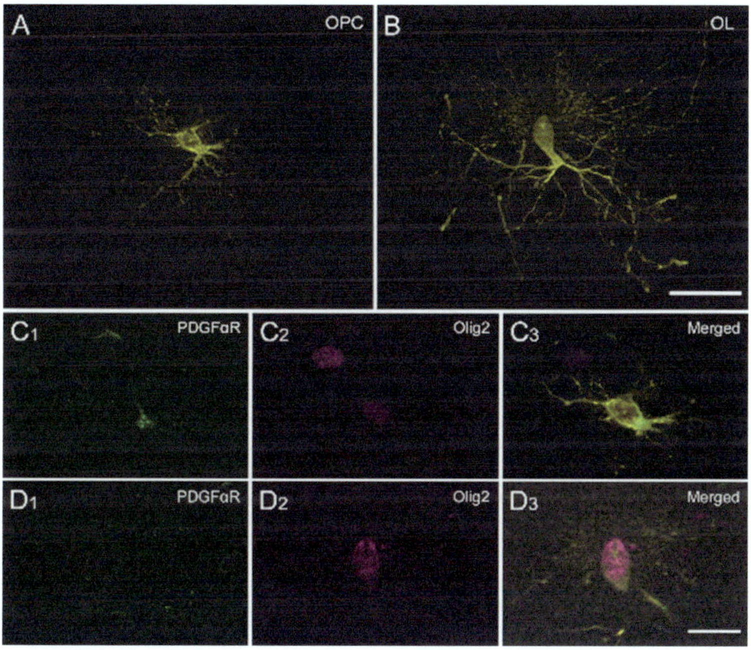

Fig. 14.3 Identification of OPCs and OLs by intracellular labeling and immunostaining for PDGFαR and Olig2 in the hippocampus. (**a**, **b**) Intracellular injection of lucifer yellow (LY) into OPC (**a**) and OL (**b**) in the mouse hippocampus. (**c**) An Olig2+/PDGFαR+ OPC shows short multi-branched processes. (**d**) An Olig2+/PDGFαR- OL shows long extended processes with ramified structure. *Scale bars* in **b** = 20 μm (applies to **a** and **b**), in $\mathbf{d}_3$ = 10 μm (applies to $\mathbf{c}_{1-3}$ and $\mathbf{d}_{1-3}$) (Modified and reproduced from Yamada and Jinno (2014), with permission of the publisher)

the rather controversial effects of aging on oligodendrogenesis. Namely, in the murine spinal cord, oligodendrogenesis was not only preserved, but it also increased during aging (Lasiene et al. 2009). In the rostral migratory stream of mice, the number of proliferative OPCs and new OLs remained unchanged during aging (Capilla-Gonzalez et al. 2013). Here we observed that the number of BrdU+ mitotic OPCs in the Ammon's horn were not compromised with age both in the dorsal and ventral hippocampus. It should also be noted that the number of BrdU+ newly generated OLs in the Ammon's horn significantly increased with age in the dorsal hippocampus, but remained unchanged in the ventral hippocampus. Together, these findings suggest that the number of OLs in the dorsal Ammon's horn may be compensatory maintained by increased generation of OLs.

14.7 Conclusion

Despite the increased number of publications in the field of gerontology, age-related changes in the topography of the hippocampus still remains largely unanswered. As this review summarizes here, the waning of adult neurogenesis and oligodendrogenesis during aging is more relevant in the ventral hippocampus than in the dorsal hippocampus. Because the ventral hippocampus mainly contributes to regulation of emotion, while the dorsal hippocampus has a preferential role in memory, these findings may provide some key to understanding various psychiatric problems seen in elderly people without dementia.

References

Aizawa K, Ageyama N, Terao K, Hisatsune T (2011) Primate-specific alterations in neural stem/progenitor cells in the aged hippocampus. Neurobiol Aging 32:140–150
Alonso G (2001) Proliferation of progenitor cells in the adult rat brain correlates with the presence of vimentin-expressing astrocytes. Glia 34:253–266
Altman J, Das GD (1967) Postnatal neurogenesis in the guinea-pig. Nature 214:1098–1101
Ambrogini P, Cuppini R, Cuppini C, Ciaroni S, Cecchini T, Ferri P, Sartini S, Del Grande P (2000) Spatial learning affects immature granule cell survival in adult rat dentate gyrus. Neurosci Lett 286:21–24
Angevine JB Jr (1965) Time of neuron origin in the hippocampal region. An autoradiographic study in the mouse. Exp Neurol Suppl 2:1–70
Artegiani B, Calegari F (2012) Age-related cognitive decline: can neural stem cells help us? Aging (Albany NY) 4:176–186
Balu DT, Lucki I (2009) Adult hippocampal neurogenesis: regulation, functional implications, and contribution to disease pathology. Neurosci Biobehav Rev 33:232–252
Banasr M, Soumier A, Hery M, Mocaër E, Daszuta A (2006) Agomelatine, a new antidepressant, induces regional changes in hippocampal neurogenesis. Biol Psychiatry 59:1087–1096
Bannerman DM, Grubb M, Deacon RM, Yee BK, Feldon J, Rawlins JN (2003) Ventral hippocampal lesions affect anxiety but not spatial learning. Behav Brain Res 139:197–213
Barres BA, Schmid R, Sendtner M, Raff MC (1993) Multiple extracellular signals are required for long-term oligodendrocyte survival. Development 118:283–295
Bartzokis G (2004) Age-related myelin breakdown: a developmental model of cognitive decline and Alzheimer's disease. Neurobiol Aging 25:5–18
Bergles DE, Roberts JD, Somogyi P, Jahr CE (2000) Glutamatergic synapses on oligodendrocyte precursor cells in the hippocampus. Nature 405:187–191
Binamé F, Sakry D, Dimou L, Jolivel V, Trotter J (2013) NG2 regulates directional migration of oligodendrocyte precursor cells via Rho GTPases and polarity complex proteins. J Neurosci 33:10858–10874
Brody H (1955) Organization of the cerebral cortex. III. A study of aging in the human cerebral cortex. J Comp Neurol 102:511–516
Bronson RT, Lipman RD, Harrison DE (1993) Age-related gliosis in the white matter of mice. Brain Res 609:124–128
Brown JP, Couillard-Després S, Cooper-Kuhn CM, Winkler J, Aigner L, Kuhn HG (2003) Transient expression of doublecortin during adult neurogenesis. J Comp Neurol 467:1–10
Cameron HA, McKay RD (1999) Restoring production of hippocampal neurons in old age. Nat Neurosci 2:894–897

Cameron HA, Woolley CS, McEwen BS, Gould E (1993) Differentiation of newly born neurons and glia in the dentate gyrus of the adult rat. Neuroscience 56:337–344

Capilla-Gonzalez V, Cebrian-Silla A, Guerrero-Cazares H, Garcia-Verdugo JM, Quiñones-Hinojosa A (2013) The generation of oligodendroglial cells is preserved in the rostral migratory stream during aging. Front Cell Neurosci 7:147

Celis JE, Celis A (1985) Cell cycle-dependent variations in the distribution of the nuclear protein cyclin proliferating cell nuclear antigen in cultured cells: subdivision of S phase. Proc Natl Acad Sci U S A 82:3262–3266

Chen L, Lu W, Yang Z, Yang S, Li C, Shi X, Tang Y (2011) Age-related changes of the oligodendrocytes in rat subcortical white matter. Anat Rec (Hoboken) 294:487–493

Claiborne BJ, Amaral DG, Cowan WM (1986) A light and electron microscopic analysis of the mossy fibers of the rat dentate gyrus. J Comp Neurol 246:435–458

Colombo M, Fernandez T, Nakamura K, Gross CG (1998) Functional differentiation along the anterior-posterior axis of the hippocampus in monkeys. J Neurophysiol 80:1002–1005

Curcio CA, Coleman PD (1982) Stability of neuron number in cortical barrels of aging mice. J Comp Neurol 212:158–172

Dawirs RR, Hildebrandt K, Teuchert-Noodt G (1998) Adult treatment with haloperidol increases dentate granule cell proliferation in the gerbil hippocampus. J Neural Transm 105:317–327

Dong HW, Swanson LW, Chen L, Fanselow MS, Toga AW (2009) Genomic-anatomic evidence for distinct functional domains in hippocampal field CA1. Proc Natl Acad Sci U S A 106:11794–11799

Drapeau E, Mayo W, Aurousseau C, Le Moal M, Piazza PV, Abrous DN (2003) Spatial memory performances of aged rats in the water maze predict levels of hippocampal neurogenesis. Proc Natl Acad Sci U S A 100:14385–14390

Dupret D, Fabre A, Döbrössy MD, Panatier A, Rodríguez JJ, Lamarque S, Lemaire V, Oliet SH, Piazza PV, Abrous DN (2007) Spatial learning depends on both the addition and removal of new hippocampal neurons. PLoS Biol 5:e214

Edgar N, Sibille E (2012) A putative functional role for oligodendrocytes in mood regulation. Transl Psychiatry 2:e109

Fanselow MS, Dong HW (2010) Are the dorsal and ventral hippocampus functionally distinct structures? Neuron 65:7–19

Farioli-Vecchioli S, Saraulli D, Costanzi M, Pacioni S, Cinà I, Aceti M, Micheli L, Bacci A, Cestari V, Tirone F (2008) The timing of differentiation of adult hippocampal neurons is crucial for spatial memory. PLoS Biol 6:e246

Feng L, Hatten ME, Heintz N (1994) Brain lipid-binding protein (BLBP): a novel signaling system in the developing mammalian CNS. Neuron 12:895–908

Ferland RJ, Gross RA, Applegate CD (2002) Differences in hippocampal mitotic activity within the dorsal and ventral hippocampus following flurothyl seizures in mice. Neurosci Lett 332:131–135

Francis F, Koulakoff A, Boucher D, Chafey P, Schaar B, Vinet MC, Friocourt G, McDonnell N, Reiner O, Kahn A, McConnell SK, Berwald-Netter Y, Denoulet P, Chelly J (1999) Doublecortin is a developmentally regulated, microtubule-associated protein expressed in migrating and differentiating neurons. Neuron 23:247–256

Gaarskjaer FB (1978) Organization of the mossy fiber system of the rat studied in extended hippocampi. I. Terminal area related to number of granule and pyramidal cells. J Comp Neurol 178:49–72

Gerdes J, Lemke H, Baisch H, Wacker HH, Schwab U, Stein H (1984) Cell cycle analysis of a cell proliferation-associated human nuclear antigen defined by the monoclonal antibody Ki-67. J Immunol 133:1710–1715

Gleeson JG, Lin PT, Flanagan LA, Walsh CA (1999) Doublecortin is a microtubule-associated protein and is expressed widely by migrating neurons. Neuron 23:257–271

Hattiangady B, Shetty AK (2008) Aging does not alter the number or phenotype of putative stem/progenitor cells in the neurogenic region of the hippocampus. Neurobiol Aging 29:129–147

Hayakawa N, Kato H, Araki T (2007) Age-related changes of astorocytes, oligodendrocytes and microglia in the mouse hippocampal CA1 sector. Mech Ageing Dev 128:311–316
Hayashi Y, Nihonmatsu-Kikuchi N, Yu X, Ishimoto K, Hisanaga SI, Tatebayashi Y (2011) A novel, rapid, quantitative cell-counting method reveals oligodendroglial reduction in the frontopolar cortex in major depressive disorder. Mol Psychiatry 16:1155–1158
Hill RA, Patel KD, Medved J, Reiss AM, Nishiyama A (2013) NG2 cells in white matter but not gray matter proliferate in response to PDGF. J Neurosci 33:14558–14566
Ishizuka N (2001) Laminar organization of the pyramidal cell layer of the subiculum in the rat. J Comp Neurol 435:89–110
Jayatissa MN, Bisgaard C, Tingström A, Papp M, Wiborg O (2006) Hippocampal cytogenesis correlates to escitalopram-mediated recovery in a chronic mild stress rat model of depression. Neuropsychopharmacology 31:2395–2404
Jinno S (2011a) Topographic differences in adult neurogenesis in the mouse hippocampus: a stereology-based study using endogenous markers. Hippocampus 21:467–480
Jinno S (2011b) Decline in adult neurogenesis during aging follows a topographic pattern in the mouse hippocampus. J Comp Neurol 519:451–466
Jinno S, Kosaka T (2006) Cellular architecture of the mouse hippocampus: a quantitative aspect of chemically defined GABAergic neurons with stereology. Neurosci Res 56:229–245
Jinno S, Kosaka T (2010) Stereological estimation of numerical densities of glutamatergic principal neurons in the mouse hippocampus. Hippocampus 20:829–840
Kaplan MS, Hinds JW (1977) Neurogenesis in the adult rat: electron microscopic analysis of light radioautographs. Science 197:1092–1094
Kempermann G, Jessberger S, Steiner B, Kronenberg G (2004) Milestones of neuronal development in the adult hippocampus. Trends Neurosci 27:447–452
Kessing LV, Nilsson FM (2003) Increased risk of developing dementia in patients with major affective disorders compared to patients with other medical illnesses. J Affect Disord 73:261–269
Kohama SG, Rosene DL, Sherman LS (2012) Age-related changes in human and non-human primate white matter: from myelination disturbances to cognitive decline. Age (Dordr) 34:1093–1110
Korczyn AD, Halperin I (2009) Depression and dementia. J Neurol Sci 283:139–142
Lasiene J, Matsui A, Sawa Y, Wong F, Horner PJ (2009) Age-related myelin dynamics revealed by increased oligodendrogenesis and short internodes. Aging Cell 8:201–213
Lin SC, Huck JH, Roberts JD, Macklin WB, Somogyi P, Bergles DE (2005) Climbing fiber innervation of NG2-expressing glia in the mammalian cerebellum. Neuron 46:773–785
Liu Y, Fujise N, Kosaka T (1996) Distribution of calretinin immunoreactivity in the mouse dentate gyrus. I. General description. Exp Brain Res 108:389–403
Maldonado PP, Vélez-Fort M, Levavasseur F, Angulo MC (2013) Oligodendrocyte precursor cells are accurate sensors of local K+ in mature gray matter. J Neurosci 33:2432–2442
Merkley CM, Jian C, Mosa A, Tan YF, Wojtowicz JM (2014) Homeostatic regulation of adult hippocampal neurogenesis in aging rats: long-term effects of early exercise. Front Neurosci 8:174
Miyamoto N, Pham LD, Hayakawa K, Matsuzaki T, Seo JH, Magnain C, Ayata C, Kim KW, Boas D, Lo EH, Arai K (2013) Age-related decline in oligodendrogenesis retards white matter repair in mice. Stroke 44:2573–2578
Moser E, Moser MB, Andersen P (1993) Spatial learning impairment parallels the magnitude of dorsal hippocampal lesions, but is hardly present following ventral lesions. J Neurosci 13:3916–3925
Noble M, Murray K, Stroobant P, Waterfield MD, Riddle P (1988) Platelet derived growth factor promotes division and motility and inhibits premature differentiation of the oligodendrocyte/type-2 astrocyte progenitor cell. Nature 333:560–562
Olariu A, Cleaver KM, Cameron HA (2007) Decreased neurogenesis in aged rats results from loss of granule cell precursors without lengthening of the cell cycle. J Comp Neurol 501:659–667

Ownby RL, Crocco E, Acevedo A, John V, Loewenstein D (2006) Depression and risk for Alzheimer disease: systematic review, meta-analysis, and metaregression analysis. Arch Gen Psychiatry 63:530–538
Pakkenberg B, Gundersen HJ (1997) Neocortical neuron number in humans: effect of sex and age. J Comp Neurol 384:312–320
Pannese E (2011) Morphological changes in nerve cells during normal aging. Brain Struct Funct 216:85–89
Peters A, Kemper T (2012) A review of the structural alterations in the cerebral hemispheres of the aging rhesus monkey. Neurobiol Aging 33:2357–2372
Peters A, Sethares C (1993) Aging and the meynert cells in rhesus monkey primary visual cortex. Anat Rec 236:721–729
Peters A, Sethares C (2002) The effects of age on the cells in layer 1 of primate cerebral cortex. Cereb Cortex 12:27–36
Pitkänen A, Pikkarainen M, Nurminen N, Ylinen A (2000) Reciprocal connections between the amygdala and the hippocampal formation, perirhinal cortex, and postrhinal cortex in rat. A review. Ann N Y Acad Sci 911:369–391
Rao MS, Shetty AK (2004) Efficacy of doublecortin as a marker to analyse the absolute number and dendritic growth of newly generated neurons in the adult dentate gyrus. Eur J Neurosci 19:234–246
Regenold WT, Phatak P, Marano CM, Gearhart L, Viens CH, Hisley KC (2007) Myelin staining of deep white matter in the dorsolateral prefrontal cortex in schizophrenia, bipolar disorder, and unipolar major depression. Psychiatry Res 151:179–188
Richardson WD, Young KM, Tripathi RB, McKenzie I (2011) NG2-glia as multipotent neural stem cells: fact or fantasy? Neuron 70:661–673
Sakry D, Karram K, Trotter J (2011) Synapses between NG2 glia and neurons. J Anat 219:2–7
Sandell JH, Peters A (2002) Effects of age on the glial cells in the rhesus monkey optic nerve. J Comp Neurol 445:13–28
Schweitzer I, Tuckwell V, O'Brien J, Ames D (2002) Is late onset depression a prodrome to dementia? Int J Geriatr Psychiatry 17:997–1005
Seki T, Arai Y (1995) Age-related production of new granule cells in the adult dentate gyrus. Neuroreport 6:2479–2482
Sheng JG, Mrak RE, Rovnaghi CR, Kozlowska E, Van Eldik LJ, Griffin WS (1996) Human brain S100 beta and S100 beta mRNA expression increases with age: pathogenic implications for Alzheimer's disease. Neurobiol Aging 17:359–363
Shors TJ, Miesegaes G, Beylin A, Zhao M, Rydel T, Gould E (2001) Neurogenesis in the adult is involved in the formation of trace memories. Nature 410:372–376
Sim FJ, Zhao C, Penderis J, Franklin RJ (2002) The age-related decrease in CNS remyelination efficiency is attributable to an impairment of both oligodendrocyte progenitor recruitment and differentiation. J Neurosci 22:2451–2459
Small SA, Schobel SA, Buxton RB, Witter MP, Barnes CA (2011) A pathophysiological framework of hippocampal dysfunction in ageing and disease. Nat Rev Neurosci 12:585–601
Snyder JS, Radik R, Wojtowicz JM, Cameron HA (2009) Anatomical gradients of adult neurogenesis and activity: young neurons in the ventral dentate gyrus are activated by water maze training. Hippocampus 19:360–370
Sturrock RR (1987) Age-related changes in the number of myelinated axons and glial cells in the anterior and posterior limbs of the mouse anterior commissure. J Anat 150:111–127
Sun Y, Xu CC, Li J, Guan XY, Gao L, Ma LX, Li RX, Peng YW, Zhu GP (2013) Transplantation of oligodendrocyte precursor cells improves locomotion deficits in rats with spinal cord irradiation injury. PLoS One 8:e57534
Tanaka J, Okuma Y, Tomobe K, Nomura Y (2005) The age-related degeneration of oligodendrocytes in the hippocampus of the senescence-accelerated mouse (SAM) P8: a quantitative immunohistochemical study. Biol Pharm Bull 28:615–618

van Groen T, Wyss JM (2003) Connections of the retrosplenial granular b cortex in the rat. J Comp Neurol 463:249–263

Walter J, Keiner S, Witte OW, Redecker C (2011) Age-related effects on hippocampal precursor cell subpopulations and neurogenesis. Neurobiol Aging 32:1906–1914

Wigley R, Hamilton N, Nishiyama A, Kirchhoff F, Butt AM (2007) Morphological and physiological interactions of NG2-glia with astrocytes and neurons. J Anat 210:661–670

Wilkins A, Majed H, Layfield R, Compston A, Chandran S (2003) Oligodendrocytes promote neuronal survival and axonal length by distinct intracellular mechanisms: a novel role for oligodendrocyte-derived glial cell line-derived neurotrophic factor. J Neurosci 23:4967–4974

Witter MP, Van Hoesen GW, Amaral DG (1989) Topographical organization of the entorhinal projection to the dentate gyrus of the monkey. J Neurosci 9:216–228

Wu Y, Zhang AQ, Yew DT (2005) Age related changes of various markers of astrocytes in senescence-accelerated mice hippocampus. Neurochem Int 46:565–574

Yamada J, Jinno S (2014) Age-related differences in oligodendrogenesis across the dorsal-ventral axis of the mouse hippocampus. Hippocampus 24:1017–1029

Chapter 15
Knowledge of Signal Transduction Provides an Approach to Attacking Memory Decline

Shuichi Yanai and Shogo Endo

Abstract Memory, the basis of higher cognitive functions, has been a major topic of interest for a very long time. One of the major setbacks most of us face as we age is a decline in declarative memory, which mainly depends on a brain structure called the hippocampus. Because memory decline has such a negative impact on quality of our life, intense efforts are being directed toward finding therapeutic interventions to ameliorate or even prevent memory decline. A major focus is on molecular targets to improve or maintain memory. Through the knowledge obtained from extensive study of amnesic patients and rapid progress in modern biochemical and neurosciences, the molecular mechanisms underpinning learning and memory and neural plasticity in the central nervous system have been revealed. In this review, we present a brief summary of memory characteristics in psychological terms, followed by a discussion of the basic concept of the memory system, its underlying cellular mechanisms, and age-related decline in memory. Anti-dementia drugs developed on the basis of the cholinergic hypothesis are also discussed. Finally, we assess the use of phosphodiesterase inhibitors, which control the cAMP-PKA-CREB signal transduction pathway, as potential candidates for treating memory decline beyond the existing cholinergic-based strategies.

Keywords Memory • Learning • Attention • cAMP • PKA • CREB • ICER • ACh • PDE

15.1 Introduction

Most of us experience decline in memory function during aging, even before we feel the effects of diminished physical ability. This decline in memory is mostly the type of memory called declarative memory, which mainly depends on a brain structure called the hippocampus (Squire 1992, 2004; Squire and Zola 1996; Tulving and Markowitsch 1998; Eichenbaum 2000). Memory decline due to

S. Yanai, Ph.D. • S. Endo, Ph.D. (✉)
Memory Neuroscience, Aging Neuroscience Research Team, Tokyo Metropolitan Institute of Gerontology, Sakae-cho 35-2, Itabashi, Tokyo 173-0015, Japan
e-mail: sendo@tmig.or.jp

© Springer Japan 2015

N. Mori, I. Mook-Jung (eds.), *Aging Mechanisms*,
DOI 10.1007/978-4-431-55763-0_15

aging has great impact on healthy aging, because memory is the basis for a variety of other higher cognitive functions, including thought, language, and emotion. Higher cognitive functions are an essential part of our life, as they are what make a biological man a human being, one having a personality and dignity (Haas and Canli 2008; Passmore et al. 2012).

Thanks to modern medical achievements, aging populations continue to expand around the world. Along with this growth comes the increasingly important task of addressing issues common to an aging population—such as memory decline and impairment. Therapeutic interventions aimed at overcoming and delaying the onset of memory decline are paramount. To attack this major need, researchers have focused their attention on identifying critical molecular targets related to memory to guide the development of future memory-enhancing drugs or drugs to prevent memory decline. Despite the existence of a huge amount of basic research on learning and memory, there are still no clear answers to basic questions about memory mechanisms. How is sensory information acquired, stored as memory, maintained, and recalled? How do we lose information stored as memories, or alternatively, lose access to information stored as memories?

Modern psychology has characterized a variety of features of memory, including the learning curve, forgetting curve, and the spacing effect, all first studied in detail by Hermann Ebbinghaus in the nineteenth century (Ebbinghaus 1885). In the last century, Donald Hebb proposed that the central nervous system (CNS) possessed an essential characteristic, called neuronal plasticity (flexibility), that could theoretically support memory formation (Hebb 1949). This hypothesized flexibility of CNS synapses was verified during the 1970s and 1980s by using electrophysiological techniques in mammals (Bliss and Lømo 1973; Ito et al. 1982). This discovery of neuronal plasticity in the CNS was a great catalyst to initiate investigations of molecules involved in neuronal plasticity and memory (Milner et al. 1998; Kandel 2001). Hundreds of molecules are now known to be involved in hippocampal long-term potentiation (LTP) (Miller and Mayford 1999) and cerebellar long-term depression (LTD) (Ito 2011; Ito et al. 2014), two examples of neuronal plasticity.

Neurological diseases such as schizophrenia, bipolar disorder, Alzheimer's disease, and Parkinson's disease impair particular cognitive functions, besides also influencing learning and memory in a variety of ways (for recent reviews, see Danion et al. 2007; Pfennig et al. 2007; Kidd 2008; Tröster 2008). Modern biomedical research on neurological diseases has developed to the point of identifying causal relationships between a particular disease and gene(s), gene product(s), transmitter(s), and brain region(s) (neuronal circuitry). A huge number of studies have produced basic information on several diseases and has also revealed molecular and cellular mechanisms for memory. Based on these results, a variety of genetically modified animal models have been developed, including ones using mice (Hsiao et al. 1996; Oddo et al. 2003); *Drosophila* (Chakraborty et al. 2011; Mhatre et al. 2014); and *Caenorhabditis elegans* (Link et al. 2003). These have begun to elucidate the mechanisms underlying higher cognitive functions, including memory.

In this review, first, we present a summary of memory characteristics in psychological terms and also discuss neuronal plasticity as cellular mechanisms for memory. Second, we focus attention on the memory system in mammals that critically depends on the hippocampus. In this context, we also summarize what is known about the underlying cellular and molecular mechanisms in the memory system. Finally, we describe potential therapeutic interventions for ameliorating memory decline through the application of knowledge on signal transduction mechanisms underlying memory.

15.2 Memory and Neuronal Plasticity in Psychology and Physiology

Traditionally, memory is thought to be essential function of thought and other cognitive function. For this reason, memory has been a major research target for philosopher and psychologist. We take a brief look at the memory from the view of psychology and discuss the classification of memory. We then summarize neuronal plasticity, the cellular substrate of memory, which alter the information processing depending on the input.

15.2.1 Classification of Memory

To date, several classifications of memory have been proposed by psychologists. Among them, two types are characterized primarily by their duration and content. These two types of classification are described and discussed below.

15.2.1.1 Sensory, Short-Term, and Long-Term Memory

The most simple and popular classification might be one that is based on the duration information persists in memory. Atkinson and Shiffrin (1968) proposed a multi-store model of memory, which consists of two distinct memory stores: short-term and long-term. A third type of memory, termed sensory memory, was subsequently added. Now three memory stores are widely recognized in most basic models of memory operations (Fig. 15.1).

Cognitive psychology often describes the five senses and the brain as an information processing system. Information from the outside world enters the human information processing system through a variety of what could be called sensory registers (Fig. 15.1). Because huge amounts of information impinge on the senses continuously and because most of this cannot be processed immediately, information that we do not attend to is held in sensory memory for a very short period of

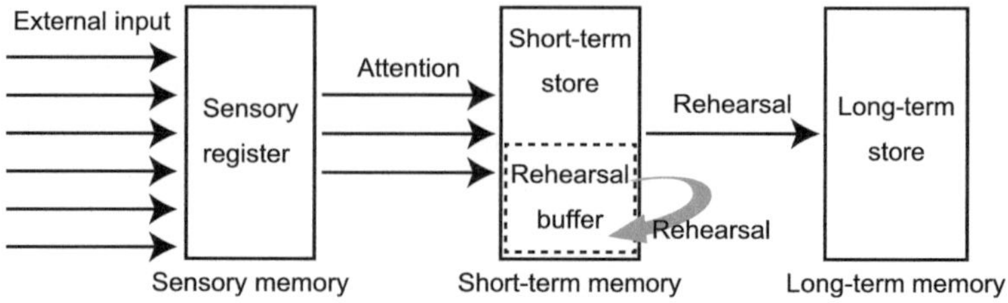

Fig. 15.1 Multi-store model of memory proposed by Atkinson and Shiffrin (1968). Huge amounts of information enter the information processing system through sensory registers (besides visual and auditory sensory registers, likely registers for the other three senses also exist). Most sensory memory is lost almost immediately; however, selectively attended to information is transferred to a second store, short-term memory, which has a somewhat longer duration. Short-term memory can be maintained longer, however, by actively repeating items in it within a rehearsal buffer. Finally, information is transferred to long-term memory. Long-term memory has an unlimited capacity and can persist as long as one's entire life

time and then dissipates rapidly. The duration of sensory memory stored in the sensory registers differs depending on the modality of the information. For example, the duration of visual sensory memory is less than 500 msec (Sperling 1960), but for auditory sensory information, the duration is about 4 sec (Glucksberg and Cowan 1970).

Selectively attended sensory memory is then transferred to another temporary store called short-term store (Fig. 15.1). The capacity of this short-term memory is about seven pieces of information (Miller 1956), and it persists for approximately 20 sec (Peterson and Peterson 1959). Information that enters the short-term store fades away as soon as it is no longer attended to. However, information at this stage can be maintained by actively repeating it within "a rehearsal buffer", which can then be transferred to the long-term store. It is believed that long-term memory has a virtually unlimited capacity and can last for several hours or as long as a lifetime. Information held in the long-term store can be recalled at a later time, even if one is no longer selectively attending to its original source or actively rehearsing it. The form of these long-term memories can be characterized in different ways, and the brain regions primarily involved have now been described.

15.2.1.2 Brain Regions Responsible for Declarative and Non-declarative Memories

Some of the best evidence for the classification of memory has emerged from the study of amnesic patients who became amnesic because of specific brain damage. One such patient, known in the literature as H.M., became densely amnesic after receiving in 1953 a bilateral lobectomy of the medial temporal lobes for the treatment of intractable epilepsy (Scoville and Milner 1957; Milner et al. 1998). His resection included the hippocampal formation and adjacent medial temporal structures. H.M. became the most studied amnesic patient in history and was influential not only because of the knowledge he provided about memory impairment and amnesia, but also because results of these studies led to a better understanding of how specific brain regions are involved in memory formation and how memory can be classified in different ways. H.M. was severely impaired on memory tests that assessed the recollection of facts and events, including names, faces, and places. On the other hand, his motor skill learning and retention of the skill were preserved (Milner 1962). These observations suggest that memory is not a single entity but instead consists of several separate components that depend on different brain systems.

Based on knowledge gained by the study of amnesic patients, a second classification of memory has been appreciated. According to Squire (1992), long-term memory can be roughly divided into declarative and non-declarative memory (Fig. 15.2). The key distinction between these two types of memories is twofold: (1) the ability or inability to "declare" the retained information, i.e., put the memory into words; and (2) the presence or absence of conscious recollection in order to express the memory. Declarative memory in humans is memory that can be described using words and that requires conscious recollection of related information (i.e., recall of facts, faces, places). Much evidence has supported the hypothesis that the medial temporal lobe, including the hippocampus, plays an important role in declarative memory (Squire 1992, 2004; Squire and Zola 1996; Tulving and Markowitsch 1998; Eichenbaum 2000. It should be noted that there are species differences in the form of memory in which the hippocampus is involved. While it is still being studied and debated, the human hippocampus is often associated with episodic memory (Squire and Zola 1998; Tulving and Markowitsch 1998; Eichenbaum 2000), whereas the rodent hippocampus is often associated with spatial navigation and memory for the environment (O'Keefe and Nadel 1978; Olton et al. 1979).

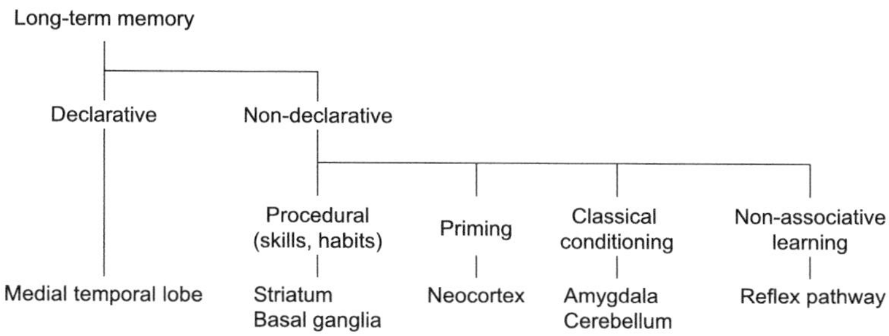

Fig. 15.2 Classification of long-term memory. Declarative memory is a type of long-term memory that can best be described using words (e.g., a parrot is a kind of bird; Paris is in France) and that requires the conscious recollection of facts and events. On the other hand, non-declarative memory is a type of long-term memory that cannot be described simply in words and that does not require conscious recollection for the incidence(s) of the learning for its expression (e.g., playing a good game of table tennis). Much evidence suggests that memory is not a monolithic entity but consists of several separate components that depend on different brain systems (Revised from Squire (1992), Squire and Zola (1996))

By contrast, non-declarative memory is memory that cannot be described easily using linguistic terms and does not require awareness of conscious recollection in order to be expressed. Squire perhaps characterizes this type of memory best: "Nondeclarative memory refers to a heterogeneous collection of distinct learning and memory abilities where performance changes but without affording access to the experience or experiences that caused the change" (Squire 1992). Thus, non-declarative memory can be further divided into four forms: procedural memory, priming, classical conditioning, and non-associative learning. Evidence obtained from amnesic patients and lesion studies using nonhuman primates revealed that non-declarative memory depends on multiple brain systems. Medial temporal lobe regions have little, if any, involvement in declarative memory. Many reviews are available that discuss in detail the responsible brain regions for declarative and non-declarative memory (Squire 1992, 2004; Squire and Zola 1996).

15.2.2 Neuronal Plasticity

By the end of the nineteenth century, it was widely believed that the number of neurons in the adult brain did not increase with advancing age. This notion constrained theories of memory formation and how neurons are involved. Within this working hypothesis of an absence of newborn neurons after brain maturation, formation of new memory in the brain could not be achieved by a simple change in neuron number. How do memories continuously accumulate throughout a lifetime without increasing the number of neurons? Hebbian theory (1949) was the answer. In his book, Hebb postulated how memories might be established by strengthening the connections (synapses) between *existing* neurons to enhance the effectiveness of signal transduction (communication) among a network of neurons. Although the theory is widely believed now to be the most viable mechanism for biological instantiation of memory, it was nearly 25 years after it was advanced before a Hebb-like synapse was discovered in the mammalian brain. This was largely because no appropriate electrophysiological equipment existed to test his theory experimentally in mammalian brain.

Partly based on the important role of the hippocampus in memory revealed by the study of amnesic patients, electrophysiologists directed their attention to the study of the mammalian hippocampus. Bliss and Lømo (1973) discovered long-lasting electrophysiological changes in synaptic efficacy at perforant path–granule cell synapses in the rabbit hippocampus. Because this phenomenon corroborated Hebbian theory, this type of synaptic plasticity was thought to underlie learning and memory. Since its initial discovery by Bliss and Lømo, long-lasting synaptic plasticity has received much focus from many scientists who have been fascinated with the ability of synapses to respond to specific patterns of activation, effecting various changes in synaptic efficacy.

15.2.2.1 Hippocampal Long-Term Potentiation (LTP)

LTP is a form of persistent, activity-dependent synaptic plasticity that is induced experimentally by brief, high-frequency stimulation of afferent fibers. Excellent reviews for the mechanisms underlying hippocampal LTP are available (Bliss and Collingridge 1993; Lynch 2004; Malenka and Bear 2004). The NMDA receptor (NMDA-R) plays essential roles in LTP induction in area CA1 of the hippocampus. NMDA-R-dependent LTP requires synaptic activation of NMDA-Rs that coincide with postsynaptic depolarization. The brief rise of Ca^{2+} through NMDA-Rs is a critical and sufficient step to initiate the signal for LTP induction (Lynch et al. 1979; Malenka et al. 1992).

Activation of kinase pathways also plays essential roles in LTP (Malinow et al. 1988; Soderling and Derkach 2000). Thus, "leaky" Ca^{2+} influx through hippocampal NMDA-Rs leads to a functional decline of the hippocampus to control memory formation. This is one reason why hippocampal NMDA-Rs are a prime target for drug development aimed at preventing or ameliorating memory decline (Parsons et al. 1999; Robinson and Keating 2006). For example, memantine is an NMDA-R antagonist approved for the treatment of moderate-to-severe Alzheimer's disease, and its mechanism of action is to prevent leaky NMDA-Rs in dementia (Van Marum 2009).

In normal aging, reduced hippocampal LTP has been reported, and this decline is associated with age-related defects in spatial memory in human and rodents (Barnes 1979; Bach et al. 1999; Huang and Kandel 2006). These facts also support the essential role of neuronal plasticity, especially LTP, in the formation and maintenance of memory.

15.2.2.2 Long-Term Depression (LTD)

In addition to potentiation of synaptic transmission, a decrease in efficacy of synaptic transmission is also observed in the CNS (Malenka 1994; Malenka and Bear 2004). LTD of parallel fiber–Purkinje cell synapses was first discovered in the cerebellum by Masao Ito, then at Tokyo University (Ito et al. 1982). Low-frequency stimulus-induced cerebellar LTD seems to control cerebellum-dependent memories, including motor memories (Llinás et al. 1997; Ito 2011; Ito et al. 2014).

Besides the cerebellum, LTD also can be induced in the hippocampus (Dudek and Bear 1992). Hippocampal LTD involves the phosphatase pathway (Mulkey et al. 1994), in contrast to the important role of kinase pathways for hippocampal LTP (Malinow et al. 1988; Soderling and Derkach 2000). Controlling the efficiency of synaptic transmission by adjusting basal levels, enhanced levels (LTP), and depressed levels (LTD) gives the CNS a myriad of different ways to affect synaptic transmission compared to a situation in which synaptic transmission might only be controlled by adjusting only basal levels and enhanced levels of synaptic transmission.

Even though LTP and LTD are thought to be essential parts of the neuronal plasticity that produce memory, there are some reports that suggest certain genetically modified mice in which hippocampal LTP is absent have normal memory (Zamanillo et al. 1999; Leiva et al. 2009).

15.3 Aging and Memory Decline

Aging is typically associated with a broad range of cognitive declines related to memory function. Imaging studies of the human brain have revealed that structural and functional declines in hippocampus are prominent during the course of aging. These changes may explain why age-related decline in memory is observed mainly in hippocampal-dependent declarative memory (Squire 1992). In fact, non-declarative memories, which don't depend on the integrity of the hippocampus, remain relatively stable (Hedden and Gabrieli 2004). For these reasons, in rodents, behavioral tests that assess hippocampal-dependent memory are frequently used to evaluate pharmacological interventions aimed at treating human dementia.

In addition to age-related decline in hippocampal-dependent memory, attentional processes that depend on the prefrontal cortex are vulnerable to aging (Prakash et al. 2009; Hedden et al. 2012). Attention plays an important role not only in memory but also in other higher cognitive functions such as executive function.

Memory decline experienced by the elderly is referred to as age-associated memory impairment, which is clearly distinguished from the memory impairment cases observed in dementia. In fact, most of the elderly do not suffer from dementia, but instead are in what can be called a state of "normal" memory decline (Plassman et al. 2007). Several risk factors have been identified for pathological memory impairment, although biological mechanisms are not completely understood yet (Scalco and Van Reekum 2006; Patterson et al. 2008). Memory decline associated with normal aging does not seem to be caused by an individual risk factor, but is caused instead by a complex mixture of several risk factors (Fig. 15.3).

15.4 Attacking the Problem of Memory Decline Through Application of Knowledge About Signal Transduction

Research on molecular mechanisms underlying memory and neuronal plasticity has provided a variety of molecules and mechanisms to control memories. Some regulate memory in positive ways and the other in negative ways (Endo 2012). Among those molecules, neuronal transmitter systems and signal transduction pathways are of the attention to attack memory decline and memory-related diseases. We take a look at cholinergic system as a representative of neuronal

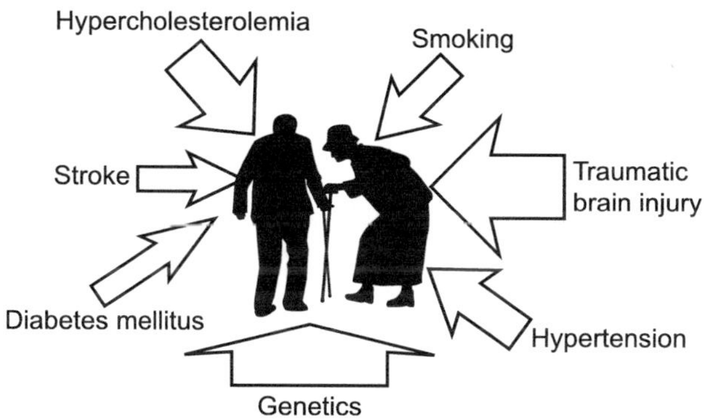

Fig. 15.3 Risk factors for pathological memory impairment. Several factors may increase the risk of developing pathological memory impairment. These risk factors include genetics, lifestyle, and age. Although inherited genes cannot be controlled, some risk factors can be changed. The thickness of arrows indicates roughly the relative risk factors for Alzheimer's disease (Details are described in Scalco and Van Reekum (2006) and in Patterson et al. (2008))

transmitter systems and, cAMP and its degradation enzyme, phosphodiesterases, as representatives of signal transduction system.

15.4.1 Acetylcholine

Acetylcholine (ACh) was the first neurotransmitter identified as a chemical signal of synaptic transmission from one neuron to the other. The role of cholinergic function in cognition was initially suggested by observations that anti-cholinergic drugs had amnesic effects (Longo 1966; Drachman and Leavitt 1974). The importance of cholinergic neurotransmission led to the "cholinergic hypothesis" for several essential brain functions, including memory, and also for the notion that at least some memory impairment is due to a disruption in cholinergic function.

The cholinergic hypothesis postulates that decreased synaptic transmission, mediated by ACh, resulting from a loss of cholinergic neurons in the *nucleus basalis magnocellularis*, leads to cognitive impairment (Appel 1981; Gibson et al. 1981; Bartus et al. 1982). Based on the cholinergic hypothesis, strategies aimed at increasing synaptic levels of ACh have been widely explored in efforts to develop anti-dementia drugs. One such strategy is to restore adequate ACh concentrations in cholinergic synapses by blocking ACh degradation by inhibition of acethycholinesterase (AChE). Two AChE inhibitors, donepezil and galantamine, were developed for treating memory impairment (Rogers et al. 1998; Tariot et al. 2000). In Europe, AChE inhibitors are prescribed to treat mild and moderate stages of Alzheimer's disease only. In Japan and the United States, donepezil is approved for all stages of the disorder.

Another anti-dementia drug based on the cholinergic hypothesis is rivastigmine, an inhibitor of AChE and butyrylcholinesterase (BChE). AChE specifically hydrolyzes the neurotransmitter ACh, while BChE non-specifically hydrolyzes choline esters in addition to ACh. Rivastigmine is reported to increase intracerebral ACh concentrations more than donepezil (Bullock et al. 2006). There are no significant differences among these AChE inhibitors with respect to efficacy on cognitive function improvement. However, rivastigmine has the possible advantage in being effective for treating age-related cognitive impairment (Birks 2006).

Among the three anti-dementia drugs (donepezil, galantamine, and rivastigmine) that were developed based on the cholinergic hypothesis, donepezil has been extensively studied and considered to be the first-line treatment in patients with mild-to-moderate Alzheimer's disease (Pepeu and Giovannini 2009). Recent studies using a mouse model of Alzheimer's disease has provided new insights on the effects of donepezil. Donepezil is effective in alleviating age-related attentional deficits (Romberg et al. 2011) and in reducing soluble amyloid β (Aβ) protein and the number of plaque depositions (Dong et al. 2009). However, to our knowledge, no clinical evidence has been reported showing that donepezil ameliorates age-related attentional deficits and the pathology associated with Alzheimer's disease.

15.4.2 cAMP, ATP, PKA, CREB

A number of lines of evidence have revealed that cyclic nucleotides are ubiquitously distributed in mammals, and these regulate intracellular signal transduction (Florian et al. 2006; Ota et al. 2008). This occurs via well-defined pathways. When cells receive stimuli such as neurotransmitters or hormones, cyclic adenosine monophosphate (cAMP) is synthesized from adenosine triphosphate (ATP) by adenylate cyclase. Then, cAMP activates cAMP-dependent protein kinase A (PKA), which phosphorylates other enzymes or transcription factors such as cAMP response element binding protein (CREB) in the nucleus.

Phosphorylation controls CREB activity. Phosphorylation of CREB bound to cAMP response element (CRE) on genes triggers the recruitment of other transcriptional components and then initiates transcription of downstream genes (Dash et al. 1990; Kaang et al. 1993). CREB-dependent gene activation leads also to the expression of immediate-early genes (IEGs), including c-fos, c-jun, and zif-256 (Franceschi and Xiao 2003). These IEGs are transcription factors, and together with CREB, initiate and control gene expression. The cascade of gene activation and resulting gene products are thought to be essential for the late phase of LTP induction and thus are thought to represent the molecular mechanism underpinning memory (Bartsch et al. 1998).

The importance of CREB in both the transition to and consolidation of short-term memory to long-term memory has been demonstrated (Bartsch et al. 1995) by using genetically modified mice in which CREB function is disrupted (Kida et al. 2002; Korzus et al. 2004). Also, the upregulation of CREB transcriptional activity in genetically modified mice enhances memory (Suzuki et al. 2011). These results support the hypothesis that the cAMP-CREB system is an essential part of normal memory function.

In concert with the CREB system inducible cAMP early repressor (ICER), which suppresses the function of CREB in transcription, plays important roles in memory (Kojima et al. 2008; Borlikova and Endo 2009). The CREB system works as an "accelerator", while the ICER system works as a "brake" for memory function (Borlikova and Endo 2009). As observed in other physiological systems, this "brake" acts to suppress memory formation in order to avoid the excess memories. Excessively strong memories that cannot be erased easily might contribute to psychological symptoms such as post-traumatic stress disorder (PTSD) in humans (Dell'Osso et al. 2011; Sekiguchi et al. 2013).

The cAMP and CREB pathways are involved in a variety of functions, including memory and age-related physiological decline. For example, the expression of sirtuin 1 (SIRT-1)—a "longevity" gene activated by caloric restriction—is suppressed in CREB-1-deficient mice (Fusco et al. 2012). Thus, CREBs are potentially good targets for drugs developed to improve declining physiological functions associated with aging. However, pharmacological agents directly modulating the function of CREB have not been developed to date.

15.4.3 Phosphodiesterases

Phosphodiesterases (PDEs) are enzymes that hydrolyze cyclic nucleotides by breaking their phosphodiester bond (Burgers et al. 1979; Goldberg et al. 1980). Since the first isolation of PDE in the early 1970s (Uzunov and Weiss 1972; Strada et al. 1974), 11 major PDE families have been identified in mammalian tissue and characterized based on their pharmacology, substrate specificity, and tissue distribution (Table 15.1). PDE-specific inhibitors increase the intracellular concentration of cAMP and/or cyclic guanosine monophosphate (cGMP) and potentiate cAMP/cGMP signal transduction. Thus, the use of PDE inhibitors is considered to be a novel pharmacological intervention for treating cognitive impairment beyond the existing cholinergic-based strategies (Reneerkens et al. 2009; Terry et al. 2011).

Among the 11 types of PDEs, PDE6 and PDE11 are not expressed in the CNS (Table 15.1). Therefore, these two types of PDEs are likely irrelevant targets for affecting memory enhancement. On the other hand, a considerable number of studies have demonstrated that the PDE4-selective inhibitor rolipram restores cognitive function in an animal model of cognitive impairment (For reviews, see Rose et al. 2005; Richter et al. 2013). Because of a number of studies that have reported beneficial effects of rolipram, a specific PDE4 inhibitor, on cognitive impairment, rolipram underwent clinical trial in the United States. Unfortunately, clinical development of PDE4 inhibitors has been terminated because of its potent

Table 15.1 Localization, substrate specificity and inhibitors of phosphodiesterase (PDE)

PDE	Localization		Substrate specificity	Typical inhibitor
	Peripheral	CNS		
PDE1	+	+	cAMP/cGMP	IC224, vinpocetine
PDE2	+	+	cAMP/cGMP	BAY60-7550, EHNA
PDE3	+	+	cAMP/cGMP	Amrinone, cilostamide, cilostazol, milrinone
PDE4	+	+	cAMP	Cilomilast, roflumilast, rolipram
PDE5	+	+	cGMP	Sildenafil, tadalafil, verdenafil, zaprinast
PDE6	+	–	cGMP	(Sildenafil)[a]
PDE7	+	+	cAMP	BRL50481, IC242
PDE8	+	+	cAMP	Dipyridamole
PDE9	+	+	cGMP	BAY73-6691
PDE10	+	+	cAMP/cGMP	Papaverine
PDE11	+	–	cAMP/cGMP	Dipyridamole, (tadalafil)[a]

[a]Sildenafil and tadalafil, PDE5 inhibitors, also inhibit PDE6 and PDE11, respectively.
CNS central nervous system, *EHNA* erythro-9-(2-hydroxy-3-nonyl)adenine, *cAMP* cyclic adenosine monophosphate, *cGMP* cyclic guanosine monophosphate.

emetic effects. This appears to be caused by their inhibition of PDE4D in the area postrema (Tenor et al. 2011). In order to be clinically useful as an anti-dementia drug, it is important for a drug to have few or no side effects.

Recently, PDE3 inhibitors have been considered as the next potential therapeutic intervention for cognitive impairment. In studies using a mouse model of Alzheimer's disease in which Aβ25-35 is injected, administration of the selective PDE3 inhibitor cilostazol significantly improved performance in the Y-maze and passive avoidance (Hiramatsu et al. 2010) and in the Morris water maze (Park et al. 2011). Surprisingly in non-genetically modified young mice, cilostazol also enhanced performance in the Morris water maze and contextual fear conditioning (Yanai et al. 2014). Thus, cilostazol might be a potential candidate for a novel anti-dementia drug.

Bolstering these results in animals, a retrospective study of mild dementia patients taking donepezil showed that adding cilostazol to the therapy slowed cognitive decline significantly (Ihara et al. 2014). Based on these findings, a clinical trial is now underway in Japan to gain approval for the use of cilostazol in treating mild cognitive impairment. Because cilostazol has been prescribed as an anti-platelet agent for more than two decades already (O'Donnell et al. 2009), drug safety and proper use are well established.

Additional recent studies have demonstrated that several other PDE inhibitors improve or enhance memory and cognitive function in rodent models. Some examples include the drugs BAY60-7550 (PDE2 inhibitor, Sierksma et al. 2013); S14 (PDE7 inhibitor, Perez-Gonzalez et al. 2013); and BAY73-6691 (PDE9 inhibitor, Van der Staay et al. 2008). Taken together, PDE–based strategies might represent the next approach to developing anti-dementia drugs (Reneerkens et al. 2009; Terry et al. 2011).

15.5 Summary and Conclusion

Studies on the molecular mechanisms underlying memory have revealed that a variety of signal transduction pathways are involved. Among the four types of anti-dementia drugs that have been developed to date, donepezil is considered to be the first-line treatment for patients with mild-to-moderate Alzheimer's disease. However, the ameliorative effect of donepezil on memory impairment can begin to abate as quickly as 6–18 months after starting drug treatment (Riepe et al. 2007). Considering that the aging of the world population is accelerating, it is imperative that a viable therapeutic intervention is established soon, one that will exert its anti-dementia effects for an extended period of time.

This potentially disastrous situation might be addressed by employing combination therapy; using two or more targeted drugs has been demonstrated to be an effective strategy for treating some diseases (Croom and Dhillon 2011). The combined use of a PDE inhibitor and an AChE inhibitor has already been shown to enhance, to some extent, cognitive function both in animal models (Lee et al. 2007) and in patients with moderate Alzheimer's disease (Arai and Takahashi 2009). In addition, these inhibitors are known to affect memory processes through different ways. PDE inhibitors improve memory consolidation, and AChE inhibitors enhance acquisition of memory (Prickaerts et al. 2005). Hence, concurrent administration of a PDE inhibitor with another anti-dementia drug, such as donepezil, may offer a new pharmacological approach for viably treating cognitive disorders in age-related neurological disease. Along with the knowledge of signal transduction already gained for the four anti-dementia drugs considered here, further research will continue to attack the problem of memory decline in aging and age-related neurological disease. Successes in this realm will achieve much more than a better quality of life for older persons. They will also allow generations of older ones to continue their existence as valued human beings, ones with personality and dignity, rather than ones with failing biological processes lacking even a hint of their former younger selves.

Acknowledgement This work was supported in part by Japan Foundation of Aging and Health, the Naito Foundation, JSPS KAKENHI (24730642, 25293331, 25560382, 26115532, 15H03103), and Otsuka Pharmaceutical Co., Ltd.

References

Appel SH (1981) A unifying hypothesis for the cause of amyotrophic lateral sclerosis, parkinsonism, and Alzheimer disease. Ann Neurol 10:499–505

Arai H, Takahashi T (2009) A combination therapy of donepezil and cilostazol for patients with moderate Alzheimer disease: pilot follow-up study. Am J Geriatr Psychiatry 17:353–354

Atkinson RC, Shifrin RM (1968) Human memory. A proposed system and its control processes. In: Spence KW, Spence JT (eds) The psychology of learning and motivation: advances in research and theory, vol 2. Academic, New York, pp 85–195

Bach ME, Barad M, Son H, Zhuo M, Lu YF, Shih R, Mansuy I, Hawkins RD, Kandel ER (1999) Age-related defects in spatial memory are correlated with defects in the late phase of hippocampal long-term potentiation in vitro and are attenuated by drugs that enhance the cAMP signaling pathway. Proc Natl Acad Sci U S A 96:5280–5285

Barnes CA (1979) Memory deficits associated with senescence: a neurophysiological and behavioral study in the rat. J Comp Physiol Psychol 93:74–104

Bartsch D, Ghirardi M, Skehel PA, Karl KA, Herder SP, Chen M, Bailey CH, Kandel ER (1995) Aplysia CREB2 represses long-term facilitation: relief of repression converts transient facilitation into long-term functional and structural change. Cell 83:979–992

Bartsch D, Casadio A, Karl KA, Serodio P, Kandel ER (1998) CREB1 encodes a nuclear activator, a repressor, and a cytoplasmic modulator that form a regulatory unit critical for long-term facilitation. Cell 95:211–223

Bartus RT, Dean RL 3rd, Beer B, Lippa AS (1982) The cholinergic hypothesis of geriatric memory dysfunction. Science 217:408–414

Birks J (2006) Cholinesterase inhibitors for Alzheimer's disease. Cochrane Database Syst Rev 1: CD005593

Bliss TV, Collingridge GL (1993) A synaptic model of memory: long-term potentiation in the hippocampus. Nature 361:31–39

Bliss TV, Lømo T (1973) Long-lasting potentiation of synaptic transmission in the dentate area of the anaesthetized rabbit following stimulation of the perforant path. J Physiol 232:331–356

Borlikova G, Endo S (2009) Inducible cAMP early repressor (ICER) and brain functions. Mol Neurobiol 40:73–86

Bullock R, Bergman H, Touchon J, Gambina G, He Y, Nagel J, Lane R (2006) Effect of age on response to rivastigmine or donepezil in patients with Alzheimer's disease. Curr Med Res Opin 22:483–494

Burgers PM, Eckstein F, Hunneman DH (1979) Stereochemistry of hydrolysis by snake venom phosphodiesterase. J Biol Chem 254:7476–7478

Chakraborty R, Vepuri V, Mhatre SD, Paddock BE, Miller S, Michelson SJ, Delvadia R, Desai A, Vinokur M, Melicharek DJ, Utreja S, Khandelwal P, Ansaloni S, Goldstein LE, Moir RD, Lee JC, Tabb LP, Saunders AJ, Marenda DR (2011) Characterization of a Drosophila Alzheimer's disease model: pharmacological rescue of cognitive defects. PLoS One 6:e20799

Croom KF, Dhillon S (2011) Bevacizumab: a review of its use in combination with paclitaxel or capecitabine as first-line therapy for HER2-negative metastatic breast cancer. Drugs 71:2213–2229

Danion JM, Huron C, Vidailhet P, Berna F (2007) Functional mechanisms of episodic memory impairment in schizophrenia. Can J Psychiatry 52:693–701

Dash PK, Hochner B, Kandel ER (1990) Injection of the cAMP-responsive element into the nucleus of Aplysia sensory neurons blocks long-term facilitation. Nature 345:718–721

Dell'Osso L, Carmassi C, Massimetti G, Daneluzzo E, Di Tommaso S, Rossi A (2011) Full and partial PTSD among young adult survivors 10 months after the L'Aquila 2009 earthquake: gender differences. J Affect Disord 131:79–83

Dong H, Yuede CM, Coughlan CA, Murphy KM, Csernansky JG (2009) Effects of donepezil on amyloid-beta and synapse density in the Tg2576 mouse model of Alzheimer's disease. Brain Res 1303:169–178

Drachman DA, Leavitt J (1974) Human memory and the cholinergic system. A relationship to aging? Arch Neurol 30:113–121

Dudek SM, Bear MF (1992) Homosynaptic long-term depression in area CA1 of hippocampus and effects of N-methyl-D-aspartate receptor blockade. Proc Natl Acad Sci U S A 89:4363–4367

Ebbinghaus H (1885) Memory: a contribution to experimental psychology (trans: Ruger HA and Bussenius CE, 1913). Dover, New York

Eichenbaum H (2000) A cortical-hippocampal system for declarative memory. Nat Rev Neurosci 1:41–50

Endo S (2012) Potential therapeutic targets for memory impairments and dementia: clues obtained from memory-enhanced mice. In: Thakur MH, Rattan SIS (eds) Brain aging, therapeutic interventions. Springer, Dordrecht, pp 219–238

Florian C, Mons N, Roullet P (2006) CREB antisense oligodeoxynucleotide administration into the dorsal hippocampal CA3 region impairs long- but not short-term spatial memory in mice. Learn Mem 13:465–472

Franceschi RT, Xiao G (2003) Regulation of the osteoblast-specific transcription factor, Runx2: responsiveness to multiple signal transduction pathways. J Cell Biochem 88:446–454

Fusco S, Ripoli C, Podda MV, Ranieri SC, Leone L, Toietta G, McBurney MW, Schütz G, Riccio A, Grassi C, Galeotti T, Pani G (2012) A role for neuronal cAMP responsive-element binding (CREB)-1 in brain responses to calorie restriction. Proc Natl Acad Sci U S A 109:621–626

Gibson GE, Peterson C, Jenden DJ (1981) Brain acetylcholine synthesis declines with senescence. Science 213:674–676

Glucksberg S, Cowen GN Jr (1970) Memory for nonattended auditory material. Cogn Psychol 1:149–156

Goldberg ND, Walseth TF, Stephenson JH, Krick TP, Graff G (1980) 18O-Labeling of guanosine monophosphate upon hydrolysis of cyclic guanosine 3′:5′-monophosphate by phosphodiesterase. J Biol Chem 255:10344–10347

Haas BW, Canli T (2008) Emotional memory function, personality structure and psychopathology: a neural system approach to the identification of vulnerability markers. Brain Res Rev 58:71–84

Hebb DO (1949) The organization of behavior. Wiley, New York

Hedden T, Gabrieli JD (2004) Insights into the ageing mind: a view from cognitive neuroscience. Nat Rev Neurosci 5:87–96

Hedden T, Van Dijk KR, Shire EH, Sperling RA, Johnson KA, Buckner RL (2012) Failure to modulate attentional control in advanced aging linked to white matter pathology. Cereb Cortex 22:1038–1051

Hiramatsu M, Takiguchi O, Nishiyama A, Mori H (2010) Cilostazol prevents amyloid β peptide (25–35)-induced memory impairment and oxidative stress in mice. Br J Pharmacol 161:1899–1912

Hsiao K, Chapman P, Nilsen S, Eckman C, Harigaya Y, Younkin S, Yang F, Cole G (1996) Correlative memory deficits, Abeta elevation, and amyloid plaques in transgenic mice. Science 274:99–102

Huang YY, Kandel ER (2006) Age-related enhancement of a protein synthesis-dependent late phase of LTP induced by low frequency paired-pulse stimulation in hippocampus. Learn Mem 13:298–306

Ihara M, Nishino M, Taguchi A, Yamamoto Y, Hattori Y, Saito S, Takahashi Y, Tsuji M, Kasahara Y, Takata Y, Okada M (2014) Cilostazol add-on therapy in patients with mild dementia receiving donepezil: a retrospective study. PLoS One 9:e89516

Ito M (2011) The cerebellum: brain for an implicit self. FT Press, Upper Saddle River

Ito M, Sakurai M, Tongroach P (1982) Climbing fibre induced depression of both mossy fibre responsiveness and glutamate sensitivity of cerebellar Purkinje cells. J Physiol 324:113–134

Ito M, Yamaguchi K, Nagao S, Yamazaki T (2014) Long-term depression as a model of cerebellar plasticity. Prog Brain Res 210:1–30

Kaang BK, Kandel ER, Grant SG (1993) Activation of cAMP-responsive genes by stimuli that produce long-term facilitation in Aplysia sensory neurons. Neuron 10:427–435

Kandel ER (2001) The molecular biology of memory storage: a dialogue between genes and synapses. Science 294:1030–1038

Kida S, Josselyn SA, Peña de Ortiz S, Kogan JH, Chevere I, Masushige S, Silva AJ (2002) CREB required for the stability of new and reactivated fear memories. Nat Neurosci 5:348–355

Kidd PM (2008) Alzheimer's disease, amnestic mild cognitive impairment, and age-associated memory impairment: current understanding and progress toward integrative prevention. Altern Med Rev 13:85–115

Kojima N, Borlikova G, Sakamoto T, Yamada K, Ikeda T, Itohara S, Niki H, Endo S (2008) Inducible cAMP early repressor acts as a negative regulator for kindling epileptogenesis and long-term fear memory. J Neurosci 28:6459–6472

Korzus E, Rosenfeld MG, Mayford M (2004) CBP histone acetyltransferase activity is a critical component of memory consolidation. Neuron 42:961–972

Lee JH, Park SY, Shin YW, Kim CD, Lee WS, Hong KW (2007) Concurrent administration of cilostazol with donepezil effectively improves cognitive dysfunctionwith increased neuroprotection after chronic cerebral hypoperfusion in rats. Brain Res 1185:246–255

Leiva J, Palestini M, Infante C, Goldschmidt A, Motles E (2009) Copper suppresses hippocampus LTP in the rat, but does not alter learning or memory in the Morris water maze. Brain Res 1256:69–75

Link CD, Taft A, Kapulkin V, Duke K, Kim S, Fei Q, Wood DE, Sahagan BG (2003) Gene expression analysis in a transgenic Caenorhabditis elegans Alzheimer's disease model. Neurobiol Aging 24:397–413

Llinás R, Lang EJ, Welsh JP (1997) The cerebellum, LTD, and memory: alternative views. Learn Mem 3:445–455

Longo VG (1966) Behavioral and electroencephalographic effects of atropine and related compounds. Pharmacol Rev 18:965–996

Lynch MA (2004) Long-term potentiation and memory. Physiol Rev 84:87–136

Lynch G, Browning M, Bennett WF (1979) Biochemical and physiological studies of long-term synaptic plasticity. Fed Proc 38:2117–2122

Malenka RC (1994) Synaptic plasticity in the hippocampus: LTP and LTD. Cell 78:535–538

Malenka RC, Bear MF (2004) LTP and LTD: an embarrassment of riches. Neuron 44:5–21

Malenka RC, Lancaster B, Zucker RS (1992) Temporal limits on the rise in postsynaptic calcium required for the induction of long-term potentiation. Neuron 9:121–128

Malinow R, Madison DV, Tsien RW (1988) Persistent protein kinase activity underlying long-term potentiation. Nature 335:820–824

Mhatre SD, Satyasi V, Killen M, Paddock BE, Moir RD, Saunders AJ, Marenda DR (2014) Synaptic abnormalities in a Drosophila model of Alzheimer's disease. Dis Model Mech 7:373–385

Miller GA (1956) The magical number even, plus or minus two: some limits of our capacity for processing information. Psychol Rev 63:81–97

Miller S, Mayford M (1999) Cellular and molecular mechanisms of memory: the LTP connection. Curr Opin Genet Dev 9:333–937

Milner B (1962) Les troubles de la memoire accompagnant des lesions hippocampiques bilaterales. In: Physiologie de l'hippocampe. Le Centre National de la Recherche Scientifique, Paris, pp 257–282

Milner B, Squire LR, Kandel ER (1998) Cognitive neuroscience and the study of memory. Neuron 20:445–468

Mulkey RM, Endo S, Shenolikar S, Malenka RC (1994) Involvement of a calcineurin/inhibitor-1 phosphatase cascade in hippocampal long-term depression. Nature 369:486–488

O'Donnell ME, Badger SA, Sharif MA, Young IS, Lee B, Soong CV (2009) The vascular and biochemical effects of cilostazol in patients with peripheral arterial disease. J Vasc Surg 49:1226–1234

O'Keefe J, Nadel L (1978) The hippocampus as a cognitive map. Oxford University Press, Oxford

Oddo S, Caccamo A, Shepherd JD, Murphy MP, Golde TE, Kayed R, Metherate R, Mattson MP, Akbari Y, LaFerla FM (2003) Triple-transgenic model of Alzheimer's disease with plaques and tangles: intracellular Abeta and synaptic dysfunction. Neuron 39:409–421

Olton DS, Becker JT, Handelmann GE (1979) Hippocampus, space, and memory. Behav Brain Sci 2:313–365

Ota KT, Pierre VJ, Ploski JE, Queen K, Schafe GE (2008) The NO-cGMP-PKG signaling pathway regulates synaptic plasticity and fear memory consolidation in the lateral amygdala via activation of ERK/MAP kinase. Learn Mem 15:792–805

Park SH, Kim JH, Bae SS, Hong KW, Lee DS, Leem JY, Choi BT, Shin HK (2011) Protective effect of the phosphodiesterase III inhibitor cilostazol on amyloid β-induced cognitive deficits associated with decreased amyloid β accumulation. Biochem Biophys Res Commun 408:602–608

Parsons CG, Danysz W, Quack G (1999) Memantine is a clinically well tolerated N-methyl-D-aspartate (NMDA) receptor antagonist – a review of preclinical data. Neuropharmacology 38:735–767

Passmore MJ, Ho A, Gallagher R (2012) Behavioral and psychological symptoms in moderate to severe Alzheimer's disease: a palliative care approach emphasizing recognition of personhood and preservation of dignity. J Alzheimers Dis 29:1–13

Patterson C, Feightner JW, Garcia A, Hsiung GY, MacKnight C, Sadovnick AD (2008) Diagnosis and treatment of dementia: 1. Risk assessment and primary prevention of Alzheimer disease. CMAJ 178:548–556

Pepeu G, Giovannini MG (2009) Cholinesterase inhibitors and beyond. Curr Alzheimers Res 6:86–96

Perez-Gonzalez R, Pascual C, Antequera D, Bolos M, Redondo M, Perez DI, Pérez-Grijalba V, Krzyzanowska A, Sarasa M, Gil C, Ferrer I, Martinez A, Carro E (2013) Phosphodiesterase 7 inhibitor reduced cognitive impairment and pathological hallmarks in a mouse model of Alzheimer's disease. Neurobiol Aging 34:2133–2145

Peterson LR, Peterson MJ (1959) Short-term retention of individual verbal items. J Exp Psychol 58:193–198

Pfennig A, Littmann E, Bauer M (2007) Neurocognitive impairment and dementia in mood disorders. J Neuropsychiatry Clin Neurosci 19:373–382

Plassman BL, Langa KM, Fisher GG, Heeringa SG, Weir DR, Ofstedal MB, Burke JR, Hurd MD, Potter GG, Rodgers WL, Steffens DC, Willis RJ, Wallace RB (2007) Prevalence of dementia in the United States: the aging, demographics, and memory study. Neuroepidemiology 29:125–132

Prakash RS, Erickson KI, Colcombe SJ, Kim JS, Voss MW, Kramer AF (2009) Age-related differences in the involvement of the prefrontal cortex in attentional control. Brain Cogn 71:328–335

Prickaerts J, Sik A, van der Staay FJ, de Vente J, Blokland A (2005) Dissociable effects of acetylcholinesterase inhibitors and phosphodiesterase type 5 inhibitors on object recognition memory: acquisition versus consolidation. Psychopharmacology (Berl) 177:381–390

Reneerkens OA, Rutten K, Steinbusch HW, Blokland A, Prickaerts J (2009) Selective phosphodiesterase inhibitors: a promising target for cognition enhancement. Psychopharmacology (Berl) 202:419–443

Richter W, Menniti FS, Zhang HT, Conti M (2013) PDE4 as a target for cognition enhancement. Expert Opin Ther Targets 17:1011–1027

Riepe MW, Kohler J, Horn R (2007) Donepezil in Alzheimer's disease: a clinical observational study evaluating individual treatment response. Curr Med Res Opin 23:1829–1835

Robinson DM, Keating GM (2006) Memantine: a review of its use in Alzheimer's disease. Drugs 66:1515–1534

Rogers SL, Doody RS, Mohs RC, Friedhoff LT (1998) Donepezil improves cognition and global function in Alzheimer disease: a 15-week, double-blind, placebo-controlled study. Donepezil Study Group. Arch Intern Med 158:1021–1031

Romberg C, Mattson MP, Mughal MR, Bussey TJ, Saksida LM (2011) Impaired attention in the 3xTgAD mouse model of Alzheimer's disease: rescue by donepezil (Aricept). J Neurosci 31:3500–3507

Rose GM, Hopper A, De Vivo M, Tehim A (2005) Phosphodiesterase inhibitors for cognitive enhancement. Curr Pharm Des 11:3329–3334

Scalco MZ, Van Reekum R (2006) Prevention of Alzheimer disease. Encouraging evidence. Can Fam Physician 52:200–207

Scoville WB, Milner B (1957) Loss of recent memory after bilateral hippocampal lesions. J Neurol Neurosurg Psychiatry 20:11–21

Sekiguchi A, Sugiura M, Taki Y, Kotozaki Y, Nouchi R, Takeuchi H, Araki T, Hanawa S, Nakagawa S, Miyauchi CM, Sakuma A, Kawashima R (2013) Brain structural changes as vulnerability factors and acquired signs of post-earthquake stress. Mol Psychiatry 18:618–623

Sierksma AS, Rutten K, Sydlik S, Rostamian S, Steinbusch HW, Van den Hove DL, Prickaerts J (2013) Chronic phosphodiesterase type 2 inhibition improves memory in the APPswe/PS1dE9 mouse model of Alzheimer's disease. Neuropharmacology 64:124–136

Soderling TR, Derkach VA (2000) Postsynaptic protein phosphorylation and LTP. Trends Neurosci 23:75–80

Sperling G (1960) The information available in brief visual presentations. Psychol Monogr Gen Appl 74:1–29

Squire LR (1992) Memory and the hippocampus: a synthesis from findings with rats, monkeys, and humans. Psychol Rev 99:195–231

Squire LR (2004) Memory systems of the brain: a brief history and current perspective. Neurobiol Learn Mem 82:171–177

Squire LR, Zola SM (1996) Structure and function of declarative and nondeclarative memory systems. Proc Natl Acad Sci U S A 93:13515–13522

Squire LR, Zola SM (1998) Episodic memory, semantic memory, and amnesia. Hippocampus 8:205–211

Strada SJ, Uzunov P, Weiss B (1974) Ontogenic development of a phosphodiesterase activator and the multiple forms of cyclic AMP phosphodiesterase of rat brain. J Neurochem 23:1097–1103

Suzuki A, Fukushima H, Mukawa T, Toyoda H, Wu LJ, Zhao MG, Xu H, Shang Y, Endoh K, Iwamoto T, Mamiya N, Okano E, Hasegawa S, Mercaldo V, Zhang Y, Maeda R, Ohta M, Josselyn SA, Zhuo M, Kida S (2011) Upregulation of CREB-mediated transcription enhances both short- and long-term memory. J Neurosci 31:8786–8802

Tariot PN, Solomon PR, Morris JC, Kershaw P, Lilienfeld S, Ding C (2000) A 5-month, randomized, placebo-controlled trial of galantamine in AD. The Galantamine USA-10 Study Group. Neurology 54:2269–2276

Tenor H, Hatzelmann A, Beume R, Lahu G, Zech K, Bethke TD (2011) Pharmacology, clinical efficacy, and tolerability of phosphodiesterase-4 inhibitors: impact of human pharmacokinetics. In: Francis SH, Conti M, Houslay MD (eds) Phosphodiesterase as drug targets. Springer, New York, pp 85–119

Terry AV Jr, Callahan PM, Hall B, Webster SJ (2011) Alzheimer's disease and age-related memory decline (preclinical). Pharmacol Biochem Behav 99:190–210

Tröster AI (2008) Neuropsychological characteristics of dementia with Lewy bodies and Parkinson's disease with dementia: differentiation, early detection, and implications for "mild cognitive impairment" and biomarkers. Neuropsychol Rev 18:103–119

Tulving E, Markowitsch HJ (1998) Episodic and declarative memory: role of the hippocampus. Hippocampus 8:198–204

Uzunov P, Weiss B (1972) Separation of multiple molecular forms of cyclic adenosine-3′,5-′-monophosphate phosphodiesterase in rat cerebellum by polyacrylamide gel electrophoresis. Biochim Biophys Acta 284:220–226

Van der Staay FJ, Rutten K, Bärfacker L, Devry J, Erb C, Heckroth H, Karthaus D, Tersteegen A, Van Kampen M, Blokland A, Prickaerts J, Reymann KG, Schröder UH, Hendrix M (2008) The novel selective PDE9 inhibitor BAY 73-6691 improves learning and memory in rodents. Neuropharmacology 55:908–918

Van Marum RJ (2009) Update on the use of memantine in Alzheimer's disease. Neuropsychiatr Dis Treat 5:237–247

Yanai S, Semba Y, Ito H, Endo S (2014) Cilostazol improves hippocampus-dependent long-term memory in mice. Psychopharmacology (Berl) 231:2681–2693

Zamanillo D, Sprengel R, Hvalby O, Jensen V, Burnashev N, Rozov A, Kaiser KM, Köster HJ, Borchardt T, Worley P, Lübke J, Frotscher M, Kelly PH, Sommer B, Andersen P, Seeburg PH, Sakmann B (1999) Importance of AMPA receptors for hippocampal synaptic plasticity but not for spatial learning. Science 284:1805–1811

Chapter 16
Critical Roles of Oxidative Signals in Age-Related Decline of Cerebellar Synaptic Plasticity

Sho Kakizawa and Nozomu Mori

Abstract A strong correlation between increasing age and the accumulation of oxidative modifications of functional molecules, such as proteins and lipids, has been observed in biological systems, including the central nervous system. Thus, oxidative signals, especially reactive oxygen species (ROS), are thought to be primary factors affecting age-related decline in brain functions. However, the molecular mechanism of the inhibitory action of oxidative signals is not yet fully understood. In this review article, we introduce our novel hypothesis on the molecular mechanism of aging in the nervous system: oxidative signals impair neuronal function through the inhibition of protein S-nitrosylation by nitric oxide (NO). This idea is based on the fact that the thiol group in cysteine residues is the common target of NO and oxidative signals. Actually, S-nitrosylation-dependent synaptic plasticity in the cerebellar cortex as well as NO-induced S-nitrosylation of cerebellar proteins are abolished by ROS treatment and aging. Furthermore, the functions of some proteins involved in synaptic plasticity, such as NSF, stargazin and ryanodine receptor 1, are demonstrated to be regulated in dependent on S-nitrosylation. Taken together, these studies open a new avenue in the study of the molecular mechanism of brain aging.

Keywords Aging • Brain • S-nitrosylation • Oxidative stress • Posttetanic potentiation • Long term potentiation • Cerebellum • Purkinje cell • Parallel fiber

---

S. Kakizawa
Department of Biological Chemistry, Graduate School of Pharmaceutical Sciences, Kyoto University, Kyoto 606-8501, Japan

Department of Anatomy and Neurobiology, Graduate School of Biomedical Sciences, Nagasaki University, 1-12-4 Sakamoto, Nagasaki 852-8523, Japan
e-mail: sho-kaki@pharm.kyoto-u.ac.jp

N. Mori (✉)
Department of Anatomy and Neurobiology, Graduate School of Biomedical Sciences, Nagasaki University, 1-12-4 Sakamoto, Nagasaki 852-8523, Japan
e-mail: morinosm@nagasaki-u.ac.jp

© Springer Japan 2015

N. Mori, I. Mook-Jung (eds.), *Aging Mechanisms*,
DOI 10.1007/978-4-431-55763-0_16

16.1 Introduction

Aging is associated with a general decline in physiological function in biological systems, including those in the nervous system, such as impairments of learning and memory. In fact, cells in all regions of the nervous system are affected by aging, as indicated by the decline of sensory, motor and cognitive functions with time (Barnes 1988, 2003; Finch 2003; Hofer et al. 2003; Landfield 1988). These nervous system dysfunctions are thought to occur when cells fail to respond adaptively to age-related increases in oxidative, metabolic and ionic stress. Among these stressors, the "oxidative stress" hypothesis of aging is widely accepted due to the strong correlation between increasing age and the accumulation of oxidative damage of cellular macromolecules observed in a wide range of studies (for a review, see Beckman and Ames 1998). In this hypothesis, oxidative stress is thought to induce damage to the cell via the oxidation of cellular components such as membrane lipids, proteins and DNA. For example, enhanced lipid peroxidation and protein oxidation are observed in aged rat brains (Calabrese et al. 2004; Cini and Moretti 1995; Devi and Kiran 2004; Forster et al. 1996; Murray and Lynch 1998). Correspondingly, several studies have shown that behavioral deficits of aged animals are associated with increases in oxidative stress (Butterfield et al. 2006; Cantuti-Castelvetri et al. 2000; Fukui et al. 2001).

Oxidative stress refers to the imbalance between the cellular production of reactive oxygen species (ROS), such as superoxide, hydroxyl radical and hydrogen peroxide (H_2O_2), for example, and the antioxidant mechanisms that remove them (Halliwell 1992). In fact, many studies indicate that the redox environment of the brain can be altered in favor of oxidation by an increased production of ROS and/or by a decreased activity of antioxidant defenses. However, the mechanisms through which oxidative signals induce decline in neuronal functions are not yet fully understood. This decline occurs, at least partly, due to the inconsistency of the effects of oxidative signals on the nervous system. For example, the effects of oxidative signals on synaptic plasticity in studies where ROS were exogenously applied to hippocampal slices induced paradoxical effects (Klann and Thiels 1999; Serrano and Klann 2004). In some studies, ROS are suggested to be essential for long-term potentiation (LTP) in hippocampal slices (Kamsler and Segal 2003a; Knapp and Klann 2002). Conversely, inhibitory effects of ROS on the synaptic plasticity are also reported in other studies (Auerbach and Segal 1997; Kamsler and Segal 2003a, b; Pellmar et al. 1991; Watson et al. 2002). In these studies, different protocols were applied for the induction of synaptic plasticity, and the species and ages of the animals studied were diverse. In addition, the target molecules or signaling pathways of oxidation in the course of the induction of the synaptic plasticity were not identified in these experiments. Therefore, the identification of the target molecules of oxidation in synaptic plasticity could provide critical insight concerning how oxidative stress results in deficits in synaptic plasticity and brain function in aged animals.

16.2 Effects of Oxidative Signals on NO-Dependent LTP in the Cerebellum

Among the excitatory synapses in the central nervous system, the parallel fiber to Purkinje cell synapse (PF synapse) in the cerebellar cortex is a good model for examining molecular mechanisms in synaptic plasticity (Ito 2006; Jörntell 2014). The cerebellum, a structure located at the back of the brain stem, consists of two parts: the cerebellar cortex and cerebellar nuclei. The cerebellar cortex consists of three layers: the molecular layer, the Purkinje cell layer and the granule cell layer (granular layer), from the surface to the inner portion. Purkinje cells (PCs), the sole output of the cerebellar cortex, receive two types of glutamatergic inputs. These inputs are climbing fibers (CF) from the inferior olive, and PF, the axon of cerebellar granule cells (Fig. 16.1). In the mature cerebellum, a single CF makes

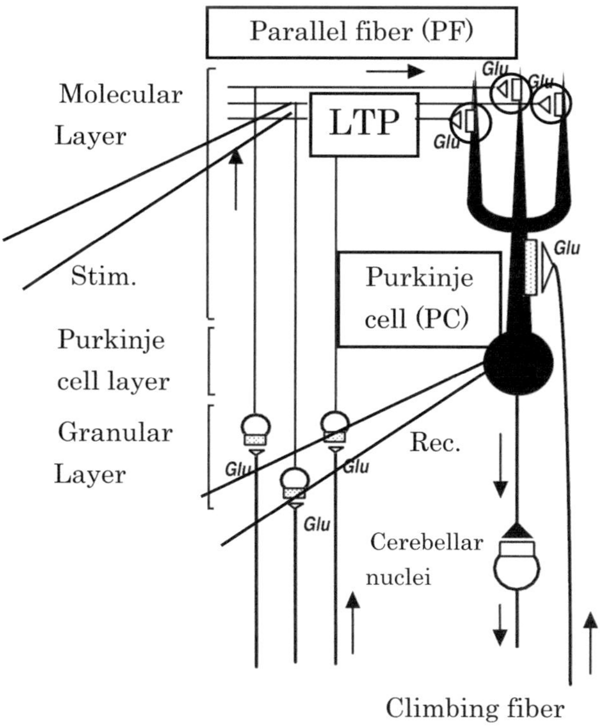

Fig. 16.1 Excitatory synaptic inputs to cerebellar Purkinje cell. The cerebellar cortex consists of three layers: the molecular layer, Purkinje cell layer and granular layer (also called granule cell layer). Purkinje cells (PCs), the sole output from the cerebellar cortex, receives two types of excitatory inputs. One input is the climbing fiber (CF), which originates from the inferior olive, and the other is the parallel fiber (PF), the axon of the granule cell. From the presynaptic terminals of CF and PF, the neurotransmitter glutamate (Glu) is released. At the PF-to-PC synapse, a form of synaptic plasticity, long-term potentiation (LTP), is induced by the repetitive stimulation of PF. This LTP is dependent on NO/S-nitrosylation signals (see text for detail)

an excitatory synapse onto the proximal dendritic region of the PC. On the other hand, each PC is innervated by hundreds thousands of PFs making synapses on the distal dendritic region. At the PF synapse, various types of synaptic plasticity have been characterized so far, including postsynaptic forms of LTP and long-term depression (LTD) as well as a presynaptically expressed LTP (Evans 2007; Ito 2006; Jörntell 2014). The LTD at the PF synapse is one of the most extensively studied plasticities in the central nervous system, and its involvement in motor learning tasks, such as eyeblink conditioning and vestibulo-ocular reflex (VOR), is indicated in multiple studies (Ito 2002, 2013). On the other hand, the functional roles of postsynaptic and presynaptic LTP at PF are not yet fully understood, although the possible involvement of postsynaptic LTP in fear conditioning has been suggested (Sacchetti et al. 2004).

Among them, the postsynaptically expressed LTP at the PF synapse is demonstrated to be dependent on NO signals (Kakegawa and Yuzaki 2005; Lev-Ram et al. 2002; Namiki et al. 2005). Nitric oxide is a gaseous messenger in biological systems and is produced from L-arginine by three distinct NO synthases (NOSs) (Alderton et al. 2001; Stuehr et al. 2004). Two of these NO synthases, neuronal (nNOS) and endothelial (eNOS), are calcium dependent, whereas inducible NOS (iNOS) is calcium independent. This NO-dependent LTP at the PF synapse (NO-LTP) is shown to be induced by the repetitive stimulation of PF (Kakegawa and Yuzaki 2005; Lev-Ram et al. 2002; Namiki et al. 2005). In addition, granule cells, the origin of PF, express high levels of nNOS (also called NOS1), whereas the immunohistochemical signal of nNOS is not observed in PCs. These observations suggest that NO, which is essential for the induction of the LTP at the PF synapse, derives from PF. Nitric oxide has two downstream signaling pathways: the activation of soluble guanylyl cyclase (sGC) and S-nitrosylation (the term "S-nitrosation" is also used for this modification of proteins Iyer et al. 2014) (Fig. 16.2). The activation of sGC induces the elevation of cyclic guanosine monophosphate (cGMP) in cytoplasm and the subsequent activation of protein kinase G, which regulates the functions of various proteins through protein phosphorylation. On the other hand, S-nitrosylation of cysteine residues in various proteins results in the modification of protein functions including those of enzymes and ionic channels (Calabrese et al. 2007; Hess et al. 2005; Jaffrey et al. 2001; Nakamura and Lipton

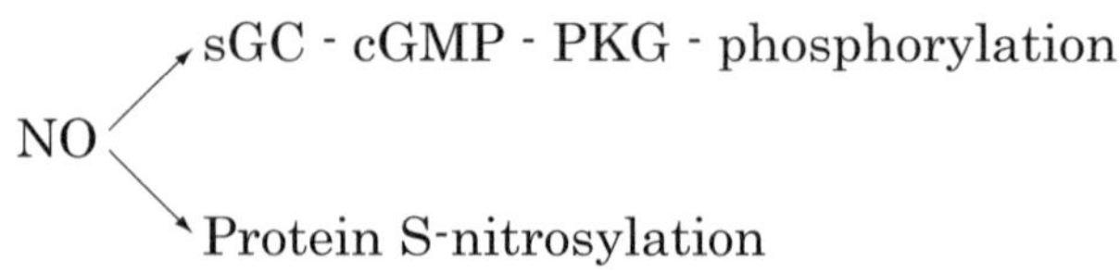

Fig. 16.2 The two downstream signaling pathways of nitric oxide. Nitric oxide (NO) has two downstream signaling pathways. One pathway involves the activation of soluble guanylyl cyclase (sGC) and results in the elevation of cyclic GMP (cGMP) levels in the cytoplasm (*upper*). cGMP subsequently activates protein kinase G (PKG), which phosphorylates a wide range of proteins. The other pathway involves protein S-nitrosylation (*lower*). The functions of some protein are known to be regulated by S-nitrosylation

2007). The NO-LTP at the PF synapse is not impaired by 1H-[1,2,4] Oxadiazolo [4,3-a] quinoxalin-1-one (ODQ), an inhibitor of sGC, but is inhibited by ascorbic acid, a reductant which reduces S-nitrosylated thiol, when applied to the PC using a micropipette (Kakizawa et al. 2012b; Namiki et al. 2005). These results indicate that the S-nitrosylation of the proteins in PCs is essential for the induction of PF-LTP. Moreover, the LTP is also abolished by the preincubation of the cerebellar slices with the N-ethyl-maleimide (NEM), a membrane-permeable thiol blocker, or glutathione (reduced form; GSH) and N-acetyl-cysteine (NAC), membrane impermeable thiol donors (Kakizawa et al. 2012a). These results indicate critical roles of thiol groups in cysteine residues in NO-LTP. Furthermore, because GSH and NAC are membrane impermeable, the results also indicate that NO released from PF into the synaptic cleft significantly contributes to the induction of the LTP. Taken together, the NO-LTP is indicated to be dependent on the S-nitrosylation of proteins in PCs.

Thiol groups in cysteine residues are the targets of S-nitrosylation by NO as well as the targets of oxidation and the subsequent formation of disulfide bonds by oxidative signals including ROS. In addition, the accumulation of protein oxidation is reported by many studies, as described above. Therefore, it is highly possible that oxidative signals inhibit NO-induced protein S-nitrosylation through the formation of disulfide bonds. When a thiol group is in a reduced state (R-SH), it has the potential to be S-nitrosylated by NO signals. However, the same thiol is difficult to S-nitrosylate when it is already oxidized and involved in a disulfide bond (R-S-S-R′) because the disulfide bond is very stable in general (Fig. 16.3).

If this hypothesis is true, protein S-nitrosylation as well as S-nitrosylation-dependent biological events could be inhibited by endogenous and exogenous oxidative signals. Kakizawa et al. examined this hypothesis and experimentally demonstrated the inhibitory effects of oxidative signals on the S-nitrosylation of cerebellar proteins and NO-LTP which is indicated to be dependent on protein S-nitrosylation in PCs (Kakizawa et al. 2012a). They first demonstrated that NO-LTP, induced by burst stimulation of PF, was blocked by the pretreatment of

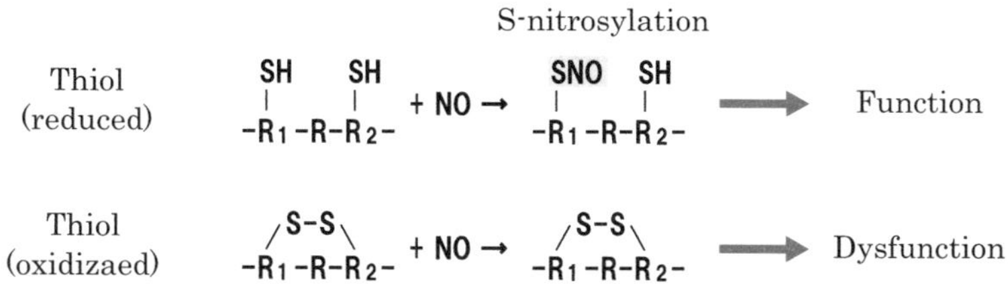

Fig. 16.3 Inhibition of protein S-nitrosylation by oxidative signals through the formation of a disulfide bond. Thiol groups in cysteine residue are targets of both S-nitrosylation by nitric oxide (NO) and oxidation (formation of a disulfide bond) by oxidative signals. When a thiol is in a reduced state (*upper*; R_1-SH), the group can be S-nitrosylated by NO (R_1-SNO) and exert its biological function. Conversely, when the same thiol is already involved in a disulfide bond (*lower*; R_1-S-S-R_2), NO cannot S-nitrosylate the group and induce its biological function because the disulfide bond is very stable and difficult to break

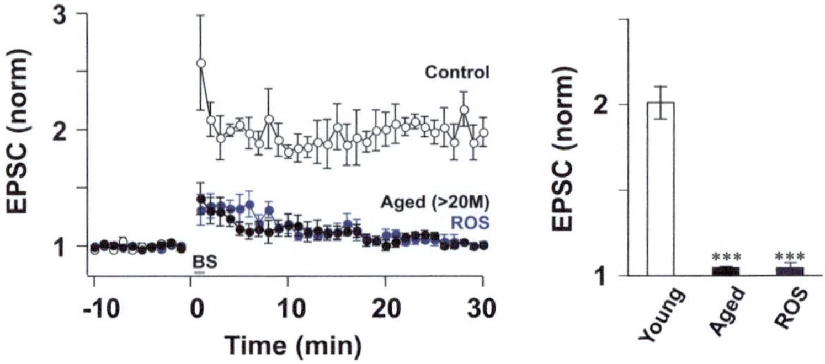

Fig. 16.4 Inhibitory effects of ROS and aging on NO-dependent LTP at the PF synapse. When the cerebellar slices from young mice (1-month-old) are not pretreated with ROS (Control), burst stimulation (BS) to parallel fibers (PF) induces long-term potentiation (LTP) at the PF synapse. Conversely, when slices from young mice are pretreated with ROS (ROS), or when cerebellar slices are obtained from aged mice (older than 20 months old (Aged >20 M)), the LTP is almost completely abolished. (*Left*) Normalized amplitude of excitatory postsynaptic current (EPSC (norm)) at the PF synapse before and after the BS). The value is normalized with the averaged value over 1–10 min before the BS, and results are expressed as the Mean ± S.E.M. (*Right*). The averaged amplitudes of PF-EPSC during the 21–30 min after the BS. Mean ± S.E.M. ***$p < 0.001$, compared with the value for the control

the cerebellar slices from young 1-month-old mice with oxidative agents (e.g., hydrogen peroxide) or a thiol blocker (NEM, for example) (Fig. 16.4). The pre-treatments did not affect basic electrophysiological properties (e.g., input-output relations and paired-pulse ratio) of the excitatory postsynaptic current of PF (PF-EPSC). In addition, the application of the oxidative reagent itself did not induce LTP at the PF synapse, and the results exclude the possibility that NO-LTP is occluded in the cerebellar slices pretreated with the oxidizing reagents. Therefore, the oxidative signals were revealed to specifically inhibit the signaling pathways essential for the induction of NO-LTP at the PF synapse.

In addition to the inhibitory effects of oxidative reagents, which are exogenous oxidative signals, the possible involvement of endogenous oxidative signals in NO-LTP inhibition was also indicated in that study (Kakizawa et al. 2012a). NO-LTP was impaired in the cerebellar slices from aged mice older than 20 months, whereas no significant changes in basic electrophysiological properties of PF-EPSC were observed. These results suggest the inhibition of NO-LTP by endogenous oxidative signals in aged mice (Fig. 16.4).

Do oxidative signals inhibit NO-LTP through the inhibition of protein S-nitrosylation? To examine this issue, the S-nitrosylation levels of cerebellar proteins were quantitatively estimated by biochemical analysis, biotin-switch assay in which S-nitrosylated thiols are displaced with biotin labels (Jaffrey et al. 2001) (Fig. 16.5). Incubation of the cerebellar slices from young mice with NO donor, NOC7, resulted in the increase in S-nitrosylation levels of the cerebellar proteins. On the other hand, pretreatment of the cerebellar slices with oxidative

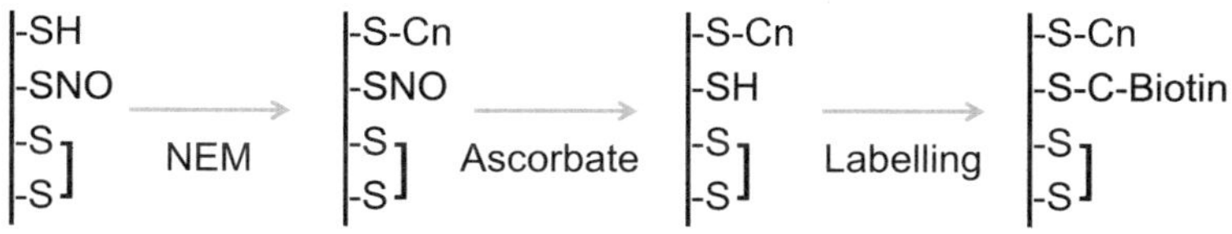

Fig. 16.5 The biotin-switch assay. S-nitrosylated thiol was labeled and detected in a biotin-switch assay. First, reduced thiols (−SH) are specifically blocked by N-ethylmaleimide (NEM). Then, S-nitrosylated thiol (−SNO) was reduced by ascorbic acid (ascorbate), and subsequently labeled by biotin (-S-C-Biotin). The biotinylated thiol was detected using streptavidin conjugated with horseradish peroxidase (HRP)

reagents, such as H_2O_2, severely impaired the NO-induced elevation of the protein S-nitrosylation levels (Kakizawa et al. 2012a) (Fig. 16.6). The possible involvement of endogenous oxidative signals in the inhibition of protein S-nitrosylation is also suggested. In the cerebellar slices from mice older than 20 months old, the elevation of protein S-nitrosylation levels induced by NO-donor application is severely impaired. Again, the result supports the idea that endogenous oxidation signals also inhibit protein S-nitrosylation (Fig. 16.6). Taken together, these results strongly indicate that the inhibitory action of oxidizing signals on the induction of the NO-LTP is mediated by the impairment of protein S-nitrosylation induced by acute NO signals.

16.3 Target Proteins of S-Nitrosylation in the Central Nervous System

In the notion that oxidative signals affect biological events through the inhibition of NO-induced protein S-nitrosylation, oxidative signals could impair the protein S-nitrosylation in a "competitive" manner because thiol groups are targets of S-nitrosylation as well as oxidation. Therefore, it is quite informative to specify S-nitrosylated /oxidized proteins to reveal the molecular mechanisms of the inhibitory actions of the oxidative signals on the biological systems. For a long time, S-nitrosylation site specification has been laboriously tackled on a protein-by-protein basis (Hess et al. 2005). Moreover, a high-throughput proteomic approach that enables the simultaneous identification of S-nitrosylated-Cys sites and their cognate proteins in complex biological mixtures has only recently been developed (Hao et al. 2006). This approach, termed SNOSID (SNO Site Identification), is a modification of the biotin-switch technique of S-nitrosylated cysteines (Jaffrey et al. 2001), including the biotinylation of protein SNO-Cys residues, trypsinolysis, affinity purification of biotinylated-peptides, and amino acid sequencing by liquid chromatography tandem mass spectrometry (MS) (Hao et al. 2006). Studies using these techniques have identified several proteins which undergo S-nitrosylation, and some of these proteins are indicated to be involved in neuronal functions, including synaptic plasticity, in the central nervous system. These proteins could be

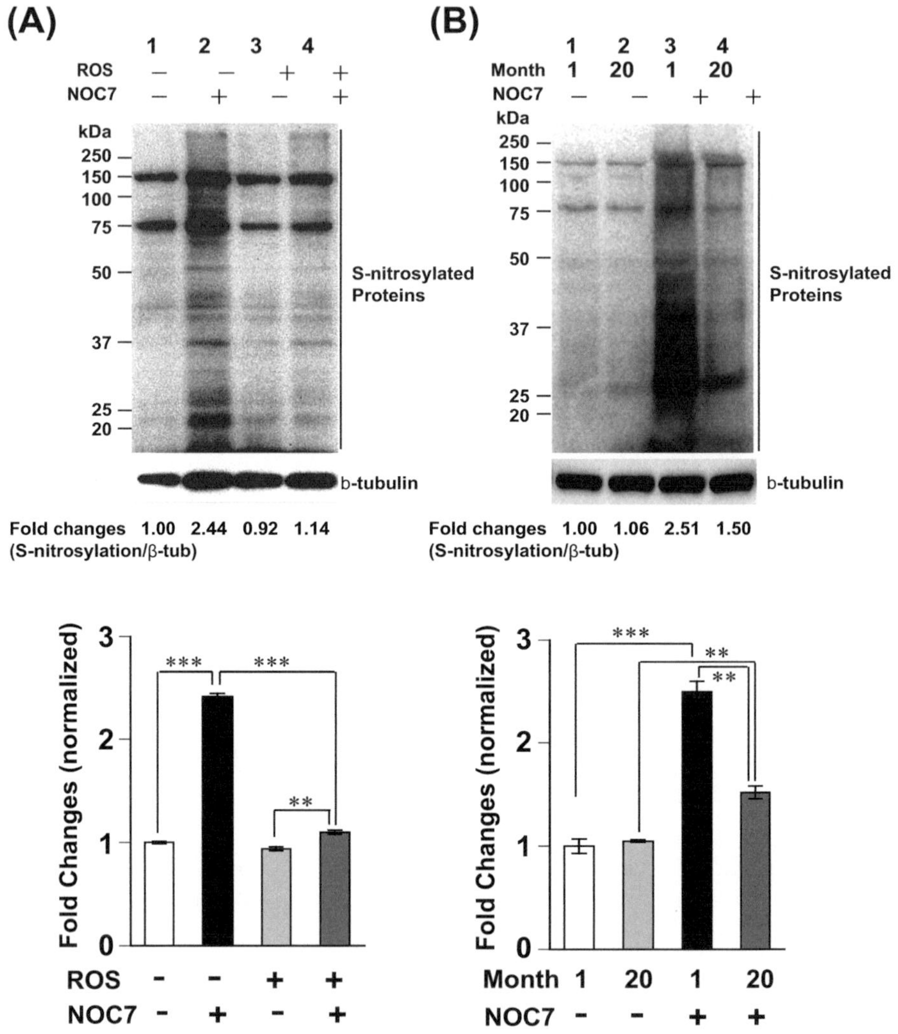

Fig. 16.6 Inhibition of NO-induced protein S-nitrosylation by pretreatment with ROS or aging. The overall protein S-nitrosylation in cerebellar slices treated with NOC7 was resolved using a biotin-switch assay (see Fig. 16.5). (**a**) (*upper*) Effects of H_2O_2 on NO-induced protein S-nitrosylation in the cerebellar slices from 1-month-old mice. Lane 1, vehicle-pretreated cerebellar slice; Lane 2, cerebellar slice treated with NOC7, an NO donor; Lane 3, H_2O_2-pretreated cerebellar slice; Lane 4, NOC7-treated cerebellar slice after the pretreatment with H_2O_2. The molecular weights of marker proteins are given to the *left*. The values shown at the *bottom* are overall protein S-nitrosylation levels, calibrated with β-tubulin levels and then normalized with the value for the vehicle-pretreated group (lane 1). (*lower*) Summary of the fold changes in S-nitrosylation levels. Mean ± S.E.M. (**b**) (*upper*) NO-induced protein S-nitrosylation in the cerebellar slices from young (1-month-old) or aged (older than 20-month-old) mice. Lane 1, cerebellar slice from a young mouse treated with vehicle; Lane 2, cerebellar slice from an aged mouse treated with vehicle; Lane 3, cerebellar slice from a young mouse treated with NOC7; Lane 4, cerebellar slice from an aged mouse treated with NOC7. Molecular weights of marker proteins

potential candidates of the targets of the oxidative signals, and the oxidation of these target-proteins is expected to inhibit S-nitrosylation-dependent biological events (e.g., NO-LTP) through the inhibition of the S-nitrosylation of the proteins. In this section, we highlight representative proteins that are demonstrated to undergo S-nitrosylation and play key roles in the modulation of neuronal functions in a manner dependent on protein S-nitrosylation.

NSF

N-ethylmaleimide sensitive factor (NSF) is thought to be a target for NO-mediated S-nitrosylation. In fact, its name originates from its sensitivity to N-ethylmaleimide, which is known to alkylate the sulfhydryl groups of cysteine residues in proteins (Block et al. 1988). Actually, NO-mediated S-nitrosylation of NSF was identified as a regulator of the exocytosis of endothelial granules (Matsushita et al. 2003).

The essential role of NSF S-nitrosylation in the regulation of synaptic plasticity is also suggested (Huang et al. 2005). NSF is revealed to regulate excitatory synaptic transmission by stabilizing or recycling alpha-amino-3-hydroxy-5-methyl-4-isoxazole propionic acid receptor (AMPA-R) subunit GluR2 (Osten et al. 1998; Song et al. 1998) and disassemble the GluR2/PICK1 protein complex that regulates cerebellar LTD (Steinberg et al. 2004; Xia et al. 2000). Huang et al. (2005) reported that NSF is physiologically S-nitrosylated by endogenous, neuronally derived NO (Huang et al. 2005). The S-nitrosylation of NSF enhances its binding to the GluR2 subunit, and the surface expression of GluR2 is regulated by NO and NSF-GluR2 interactions. These results strongly suggest that the S-nitrosylation of NSF is a physiological mediator for the surface expression of GluR2 during the induction of synaptic plasticity.

16.3.1 Stargazin

Stargazin has also emerged as a principal regulator of AMPAR surface expression. Selvakumar et al. (2009) demonstrated that stargazin is physiologically S-nitrosylated, resulting in increased surface expression of the receptor in HEK 293 cells and primary neurons (Selvakumar et al. 2009). The S-nitrosylation of stargazin enhances its binding to the AMPAR subunit GluR1, resulting in increased surface expression of the AMPAR.

Fig. 16.6 (continued) and the values shown at the *bottom* are the same as (**a**). (*lower*) Summary of the fold changes in S-nitrosylation levels. Mean ± S.E.M. $**p < 0.01$ and $***p < 0.001$ indicate significant differences different

16.3.2 Type 1 Ryanodine Receptors

Ryanodine receptors (RyRs) located in the sarcoplasmic / endoplasmic reticulum (SR/ER) membrane are required for the intracellular Ca^{2+} release that is involved in a wide range of cellular functions. In addition to Ca^{2+}-induced Ca^{2+} release (CICR) in cardiac cells and voltage-induced Ca^{2+} release in skeletal muscle cells, NO-induced Ca^{2+} release (NICR) was recently identified in neuronal cells such as cerebellar Purkinje cells and cortical neuronal cells (Kakizawa et al. 2012b).

Thus far, three isoforms of the RyRs have been identified and cloned: type 1 RyR (RyR1) was first detected in skeletal muscle (Takeshima et al. 1989), type 2 RyR (RyR2) was first found in cardiac muscles (Nakai et al. 1990) and type 3 RyR (RyR3) was found in the brain (Giannini et al. 1992; Hakamata et al. 1992). They are encoded by three distinct genes and share ~70 % sequence homology (Rossi and Sorrentino 2002). Because the RyR has approximately 100 cysteine residues per subunit and almost half of the thiol groups are kept in a reduced state under the resting condition, RyR is thought to be one of the substrate proteins for S-nitrosylation. Actually, the open probability of RyR1 measured in lipid bilayers is increased by NO donors (Eu et al. 2000; Stoyanovsky et al. 1997; Sun et al. 2003; Xu et al. 1998), and this response is accompanied by an increase in the S-nitrosylation of the Ca^{2+} release channels.

S-nitrosylation-dependent activation of RyR1 is also observed in neuronal cells. In cerebellar Purkinje cells, which express significantly higher levels of RyR1, the application of an NO donor induced prominent increases in Ca^{2+} levels. Because this NO-induced Ca^{2+} increase is impaired by dantrolene, an inhibitor of RyR1, or the systemic deletion of the *RyR1* gene, the Ca^{2+} increase is revealed to be due to Ca^{2+} release from the ER and mediated by RyR (Kakizawa et al. 2012b). In addition, the NO-induced Ca^{2+} release is not inhibited by ODQ, a sGC inhibitor, but is inhibited by ascorbic acid. These results indicate that the NO-induced Ca^{2+} release is indicated to be dependent on protein S-nitrosylation. Furthermore, a single-site C3635A-mutation in rabbit RyR1 abolished NO-induced Ca^{2+} release and S-nitrosylation of the mutated channels expressed in HEK 293 cells. Taken together, S-nitrosylation of RyR1 is revealed to be essential for the activation of the channel and the resultant cellular response, NO-induced Ca^{2+} release (Fig. 16.7). Thus far, the possible involvement of NICR in NO-LTP and NO-induced neuronal cell death is suggested by pharmacological studies (Kakizawa et al. 2012b).

16.4 Perspectives

In this review article, we propose a possible molecular mechanism of the inhibitory action of oxidative signals on biological events. Specifically, we propose that the inhibitory action of an oxidizing signal could be mediated through the impairment of protein S-nitrosylation. Because thiol groups in cysteine residues are the targets of S-nitrosylation by NO as well as the targets of oxidation and the subsequent formation of disulfide bond, NO and oxidative signals, such as H_2O_2, could

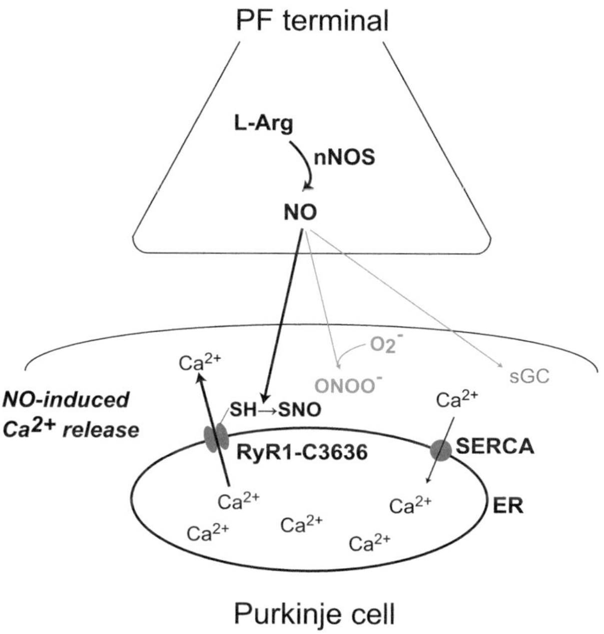

Fig. 16.7 A schematic diagram of signaling pathways for nitric oxide-induced Ca^{2+} release (NICR) in cerebellar Purkinje cells. Nitric oxide (NO) is produced at the parallel-fiber terminal in response to neuronal activity and diffuses into the cerebellar Purkinje cell. Then, type 1 ryanodine receptors (RyR1) located in the endoplasmic reticulum (ER) membrane are S-nitrosylated and activated, resulting in Ca^{2+} release from the ER. Cyclic GMP and peroxynitrite produced by the reaction of NO with superoxide are unlikely to be involved in NICR. *nNOS* neuronal NO synthase, *SERCA* sarco/endoplasmic reticulum Ca^{2+} ATPase

compete the binding to thiol groups. When a thiol is in a reduced state (R-SH), NO could S-nitrosylate the thiol (formation of R-SNO) and induce its effects. On the other hand, when a thiol is already involved in a disulfide bond (R-S-S-R′), NO can no longer modify the groups because the disulfide bond is very stable. In accordance with this notion, in cerebellar slices pretreated with an oxidizing reagent (e.g., H_2O_2), NO-induced S-nitrosylation of cerebellar proteins is severely impaired. Furthermore, the treatment also inhibits S-nitrosylation-dependent LTP at the PF-PC synapse (NO-LTP).

These findings indicate that the S-nitrosylation of functional protein(s) is essential for the induction of the NO-LTP. As described above, some functional proteins such as NSF, stargazin and RyR1 are candidates for S-nitrosylation and oxidation (formation of disulfide bond) because these proteins are indicated to be involved in synaptic plasticity and the functions of these molecules are regulated in a manner dependent on S-nitrosylation. The inhibitory actions of oxidative signals on the molecular functions of these proteins are expected to be clarified in future studies.

However, it should also be noted that the target cysteines of S-nitrosylation and oxidation are not necessarily the same. For example, phosphatase with sequence homology to tensin (PTEN) is known to be oxidized by high concentrations of H_2O_2

(>0.5 mM), which results in disulfide bond formation between Cys-71 and Cys-124 (Lee et al. 2002). On the other hand, the protein is selectively S-nitrosylated by low concentrations of NO donor (1–10 μM SNOC) at a specific cysteine residue, Cys-83, and the nitrosylation of the cysteine is reported to regulate the enzymatic activity of the protein (Numajiri et al. 2011). Whether NO-dependent-S-nitrosylation is abolished by oxidative signals is probably influenced by the 3-D topology of the protein as well as the concentration of NO and oxidative signals.

In this review article, we discussed a possible molecular mechanism for age-related decline in brain function, focusing on S-nitrosylation-dependent LTP (NO-LTP) at the PF synapse in the cerebellum as a model. In addition to NSF, stargazin and RyR1, which have been already demonstrated to have S-nitrosylation-dependent functions involved in synaptic plasticity, it is also possible that molecules involved in the structural plasticity of dendritic spines in hippocampal and cortical regions could be S-nitrosylated or oxidized. The identification of the candidate molecules involved in S-nitrosylation-dependent structural plasticity and the observation of an inhibitory action of oxidative signals on plasticity may further clarify the molecular mechanisms of aging in brain systems (Fig. 16.8).

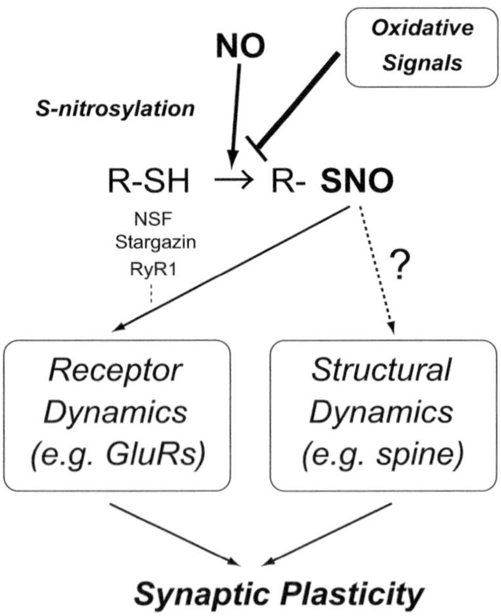

Fig. 16.8 A schematic diagram of possible mechanisms through which oxidative signals inhibit synaptic plasticity. A thiol group in a reduced state (R-SH) is S-nitrosylated (R-SNO) by nitric oxide (NO). S-nitrosylation of some molecules, including N-ethylmaleimide sensitive factor (NSF), stargazin and type 1 ryanodine receptor (RyR1), regulates the dynamics of postsynaptic receptors, e.g., glutamate receptors (GluRs), and subsequently induces synaptic plasticity. It is also possible that molecules involved in the structural dynamics of dendritic spines are S-nitrosylated and induce synaptic plasticity, although the candidate molecule(s) has yet to be identified. On the other hand, oxidative signals could inhibit S-nitrosylation through the mechanism shown in Fig. 16.3 and could therefore abolish synaptic plasticity

References

Alderton WK, Cooper CE, Knowles RG (2001) Nitric oxide synthases: structure, function and inhibition. Biochem J 357:593–615

Auerbach JM, Segal M (1997) Peroxide modulation of slow onset potentiation in rat hippocampus. J Neurosci 17:8695–8701

Barnes CA (1988) Aging and the physiology of spatial memory. Neurobiol Aging 9:563–568

Barnes CA (2003) Long-term potentiation and the ageing brain. Philos Trans R Soc Lond B Biol Sci 358:765–772

Beckman KB, Ames BN (1998) The free radical theory of aging matures. Physiol Rev 78:547–581

Block MR, Glick BS, Wilcox CA, Wieland FT, Rothman JE (1988) Purification of an N-ethylmaleimide-sensitive protein catalyzing vesicular transport. Proc Natl Acad Sci U S A 85:7852–7856

Butterfield DA, Perluigi M, Sultana R (2006) Oxidative stress in Alzheimer's disease brain: new insights from redox proteomics. Eur J Pharmacol 545:39–50

Calabrese V, Scapagnini G, Ravagna A, Colombrita C, Spadaro F, Butterfield DA, Giuffrida Stella AM (2004) Increased expression of heat shock proteins in rat brain during aging: relationship with mitochondrial function and glutathione redox state. Mech Ageing Dev 125:325–335

Calabrese V, Mancuso C, Calvani M, Rizzarelli E, Butterfield DA, Stella AM (2007) Nitric oxide in the central nervous system: neuroprotection versus neurotoxicity. Nat Rev Neurosci 8:766–775

Cantuti-Castelvetri I, Shukitt-Hale B, Joseph JA (2000) Neurobehavioral aspects of antioxidants in aging. Int J Dev Neurosci 18:367–381

Cini M, Moretti A (1995) Studies on lipid peroxidation and protein oxidation in the aging brain. Neurobiol Aging 16:53–57

Devi SA, Kiran TR (2004) Regional responses in antioxidant system to exercise training and dietary vitamin E in aging rat brain. Neurobiol Aging 25:501–508

Eu JP, Sun JH, Xu L, Stamler JS, Meissner G (2000) The skeletal muscle calcium release channel: coupled O-2 sensor and NO signaling functions. Cell 102:499–509

Evans GJ (2007) Synaptic signalling in cerebellar plasticity. Biol Cell 99:363–378

Finch CE (2003) Neurons, glia, and plasticity in normal brain aging. Neurobiol Aging 24(Suppl 1): S123–S127; discussion S131

Forster MJ, Dubey A, Dawson KM, Stutts WA, Lal H, Sohal RS (1996) Age-related losses of cognitive function and motor skills in mice are associated with oxidative protein damage in the brain. Proc Natl Acad Sci U S A 93:4765–4769

Fukui K, Onodera K, Shinkai T, Suzuki S, Urano S (2001) Impairment of learning and memory in rats caused by oxidative stress and aging, and changes in antioxidative defense systems. Ann N Y Acad Sci 928:168–175

Giannini G, Clementi E, Ceci R, Marziali G, Sorrentino V (1992) Expression of a ryanodine receptor-Ca2+ channel that is regulated by TGF-beta. Science 257:91–94

Hakamata Y, Nakai J, Takeshima H, Imoto K (1992) Primary structure and distribution of a novel ryanodine receptor/calcium release channel from rabbit brain. FEBS Lett 312:229–235

Halliwell B (1992) Reactive oxygen species and the central nervous system. J Neurochem 59:1609–1623

Hao G, Derakhshan B, Shi L, Campagne F, Gross SS (2006) SNOSID, a proteomic method for identification of cysteine S-nitrosylation sites in complex protein mixtures. Proc Natl Acad Sci U S A 103:1012–1017

Hess DT, Matsumoto A, Kim SO, Marshall HE, Stamler JS (2005) Protein S-nitrosylation: purview and parameters. Nat Rev Mol Cell Biol 6:150–166

Hofer SM, Berg S, Era P (2003) Evaluating the interdependence of aging-related changes in visual and auditory acuity, balance, and cognitive functioning. Psychol Aging 18:285–305

Huang Y, Man HY, Sekine-Aizawa Y, Han Y, Juluri K, Luo H, Cheah J, Lowenstein C, Huganir RL, Snyder SH (2005) S-nitrosylation of N-ethylmaleimide sensitive factor mediates surface expression of AMPA receptors. Neuron 46:533–540
Ito M (2002) Historical review of the significance of the cerebellum and the role of Purkinje cells in motor learning. Ann N Y Acad Sci 978:273–288
Ito M (2006) Cerebellar circuitry as a neuronal machine. Prog Neurobiol 78:272–303
Ito M (2013) Error detection and representation in the olivo-cerebellar system. Front Neural Circ 7:1
Iyer AK, Rojanasakul Y, Azad N (2014) Nitrosothiol signaling and protein nitrosation in cell death. Nitric Oxide 42C:9–18
Jaffrey SR, Erdjument-Bromage H, Ferris CD, Tempst P, Snyder SH (2001) Protein S-nitrosylation: a physiological signal for neuronal nitric oxide. Nat Cell Biol 3:193–197
Jörntell H (2014) Cerebellar synaptic plasticity and the credit assignment problem. Cerebellum
Kakegawa W, Yuzaki M (2005) A mechanism underlying AMPA receptor trafficking during cerebellar long-term potentiation. Proc Natl Acad Sci U S A 102:17846–17851
Kakizawa S, Shibazaki M, Mori N (2012a) Protein oxidation inhibits NO-mediated signaling pathway for synaptic plasticity. Neurobiol Aging 33:535–545
Kakizawa S, Yamazawa T, Chen Y, Ito A, Murayama T, Oyamada H, Kurebayashi N, Sato O, Watanabe M, Mori N et al (2012b) Nitric oxide-induced calcium release via ryanodine receptors regulates neuronal function. EMBO J 31:417–428
Kamsler A, Segal M (2003a) Hydrogen peroxide modulation of synaptic plasticity. J Neurosci 23:269–276
Kamsler A, Segal M (2003b) Paradoxical actions of hydrogen peroxide on long-term potentiation in transgenic superoxide dismutase-1 mice. J Neurosci 23:10359–10367
Klann E, Thiels E (1999) Modulation of protein kinases and protein phosphatases by reactive oxygen species: implications for hippocampal synaptic plasticity. Prog Neuropsychopharmacol Biol Psychiatry 23:359–376
Knapp LT, Klann E (2002) Potentiation of hippocampal synaptic transmission by superoxide requires the oxidative activation of protein kinase C. J Neurosci 22:674–683
Landfield PW (1988) Hippocampal neurobiological mechanisms of age-related memory dysfunction. Neurobiol Aging 9:571–579
Lee SR, Yang KS, Kwon J, Lee C, Jeong W, Rhee SG (2002) Reversible inactivation of the tumor suppressor PTEN by H_2O_2. J Biol Chem 277:20336–20342
Lev-Ram V, Wong ST, Storm DR, Tsien RY (2002) A new form of cerebellar long-term potentiation is postsynaptic and depends on nitric oxide but not cAMP. Proc Natl Acad Sci U S A 99:8389–8393
Matsushita K, Morrell CN, Cambien B, Yang SX, Yamakuchi M, Bao C, Hara MR, Quick RA, Cao W, O'Rourke B et al (2003) Nitric oxide regulates exocytosis by S-nitrosylation of N-ethylmaleimide-sensitive factor. Cell 115:139–150
Murray CA, Lynch MA (1998) Evidence that increased hippocampal expression of the cytokine interleukin-1 beta is a common trigger for age- and stress-induced impairments in long-term potentiation. J Neurosci 18:2974–2981
Nakai J, Imagawa T, Hakamat Y, Shigekawa M, Takeshima H, Numa S (1990) Primary structure and functional expression from cDNA of the cardiac ryanodine receptor/calcium release channel. FEBS Lett 271:169–177
Nakamura T, Lipton SA (2007) Molecular mechanisms of nitrosative stress-mediated protein misfolding in neurodegenerative diseases. Cell Mol Life Sci 64:1609–1620
Namiki S, Kakizawa S, Hirose K, Iino M (2005) NO signalling decodes frequency of neuronal activity and generates synapse-specific plasticity in mouse cerebellum. J Physiol 566:849–863
Numajiri N, Takasawa K, Nishiya T, Tanaka H, Ohno K, Hayakawa W, Asada M, Matsuda H, Azumi K, Kamata H et al (2011) On-off system for PI3-kinase-Akt signaling through S-nitrosylation of phosphatase with sequence homology to tensin (PTEN). Proc Natl Acad Sci U S A 108:10349–10354

Osten P, Srivastava S, Inman GJ, Vilim FS, Khatri L, Lee LM, States BA, Einheber S, Milner TA, Hanson PI et al (1998) The AMPA receptor GluR2 C terminus can mediate a reversible, ATP-dependent interaction with NSF and alpha- and beta-SNAPs. Neuron 21:99–110

Pellmar TC, Hollinden GE, Sarvey JM (1991) Free radicals accelerate the decay of long-term potentiation in field CA1 of guinea-pig hippocampus. Neuroscience 44:353–359

Rossi D, Sorrentino V (2002) Molecular genetics of ryanodine receptors Ca^{2+}-release channels. Cell Calcium 32:307–319

Sacchetti B, Scelfo B, Tempia F, Strata P (2004) Long-term synaptic changes induced in the cerebellar cortex by fear conditioning. Neuron 42:973–982

Selvakumar B, Huganir RL, Snyder SH (2009) S-nitrosylation of stargazin regulates surface expression of AMPA-glutamate neurotransmitter receptors. Proc Natl Acad Sci U S A 106:16440–16445

Serrano F, Klann E (2004) Reactive oxygen species and synaptic plasticity in the aging hippocampus. Ageing Res Rev 3:431–443

Song I, Kamboj S, Xia J, Dong H, Liao D, Huganir RL (1998) Interaction of the N-ethylmaleimide-sensitive factor with AMPA receptors. Neuron 21:393–400

Steinberg JP, Huganir RL, Linden DJ (2004) N-ethylmaleimide-sensitive factor is required for the synaptic incorporation and removal of AMPA receptors during cerebellar long-term depression. Proc Natl Acad Sci U S A 101:18212–18216

Stoyanovsky D, Murphy T, Anno PR, Kim YM, Salama G (1997) Nitric oxide activates skeletal and cardiac ryanodine receptors. Cell Calcium 21:19–29

Stuehr DJ, Santolini J, Wang ZQ, Wei CC, Adak S (2004) Update on mechanism and catalytic regulation in the NO synthases. J Biol Chem 279:36167–36170

Sun JH, Xu L, Eu JP, Stamler JS, Meissner G (2003) Nitric oxide, NOC-12, and S-nitrosoglutathione modulate the skeletal muscle calcium release channel/ryanodine receptor by different mechanisms – an allosteric function for O-2 in S-nitrosylation of the channel. J Biol Chem 278:8184–8189

Takeshima H, Nishimura S, Matsumoto T, Ishida H, Kangawa K, Minamino N, Matsuo H, Ueda M, Hanaoka M, Hirose T et al (1989) Primary structure and expression from complementary DNA of skeletal muscle ryanodine receptor. Nature 339:439–445

Watson JB, Khorasani H, Persson A, Huang KP, Huang FL, O'Dell TJ (2002) Age-related deficits in long-term potentiation are insensitive to hydrogen peroxide: coincidence with enhanced autophosphorylation of Ca^{2+}/calmodulin-dependent protein kinase II. J Neurosci Res 70:298–308

Xia J, Chung HJ, Wihler C, Huganir RL, Linden DJ (2000) Cerebellar long-term depression requires PKC-regulated interactions between GluR2/3 and PDZ domain-containing proteins. Neuron 28:499–510

Xu L, Eu JP, Meissner G, Stamler JS (1998) Activation of the cardiac calcium release channel (ryanodine receptor) by poly-S-nitrosylation. Science 279:234–237

Chapter 17
Brain Aging Using Large Brain MRI Database

Yasuyuki Taki

Abstract Now we confront a super aging society in Japan. In the situation, it is important to preserve our cognitive function for entire life by preventing us from pathological brain aging. To perform the aim, we have built a large brain magnetic resonance imaging (MRI) database from around 3000 subjects aged from 5 to 80 in order to reveal how brain develops and ages. We have also collected several cognitive functions, lifestyle such as eating and sleeping habits, and genetic data. Using the database, we have revealed normal brain development and aging, and also have revealed what factors affect brain development and aging. For example, there were significant negative correlation between alcohol drinking and gray matter volume of front-parietal region, and body mass index and gray matter volume of the hippocampus in cross-sectional analysis. In addition, having intellectual curiosity showed significant negative correlation with regional gray matter volume decline rate in the temporo-parietal region. These findings help understanding the mechanism of brain development and aging as well as performing differential diagnosis or diagnosis at an early stage of several diseases/disorders such as autism and Alzheimer's disease. In addition, I will introduce you the Tohoku Medical Megabank Project, in which we will build brain MRI database of around 30,000 healthy subjects. By performing the project, we aim to build a system of preventive medicine for several diseases/disorders such as Alzheimer's disease.

Keywords Brain development • Brain aging • Magnetic resonance imaging • Database • Preventive medicine • Normal subject

Y. Taki (✉)
Department of Nuclear Medicine & Radiology, Institute of Development, Aging and Cancer, Tohoku University, 4-1 Seiryo-cho, Aoba-ku, Sendai 980-8575, Japan

Division of Medical Neuroimage Analysis, Department of Community Medical Supports, Tohoku Medical Megabank Organization, Tohoku University, Sendai, Japan

Division of Developmental Cognitive Neuroscience, Institute of Development, Aging and Cancer, Tohoku University, Sendai, Japan
e-mail: ytaki@idac.tohoku.ac.jp

© Springer Japan 2015

N. Mori, I. Mook-Jung (eds.), *Aging Mechanisms*,
DOI 10.1007/978-4-431-55763-0_17

17.1 Introduction

Now we confront a super aging society in Japan. In the situation, it is important to preserve our cognitive function for entire life by preventing us from pathological brain aging. Recently, the importance of human neuroimaging database was recognized greatly. The normal brain structure and function database can be used as the references not only for neuroimaging study for humans but also for early diagnosis and computer aided automated diagnosis of the brain diseases. Most remarkable recently developed method for brain image analysis is a voxel-based morphometry (VBM). It includes anatomical standardization of the brain to a standard brain, brain tissue segmentation and finally voxel based statistical analysis based on general linear model. This technique enables us to extract brain regions which show correlations between tissue volume and variables, such as age, sex and other subject's characteristics. We can analyze not only age-related normal changes but also diseased brain, such as dementia and schizophrenia. It has been believed that functional imaging precede structural imaging to detect early pathological findings of the diseases. However, recent development of high resolution structural imaging and sophisticated analytical technique enable us to detect the brain disease at very early stage. Now we have collected over 3000 brain MRI of healthy Japanese aged from 5 to 80 and constructed an MRI database together with their characteristics such as age, sex, blood pressure, present and past disease history and cognitive functions. This is a largest brain MRI database in Japan and one of the largest one in the world.

17.2 Imaging Studies of Brain Aging

17.2.1 Correlation Between Baseline Regional Gray Matter Volume and Global Gray Matter Volume Decline Rate

Evaluating whole-brain or global gray matter volume decline rate is important in distinguishing neurodegenerative diseases from normal aging and in anticipating cognitive decline over a given period in non-demented subjects. Whether a significant negative correlation exists between baseline regional gray matter volume of several regions and global gray matter volume decline in the subsequent time period in healthy subjects has not yet been clarified. Therefore, we analyzed the correlation between baseline regional gray matter volumes and the rate of global gray matter volume decline in the period following baseline using magnetic resonance images of the brains of 381 healthy subjects by applying a longitudinal design over 6 years using voxel-based morphometry.

All subjects were Japanese individuals recruited from our previous brain-imaging project. We selected participants who had lived in Sendai City at the time of the previous study, whose collected data had no missing values and who had no serious medical problems from an initial 1604 eligible persons. All participants were screened with a mail-in health questionnaire and underwent telephone and personal interviews. Persons who reported a history of any malignant tumor, head trauma with loss of consciousness for >5 min, cerebrovascular disease, epilepsy, any psychiatric disease, or claustrophobia were excluded from the study. All subjects were screened for dementia using the Mini-Mental State Examination (MMSE), An experienced neuroradiologist examined the MR scans for any tumors and cerebrovascular disease. The final sample consisted of 381 participants (40.1 % of the eligible cohort: 158 men, 223 women). All images were collected using the same 0.5 T MR scanner, including baseline images using MP-RAGE pulse sequences. After the image acquisition, all MR images were analyzed using statistical parametric mapping 2 in Matlab. We calculated gray matter volume, white matter volume using fully automated techniques. To normalize the head size of each subject, we defined the gray matter ratio (GMR) as the percentage of gray matter volume divided by the intracranial volume. Next, to reveal the annualized rate of change in GMR with age, we determined the annual percentage change in GMR (APC_{GMR}) for each subject. We determined regional gray matter volume using voxel-based morphometry as shown in Fig. 17.1. To investigate the correlation between baseline regional gray matter volume and APC_{GMR}, we performed a multiple regression analysis with age, gender, intracranial volume, and APC_{GMR} as independent valuables and baseline regional gray matter volume as a dependent valuable. We used the random field theory method to correct for the Familywise Error Rate (FWE); any resulting P-value less than 0.05 was considered significant. Next, we tested whether the gray matter regional volume that showed the significant

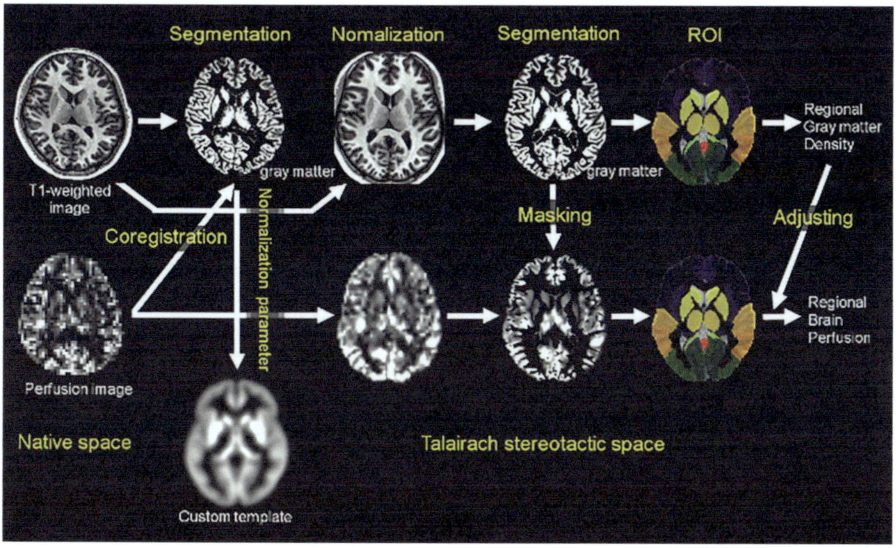

Fig. 17.1 Schematic of the image analysis

negative correlation with APC_{GMR} at baseline could predict whether the APC_{GMR} was above or below the APC_{GMR} mean by applying a standard (not stepwise) linear discriminant analysis in SPSS11.5. For the discriminant analysis we used the mean gray matter volume over a cluster in each region, and the regional gray matter volume as defined by multiple regression analysis. We set the significance level at $P < 0.05$.

As a result, the gray matter regions showing significant negative correlation with APC_{GMR} adjusted for age, gender, and intracranial volume are shown in Fig. 17.2. Baseline regional gray matter volumes of the right PCC/precuneus and the left hippocampus showed significant negative correlations with APC_{GMR} after adjusting for age, gender, and intracranial volume (right PCC/precuneus, $t = 5.42$, $P = 0.020$; left hippocampus, $t = 5.29$, $P = 0.035$) (Taki et al. 2011a). Therefore, we used the gray matter regions of the right PCC/precuneus and the left hippocampus in the next discriminant analysis. Baseline regional gray matter volume of both the right PCC/precuneus and the left hippocampus significantly distinguished whether APC_{GMR} was above or below the APC_{GMR} mean. The F-value, p-value, and discriminant function coefficient were 13.51, <0.001, and 0.833 in the right PCC/precuneus, and 5.71, 0.017, and 0.350 in the left hippocampus, respectively. Overall, 58.4 % of the APC_{GMR} (55.8 % of APC_{GMR} below the mean of APC_{GMR} and 60.9 % of APC_{GMR} above the mean of APC_{GMR}) was correctly distinguished using the discriminant function.

This study provides the first longitudinal findings showing that baseline regional gray matter volumes in the right PCC/precuneus, and the left hippocampus show a significant negative correlation with the rate of global gray matter volume decline in the following period, as represented by APC_{GMR}, adjusting for age, gender, and intracranial volume. In addition, baseline regional gray matter volumes of both the right PCC/precuneus and the left hippocampus significantly distinguished whether the APC_{GMR} was above or below the APC_{GMR} mean. These results indicate that subjects who had smaller baseline regional gray matter volumes in those regions showed higher rate of global gray matter volume decline in the following period.

In summary, using a longitudinal design over 6 years in 381 community-dwelling healthy individuals, we examined the correlation between baseline regional gray matter volume and the rate of global gray matter volume decline in the following period. We found a significant negative correlation between APC_{GMR} and the baseline regional gray matter volumes of the right PCC/precunei and the left hippocampus after adjusting for age and gender. In addition, baseline regional gray matter volume of both the right PCC/precuneus and the left hippocampus significantly distinguished whether the APC_{GMR} was above or below the APC_{GMR} mean. Our results suggest that baseline regional gray matter volume predicts the rate of global gray matter volume decline in the following period in healthy subjects. Our study may contribute to distinguishing neurodegenerative diseases from normal aging, and to predicting cognitive decline.

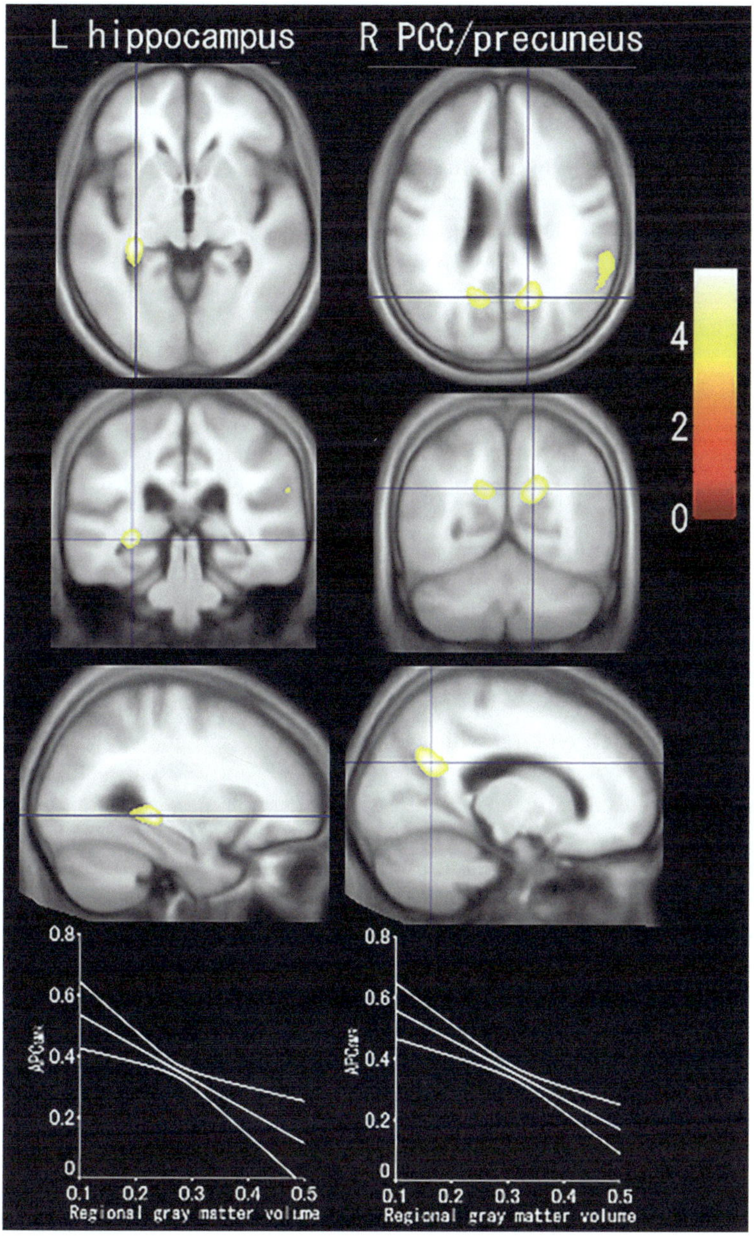

Fig. 17.2 Gray matter regions showing significant negative correlations with annual percent change of the gray matter ratio (APC_{GMR}) adjusted for age, gender, and intracranial volume

17.2.2 Correlation Between Degree of White Matter Hyperintensities and Global Gray Matter Volume Decline Rate

Whether the degree of white matter hyperintensities (WMHs) shows a significant correlation with the rate of global gray matter volume decline over a period following initial baseline measurement remains unclear. The purpose of the present study was to reveal the relationship between the degree of WMHs at baseline and the rate of global gray matter volume decline by applying a longitudinal design.

Using a 6-year longitudinal design and magnetic resonance images of the brains of 160 healthy individuals aged over 50 years and living in the community, we analyzed the correlation between degree of WMHs using Fazekas scaling at baseline and rate of global gray matter volume decline 6 years later. To obtain the rate of global gray matter volume decline, we calculated global gray matter volume and intracranial volume at baseline and at follow-up using a fully automated method.

As a result, the annual percentage change in the gray matter ratio (GMR, APC_{GMR}), in which GMR represents the percentage of gray matter volume in the intracranial volume, showed a significant positive correlation with the degree of deep WMHs and periventricular WMHs at baseline, after adjusting for age, gender, present history of hypertension, and diabetes mellitus (Taki et al. 2011b).

The degree of WMHs, both DWMH and PVWMH, at baseline showed a significant positive correlation with the rate of global gray matter volume decline, represented by APC_{GMR}, adjusting for age, gender, and present history of hypertension and diabetes mellitus in healthy subjects using longitudinal analysis. To our knowledge, we are the first to show the correlation between the degree of WMHs at baseline and the rate of subsequent global gray matter volume decline in healthy elderly individuals. Our result is partially consistent with recent studies that showed a significant positive correlation between the degree or load of WMHs and decreases in gray matter volume in healthy elderly people, although those studies were conducted using cross-sectional design. However, another recent study using longitudinal analysis has shown that WMH is not a predictor of brain atrophy rate in elderly subjects. The inconsistency between the findings of the recent study and the present study may have arisen from differences in the volume that was measured. In the present study, we focused on the rate of decline of gray matter volume, not whole-brain volume, because gray matter volume is significantly correlated with several cognitive functions. Our results suggest that the rate of global gray matter volume decline could be predicted using the degree of WMHs at baseline, evaluated by simple visual scaling.

In summary, using a longitudinal design over 6 years in 160 community-dwelling healthy individuals, the degree of WMHs was measured at baseline, and the rate of global gray matter volume decline was obtained. As a result, APC_{GMR} showed a significant positive correlation with the degree of deep WMHs and periventricular WMHs at baseline adjusting for age, gender, and present history of hypertension and diabetes mellitus. Our results suggest that degree of WMHs at baseline predicts the rate of subsequent gray matter volume decline and also suggests that simple visual scaling of WMHs could contribute to the prediction of the rate of global gray matter volume decline.

17.2.3 Correlation Between Gray Matter Density-Adjusted Brain Perfusion and Age Using Brain MR Images of 202 Healthy Children

In understanding brain aging, the knowledge of brain maturation is very important, for the relationship between brain maturation and brain aging is regarding as a "mirror pattern". In detail, brain regions that mature earlier such as occipital regions are robust in brain aging, whereas brain regions that mature rather late such as prefrontal regions are vulnerable for aging. Brain development continues through childhood and adolescence. Recently, it has been revealed that human brain development is a structurally and functionally non-linear process. However, despite this growing wealth of knowledge about maturational changes in brain structure in children, the trajectory of brain perfusion with age in healthy children is not yet well documented.

Recently, arterial spin-labeling (ASL) perfusion magnetic resonance imaging (MRI) has been developed for evaluating brain perfusion. We examined the correlation between brain perfusion and age using pulsed ASL MRI in a large number of healthy children.

We collected data on brain structural and ASL perfusion MRI in 202 healthy children aged 5–18 years. Structural MRI data were segmented and normalized, applying a voxel-based morphometric analysis. Perfusion MRI was normalized using the normalization parameter of the corresponding structural MRI. We calculated brain perfusion with an adjustment for gray matter density (BP-GMD) by dividing normalized ASL MRI by normalized gray matter segments in 22 regions. Next, we analyzed the correlation between BP-GMD and age in each region by estimating linear, quadratic, and cubic polynomial functions, using the Akaike information criterion.

As a result, the correlation between BP-GMD and age showed an inverted U shape followed by a U-shaped trajectory in most regions (Taki et al. 2011c) as shown in Fig. 17.3. In addition, age at which BP-GMD was highest was different among the lobes and gray matter regions, and the BP-GMD association with age increased from the occipital to the frontal lobe via the temporal and parietal lobes.

In the frontal lobe, all gray matter regions showed an inverted U-shaped trajectory for the correlation between BP-GMD and age, and the best fit was a negative quadratic or positive cubic polynomial function. The estimated age at which BP-GMD was highest was earlier in the precentral gyrus, cingulate gyrus, and

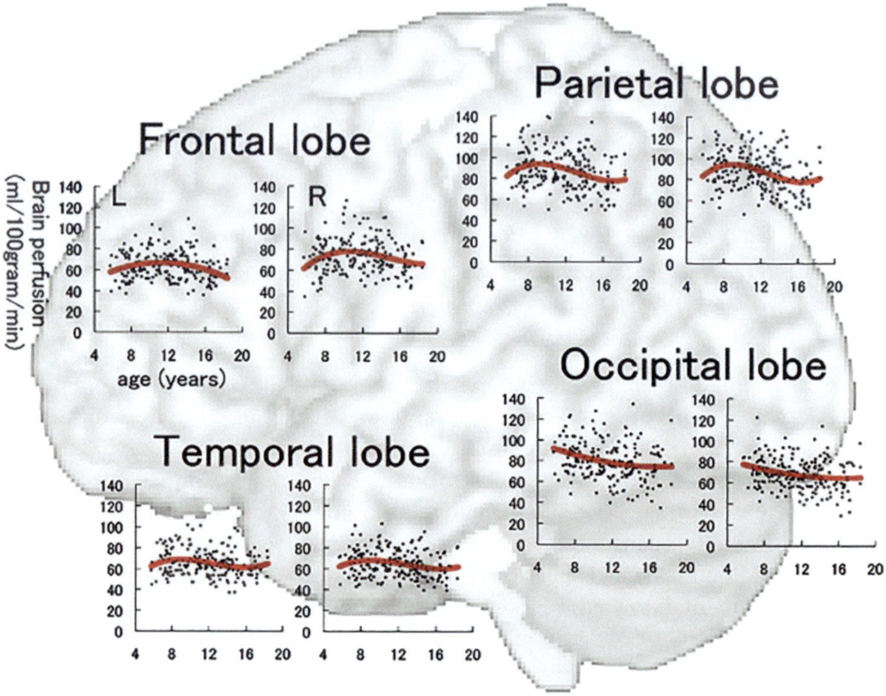

Fig. 17.3 Correlation between brain perfusion, adjusted for gray matter density, and age in the frontal lobe, parietal lobe, occipital lobe, and temporal lobe in each hemisphere

anterior cingulate cortex than in the superior, middle, and inferior frontal gyri as shown in Fig. 17.4.

We demonstrated a correlation between BP-GMD and age using ASL brain perfusion MRI in a large number of healthy children over a wide age range. As a result, the trajectory of the correlation between BP-GMD and age showed an inverted U-shaped second-order polynomial function in most regions in the frontal lobe, a third-order polynomial function in the parietal and temporal lobes, and a U-shaped second-order and negative linear correlation in the occipital lobe. Our results indicate that higher-order association cortices mature after the lower-order cortices in terms of brain perfusion. As a result, the trajectory of the correlation between BP-GMD and age showed an inverted U shape followed by a U-shaped trajectory in most regions. In addition, the age at which BP-GMD was highest was different among the lobes and gray matter regions, showing a progression from the occipital lobe to the frontal lobe, via the temporal and parietal lobes. Our results indicate that higher-order association cortices mature after the lower-order cortices mature. This may help not only clarify the mechanisms of normal brain maturation from the viewpoint of brain perfusion, but also distinguish normal from developmental disorders that show abnormal brain perfusion patterns.

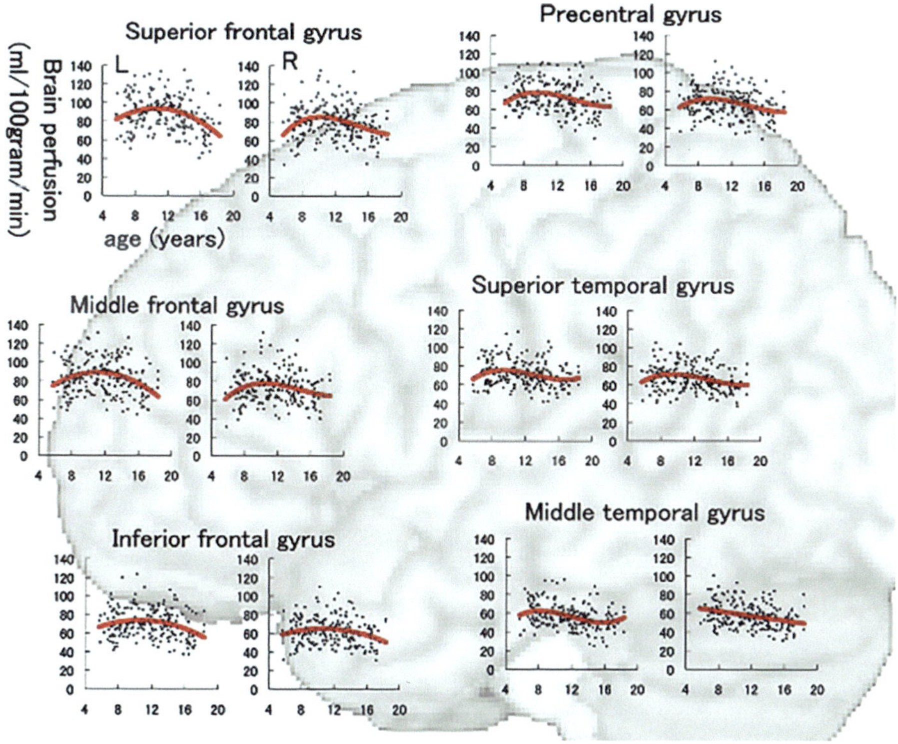

Fig. 17.4 Correlation between brain perfusion, adjusted for gray matter density, and age in the precentral gyrus, superior frontal gyrus, middle frontal gyrus, inferior frontal gyrus, superior temporal gyrus, and middle temporal gyrus

17.2.4 Risk Factors for Brain Volume Decrease

17.2.4.1 Hypertension

By VBM technique, we analyzed correlation between regional gray matter volume and subject's characteristics. We found that total gray matter volume negatively correlated with systolic blood pressure (Taki et al. 2004). Figure 17.5 represents the brain regions that showed negative correlation between gray matter volume and systolic blood pressure. These regions mainly distributed watershed regions between major cerebral arteries.

17.2.4.2 Alcohol Drinking

We also tested the correlation between gray matter ratio and life time alcohol intake. There was a strong negative correlation between the log transformed

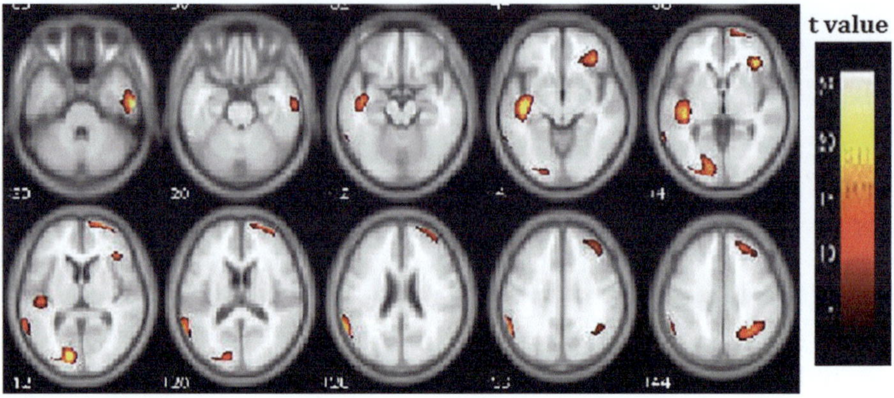

Fig. 17.5 Brain regions that showed negative correlation between gray matter volume and systolic blood pressure

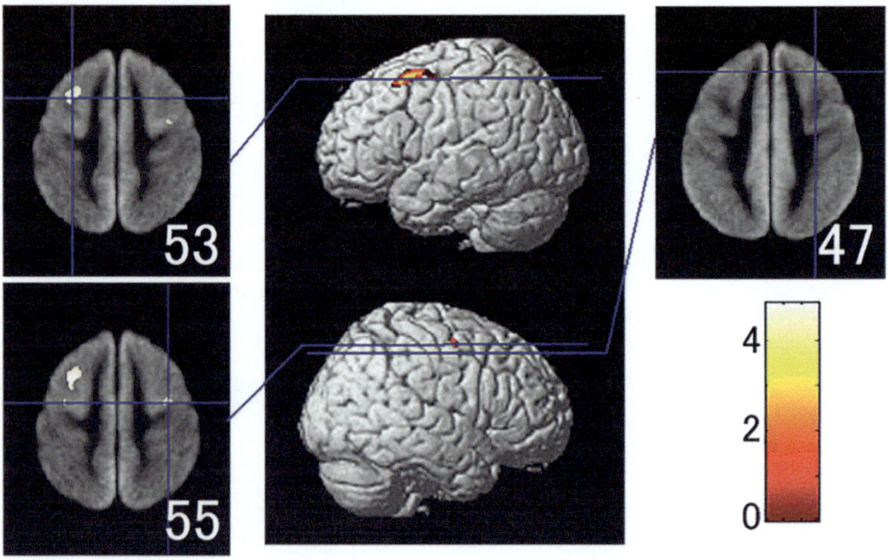

Fig. 17.6 Brain regions that showed negative correction between gray matter volume and life time alcohol intake

lifetime alcohol intake and the gray matter ratio (Taki et al. 2006). Figure 17.6 shows the gray matter regions that had a significant negative correlation between the lifetime alcohol intake and the regional gray matter volume. The gray matter volume of the bilateral middle frontal gyri showed a significant negative correlation with the log transformed lifetime alcohol intake.

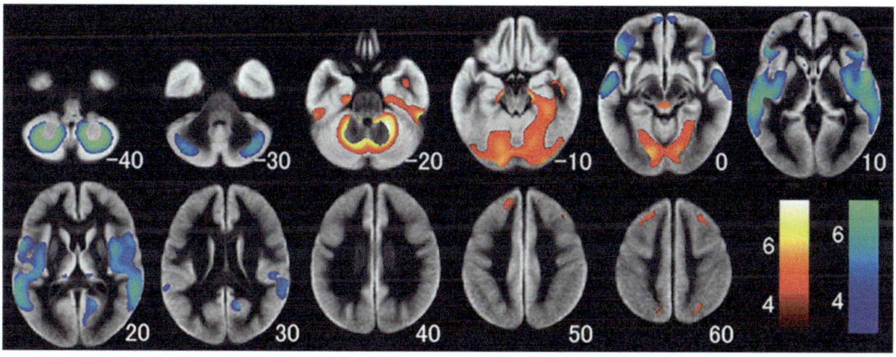

Fig. 17.7 Brain regions that showed correlations between gray matter volume and body mass index (BMI). *Red* and *blue* color indicated negative and positive correlations, respectively

17.2.4.3 Obesity

We tested correlation between gray matter ratio and obesity. As an indicator for obesity, body mass index (BMI) was used. Volumetric analysis revealed that there are significant negative correlation between BMI and the gray matter ratio, which represents the percentage of gray matter volume in the intracranial volume, in men ($p < 0.001$, adjusting for age, systolic blood pressure, and lifetime alcohol intake) whereas not in women. VBM revealed that regional gray matter volumes of the bilateral medial temporal lobe, occipital lobe, frontal lobe, and anterior lobe of the cerebellum show significant negative correlation with BMI, and those of posterior lobe of the cerebellum, perisylvian regions of the bilateral frontal and temporal lobes, and bilateral orbitofrontal gyri show significant positive correlation with BMI in men as shown in Fig. 17.7 (Taki et al. 2008).

17.3 Conclusion

We constructed a large scale brain MRI database for healthy Japanese and clarified age-related volume changes of the human brain and their risk factors. Several factors such as hypertension, alcohol drinking, and obese are related with gray matter volume reduction of several regions. In addition, we have shown that several factors such as baseline gray matter volume structure and white matter lesions predict the global gray matter volume decline rate. These results may contribute to the understanding of normal brain aging, as well as age-related brain diseases, such as dementia.

References

Taki Y, Goto R, Evans A, Zijdenbos A, Neelin P, Lerch J, Sato K, Ono S, Kinomura S, Nakagawa M, Sugiura M, Watanabe J, Kawashima R, Fukuda H (2004) Voxel-based morphometry of human brain with age and cerebrovascular risk factors. Neurobiol Aging 25:455–463

Taki Y, Kinomura S, Sato K, Goto R, Inoue K, Okada K, Ono S, Kawashima R, Fukuda H (2006) Both global gray matter volume and regional gray matter volume negatively correlate with lifetime alcohol intake in non-alcohol-dependent Japanese men. A volumetric analysis and a voxel-based morphometry. Alcohol: Clin Exp Res 30:1045–1050

Taki Y, Kinomura S, Sato K, Inoue K, Goto R, Okada K, Uchida S, Kawashima R, Fukuda H (2008) Relationship between body mass index and gray matter volumes in 1428 healthy individuals. Obes Res 16:119–124

Taki Y, Kinomura S, Sato K, Goto R, Wu K, Kawashima R, Fukuda H (2011a) Correlation between baseline regional gray matter volume and global gray matter volume decline rate. Neuroimage 54:743–749

Taki Y, Kinomura S, Sato K, Goto R, Wu K, Kawashima R, Fukuda H (2011b) Correlation between degree of white matter hyperintensities and global gray matter volume decline rate. Neuroradiology 53:397–403

Taki Y, Hashizume H, Sassa Y, Takeuchi H, Wu K, Asano M, Asano K, Fukuda H, Kawashima R (2011c) Correlation between gray matter density-adjusted brain perfusion and age using brain MR images of 202 healthy children. Hum Brain Mapp 32:1973–1985

Part VI
Aged Brain: Mechanisms of Neurodegeneration

Chapter 18
Brain Aging as a Cause of Alzheimer's Disease

Toshiharu Suzuki, Ayano Kimura, Kyoko Chiba, Tadashi Nakaya, and Saori Hata

Abstract Alzheimer's disease (AD) is the most common form of senile dementia. Identification of genes causally associated with familial Alzheimer's disease (FAD) advanced our understanding of the molecular mechanisms of AD pathogenesis. However, FAD is much less common than sporadic Alzheimer's disease (SAD), which constitutes the majority of cases. Despite its similar pathology (albeit at a later age of onset), SAD is not linked to mutations in FAD-associated genes. In both FAD and SAD, the generation and oligomerization of amyloid β (Aβ) peptide play central roles in neurotoxicity, but it remains unclear how qualitative and quantitative alterations in Aβ occur in SAD patients in the absence of causative mutations. The predominant risk factor for SAD is aging, suggesting that some as-yet-unknown alterations in the aged brain augment the amyloidogenic metabolism of APP and promote the neural toxicity of Aβ oligomers. In this chapter, we discuss potential biochemical changes in amyloid β precursor protein (APP) and proteins related to APP metabolism and function in the aged brain. APP axonal transport, membrane microlocalization and metabolism, including generation of Aβ in neurons, are regulated by interactions with several cytoplasmic proteins and phosphorylation of the APP cytoplasmic region. Age-related decline or aberration in the regulation of APP transport, localization and metabolism may induce generation of altered Aβ. Here, we focus on APP phosphorylation at threonine 668 in the cytoplasmic domain and the roles of APP regulatory proteins, including X11-like (X11L), JIP1, kinesin-1, and Alcadein, on the regulation of APP metabolism and intracellular trafficking.

Keywords Brain aging • Alzheimer's disease • APP • Amyloid β peptide • X11-like • JIP1 • Alcadein • Protein phosphorylation • Kinesin • p3-Alc

---

T. Suzuki (✉) • A. Kimura • K. Chiba • T. Nakaya • S. Hata
Laboratory of Neuroscience, Graduate School of Pharmaceutical Sciences, Hokkaido University, Sapporo 060-0812, Japan
e-mail: tsuzuki@pharm.hokudai.ac.jp

© Springer Japan 2015

N. Mori, I. Mook-Jung (eds.), *Aging Mechanisms*,
DOI 10.1007/978-4-431-55763-0_18

18.1 Dementia and the Growing Population of Aged People

In almost all developed countries, the elderly population is increasing rapidly, and the relative size of the younger population is shrinking. Dementia is a major illness that critically impairs the health of older people, threatening their ability to live independently. The latest WHO report suggested that over 35 million people worldwide suffer from dementia, and this number is expected to triple by 2050 (WHO and Alzheimer's Disease International 2012). In countries with a growing aged population, the expanding costs of treating and caring for people with dementia has become a serious social and economic issue.

Alzheimer's disease (AD) is the most common form of senile dementia, accounting for 60–70 % of cases (Thies et al. 2013). Contributions by many AD researchers identified three causative genes involved in the pathogenesis of familial AD (FAD): amyloid β precursor protein (APP) and presenilins 1 and 2 (PS1 and PS2) (Haass and Selkoe 2007). Patients carrying pathogenic mutations in the APP or presenilin (PS) genes suffer from early-onset AD, as distinct from late-onset or sporadic AD (LOAD or SAD). Since their discovery of these genes in the late 1980s to early 1990s, the accumulated findings regarding their functions and the effects of causative mutations advanced our understanding of the molecular mechanisms of AD pathogenesis in both forms of the disease.

FAD is relatively rare; the majority of the AD patients have SAD, which is not linked to mutations in genes causally associated with FAD. Nonetheless, SAD exhibits pathology similar to that of typical FAD: neurodegeneration accompanied by extracellular senile plaques and intracellular neurofibrillary tangles. In both FAD and SAD, generation and oligomerization of amyloid β (Aβ) peptide are believed to play central roles in the expression of neurotoxicity; thus, Aβ is thought to be a primary trigger of neurodegeneration (Haass and Selkoe 2007). However, it remains unclear how qualitative and quantitative alterations in Aβ generation occur in SAD patients in the absence of causative mutations. Studies aimed at identifying genes involved in SAD revealed the ε4 allele of the ApoE gene as the strongest risk factor among many candidate genes (Kanekiyo et al. 2014). Although ApoE4 has been implicated in several aspects of AD pathogenesis, including Aβ deposition and clearance (Kok et al. 2009; Mawuenyaga et al. 2010), the molecular mechanisms and pathogenic roles of this ApoE4 isoform in SAD remain incompletely understood (Kanekiyo et al. 2014). A genomics analysis of AD-related genes affected by ApoE isoforms identified several genes related to intracellular protein trafficking, including the X11L gene (*APBA2*), as risk genes for AD (Rhinn et al. 2013), suggesting a role for intracellular trafficking defects in pathogenesis.

Notwithstanding the contributions of genetic factors, the predominant risk factor for SAD is aging itself. Therefore, it is important to understand what types of alterations in the aged brain augment the amyloidogenic metabolism of APP and promote the neurotoxicity of Aβ oligomers. In this chapter, we focus on potential biochemical changes in APP, as well as in proteins linked to APP metabolism or intracellular trafficking, in the aged brain. For almost three decades, we have sought to understand the intracellular transport, metabolism, and function of APP (Suzuki et al. 2006; Suzuki and Nakaya 2008; Taru and Suzuki 2009). APP axonal transport, membrane microlocalization and metabolism, including Aβ generation in neurons, is regulated by interactions with several cytoplasmic proteins and the phosphorylation of the APP cytoplasmic region. Age-related dysfunction in APP transport and metabolism may trigger quantitative or qualitative alterations in the generation of Aβ, even in patients lacking pathogenic mutations in causative genes. This chapter describes neuron-specific APP phosphorylation at threonine 668 (Thr668) in the cytoplasmic region; APP-binding proteins that regulate intracellular metabolism, such as X11L and JIP1; and trafficking of APP. Furthermore, we describe alterations in substrate cleavage by γ-secretase, based on our analysis of the membrane proteins known as Alcadeins. Over the course of brain aging, changes in expression, localization, modification, and/or interaction of APP with regulators can augment amyloidogenic APP metabolism.

18.2 Phosphorylation of APP

APP is subject to proteolytic cleavages, initially by α- or β-secretase primarily and then by γ-secretase (Fig. 18.1). A combination of cleavages of APP by α- and γ-secretase generates the p3 fragment, which consists of 24–26 amino acids. This is

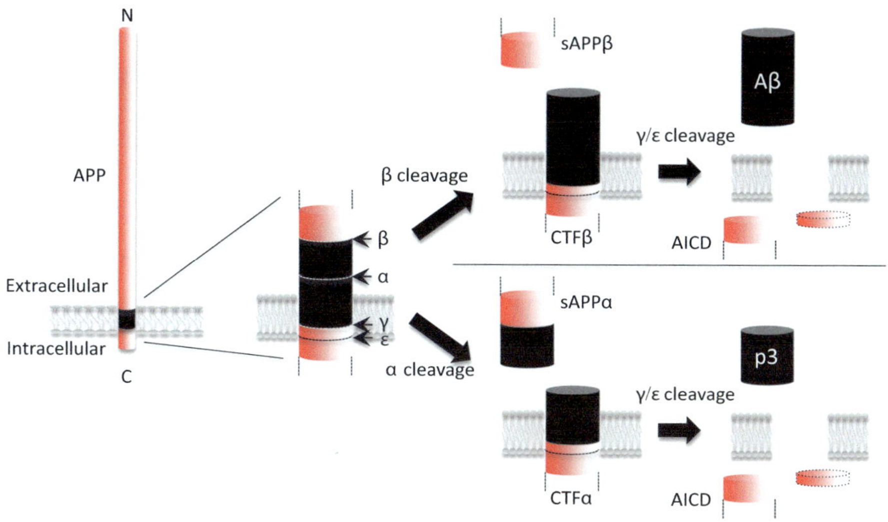

Fig. 18.1 Metabolism of APP. APP is cleaved at the juxtamembrane α- or β-site to generate membrane-associated CTFα or CTFβ, accompanied by secretion of sAPPα or sAPPβ. CTFs are further cleaved at the γ-site, followed by the ε-cleavage site. This intramembrane cleavages result in secretion of p3 or Aβ and release of AICD into the cytoplasmic milieu. Amyloidogenic (*right upper*) and non-amyloidogenic (*right lower*) pathways are shown schematically

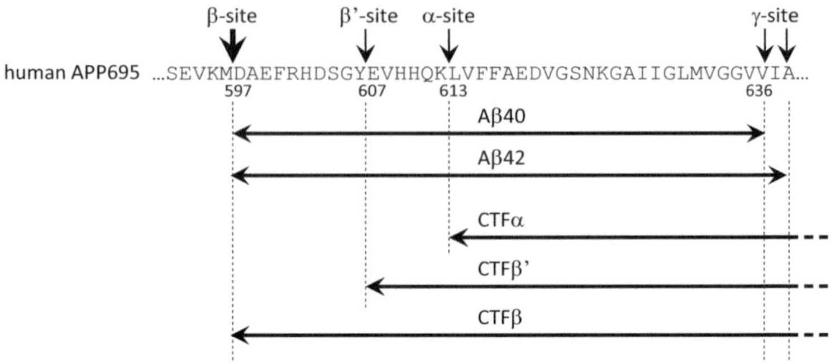

Fig. 18.2 Cleavage site of APP by α-, β-, and γ-secretases, and small peptide products of APP generated by secretase cleavages. α-site, cleavage site by α-secretase; β- and β′-sites, β-cleavage site by β-secretase; γ-site, major γ-cleavage sites by γ-secretase. Amino-acid sequences of Aβ and N-terminal amino-acid sequence of CTFs are shown

accompanied by secretion of sAPPα, a large extracellular domain, and the cytoplasmic release of the AICD fragment, followed by the generation of membrane-associated CTFα (C83), the 83 amino-acid C-terminal fragment. Because the α-secretase, which is largely composed of ADAM10 and ADAM17, cleaves APP at the α-site within the Aβ domain on the cell membrane, this primary cleavage of APP is referred to as the non-amyloidogenic pathway or amyloidolytic processing. By contrast, the β-secretase, BACE, cleaves APP at the β-site and β′ site to generate CTFβ (C99) and CTFβ′ (C89), respectively. Although cleavage of the β′-site is also amyloidolytic, the major BACE cleavage occurs at the β-site. Therefore, cleavage by BACE is generally referred to as amyloidogenic processing (Cole and Vassar 2008) (Fig. 18.2). The cleavage of APP by β-secretase occurs in the acidic environment of early and/or recycling endosomes, into which APP molecules that escape from cleavage by α-secretase on plasma membrane are incorporated (Thinakaran and Koo 2008).

All secretases and their components for APP processing are membrane proteins. BACE and α-secretase (ADAM 10/17) are type I membrane proteins with single transmembrane domains, like their substrate APP (Suzuki et al. 2006). By contrast, the γ-secretase is a complex of membrane proteins composed of PS, anterior pharynx-defective complex subunit (APH-1), presenilin enhancer-2 (PEN-2), and nicastrin (NCT). PS is subject to endoproteolytic cleavage by the active γ-secretase complex. Therefore, the regulation of microlocalization of APP and active secretases in the membrane are important determinants of whether APP is subject to amyloidogenic or non-amyloidogenic processing.

Active BACE and γ-secretase predominantly localize in lipid raft–like membrane microdomains. The microdomain containing BACE activity can be biochemically isolated as a 1 % Triton X-100-resistant membrane fraction (Riddle et al. 2001; Saito et al. 2008), whereas the fraction containing active γ-secretase complex is prepared using 1 % CHAPSO (Matsushima et al. 2012). Thus, active BACE may localize in a detergent-resistant membrane (DRM) fraction that is biochemically distinct from the fraction containing the active γ-secretase complex. Alternatively, because γ-secretase consists of multiple subunits, it may be unstable in DRMs prepared with Triton X-100.

APP is a type I membrane protein, and its short 47 amino-acid cytoplasmic region is exposed to many factors that regulate the trafficking of APP-containing membrane vesicles and APP metabolism (Suzuki et al. 2006; Suzuki and Nakaya 2008; Taru and Suzuki 2009). Several APP isoforms/splicing variants with identical cytoplasmic regions exist, including APP770, APP751, and APP695; the APP695 isoform (composed of 695 amino acids) is exclusively expressed in neuronal cells. APP is subject to *N*-glycosylation in the ER to yield immature APP (imAPP) and *O*-glycosylation in Golgi to yield mature APP (mAPP). Therefore, mAPP with both *N*- and *O*-glycans is the true substrate of secretases, which act in the late secretory pathway. mAPP is further subject to neuron-specific phosphorylation at Thr668 (numbering of amino-acid positions is based on the APP695 isoform) (Iijima et al. 2000) (Fig. 18.3). The constitutive phosphorylation level of mAPP is low compared to CTFs, and in the brain almost 50 % of CTFs are phosphorylated at Thr668 (Matsushima et al. 2012). Phosphorylation induces a conformational change throughout the cytoplasmic region (Ramelot and Nicholson 2001; Ando et al. 2001), which regulates DRM localization of APP CTFs. The C-terminal end

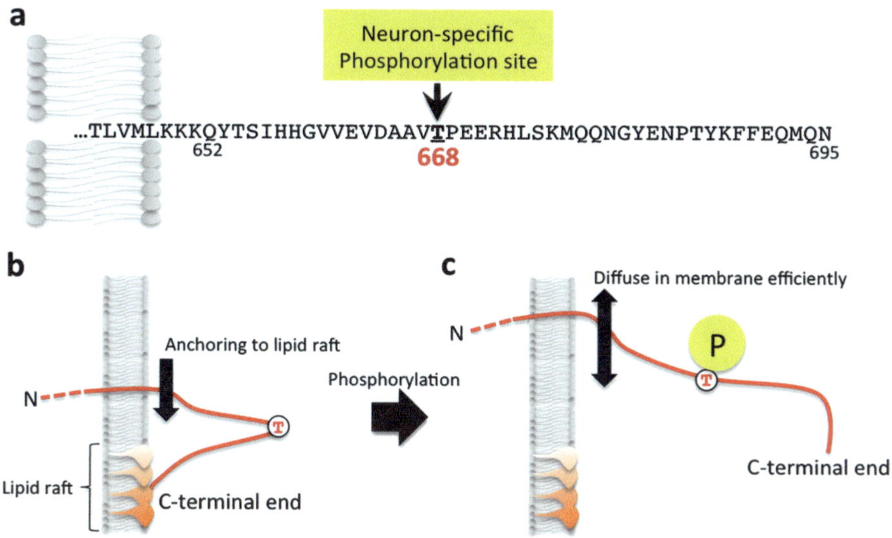

Fig. 18.3 Neuron-specific phosphorylation site in APP cytoplasmic region, and the conformational change of the cytoplasmic tail of APP that occurs upon phosphorylation. (**a**) Amino-acid sequence of the APP cytoplasmic region and the position of Thr668, which is subject to phosphorylation in neurons. (**b**) The C-terminal tail of APP CTFs exhibits lipid-binding ability. This binding decreases the membrane fluidity of APP CTFs. (**c**) Phosphorylation of APP CTFs alters the overall conformation of the cytoplasmic region, resulting in translocation of the tails of APP CTFs into the cytoplasm

of the APP cytoplasmic region tends to be anchored to membrane lipids, restricting CTF fluidity in the membrane (Beel et al. 2008). CTF phosphorylated at Thr668 loses its lipid-binding ability due to the conformational change, and thereby acquires greater mobility in the plasma membrane (Matsushima et al. 2012). Although phosphorylated and non-phosphorylated CTFs are kinetically equivalent substrates of γ-secretase, phosphorylated CTF can escape more easily from lipid raft–like membrane microdomains enriched in active γ-secretase. The phosphorylation of CTFβ is not affected for γ-cleavage of CTFβ; nevertheless, phosphorylated CTFβ is a poorer source of Aβ than non-phosphorylated CTFβ due to its freer mobility and ability to escape from lipid raft–like membrane microdomains (Matsushima et al. 2012) (Fig. 18.4).

The phosphorylation level of APP CTF decreases with brain aging in monkey, indicating that more APP CTF is available for cleavage by γ-secretase in the aged

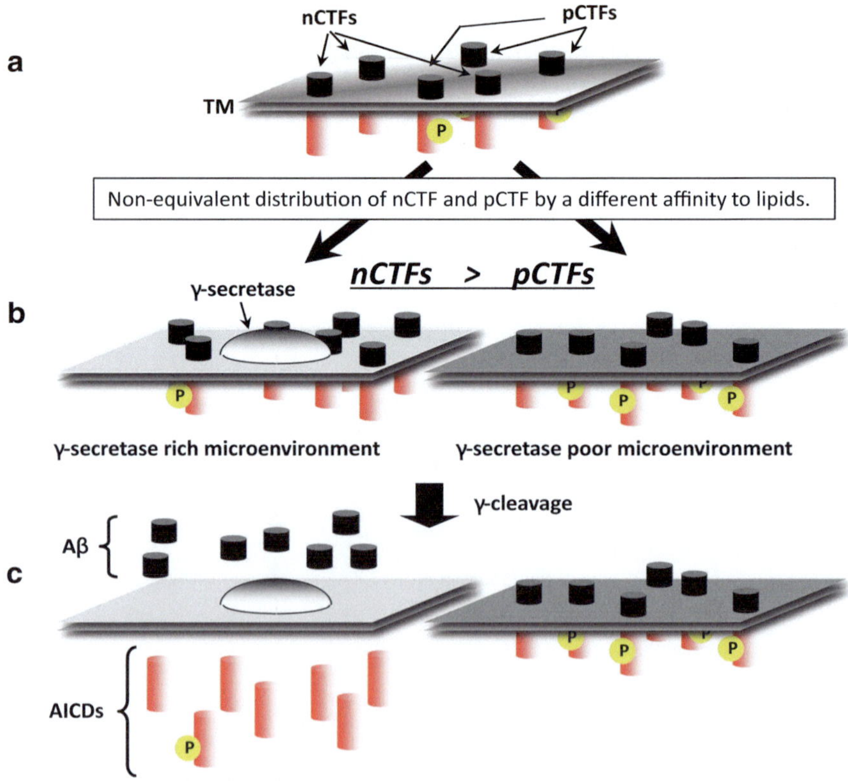

Fig. 18.4 Membrane localization of APP CTFs. (**a**) In brain neurons, phosphorylated (pCTF) and non-phosphorylated (nCTF) forms of APP CTFs are equally abundant, but nCTFs are preferentially cleaved by γ-secretase. (**b**, **c**) Because the C-terminal tails of nCTFs are prone to be anchored in membrane lipids (see Fig. 18.3), more nCTFs are localized in the lipid-rich membrane microdomain where γ-secretase is active. pCTFs are also localized in the microdomain, but without a lipid-binding C-terminal tail they are more mobile and can escape quickly

brain. This phenomenon may arise due to elevated generation of neurotoxic Aβ. In fact, deposition of Aβ peptide and Tau phosphorylation in the brain both increase over the course of aging in cynomolgus monkeys (Matsushima et al. 2012; Oikawa et al. 2010). Thus, it is possible that phosphorylation of APP CTFs at Thr668 plays an important role in regulation of membrane microlocalization of CTFs. However, it remains unclear which protein kinase phosphorylates APP at Thr668 in the brain in vivo; candidate kinases include CDK5, JNK, and GSK3β. Likewise, the protein phosphatase that dephosphorylates phosphorylated APP CTFs in neurons has not been identified. Prevention of APP CTF phosphorylation represents a therapeutic strategy to overcome Aβ generation.

18.3 Intracellular Trafficking of APP in Neurons

Neurons have highly specialized, polarized shapes consisting of a long axon and highly branched dendrites. In these cells, transport systems for proteins, RNA, and organelles are organized by a bidirectional motor system (Hancock 2014). In particular, anterograde transport in axon plays an essential role in supplying various materials to nerve termini (synapses); consequently, perturbation of axonal transport system might induce neurodegenerative diseases (Millecamps and Julien 2013). APP is subject to axonal anterograde transport by the kinesin-1 motor (Kamal et al. 2000; Araki et al. 2007), which is composed of two light chains (KLC) and two heavy chains (KHC) (Verhey and Hammond 2009; Hirokawa et al. 2009). Amyloidogenic processing, including neurotoxic Aβ generation and axonopathy, are facilitated when anterograde transport of APP cargo is disturbed (Araki et al. 2007; Stokin et al. 2005).

In the presence of JNK-interacting protein 1 (JIP1), APP predominantly associates with kinesin-1 motor. This interaction is mediated by JIP1b (Araki et al. 2007; Chiba et al. 2014); among JIP1a, JIP1b, and JIP2, the JIP1b isoform has the highest affinity for APP (Taru et al. 2002). In *Drosophila*, APP-like protein (APPL) is transported in an anterograde direction in axons, as is APP in mammals, and *Drosophila* APPL-interacting protein (APLIP1) is an ortholog of JIP1 (Taru et al. 2002; Horiuchi et al. 2005). Human APP expressed in *Drosophila* is transported to fly synapses in a manner resembling axonal transport in mammals (Yagi et al. 2000). Thus, APP cargo anterograde transport by kinesin-1, mediated by the JIP1 association, is conserved across widely divergent animals. Together, these observations indicate that APP is the cargo receptor for kinesin-1, rather than a simple cargo that is transported within vesicles. Although kinesin-1 was the first anterograde motor identified in squid giant axon (Vale et al. 1985), the cargo receptors of kinesin-1 and the mechanisms of regulation of cargo transport by this protein remain unknown. In addition to APP, apolipoprotein receptor 2 (ApoER2) and Alcadein α (Alcα) have been identified as kinesin-1 cargo receptors. Alcadein forms a tripartite complex with APP, mediated by its association with X11-like (X11L) (Araki et al. 2003). Complex formation stabilizes the intracellular metabolism of APP and Alcα; however, both APP and Alcα cargoes are transported largely independently (Araki et al. 2004, 2007).

Alcα is a cargo receptor that can directly bind KLC (Araki et al. 2007; Konecta et al. 2006; Kawano et al. 2012), whereas APP and ApoER2 require scaffold and/or adaptor proteins, such as JIP1, to associate with KLC. APP and ApoER2 contain the NP*X*Y motif, to which JIP1 binds, whereas Alcα contains an NP motif, to which X11L but not JIP1 can bind. The velocity of Alcα cargo transport by kinesin-1, measured in live axons, closely matches the speed of kinesin-1 on microtubules in vitro. Intriguingly, anterograde transport of APP by kinesin-1 is almost twice as fast as transport of Alcα cargo (Araki et al. 2007). Efficient APP cargo transport with increased velocity and higher frequency in the anterograde direction suggests the importance of cargos that APP carries to the nerve terminus, and implies that neuronal function would deteriorate if the APP cargo transport system were impaired.

In JIP1-deficient neurons, the velocity of anterograde transport of APP cargo is reduced, but the speed can be restored by expression of exogenous JIP1b. Efficiency of anterograde transport of APP cargos is also reduced in the absence of JIP1 (Chiba et al. 2014). Therefore, JIP1b is an essential factor for rapid and efficient anterograde transport of APP cargo. The JIP1 carboxyl-terminal region, consisting of 11 amino acids (C11), interacts with the tetratricopeptide repeat (TPR) motifs of KLC1, and the amino-terminus of KLC1 binds to KHC (Verhey et al. 2001). However, recent analysis revealed that the interaction between JIP1b and KLC is more complex than previously thought. A tyrosine residue in C11 is essential for high-velocity transport, but a more complex interaction regulates the association of JIP1b with KLC, resulting in efficient anterograde transport of APP cargo (Chiba et al. 2014) (Fig. 18.5). Although it remains unknown at the molecular level how JIP1b increases the velocity of APP cargo transport by kinesin-1, it is clear that JNK binding to JIP1b and the phosphorylation of APP by JNK are not involved in

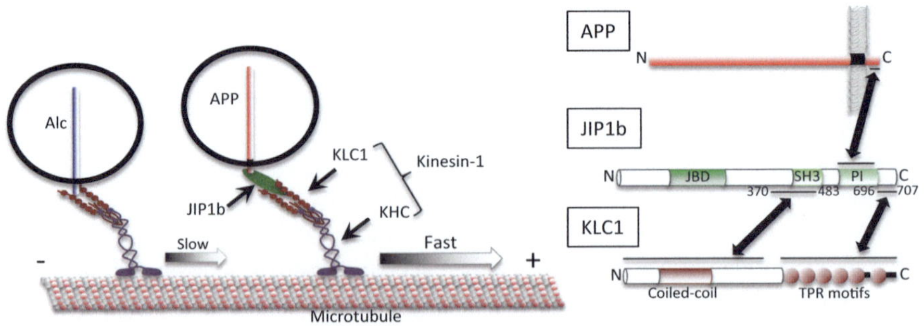

Fig. 18.5 Schematic picture of Alcα and APP cargos transport by kinesin-1. Alcα associates with kinesin-1 directly, whereas APP associates with KLC by mediation of JIP1. JIP1 mediates efficient anterograde transport of APP cargo at high velocity. Schematic interaction between JIP1b and KLC1 is also shown at *right*. Complex interactions regulate the transport activity of APP cargos. *JBD* JNK-binding domain, *SH3* Src homology domain 3, *PI* phosphotyrosine interaction domain, *TPR* tetratricopeptide repeat. KLC1 is the major KLC isoform expressed in neurons

regulating the interaction between APP and JIP1b or the association of JIP1b with KLC (Chiba et al. 2014). In the absence of JIP1 expression, the efficiency of APP cargo transport is slower, and the frequency of anterograde transport is reduced. If the efficiency of APP anterograde transport is impaired in the neurons of aged brain, as in JIP1-deficient neurons, this may affect neuronal function. The reduced efficiency of APP cargo transport could also affect the transport of contents in cargo and/or induce metabolism of APP itself, including alterations in Aβ generation. In fact, kinesin-1 levels are reduced in the brains of AD patients (Morel et al. 2012), and a specific KLC isoform modifies Aβ accumulation in brain (Morihara et al. 2014). Although it is impossible to observe alteration of APP axonal transport in living neurons in vivo, biochemical changes in cargo receptors, adaptor proteins, and motor molecules suggest that the cargo transport system is altered in the aged brain. Preservation of axonal transport system may be important for maintenance of neural function in the aged brain, and for protecting the brain against neurodegenerative diseases.

18.4 Alcadein in AD

Alcadeins (Alcs), also called calsyntenins, are evolutionarily conserved neural membrane proteins of the cadherin superfamily. The family has three members in mammals: Alcα, Alcβ, and Alcγ (Araki et al. 2003; Vogt et al. 2001). Alcs are largely similar to APP in regard to function, metabolism, and localization. Metabolically, Alcs are subject to a primary cleavage by APP α-secretase, resulting in secretion of a large extracellular domain fragment (sAlc) and generation of membrane-associated carboxyl-terminal fragments (Alc CTFs). Alc CTFs are further cleaved by γ-secretase, as is APP, resulting in secretion of a small peptide (p3-Alc) and release of the cytoplasmic-domain fragment (Alc ICD) into the cytoplasm (Araki et al. 2003, 2004, 2007; Hata et al. 2009; Takei et al. 2015). In contrast to Aβ from APP, secreted p3-Alc peptide is not prone to aggregation; consequently, unlike p3 peptide of APP, p3-Alc is sufficiently metabolically stable to be detected in cerebrospinal fluid (CSF) and blood (Fig. 18.6). Therefore, qualitative and quantitative alterations of p3-Alc can reflect metabolic changes in APP, including Aβ generation, in the aged brain (Hata et al. 2011, 2012; Konnno et al. 2011; Kamogawa et al. 2012; Omori et al. 2014). Because APP γ-cleavage is altered in the brains of SAD patients (Kakuda et al. 2012), as are the γ-cleavage sites of p3-Alcα in the CSF of SAD patients (Hata et al. 2011), dysfunction in γ-cleavage is as obvious a contributor to the pathogenesis of SAD as PS pathogenic mutations are to the pathogenesis of FAD. Recently, the molecular mechanisms by which γ-secretase cleaves APP and Alc have been revealed (Takami et al. 2009; Piao et al. 2013). Both APP and Alcs are surrogate substrates for detecting dysfunctions in γ-secretase and/or changes in the membrane environment where γ-cleavage occurs.

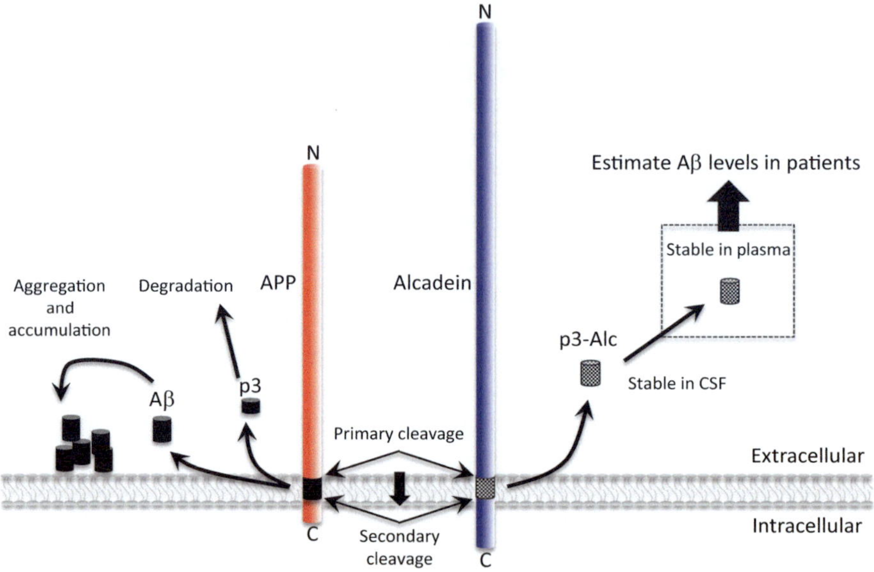

Fig. 18.6 Comparison of Alcadein proteolysis with APP proteolysis. Alcadein is cleaved by α- and γ-secretase, as is APP. The cleavages generate a small secreted peptide, p3-Alc, which is not prone to aggregation. Therefore, quantitative and qualitative alterations of p3-Alc are detectable in body fluids. Alterations of γ-cleavage appear on the C-terminal amino-acid sequence of p3-Alc as an endophenotype as can be seen in Aβ generation

The magnitudes of carboxyl-terminal alterations generated by altered cleavages of CTFs by γ-secretase are not equivalent among APP, Alcα, Alcβ, and Alcγ (Hata et al. 2009; Piao et al. 2013). This suggests that some alteration for γ-secretase or its environment in a specific brain area appears on the alteration of γ-cleavage of APP, such as an increase of Aβ42 generation, whereas other alteration is apparent in Alc but not APP. Even if we cannot detect alterations in Aβ generation by analyzing CSF and/or blood, the underlying alteration in γ-cleavage alteration might initially be detected as a change in Alc γ-cleavage, and appear subsequently in Aβ as brain aging progresses. Therefore, the use of a substrate panel would represent an effective means for detecting γ-secretase dysfunction and alterations in membrane environment that induce altered γ-cleavage of substrates in the pre-pathogenic state of AD.

18.5 Summary and Conclusion

The majority of AD patients have the sporadic type of disease, which is distinct from the much less common familial disease associated with mutations in APP and PSs. Although the brain pathologies of SAD patients are similar to those of FAD patients, as can be observed from the formation of senile plaques and neurofibrillary tangles, SAD may have various primary causes. In some cases, β- or γ-secretase activity may be altered, resulting in elevated levels of pathogenic Aβ, whereas in other cases, the ability to clear Aβ may be reduced. Currently, we do not know which events in the aged brain are the major causes of AD. Because the critical causes may differ among individual patients, we should identify alterations that might contribute to AD pathogenesis to advance the development of personalized medicine.

In this chapter, we introduced three potential mechanisms that might alter the level of pathogenic Aβ, as follows. (1) *Phosphorylation of APP at Thr668*. Phosphorylated APP CTFs, substrates of γ-secretase, are clearly capable of escaping from membrane microdomains in which γ-secretase is active. Aging generally decreases the phosphorylation level of CTFβ in monkey brain, and this reduction is significant in the brains of AD patients relative to age-matched non-demented subjects. (2) *The axonal transport system*. Dysfunction in axonal transport can induce many types of neurodegenerative diseases; thus, it is a general cause of brain disorders rather than a specific cause of AD. In this context, however, we should focus our attention on the function of APP as a cargo receptor. Although previous studies have described multiple APP functions, it is clear that APP's primary function in neurons is to serve as the cargo receptor for kinesin-1, the major anterograde motor in this cell type, suggesting that APP plays an important role in a transport of proteins. Defective transport of as-yet-unidentified proteins by APP cargo may promote neurotoxicity in the aged brain and trigger AD. (3) *Dysfunction of γ-secretase or γ-cleavage of APP, which is common in FAD patients carrying PS gene mutations*. Recent studies showed that alteration of γ-cleavage also occurs in SAD patients, even if they do not carry any mutation in PS genes. Aβ, the pathogenic product of APP γ-cleavage, shows a strong tendency to aggregate; thus, because the majority of Aβ is aggregated or precipitated in body fluids, it is difficult to determine the extent of qualitative and quantitative alteration of γ-cleavage of Aβ peptides in quality and/or quantity. Almost 100 type I membrane proteins are candidate substrates of the γ-secretase. However, the detailed mechanism of γ-cleavages has been revealed for only a few substrates, including APP, APP family proteins, Notch, and the Alcadeins. We have proposed using a substrate panel to detect γ-secretase dysfunction and alteration of γ-cleavage, and thereby identify subjects in preclinical or very early stages of AD. Classification of AD patients based on major pathogenetic events and early detection of preclinical subjects will be essential for the effective development of personalized therapies for elderly people.

Acknowledgements We are thankful to Prof. Nozomu Mori (Nagasaki University) and Prof. Inhee Mook-Jung (Seoul National University) for giving us this opportunity to write this review. This work was supported in part by Grant-in-Aid for Scientific Research 26293010 (to T.S.) and 24790062 (to S.H.) from MEXT, Japan, and in part by the Asian Core Program of the JSPS, Japan. S.H. was supported by the Bilateral Joint Research Projects of the JSPS.

References

Ando K, Iijima KI, Elliott JI, Kirino Y, Suzuki T (2001) Phosphorylation-dependent regulation of the interaction of amyloid precursor protein with FE65 affects the production of β-amyloid. J Biol Chem 276:40353–40361

Araki Y, Tomita S, Yamaguchi H, Miyagi N, Sumioka A, Kirino Y, Suzuki T (2003) Novel cadherin-related membrane proteins, Alcadeins, enhance the X11-like protein mediated stabilization of amyloid β-protein precursor metabolism. J Biol Chem 278:49448–49458

Araki Y, Miyagi N, Kato N, Yoshida T, Wada S, Nishimura M, Komano H, Yamamoto T, De Strooper B, Yamamoto K, Suzuki T (2004) Coordinated metabolism of Alcadein and amyloid β-protein precursor regulates FE65-dependent gene transactivation. J Biol Chem 279:24343–24354

Araki Y, Kawano T, Taru H, Saito Y, Wada S, Miyamoto K, Kobayashi H, Ishikawa HO, Ohsugi Y, Yamamoto T, Matsuno K, Kinjo M, Suzuki T (2007) The novel cargo Alcadein induces vesicle association of kinesin-1 motor components and activates axonal transport. EMBO J 26:1475–1486

Beel A, Mobley CK, Kim HJ, Tian F, Hadziselimovic A, Jap B, Prestegard JH, Sanders CR (2008) Structural studies of the transmembrane C-terminal domain of the amyloid precursor protein (APP). Does APP functions as a cholesterol sensor? Biochemistry 47:9428–9446

Chiba K, Araseki M, Nozawa K, Furukori K, Araki Y, Matsushima T, Nakaya T, Hata S, Saito Y, Uchida S, Okada Y, Nairn AC, Davis RJ, Yamamoto T, Kinjo M, Taru H, Suzuki T (2014) Quantitative analysis of APP axonal transport in neurons: role of JIP1 in enhanced APP anterograde transport. Mol Biol Cell 25:3569–3580

Cole SL, Vassar R (2008) The role of amyloid precursor protein processing by BACE1, the β-secretase, in Alzheimer disease pathophysiology. J Biol Chem 283:29621–29625

Haass C, Selkoe DJ (2007) Soluble protein oligomers in neurodegeneration: lessons from the Alzheimer's amyloid β-peptide. Nat Rev Mol Cell Biol 8:101–112

Hancock WO (2014) Bidirectional cargo transport: moving beyond tug of war. Nat Rev Mol Cell Biol 9:615–628

Hata S, Fujishige S, Araki Y, Kato N, Araseki M, Nishimura M, Hartmann D, Saftig P, Fahrenholz F, Taniguchi M, Urakami K, Akatsu H, Martins RN, Yamamoto K, Maeda M, Yamamoto T, Nakaya T, Gandy S, Suzuki T (2009) Alcadein cleavages by APP α- and γ-secretases generate small peptides p3-Alcs indicating Alzheimer disease-related γ-secretase dysfunction. J Biol Chem 284:36024–36033

Hata S, Fujishige S, Araki Y, Taniguchi M, Urakami K, Peskind E, Akatsu H, Araseki M, Yamamoto K, Martins NR, Maeda M, Nishimura M, Levey A, Chung KA, Montine T, Leverenz J, Fagan A, Goate A, Bateman R, Holtzman DM, Yamamoto T, Nakaya T, Gandy S, Suzuki T (2011) Alternative γ-secretase processing of γ-secretase substrates in common forms of mild cognitive impairment and Alzheimer disease: evidence for γ-secretase dysfunction. Ann Neurol 69:1026–1031

Hata S, Taniguchi M, Piao Y, Ikeuchi T, Fagan AM, Holzman DM, Bateman R, Sohrabi HR, Martins RN, Gandy S, Urakami K, Suzuki T, J-ADNI (2012) Multiple γ-secretase product peptides are coordinately increased in concentration in the CSF of a subpopulation of sporadic Alzheimer's disease subjects. Mol Neurodegener 7:16

Hirokawa N, Noda Y, Tanaka Y, Niwa S (2009) Kinesin superfamily motor proteins and intracellular transport. Nat Rev Mol Cell Biol 10:682–696

Horiuchi D, Barkus RV, Pilling AD, Gassman A, Saxton WM (2005) APLIP1, a kinesin binding JIP-1/JNK scaffold protein, influences the axonal transport of both vesicles and mitochondria in *Drosophila*. Curr Biol 15:2137–2141

Iijima K, Ando K, Takeda S, Satoh Y, Seki T, Itohara S, Greengard P, Kirino Y, Nairn AC, Suzuki T (2000) Neuron-specific phosphorylation of Alzheimer's b-amyloid precursor protein by cyclin-dependent kinase 5. J Neurochem 75:1085–1091

Kakuda N, Shoji M, Arai H, Furukawa K, Ikeuchi T, Akazawa K, Takami M, Hatsuta H, Murayama S, Hashimoto Y, Miyajima M, Arai H, Nagashima Y, Yamaguchi H, Kuwano R, Nagaike K, Ihara Y, for J-ADNI (2012) Altered γ-secretase activity in mild cognitive impairment and Alzheimer's disease. EMBO Mol Med 4:344–352

Kamal A, Stokin GB, Yang Z, Xia CH, Goldstein LS (2000) Axonal transport of amyloid precursor protein is mediated by direct binding to the kinesin light chain subunit of kinesin-1. Neuron 28:449–459

Kamogawa K, Kohara K, Tabara Y, Takita R, Miki T, Konno T, Hata S, Suzuki T (2012) Utility of plasma levels of soluble p3-Alcadeinα as a biomarker for sporadic Alzheimer's disease. J Alzheimers Dis 31:421–428

Kanekiyo T, Xu H, Bu G (2014) ApoE and Ab in Alzheimer's disease: accidental encounters or partners. Neuron 81:740–754

Kawano T, Araseki M, Araki Y, Kinjo M, Yamamoto T, Suzuki T (2012) A small peptide sequence is sufficient for initiating kinesin-1 activation through part of TPR region of KLC1. Traffic 13:834–848

Kok E, Haikonen S, Luoto T, Huhtala H, Goebeler S, Haapasalo H, Karhunen PJ (2009) Appolipoprotein E-dependent accumulation of Alzheimer disease-related lesions begins in middle age. Ann Neurol 65:650–657

Konecta A, Frischknecht R, Kinter J, Ludwig A, Steuble M, Meskenaite V, Indermuhle M, Engel M, Cen C, Mateos JM, Sreit P, Sonderegger P (2006) Calsyntenin-1 docks vescular cargo to kinesin-1. Mol Biol Cell 17:3651–3663

Konnno T, Hata S, Hamada Y, Horikoshi Y, Nakaya T, Saito Y, Yamamoto T, Yamamoto T, Maeda M, Gandy S, Akatsu H, Suzuki T, for J-ADNI (2011) Coordinate increase of γ-secretase reaction products in the plasma of some female Japanese sporadic Alzheimer's disease patients: quantitative analysis with a new ELISA system. Mol Neurodegener 6:76

Matsushima T, Saito Y, Elliott JI, Iijim-Ando K, Nishimura M, Kimura N, Hata S, Yamamoto T, Nakaya T, Suzuki T (2012) Membrane-microdomain localization of amyloid β-precursor protein (APP) C-terminal fragments is regulated by phosphorylation of the cytoplasmic Thr668 residue. J Biol Chem 287:19715–19724

Mawuenyaga KG, Sigurdson W, Ovod V, Munsell L, Kasten T, Morris JC, Yarasheski KE, Bateman RJ (2010) Decreased clearance of CNS beta-amyloid in Alzheimer's disease. Science 330:1774

Millecamps S, Julien J-P (2013) Axonal transport deficits and neurodegenerative disease. Nat Rev Neurosci 14:161–176

Morel M, Heraud C, Nicaise C, Suain V, Brion JP (2012) Levels of kinesin light chain and dynein intermediate chain are reduced in the frontal cortex in Alzheimer's disease: implications for axoplasmic transport. Acta Neuropathol 123:71–84

Morihara T, Hayashi N, Yokokoji M, Akatsu H, Silverman MA, Kimura N, Sato M, Saito Y, Suzuki T, Yanagida K, Kodama TS, Tanaka T, Okochi M, Tagami S, Kazui H, Kubo T, Hashimoto R, Itoh N, Nishitomi K, Yamaguchi-Kabata Y, Tsunoda T, Takamura H, Katayama T, Kimura R, Kamino K, Hashizume Y, Takeda M (2014) Transcriptome analysis of distinct mouse strains reveals kinesin light chain-1 splicing as an amyloid-β accumulation modifier. Proc Natl Acad Sci USA 111:2638–2643

Oikawa N, Kimura N, Yanagisawa K (2010) Alzheimer-type tau pathology in advanced age nonhuman primate brains harboring substantial amyloid deposition. Brain Res 1315:137–149

Omori C, Kaneko M, Nakajima E, Akatsu H, Waragai M, Maeda M, Morishima-Kawashima M, Saito Y, Nakaya T, Taru H, Yamamoto T, Asada T, Hata S, Suzuki T, for J-ADNI (2014) Increased levels of plasma p3-Alcα35, a major fragment of alcaseinα by γ-secretase cleavage, in Alzheimer disease. J Alzheimers Dis 39:861–870

Piao Y, Kimura A, Urano S, Saito Y, Taru H, Yamamoto T, Hata S, Suzuki T (2013) Mechanism of intramembrane cleavage of Alcadeins by γ-secretase. PLoS One 8:e62431

Ramelot TA, Nicholson LK (2001) Phosphorylation-induced structural changes in the amyloid precursor protein cytoplasmic tail detected by NMR. J Mol Biol 307:871–884

Rhinn H, Fujita R, Qiang L, Cheng R, Lee JH, Abeliovich A (2013) Integrative genomics identifies APOEe4 effectors in Alzheimer's disease. Nature 500:45–50

Riddle DR, Christie G, Hussain I, Dingwall C (2001) Compartmentalization of beta-secretase (Asp2) into low-buoyant density, noncaveolar lipid rafts. Curr Biol 11:1288–1293

Saito Y, Sano Y, Vassar R, Gandy S, Nakaya T, Yamamoto T, Suzuki T (2008) X11 proteins regulate the translocation of amyloid β-protein precursor (APP) into detergent-resistant membrane and suppress the amyloidogenic cleavage of APP by β-site-cleaving enzyme in brain. J Biol Chem 283:35763–35771

Stokin GG, Lillo C, Falzone TL, Brusch RG, Rockenstein E, Mount SL, Raman R, Davies P, Masliah E, Williams DS, Goldstein LS (2005) Axonopathy and transport deficities early in the pathogenesis of Alzheimer's disease. Science 307:1282–1288

Suzuki T, Nakaya T (2008) Regulation of amyloid β-protein precursor by phosphorylation and protein interactions. J Biol Chem 283:29633–29637

Suzuki T, Araki Y, Yamamoto T, Nakaya T (2006) Trafficking of Alzheimer's disease-related membrane proteins and its participation in disease pathogenesis. J Biochem 139:949–955

Takami M, Nagashima Y, Sano Y, Ishihara S, Morishima-Kawashima M, Funamoto S, Ihara Y (2009) Gamma-secretase: successive tripeptide and tetrapeptide release from the transmembrane domain of beta-carboxyl terminal fragment. J Neurosci 29:13042–13052

Takei N, Sobu Y, Kimura A, Urano S, Piao Y, Araki Y, Taru H, Yamamoto T, Hata S, Nakaya T, Suzuki T (2015) Cytoplasmic fragment of Alcadeinα generated by regulated intramembrane proteolysis enhances APP transport into the late-secretory pathway and facilitates APP cleavage. J Biol Chem 290:987–995

Taru H, Suzuki T (2009) Regulation of physiological function and metabolism of AβPP by AβPP binding proteins. J Alzheimer Dis 18:253–265

Taru T, Iijima K, Hase M, Kirino Y, Yagi Y, Suzuki T (2002) Interaction of Alzheimer's β-amyloid precursor family proteins with scaffold proteins of the JNK signaling cascade. J Biol Chem 277:20070–20078

Thies W, Bleiler L, Alzheimer's Association (2013) Alzheimer's disease facts and figures. Alzheimers Dement 9:208–245

Thinakaran G, Koo EH (2008) Amyloid precursor protein trafficking, processing, and function. J Biol Chem 283:29615–29619

Vale RD, Reese TS, Sheetz MP (1985) Identification of a novel force-generating protein, kinesin, involved in microtubule-based motility. Cell 42:39–50

Verhey KJ, Hammond JW (2009) Traffic control: regulation of kinesin motors. Nat Rev Mol Cell Biol 10:765–777

Verhey KJ, Meyer D, Deehan R, Blenis J, Schnapp BJ, Rapoport TA, Margolis B (2001) Cargo of kinesin identified as JIP scaffolding proteins and associated signaling molecules. J Cell Biol 152:959–970

Vogt L, Schrimpf SP, Meskenaite V, Frischknecht R, Kinter J, Leone DP, Ziegler U, Sonderegger P (2001) Calsyntenin-1, a proteolytically processed postsynaptic membrane protein with a cytoplasmic calcium-binding domain. Mol Cell Neurosci 17:151–166

World Health Organization and Alzheimer's Disease International (2012) Mental health. Dementia: a public health priority. WHO Press, Geneva. ISBN:978 92 4 156445 8

Yagi Y, Tomita S, Nakamura M, Suzuki T (2000) Overexpression of human amyloid precursor protein in *Drosophila*. Mol Cell Biol Res Commun 4:43–49

Chapter 19
Catabolism and Anabolism of Amyloid-β

Shoko Hashimoto, Per Nilsson, and Takaomi C. Saido

Abstract According to the amyloid cascade hypothesis, the initial elevation of amyloid β-peptide (Aβ) level is the primary trigger of Alzheimer's disease (AD). The steady-state levels of Aβ are determined by the balance of the production and clearance of Aβ. In familial AD, an imbalance in Aβ kinetics arises from the abnormal generation of Aβ due to mutations in the secretases, which cleave Aβ from the amyloid precursor protein (APP), or in APP itself. However, the large majority of AD is sporadic AD (SAD), which lacks a strong genetic component. In SAD, inactivations of Aβ degradation systems might instead be the cause of the disease. We previously identified neprilysin (NEP) as the major Aβ-degrading enzyme. Importantly, NEP declines in the human brain with aging, which may contribute to the increased Aβ pathology. Therefore, the up-regulation of NEP activity in the brain represents a potential therapy for the prevention of AD. To that end, we recently developed a system to overexpress NEP throughout the brain using an newly designed adeno-associated viral vector carrying the NEP gene (AAV-NEP). In addition to the Aβ-degrading enzymes, we recently found that autophagy plays an important role in Aβ metabolism. Specifically, autophagy influences the intracellular sorting and secretion of Aβ. Intriguingly, autophagy deficiency induces intracellular Aβ accumulation, which enhances autophagy deficiency-induced neurodegeneration. This result indicates that intracellular Aβ might be toxic. Indeed, Aβ clearance systems are potential therapeutic targets in AD, to prevent the disease via a treatment already in early stages.

Keywords APP • BACE1 • γ-secretase • Neprilysin • Autophagy • Aβ transport • APP knock-in model mice

---

S. Hashimoto • P. Nilsson • T.C. Saido (✉)
Laboratory for Proteolytic Neuroscience, RIKEN Brain Science Institute, 2-1 Hirosawa, Wako, Saitama 351-0198, Japan
e-mail: saido@brain.riken.jp

© Springer Japan 2015

N. Mori, I. Mook-Jung (eds.), *Aging Mechanisms*,
DOI 10.1007/978-4-431-55763-0_19

19.1 Introduction

Although the physiological function of the amyloid β (Aβ) precursor protein (APP) remains largely unknown, the Aβ peptide that is cleaved out of APP plays a central role in Alzheimer's disease (AD) because Aβ accumulation is a key event in AD. Aβ is generated from APP, which is a type I membrane protein, through proteolytic cleavages mediated by the β- and γ-secretases. APP is first cleaved by either α- or β-secretase (also termed β-site APP-cleaving enzyme 1 (BACE1)). Following cleavage by BACE1, which is a type I transmembrane aspartic protease, the C-terminal part of APP is further digested by γ-secretase which releases Aβ. The peptide that is produced from APP by α-and γ-secretase, termed P3, is highly soluble and non-amyloidogenic, while the β- and γ-cleaved Aβ peptide, which is 40–43 amino acids long, is biochemically less soluble and tends to form pathological oligomers and fibrils (Hardy and Selkoe 2002). Mutations in the genes encoding presenilin 1 (PS1) and presenilin 2 (PS2) (*PSENl* and *PSEN2*, respectively), which are the enzymatic components of γ-secretase, or in *APP* cause the majority of cases of early onset familial AD (FAD). These FAD-linked mutations increase either the total Aβ levels or the ratio of Aβ42/Aβ40 and give rise to an aggressive pathology due to the higher aggregation propensity of Aβ42 compared to Aβ40. Consistently, Aβ42 levels are increased by approximately 1.5-fold in the brains of mutant PS1 transgenic or knock-in mice that possess the PS1/I213T missense mutation (Nakano et al. 1999b). More than 100 mutations that cause familial AD have been identified thus far in *PSEN1/2* and *APP*. A larger proportion of familial AD patients (more than 70 %) have mutations in the *PSEN1* gene while mutations in the *APP* sequence account for less than 10 % of FAD. The FAD-linked mutations in APP are located around the β- or γ- cleavage sites or in the N-terminal part of Aβ. The former mutations affect the proteolytic cleavage performed by β- or γ-secretase, and the latter alter the aggregation propensity and cause amyloid angiopathy. Additionally, the recent discovery of a mutation in APP that protects against AD further strengthens the association of Aβ with AD (Jonsson et al. 2012). Moreover, we have demonstrated that some of the FAD-associated mutations in APP, such as the Dutch, Flemish, Italian and the Arctic mutations, not only affect the production of Aβ but also the resistance of Aβ to the proteolysis catalyzed by the Aβ catabolic enzyme neprilysin (NEP) (Tsubuki et al. 2003). Even though the precise neurotoxic role of Aβ is yet to be fully understood, Aβ has been shown to interfere with several receptors. Aβ binds to α7-nicotinic acetylcholine receptor, tumor necrosis factor-related apoptosis-inducing ligand (TRAIL) death receptors, and receptor tyrosine kinase Ephrin type-B receptor 2 (EphB2) which activate neurodegenerative pathways (Cisse et al. 2011; Wang et al. 2000; Liu et al. 2001; Fossati et al. 2012). In the Aβ anabolism section, we will describe the pathway of physiological and pathological Aβ generation in more detail.

Because the FAD-linked mutations in *PSEN1/2* and *APP* strongly associate Aβ with AD, the Aβ-generating enzymes BACE1 and PS1 have been focused on as therapeutic targets, and numerous attempts have been made to develop inhibitors of

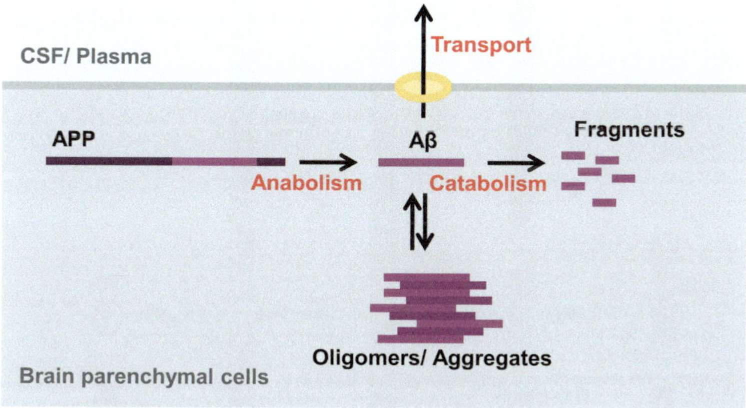

Fig. 19.1 The relationships between Aβ production, degradation and transport within and outside of the brain. The steady-state Aβ level is determined by the APP concentration, the rate constants of Aβ production, the rate of Aβ degradation in the brain, and the out-of-brain transport of Aβ

these enzymes. Clinical trials of BACE1 and γ-secretase inhibitors have faced several difficulties including the facts that these secretases have endogenous substrates other than APP. However, autosomal dominant FAD-linked mutations account for less than 1 % of all AD cases, whereas sporadic AD (SAD) accounts for the majority of AD case. SAD is also characterized by increased Aβ levels but has a much more elusive etiology than FAD, which suggests that other mechanisms contribute to Aβ accumulation in SAD. Steady-state Aβ levels are maintained by the balance between the production and clearance of Aβ (Fig. 19.1). Because an over-production of Aβ has not been observed in SAD, it might be an impairment of Aβ clearance that causes the increase in Aβ levels and the onset of the disease. The major processes involved in Aβ clearance include the proteolytic degradation of Aβ and transport of Aβ from the brain tissue to the cerebrospinal fluid or blood. In the Aβ catabolism section, we summarize the present knowledge regarding Aβ clearance, including the actions of Aβ catabolic enzymes and autophagy-mediated Aβ metabolism.

19.2 Aβ Anabolism

Aβ is produced from the cleavage of APP by two aspartic proteases, β-secretase (BACE1) and γ-secretase (Saido 2003). First, BACE1 cleaves APP at the first Asp residue of the Aβ-sequence, which gives rise to the N-terminus of the Aβ peptide. This cleavage produces two fragments, the secreted ectodomain (sAPPβ) and the membrane-bound C-terminal fragment (CTFβ). Next, CTFβ is digested at the γ-site by γ-secretase which releases the Aβ peptide from the APP intracellular domain

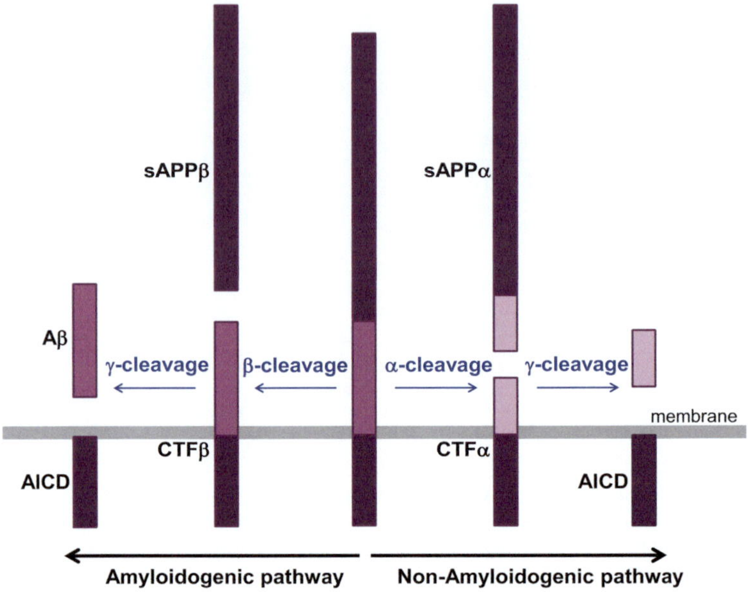

Fig. 19.2 APP processing and Aβ generation. APP is localized to the membrane and is sequentially cleaved by β- and γ-secretase to generate Aβ. Aβ formation is prevented by the non-amyloidgenic pathway, which is initiated by α-cleavage of APP

(AICD) (Wolfe et al. 1999) (Fig. 19.2). γ-secretase is a membrane-associated complex consisting of the following four different proteins: PS1/2, nicastrin, Aph1, and Pen2 (Saido 2003; Hardy and Selkoe 2002). Due to the ability of γ-secretase to cleave at multiple positions, not only the major Aβ40 peptide but also longer and shorter peptides are produced. Among these peptides, Aβ42 is known to be a primary amyloidogenic and pathogenic agent in AD due to its high aggregation propensity. Additionally, Aβ43, which is overproduced in PS1/R278I knock-in mice and increased in AD brains (Welander et al. 2009; Sandebring et al. 2013), exhibits an even greater propensity to aggregate and is more neurotoxic than Aβ42 (Saito et al. 2011).

APP is produced in the endoplasmic reticulum (ER) and transported to the cell surface by the trans-golgi network (TGN). BACE1 is localized in the TGN, and although the majority of γ-secretase is localized in the ER, a portion is transported from the ER together with APP by the TGN where Aβ is generated. Alternatively, APP located at the cell surface is endocytosed and cleaved by BACE1 and γ-secretase in the endosomes. It remains to be resolved which of these two pathways makes the greatest contribution to Aβ generation and secretion.

Inhibitors of γ-secretase are likely to reduce Aβ generation and are therefore potential drugs for AD. A substantial number of such inhibitors have been developed over the years (Ghosh et al. 2012), and some have progressed to human clinical trials. However, γ-secretase acts not only on APP but also on more than

60 other substrates, including the trans-membrane receptor and signaling protein Notch (Nakayama et al. 2011). Notch signaling is involved in somitogenesis, cell differentiation, and neural development; therefore, this might be the reason why many γ-secretase inhibitors are toxic to patients (Nakayama et al. 2011; Bolos et al. 2007). Nevertheless, several γ-secretase inhibitors, including R-flurbiprofen, exhibit low toxicity. Indeed, R-flurbiprofen decreases the levels of Aβ42 without inhibiting the activities of Notch or other substrates. Although an Aβ-lowering effect has been observed, the inhibition of cognitive decline by R-flurbiprofen is limited to mild AD patients, and no improvements have been observed in moderate AD (Green et al. 2009).

As an alternative to γ-secretase inhibitors, the inhibition of BACE1 would decrease Aβ generation. Indeed, several groups have reported that BACE1 knockout mice are nearly devoid of Aβ in addition to being healthy and fertile. Therefore, research progressed to develop inhibitors of BACE1, but the recent discovery of additional endogenous substrates of BACE1 and the side effects of BACE1 inhibitors have been discouraging. Moreover, because there are many other aspartic acid proteinases than BACE1, such as renin, neuregulin 1 and cathepsin D, a high specificity is required to establish a BACE1 inhibitor without side effects. In addition, the large active site of BACE1 is a challenge for the design of inhibitors with high specificity (Hong et al. 2000). Moreover, additional research has suggested that BACE1 knockout mice do exhibit several phenotypes, including schizophrenia endophenotypes (Savonenko et al. 2008), hyperactivity (Dominguez et al. 2005; Savonenko et al. 2008), spine density reduction (Savonenko et al. 2008), an increase in astrogenesis (Hu et al. 2013), a decrease in neurogenesis (Hu et al. 2013), neurodegeneration (Hu et al. 2010), memory deficits (Ohno et al. 2004; Laird et al. 2005), or axon guidance defects (Cao et al. 2012; Hitt et al. 2012), that are likely induced via substrates other than APP. Recently, a clinical trial of the BACE1 inhibitor LY2886721 was discontinued due to abnormal liver values, although the relationship between its toxicity and the off-target effect of the compound is unclear. Simultaneously, a Phase III study of the BACE1 inhibitor MK-8931 for the treatment of AD and mild cognitive impairment was initiated. Although the outcome of this trial has yet to be revealed, it has been difficult to establish a drug based on the inhibition of β- and γ-secretases thus far. However, a recent research field has emerged that aims to identify compounds that modify the activity of γ-secretase in an APP-specific manner rather than inhibit it which may open up for γ-secretase-based treatments.

Due to the apparent challenges to developing BACE1 and γ-secretase inhibitors, identifying new targets and designing drugs for these targets are necessary to overcome AD. Indeed, Aβ catabolism is a potential pathway that has attracted increasing attention recently and has advanced since our discovery of the major Aβ-degrading enzyme NEP (Iwata et al. 2001). The next section will focus on the research related to Aβ degrading enzymes.

19.3 Aβ Catabolism

19.3.1 Aβ Degrading Enzymes

To a large extent, AD research has focused on Aβ generation based on the proposed Aβ cascade hypothesis (Selkoe 1991) and the discovery of the FAD-linked mutations. In contrast, research on Aβ catabolism has been less in focus. However, Aβ catabolism has gained increasing attention due to its crucial role in determining the steady-state levels of Aβ, a mechanism that might be impaired in SAD. Through in vitro experiments that aimed to investigate Aβ degrading activities, many proteases or peptidases have been identified and reported to exhibit Aβ-cleaving activity. These include NEP, insulin-degrading enzyme (IDE) (Kurochkin and Goto 1994; McDermott and Gibson 1997; Qiu et al. 1998), endothelin-converting enzyme (ECE)1/2 (Eckman et al. 2001), angiotensin-converting enzyme (ACE) (Hu et al. 2001), cathepsin D (Yamada et al. 1995; Hamazaki 1996), urokinase-type plasminogen activator (Sasaki et al. 1988), and matrix metalloendopeptidase-9 (Carvalho et al. 1997). Below, we briefly discuss some of these peptidases, and subsequently present a separate section about NEP.

IDE is known as a zinc metalloendopeptidase that was originally implicated in insulin metabolism (Duckworth 1979). IDE hydrolyzes multiple peptides, including insulin, glucagon, atrial natriuretic factor, transforming growth factor-α, β-endorphin, amylin, and AICD in addition to Aβ (Duckworth et al. 1998; Selkoe 2001). Numerous reports have indicated a role for IDE in Aβ metabolism via in vitro, cell-based, and animal model studies (Farris et al. 2003). However, in mice, a genetic deficiency of IDE only increases Aβ42 levels by 1.4-fold (Miller et al. 2003). This rather small increase in Aβ levels might be explained by a difference in cellular localization; IDE is located in the cytosol, while Aβ is primarily membrane bound. Furthermore, IDE lacks a secretory signal and therefore might not appropriately interact with Aβ in the extracellular space to metabolize Aβ (Table 19.1).

ECE1 and 2 are transmembrane metalloproteases and belong to the same family as NEP. ECE1 and 2 are detected primarily in vascular endothelial cells where they process pro-endothelin-1, which is a potent vasoconstrictive peptide produced in vascular endothelial cells, but are also present in neuronal cells in the brain (Masaki 2004). Knockout of ECE1/2 in AD model mice leads to a 1.2 to 1.3-fold increase in Aβ, which indicates that ECE1 and 2 are potentially Aβ-degrading enzymes (Eckman et al. 2003) (Table 19.1).

ACE is a membrane bound zinc-dependent dipeptidyl carboxypeptidase. ACE converts angiotensin I (AngI) to AngII, which plays roles in maintaining blood pressure, body fluid, and sodium homeostasis (Guy et al. 2005). Genetic studies in humans have provided evidence for a potential relationship between ACE and AD (Kehoe et al. 1999). Patients with an insertion polymorphism within intron 16 of the ACE gene are at a greater risk to develop AD (Kehoe et al. 1999). In contrast, patients with a deletion polymorphism are at a lower risk for AD (Elkins

Table 19.1 Aβ levels in mice after knockout (KO) of different peptidases

Mouse genotype	Ab42 (fold)	Ab40 (fold)	References
NEP-KO (−/−)	2	2	Iwata et al. (2001)
NEP-KO (+/−)	1.5	1.5	Iwata et al. (2001)
ECE2-KO (−/−)	1.3	1.3	Eckman et al. (2003)
ECE2-KO (+/−)	1.2	1.2	Eckman et al. (2003)
ECE1-KO (+/−)	1.3	1.3	Eckman et al. (2003)
IDE-KO (−/−)	Not significant	1.1–1.2	Farris et al. (2003)
IDE-KO (−/−)	1.4	1.6	Miller et al. (2003)
tPA-KO (−/−)	Not significant	Not significant	(Unpublished data by Iwata et al.)
uPA-KO (−/−)	Not significant	Not significant	Ertekin-Taner et al. (2005)
ACE-KO (−/−)	Not significant	Not significant	(Unpublished data by Takaki et al.)
Plasmin (−/−)	Not significant	Not significant	Nakano et al. (1999a)

et al. 2004). Moreover, it has been shown in vitro that ACE exhibits significant inhibitory effects on the aggregation, deposition, and cytotoxicity of Aβ (Hu et al. 2001) and that purified ACE can efficiently cleave Aβ42 to Aβ40 (Zou et al. 2007). In agreement, it has been shown that ACE degrades endogenous Aβ40 and Aβ42 in cell culture (Hemming and Selkoe 2005). However, treating AD models with ACE inhibitor did not increase Aβ accumulation in vivo. Consistently, we previously showed that the knockout of ACE has no effects on Aβ levels (Takaki et al. unpublished data) (Table 19.1).

19.3.2 Neprilysin

Many potential Aβ-degrading enzymes have been reported. To identify the main Aβ-degrading enzyme in vivo, we previously established a physiologically relevant experimental system which can detect the Aβ catabolic processes in the brain (Iwata et al. 2000). In this approach, synthetic radiolabeled Aβ42 is injected into the hippocampus of anesthetized live rats, and the degradation process is analyzed by high-pressure liquid chromatography and a liquid scintillation counter. We tested the effects of Aβ degradation of more than 20 peptidase inhibitors and found that thiorphan, a well-characterized neutral endopeptidase inhibitor, was the most potent inhibitor. Hence, a neutral endopeptidase family member that is similar or identical to NEP appears to play a major role in Aβ catabolism.

NEP, which is membrane bound protein, was originally discovered in renal brush border membranes (Wong-Leung and Kenny 1968). Later, it was also found that NEP was identical to enkephalinase, which participates in the proteolytic inactivation of enkephalins (Malfroy et al. 1978). In the brain, NEP is abundant in the striatonigral pathway (Barnes et al. 1988) and plays a major role in degrading opioids and other neuropeptides (Matsas et al. 1983; Turner et al. 2001). Indeed,

NEP colocalizes with opioid receptors (Waksman et al. 1986), tachykinin and substance P (Barnes et al. 1993). NEP is also abundant in the hippocampus, especially in the dentate gyrus molecular layer and the dentate gyrus granule cell layer, while in the cortex, NEP is present in layers 2/3 and 5 (Iwata et al. 2001). In the neurons, NEP is localized presynaptically in the axons but is also present in the membrane fractions of ER, TGN and endosomes.

Knockout studies in mice have shown that NEP deficiency decreases the degradation of both exogenously administered Aβ and endogenous Aβ (Iwata et al. 2001). This leads to a twofold increase in the Aβ levels in the brains of NEP-deficient mice, which clearly indicates that NEP dominantly regulates the steady-state level of Aβ (Table 19.1). Moreover, Aβ levels in the mouse brain are inversely correlated with NEP levels; mice that are heterozygous for NEP-KO exhibit approximately 50 % increases in the levels of both Aβ40 and Aβ42 (Iwata et al. 2001; Akiyama et al. 2001). Importantly, clear decreases in NEP activity have been detected in the brains of both aged mice and in aging healthy people as well as in AD patients through biochemical and neuropathological examinations (Iwata et al. 2002). Interestingly, the amount and activity of NEP decline in age-dependent manners, especially in the outer molecular layer and the polymorph layer of the hippocampal dentate gyrus, and Aβ levels consequently increase in these brain areas (Iwata et al. 2002). The outer molecular layer of the dentate gyrus receives nerve projections from the entorhinal cortex through the perforant pathway. This pathway is markedly damaged in AD, and pathological changes in this region are closely linked to the cognitive impairments in the early stage of AD (Iwata et al. 2002). McGeer and co-workers reported that the NEP protein levels decrease in the hippocampus and the temporal lobe in Braak stage II and suggested that these reductions in NEP precedes neurodegeneration (Yasojima et al. 2001a). Additionally, NEP mRNA levels are decreased in the hippocampi and temporal cortices of early AD patients, whereas these levels are unchanged in the cerebellum (Yasojima et al. 2001a, b). Because AD pathology primarily involves the hippocampus and the association cortices and spares the cerebellum, these data suggest a close relationship between region-selective declines of NEP activity and AD pathogenesis. Further contributing to the lowered NEP activity in AD is the increased binding of 4-hydroxynonenal (HNE) to NEP (Wang et al. 2003). HNE is produced under oxidative stress possibly as a result of Aβ deposition or neuroinflammation and leads to increased inhibition of the enzymatic activity of NEP.

In an effort to understand the regulatory mechanisms of NEP, we recently found that the phosphorylation status of NEP controls the cellular localization of NEP, which affects extracellular Aβ levels (Kakiya et al. 2012). By analyzing the effects of neurotrophic factors, including brain-derived neurotrophic factor, nerve growth factor, and neurotrophins 3 and 4, on NEP localization, we established that these factors enhance the phosphorylation of the NEP intracellular domain and that this process is mediated by MEK/ERK signaling and leads to reduced NEP activity on the cell surface. In contrast, the activation of protein phosphatase-1a increases cell surface NEP activity, which lowers the Aβ levels.

What other mechanisms regulate NEP activity in the brain? In peripheral cells, such as fibroblasts, neutrophils, and bone marrow cells, NEP levels are influenced by several peptides including substance P and opioid calcitonin (Wang et al. 1997; Howell et al. 1993; Bae et al. 2002). These factors act on receptors and regulate NEP levels; therefore, ligand-receptor systems have been considered to also control NEP activity in the brain. To that end we previously screened tens of biochemical ligands, including hormones, neurotransmitters and cytokines, and found that NEP activity in primary neuronal cultures are enhanced by the addition of the neuropeptide somatostatin (Saito et al. 2005). Somatostatin activates NEP activity by increasing the localization of NEP to the cell surface, which leads to an enhanced degradation of Aβ42. As described above, the localization of NEP is regulated by phosphorylation and dephosphorylation, which are mediated by MEK-ERK and PP1a, respectively. PP1a activity is in turn controlled by DARPP32 (Greengard et al. 1999), the levels of which are reduced in the brains of somatostatin receptor-1 (*SSTR1*) and *SSTR5* double-KO mice (Rajput et al. 2011). Therefore, somatostatin possibly regulates the phosphorylation statues and hence the NEP activity through PP1a. Together, these observations suggest that SSTR agonists are potential drugs as Aβ-lowering treatments for AD. SSTRs are 7-transmembrane G-protein-coupled receptors (GPCR) (Hoyer et al. 1995), and approximately 45 % of all pharmacological products act on receptors, most of which are GPCRs. Through our recent research, we have discovered that two SSTR subtypes mediate the somatostatin-induced activation of NEP. These findings might be a step forward in the development of a somatostatin-based treatment for AD.

19.3.3 Gene Therapy Using Neprilysin

Because a decline in NEP activity might contribute to the Aβ pathology of AD, the up-regulation of NEP would be a promising therapy for the prevention of AD. Indeed, several studies have shown that gene delivery of NEP can efficiently rescue impaired Aβ catabolism (reviewed in Nilsson et al. 2010). In an initial approach, we found that the expression of NEP from a sindbis viral vector in murine primary cortical neurons leads to significant decreases in intra- and extracellular Aβ40 and Aβ42 (Hama et al. 2001). Subsequent in vivo studies showed that the overexpression of NEP by intracerebral injection of a lentiviral vector expressing human NEP under the control of a CMV promoter reduces Aβ and ameliorates neurodegeneration in the frontal cortices and hippocampi of two APP transgenic mouse lines (Marr et al. 2003). Further, mice injected with an adeno-associated viral vector carrying the NEP gene (AAV-NEP) into their brains exhibit increased presynaptic NEP throughout the brain (Iwata et al. 2004). The overexpression of NEP significantly reduces the increased levels of Aβ in NEP-KO mice and reduces the formation of amyloid plaques in APP-transgenic mice. However, intracerebroventricular (ICV) injection is not a suitable method for gene delivery to humans in terms of safety and efficiency. Additionally, ICV injection leads to the local expression of the target. Taking the broadly distributed pathology of AD into account, we therefore constructed AAV vectors that mediate neuronal gene expression throughout the brain after peripheral administration (Iwata et al. 2013). This newly designed rAAV9 vector encoding NEP was introduced into NEP-deficient mice by intracardiac administration. The injection was performed into the left ventricle of the heart because this allows direct access to the brain through the internal carotid artery. Using this method to express NEP in AD model mice resulted in the elevation of NEP activity broadly across the brain, reductions in monomeric and oligomeric Aβ, and the amelioration of the abnormal learning and memory deficits induced by the amyloid burden (Fig. 19.3). This study is the first example of an innovative therapeutic strategy for neurodegenerative diseases in which the global transduction of a target gene into the brain via extracerebral injection was achieved. The use of modified NEPs, such as mutants with higher activities or NEP variants that are targeted to the cell membrane or bind Aβ with higher specificity, might further improve the efficiency of the method. Moreover, co-treatments with somatostatin agonists in combination with NEP-AAV might synergistically increase NEP activity and reduce Aβ pathology.

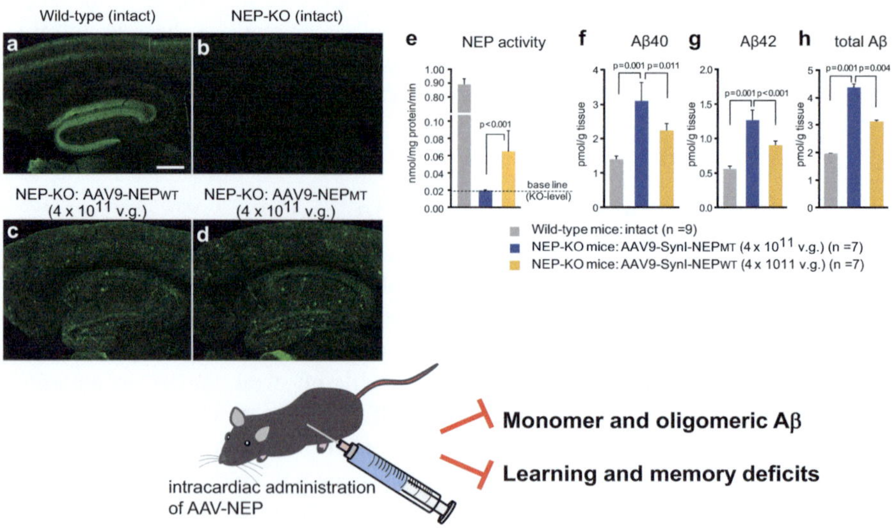

Fig. 19.3 Gene delivery of NEP decreases Aβ amyloidosis in vivo. Intracardiac injection of a rAAV9 vector encoding NEP under the control of the Synapsin I (SynI) promoter leads to a widespread gene transduction of wild-type NEP (NEP_{WT}) and non-active NEP mutant (NEP_{MT}) in the brain. Brain sections from wild-type mice (**a**), NEP-knockout (KO) (**b**), and NEP-KO mice 14 days after intracardiac injection of rAAV9 encoding wild-type (**c**) or mutant NEP (**d**) were stained for NEP. (**e**) NEP-activity in the limbic regions of wild-type mice, NEP-KO mice injected with rAAV9 encoding NEP_{MT} and NEP-KO mice injected with rAAV9 encoding NEP_{WT} 14 days after intracardiac injection. (**f–h**) The levels of Aβ40, Aβ42 and total Aβs in the limbic regions after NEP gene transfer were determined by sandwich ELISA. The data are presented as the mean ± s.e.m. of 7–9 mice. Scale bars, 200 μm

19.3.4 *Aβ Metabolism and Autophagy*

Together with the proteasome, autophagy is a protein quality control system in the cell that efficiently recycles cellular waste, such as aberrant organelles including depolarized mitochondria and toxic protein aggregates. This process generates free amino acids that are reused to synthesize new proteins. Autophagy is therefore a key mechanism in cellular metabolism and homeostasis. Increasing evidence points towards a dysfunctional protein quality control in AD that includes protein misfolding, ER stress, aberrant protein degradation, and impaired clearance of misfolded protein. Furthermore, the accumulation of autophagosomes in AD brains indicates that the autophagic system is disturbed, which might contribute to the impaired proteostasis in AD. Importantly, autophagy is also involved in the metabolism of Aβ (reviewed in Nilsson and Saido 2014). In addition to the previously established degradative role of Aβ by autophagy, we recently found that autophagy also mediates the intracellular sorting and secretion of Aβ to the extracellular space (Nilsson et al. 2013). Hence, autophagy directly influences Aβ plaque formation.

The initiation of autophagy is highly regulated by network signaling that responds to starvation, different nutrients, hypoxia, ER stress, and oxidative stress (Lipinski 2010). Oxidative stress is a particularly important part of AD pathology which includes abnormal metal ion homeostasis and the activation of glial cells that produce reactive oxygen species (ROS). Therefore, dysfunctions in the oxidative stress response system and the protein quality control system, including autophagy, might cause cell injury. Indeed, alterations in autophagy have been observed in various neurodegenerative diseases (reviewed in Nixon 2013). The main regulator that controls autophagy is mTOR complex 1 (mTORC1). Under stress conditions, such as starvation, mTORC1 is suppressed, which activates a cascade that involves the autophagy-related proteins (Atg) and leads to the formation of autophagosomes.

To investigate the contribution of autophagy to Aβ pathology, we previously generated an autophagy-deficient APP transgenic mouse by conditionally knocking out Atg7 in excitatory neurons in the mouse forebrain (Nilsson et al. 2013). Genetic manipulation of Atg7, which is one of the upstream proteins that is involved in the formation of autophagosomes, is commonly used to experimentally abrogate autophagy. Contrary to our expectations, the autophagy deficiency drastically reduced the extracellular Aβ plaque burden despite the inhibition of protein quality control mediated by autophagy (Fig. 19.4). After having thoroughly investigated the autophagy-deficient APP mice, we found that the reduction in Aβ plaque was due to the inhibition of Aβ secretion, which in turn led to intracellular Aβ accumulation in the golgi prior to Aβ plaque formation (Nilsson et al. 2014). Consistently, rapamycin, an inhibitor of mTORC1 that activates autophagy, induces Aβ secretion, while spautin-1, an inhibitor of autophagy, decreases Aβ secretion from cultured neurons.

Interestingly, the combination of autophagy-deficiency and intracellular Aβ accumulation cause neurodegeneration and severely impair the memories of the mice. Intracellular Aβ accumulation is also observed in AD brains. The

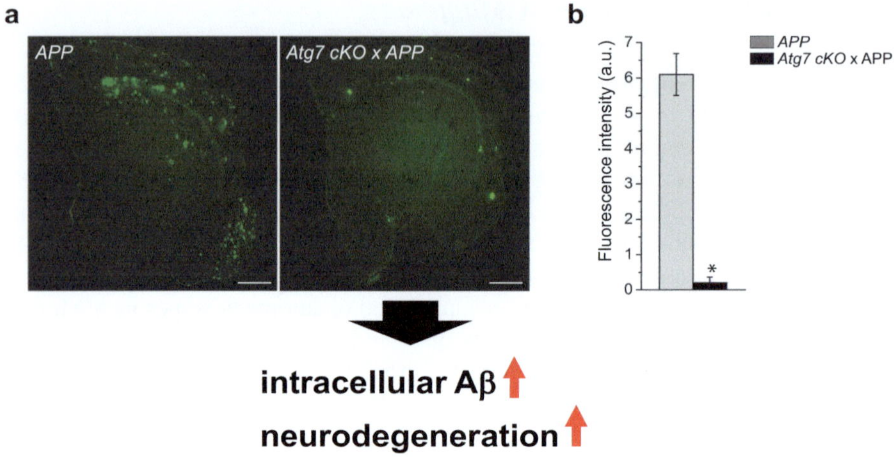

Fig. 19.4 Autophagy mediates the secretion of Aβ and thereby influences extracellular Aβ plaque formation. (**a**) Aβ plaque staining of brains of 15-month-old APP mouse and APP mouse with conditional knockout of *Atg7* (*Atg7* cKO x *APP*). (**b**) The fluorescence intensities were quantified. Note the drastic decrease in Aβ plaque load upon deletion of *Atg7*. The data are presented as the mean ± s.e.m., n = 4, $^*p < 0.005$

accumulation of Aβ in cells induces the production of ROS, which are toxic to lysosomes and mitochondria, and enhances cell death through lysosomal membrane permeabilization and the subsequent release of cathepsins (Wirths and Bayer 2012; Friedrich et al. 2010; Abramowski et al. 2012; Hedskog et al. 2012; Aits and Jaattela 2013). Further investigation is warranted to clarify how autophagy deficiency-induced intracellular Aβ causes neurodegeneration. Taking these results together, we propose that autophagy mediates Aβ sorting and secretion and that failure in the autophagic system causes neurodegeneration due to the intracellular accumulation of Aβ.

19.3.5 Aβ Clearance by Transport

In addition to degradation, secreted Aβ is cleared from the brain through the cerebrospinal fluid (CSF) and further transported to the vascular system. It has been demonstrated that ^{125}I-Aβ40 peptides injected to the mouse brain are transported to the blood with a half-time of approximately 25 min (Shibata et al. 2000). Aβ peptides are exported in a dose-dependent manner that can be saturated by high dose injections. These findings indicate that Aβ is cleared through active transport. The brain is separated from the blood by the blood brain barrier (BBB), which is localized to the brain capillaries and pia-subarachnoid membranes, and the CSF barrier, which is localized to the choroid plexus. These barriers consist of tight junctions between brain endothelial cells and epithelial cells. Therefore, Aβ needs to bind to receptors or other transport systems to cross these barriers. Thus, the system that transports Aβ from the brain parenchyma to blood consists of Aβ uptake from parenchymal to vascular endothelial cells and the subsequent export from the endothelial cells to the blood. The low-density lipoprotein receptor (LDLR) family proteins, which are cell surface receptors that recognize extracellular ligands, play a role in the active transport of Aβ from the brain to the blood or CSF. Low-density lipoprotein receptor-related protein 1 (LRP1), which is a member of the LDLR family, and very-low-density lipoprotein receptor (VLDL) are located in the membranes of endothelial cells and are involved in the uptake of Aβ from parenchymal to endothelial cells, while P-glycoprotein (also termed ATP binding cassette B1 (ABCB1)), ABCG2, and ABCC1 participate in the translocation of Aβ from endothelial cells to the blood (Yan et al. 2006; Deane et al. 2008; Cirrito et al. 2005; Xiong et al. 2009) (Fig. 19.5). P-glycoprotein, ABCG2 and ABCC1 are family members of ABC transporters.

LRP1 binds multiple ligands, including ApoE, α2-macroglobulin and Aβ, and is the main receptor for Aβ transport across the BBB (Herz 2001). The binding of Aβ to LRP1 on the abluminal side of the BBB facilitates Aβ clearance via transcytosis across the BBB into the blood (Shibata et al. 2000; Deane et al. 2004). A recent study suggested that Aβ binds to ApoE or α2-macroglobulin and that Aβ is cleared by the binding of these complexes to LRP1 (Koistinaho et al. 2004; Narita et al. 1997). Among the three different isoforms of ApoE that exist in humans, ApoE4 is a risk allele for AD. Interestingly, the binding of ApoE4 to Aβ reduces the

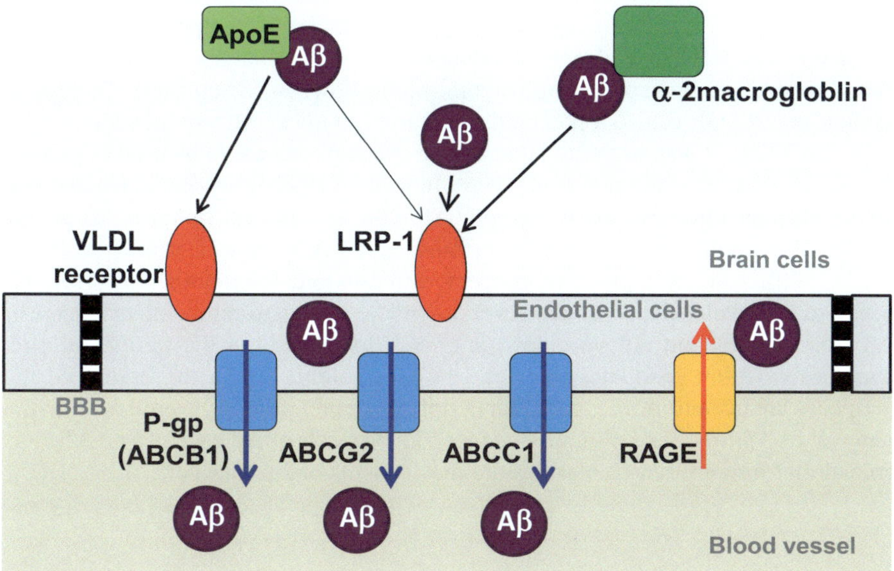

Fig. 19.5 Transport of Aβ over the BBB. Aβ is transported to the blood or CSF through active transport systems that are mediated by membrane receptors and transporters. LRP1 and VLDL are involved in the uptake of Aβ by endothelial cells, and ABC transporters facilitate export to the blood. RAGE regulates the import of Aβ (*P-gp* P-glycoprotein)

clearance of Aβ from the brain because the binding of ApoE4 to Aβ causes a shift from LRP1-mediated rapid efflux to VLDL receptor-mediated slow efflux that results in poor Aβ clearance from brain. Additionally, transport efficiency is further decreased by the binding of lipids to ApoE4. In contrast, ApoE2 and ApoE3 only moderately inhibit clearance due to the ApoE2/3-Aβ complex's ability to interact at least partially with LRP1 (Deane et al. 2008). Several groups have found that elderly people with less P-glycoprotein in their brain epithelia exhibited more Aβ plaque and more vascular Aβ (Vogelgesang et al. 2002, 2004). A recent study suggested that the deficiency of ABCC1 in APP/PS1 mice more strongly influences Aβ transport and accumulation than do deficiencies of P-glycoprotein or ABCG2 (Krohn et al. 2011). Moreover, several other Aβ transporters such as ABCA1 and ABCG4 have recently been identified, and additional research is required to clarify the whole picture of Aβ transport (Do et al. 2012; Jedlitschky et al. 2014).

In addition to the clearance of Aβ from the brain across the BBB, a system to import Aβ into the brain across the BBB also exists. The receptor for advanced glycation end products (RAGE) binds to a number of ligands in the blood, including Aβ, which it incorporates into the brain parenchyma (Yan et al. 1996). Interestingly, RAGE expression is altered depending on the levels of its ligands (Bierhaus et al. 2005). Upon the accumulation of pathogenic Aβ species in AD brains or in the brains of transgenic mouse models of amyloidosis, RAGE expression is upregulated in cerebral vessels, neurons and microglia (Yan et al. 1996; Deane et al. 2003; Miller et al. 2008). This mechanism potentially aggravates the pathology by increasing amyloidosis in the brain. The RAGE/Aβ interaction is also involved in neurovascular stress not only via mediating the import of circulating Aβ across the BBB but also by influencing the inflammatory responses of the endothelium, and by suppressing cerebral blood flow. Based on these findings, it is clear that RAGE is also a potential drug target to lower Aβ amyloidosis.

In the 1990s, it was suggested that an anti-Aβ antibody could be used to prevent Aβ fibril formation (Solomon et al. 1996). Indeed, in 1999, it was demonstrated that active immunization against full-length Aβ with an adjuvant reduced Aβ plaque burden in a PDAPP (APP/V717F-transgenic) AD mouse model (Schenk et al. 1999). After this proof of concept, anti-amyloid immunotherapies for AD began to be developed. There are several hypotheses regarding the mechanisms of Aβ clearance by the Aβ vaccines. One such hypothesis is the peripheral sink hypothesis, which postulates that the levels of soluble Aβ in the brain and the periphery are in equilibrium, such that peripheral depletion of Aβ would leads to the removal of Aβ from the brain. To understand the effects of Aβ vaccines on AD, it is crucial that future research clarifies the mechanisms of Aβ transport.

19.4 Future Perspectives

Elucidating the anabolic and catabolic systems of Aβ is of the utmost importance for the understanding of AD. Studies of mouse models of Aβ amyloidosis have provided important knowledge about Aβ metabolism. However, these previously generated models of AD are based on transgenic paradigms that overexpress APP to achieve high levels of Aβ. Due to the APP overexpression, fragments other than Aβ, such as sAPP, CTFα, CTFβ and AICD, are generated at non-physiologically high levels. These elevated levels might have affected the obtained results related to the pathways upstream and downstream of Aβ pathology and might have induced artificial phenotypes. To circumvent these intrinsic problems, we produced novel AD mouse models based on a knock-in strategy. The APP knock-in mice, which express APP at endogenous levels, exhibit a pronounced Aβ amyloidosis (Saito et al. 2014). These mice also exhibit neuroinflammation and impaired memory. Our intention is that these new mouse models will become the standard AD mouse models in AD research and will lead to better understandings of Aβ anabolism and catabolism and to the discovery of new drug targets for AD.

References

Abramowski D, Rabe S, Upadhaya AR, Reichwald J, Danner S, Staab D, Capetillo-Zarate E, Yamaguchi H, Saido TC, Wiederhold KH, Thal DR, Staufenbiel M (2012) Transgenic expression of intraneuronal Abeta42 but not Abeta40 leads to cellular Abeta lesions, degeneration, and functional impairment without typical Alzheimer's disease pathology. J Neurosci 32: 1273–1283

Aits S, Jaattela M (2013) Lysosomal cell death at a glance. J Cell Sci 126:1905–1912

Akiyama H, Kondo H, Ikeda K, Kato M, Mcgeer PL (2001) Immunohistochemical localization of neprilysin in the human cerebral cortex: inverse association with vulnerability to amyloid beta-protein (Abeta) deposition. Brain Res 902:277–281

Bae SJ, Matsunaga Y, Takenaka M, Tanaka Y, Hamazaki Y, Shimizu K, Katayama I (2002) Substance P induced preprotachykinin-a mRNA, neutral endopeptidase mRNA and substance P in cultured normal fibroblasts. Int Arch Allergy Immunol 127:316–321

Barnes K, Matsas R, Hooper NM, Turner AJ, Kenny AJ (1988) Endopeptidase-24.11 is striosomally ordered in pig brain, and in contrast to aminopeptidase N and peptidyl dipeptidase A ('angiotensin converting enzyme'), is a marker for a set of striatal efferent fibres. Neuroscience 27:799–817

Barnes K, Turner AJ, Kenny AJ (1993) An immunoelectron microscopic study of pig substantia nigra shows co-localization of endopeptidase-24.11 with substance P. Neuroscience 53:1073–1082

Bierhaus A, Humpert PM, Morcos M, Wendt T, Chavakis T, Arnold B, Stern DM, Nawroth PP (2005) Understanding RAGE, the receptor for advanced glycation end products. J Mol Med (Berl) 83:876–886

Bolos V, Grego-Bessa J, De La Pompa JL (2007) Notch signaling in development and cancer. Endocr Rev 28:339–363

Cao L, Rickenbacher GT, Rodriguez S, Moulia TW, Albers MW (2012) The precision of axon targeting of mouse olfactory sensory neurons requires the BACE1 protease. Sci Rep 2:231

Carvalho KM, Franca MS, Camarao GC, Ruchon AF (1997) A new brain metalloendopeptidase which degrades the Alzheimer beta-amyloid 1-40 peptide producing soluble fragments without neurotoxic effects. Braz J Med Biol Res 30:1153–1156

Cirrito JR, Deane R, Fagan AM, Spinner ML, Parsadanian M, Finn MB, Jiang H, Prior JL, Sagare A, Bales KR, Paul SM, Zlokovic BV, Piwnica-Worms D, Holtzman DM (2005) P-glycoprotein deficiency at the blood-brain barrier increases amyloid-beta deposition in an Alzheimer disease mouse model. J Clin Invest 115:3285–3290

Cisse M, Halabisky B, Harris J, Devidze N, Dubal DB, Sun B, Orr A, Lotz G, Kim DH, Hamto P, Ho K, Yu GQ, Mucke L (2011) Reversing EphB2 depletion rescues cognitive functions in Alzheimer model. Nature 469:47–52

Deane R, Du Yan S, Submamaryan RK, Larue B, Jovanovic S, Hogg E, Welch D, Manness L, Lin C, Yu J, Zhu H, Ghiso J, Frangione B, Stern A, Schmidt AM, Armstrong DL, Arnold B, Liliensiek B, Nawroth P, Hofman F, Kindy M, Stern D, Zlokovic B (2003) RAGE mediates amyloid-beta peptide transport across the blood-brain barrier and accumulation in brain. Nat Med 9:907–913

Deane R, Wu Z, Sagare A, Davis J, Du Yan S, Hamm K, Xu F, Parisi M, Larue B, Hu HW, Spijkers P, Guo H, Song X, Lenting PJ, Van Nostrand WE, Zlokovic BV (2004) LRP/amyloid beta-peptide interaction mediates differential brain efflux of Abeta isoforms. Neuron 43: 333–344

Deane R, Sagare A, Hamm K, Parisi M, Lane S, Finn MB, Holtzman DM, Zlokovic BV (2008) apoE isoform-specific disruption of amyloid beta peptide clearance from mouse brain. J Clin Invest 118:4002–4013

Do TM, Noel-Hudson MS, Ribes S, Besengez C, Smirnova M, Cisternino S, Buyse M, Calon F, Chimini G, Chacun H, Scherrmann JM, Farinotti R, Bourasset F (2012) ABCG2- and ABCG4-mediated efflux of amyloid-beta peptide 1-40 at the mouse blood-brain barrier. J Alzheimers Dis 30:155–166

Dominguez D, Tournoy J, Hartmann D, Huth T, Cryns K, Deforce S, Serneels L, Camacho IE, Marjaux E, Craessaerts K, Roebroek AJ, Schwake M, D'hooge R, Bach P, Kalinke U, Moechars D, Alzheimer C, Reiss K, Saftig P, De Strooper B (2005) Phenotypic and biochemical analyses of BACE1- and BACE2-deficient mice. J Biol Chem 280:30797–30806

Duckworth WC (1979) Insulin degradation by liver cell membranes. Endocrinology 104:1758–1764

Duckworth WC, Bennett RG, Hamel FG (1998) Insulin degradation: progress and potential. Endocr Rev 19:608–624

Eckman EA, Reed DK, Eckman CB (2001) Degradation of the Alzheimer's amyloid beta peptide by endothelin-converting enzyme. J Biol Chem 276:24540–24548

Eckman EA, Watson M, Marlow L, Sambamurti K, Eckman CB (2003) Alzheimer's disease beta-amyloid peptide is increased in mice deficient in endothelin-converting enzyme. J Biol Chem 278:2081–2084

Elkins JS, Douglas VC, Johnston SC (2004) Alzheimer disease risk and genetic variation in ACE: a meta-analysis. Neurology 62:363–368

Ertekin-Taner N, Ronald J, Feuk L, Prince J, Tucker M, Younkin L, Hella M, Jain S, Hackett A, Scanlin L, Kelly J, Kihiko-Ehman M, Neltner M, Hersh L, Kindy M, Markesbery W, Hutton M, De Andrade M, Petersen RC, Graff-Radford N, Estus S, Brookes AJ, Younkin SG (2005) Elevated amyloid beta protein (Abeta42) and late onset Alzheimer's disease are associated with single nucleotide polymorphisms in the urokinase-type plasminogen activator gene. Hum Mol Genet 14:447–460

Farris W, Mansourian S, Chang Y, Lindsley L, Eckman EA, Frosch MP, Eckman CB, Tanzi RE, Selkoe DJ, Guenette S (2003) Insulin-degrading enzyme regulates the levels of insulin, amyloid beta-protein, and the beta-amyloid precursor protein intracellular domain in vivo. Proc Natl Acad Sci U S A 100:4162–4167

Fossati S, Ghiso J, Rostagno A (2012) TRAIL death receptors DR4 and DR5 mediate cerebral microvascular endothelial cell apoptosis induced by oligomeric Alzheimer's Abeta. Cell Death Dis 3:e321

Friedrich RP, Tepper K, Ronicke R, Soom M, Westermann M, Reymann K, Kaether C, Fandrich M (2010) Mechanism of amyloid plaque formation suggests an intracellular basis of Abeta pathogenicity. Proc Natl Acad Sci U S A 107:1942–1947

Ghosh AK, Brindisi M, Tang J (2012) Developing beta-secretase inhibitors for treatment of Alzheimer's disease. J Neurochem 120(Suppl 1):71–83

Green RC, Schneider LS, Amato DA, Beelen AP, Wilcock G, Swabb EA, Zavitz KH, Tarenflurbil Phase 3 Study G (2009) Effect of tarenflurbil on cognitive decline and activities of daily living in patients with mild Alzheimer disease: a randomized controlled trial. JAMA 302:2557–2564

Greengard P, Allen PB, Nairn AC (1999) Beyond the dopamine receptor: the DARPP-32/protein phosphatase-1 cascade. Neuron 23:435–447

Guy JL, Lambert DW, Warner FJ, Hooper NM, Turner AJ (2005) Membrane-associated zinc peptidase families: comparing ACE and ACE2. Biochim Biophys Acta 1751:2–8

Hama E, Shirotani K, Masumoto H, Sekine-Aizawa Y, Aizawa H, Saido TC (2001) Clearance of extracellular and cell-associated amyloid beta peptide through viral expression of neprilysin in primary neurons. J Biochem 130:721–726

Hamazaki H (1996) Cathepsin D is involved in the clearance of Alzheimer's beta-amyloid protein. FEBS Lett 396:139–142

Hardy J, Selkoe DJ (2002) The amyloid hypothesis of Alzheimer's disease: progress and problems on the road to therapeutics. Science 297:353–356

Hedskog L, Zhang S, Ankarcrona M (2012) Strategic role for mitochondria in Alzheimer's disease and cancer. Antioxid Redox Signal 16:1476–1491

Hemming ML, Selkoe DJ (2005) Amyloid beta-protein is degraded by cellular angiotensin-converting enzyme (ACE) and elevated by an ACE inhibitor. J Biol Chem 280:37644–37650

Herz J (2001) The LDL receptor gene family: (un)expected signal transducers in the brain. Neuron 29:571–581

Hitt B, Riordan SM, Kukreja L, Eimer WA, Rajapaksha TW, Vassar R (2012) beta-Site amyloid precursor protein (APP)-cleaving enzyme 1 (BACE1)-deficient mice exhibit a close homolog of L1 (CHL1) loss-of-function phenotype involving axon guidance defects. J Biol Chem 287: 38408–38425

Hong L, Koelsch G, Lin X, Wu S, Terzyan S, Ghosh AK, Zhang XC, Tang J (2000) Structure of the protease domain of memapsin 2 (beta-secretase) complexed with inhibitor. Science 290: 150–153

Howell S, Caswell AM, Kenny AJ, Turner AJ (1993) Membrane peptidases on human osteoblast-like cells in culture: hydrolysis of calcitonin and hormonal regulation of endopeptidase-24.11. Biochem J 290(Pt 1):159–164

Hoyer D, Bell GI, Berelowitz M, Epelbaum J, Feniuk W, Humphrey PP, O'carroll AM, Patel YC, Schonbrunn A, Taylor JE et al (1995) Classification and nomenclature of somatostatin receptors. Trends Pharmacol Sci 16:86–88

Hu J, Igarashi A, Kamata M, Nakagawa H (2001) Angiotensin-converting enzyme degrades Alzheimer amyloid beta-peptide (A beta); retards A beta aggregation, deposition, fibril formation; and inhibits cytotoxicity. J Biol Chem 276:47863–47868

Hu X, Zhou X, He W, Yang J, Xiong W, Wong P, Wilson CG, Yan R (2010) BACE1 deficiency causes altered neuronal activity and neurodegeneration. J Neurosci 30:8819–8829

Hu X, He W, Luo X, Tsubota KE, Yan R (2013) BACE1 regulates hippocampal astrogenesis via the Jagged1-Notch pathway. Cell Rep 4:40–49

Iwata N, Tsubuki S, Takaki Y, Watanabe K, Sekiguchi M, Hosoki E, Kawashima-Morishima M, Lee HJ, Hama E, Sekine-Aizawa Y, Saido TC (2000) Identification of the major Abeta1-42-degrading catabolic pathway in brain parenchyma: suppression leads to biochemical and pathological deposition. Nat Med 6:143–150

Iwata N, Tsubuki S, Takaki Y, Shirotani K, Lu B, Gerard NP, Gerard C, Hama E, Lee HJ, Saido TC (2001) Metabolic regulation of brain Abeta by neprilysin. Science 292:1550–1552

Iwata N, Takaki Y, Fukami S, Tsubuki S, Saido TC (2002) Region-specific reduction of A beta-degrading endopeptidase, neprilysin, in mouse hippocampus upon aging. J Neurosci Res 70: 493–500

Iwata N, Mizukami H, Shirotani K, Takaki Y, Muramatsu S, Lu B, Gerard NP, Gerard C, Ozawa K, Saido TC (2004) Presynaptic localization of neprilysin contributes to efficient clearance of amyloid-beta peptide in mouse brain. J Neurosci 24:991–998

Iwata N, Sekiguchi M, Hattori Y, Takahashi A, Asai M, Ji B, Higuchi M, Staufenbiel M, Muramatsu S, Saido TC (2013) Global brain delivery of neprilysin gene by intravascular administration of AAV vector in mice. Sci Rep 3:1472

Jedlitschky G, Grube M, Mosyagin I, Kroemer HK, Vogelgesang S (2014) Targeting CNS transporters for treatment of neurodegenerative diseases. Curr Pharm Des 20:1523–1533

Jonsson T, Atwal JK, Steinberg S, Snaedal J, Jonsson PV, Bjornsson S, Stefansson H, Sulem P, Gudbjartsson D, Maloney J, Hoyte K, Gustafson A, Liu Y, Lu Y, Bhangale T, Graham RR, Huttenlocher J, Bjornsdottir G, Andreassen OA, Jonsson EG, Palotie A, Behrens TW, Magnusson OT, Kong A, Thorsteinsdottir U, Watts RJ, Stefansson K (2012) A mutation in APP protects against Alzheimer's disease and age-related cognitive decline. Nature 488:96–99

Kakiya N, Saito T, Nilsson P, Matsuba Y, Tsubuki S, Takei N, Nawa H, Saido TC (2012) Cell surface expression of the major amyloid-beta peptide (Abeta)-degrading enzyme, neprilysin, depends on phosphorylation by mitogen-activated protein kinase/extracellular signal-regulated kinase kinase (MEK) and dephosphorylation by protein phosphatase 1a. J Biol Chem 287: 29362–29372

Kehoe PG, Russ C, Mcilory S, Williams H, Holmans P, Holmes C, Liolitsa D, Vahidassr D, Powell J, Mcgleenon B, Liddell M, Plomin R, Dynan K, Williams N, Neal J, Cairns NJ, Wilcock G, Passmore P, Lovestone S, Williams J, Owen MJ (1999) Variation in DCP1, encoding ACE, is associated with susceptibility to Alzheimer disease. Nat Genet 21:71–72

Koistinaho M, Lin S, Wu X, Esterman M, Koger D, Hanson J, Higgs R, Liu F, Malkani S, Bales KR, Paul SM (2004) Apolipoprotein E promotes astrocyte colocalization and degradation of deposited amyloid-beta peptides. Nat Med 10:719–726

Krohn M, Lange C, Hofrichter J, Scheffler K, Stenzel J, Steffen J, Schumacher T, Bruning T, Plath AS, Alfen F, Schmidt A, Winter F, Rateitschak K, Wree A, Gsponer J, Walker LC, Pahnke J (2011) Cerebral amyloid-beta proteostasis is regulated by the membrane transport protein ABCC1 in mice. J Clin Invest 121:3924–3931

Kurochkin IV, Goto S (1994) Alzheimer's beta-amyloid peptide specifically interacts with and is degraded by insulin degrading enzyme. FEBS Lett 345:33–37

Laird FM, Cai H, Savonenko AV, Farah MH, He K, Melnikova T, Wen H, Chiang HC, Xu G, Koliatsos VE, Borchelt DR, Price DL, Lee HK, Wong PC (2005) BACE1, a major determinant of selective vulnerability of the brain to amyloid-beta amyloidogenesis, is essential for cognitive, emotional, and synaptic functions. J Neurosci 25:11693–11709

Lipinski MM (2010) Towards the global understanding of the autophagy regulatory network. Autophagy 6:1218–1220

Liu Q, Kawai H, Berg DK (2001) beta -Amyloid peptide blocks the response of alpha 7-containing nicotinic receptors on hippocampal neurons. Proc Natl Acad Sci U S A 98:4734–4739

Malfroy B, Swerts JP, Guyon A, Roques BP, Schwartz JC (1978) High-affinity enkephalin-degrading peptidase in brain is increased after morphine. Nature 276:523–526

Marr RA, Rockenstein E, Mukherjee A, Kindy MS, Hersh LB, Gage FH, Verma IM, Masliah E (2003) Neprilysin gene transfer reduces human amyloid pathology in transgenic mice. J Neurosci 23:1992–1996

Masaki T (2004) Historical review: endothelin. Trends Pharmacol Sci 25:219–224

Matsas R, Fulcher IS, Kenny AJ, Turner AJ (1983) Substance P and [Leu]enkephalin are hydrolyzed by an enzyme in pig caudate synaptic membranes that is identical with the endopeptidase of kidney microvilli. Proc Natl Acad Sci U S A 80:3111–3115

Mcdermott JR, Gibson AM (1997) Degradation of Alzheimer's beta-amyloid protein by human and rat brain peptidases: involvement of insulin-degrading enzyme. Neurochem Res 22:49–56

Miller BC, Eckman EA, Sambamurti K, Dobbs N, Chow KM, Eckman CB, Hersh LB, Thiele DL (2003) Amyloid-beta peptide levels in brain are inversely correlated with insulysin activity levels in vivo. Proc Natl Acad Sci U S A 100:6221–6226

Miller MC, Tavares R, Johanson CE, Hovanesian V, Donahue JE, Gonzalez L, Silverberg GD, Stopa EG (2008) Hippocampal RAGE immunoreactivity in early and advanced Alzheimer's disease. Brain Res 1230:273–280

Nakano S, Ikata T, Kinoshita I, Kanematsu J, Yasuoka S (1999a) Characteristics of the protease activity in synovial fluid from patients with rheumatoid arthritis and osteoarthritis. Clin Exp Rheumatol 17:161–170
Nakano Y, Kondoh G, Kudo T, Imaizumi K, Kato M, Miyazaki JI, Tohyama M, Takeda J, Takeda M (1999b) Accumulation of murine amyloidbeta42 in a gene-dosage-dependent manner in PS1 'knock-in' mice. Eur J Neurosci 11:2577–2581
Nakayama K, Nagase H, Koh CS, Ohkawara T (2011) gamma-Secretase-regulated mechanisms similar to notch signaling may play a role in signaling events, including APP signaling, which leads to Alzheimer's disease. Cell Mol Neurobiol 31:887–900
Narita M, Holtzman DM, Schwartz AL, Bu G (1997) Alpha2-macroglobulin complexes with and mediates the endocytosis of beta-amyloid peptide via cell surface low-density lipoprotein receptor-related protein. J Neurochem 69:1904–1911
Nilsson P, Saido TC (2014) Dual roles for autophagy: degradation and secretion of Alzheimer's disease Abeta peptide. Bioessays 36:570–578
Nilsson P, Iwata N, Muramatsu S, Tjernberg LO, Winblad B, Saido TC (2010) Gene therapy in Alzheimer's disease – potential for disease modification. J Cell Mol Med 14:741–757
Nilsson P, Loganathan K, Sekiguchi M, Matsuba Y, Hui K, Tsubuki S, Tanaka M, Iwata N, Saito T, Saido TC (2013) Abeta secretion and plaque formation depend on autophagy. Cell Rep 5:61–69.
Nilsson P, Sekiguchi M, Akagi T, Izumi S, Komori T, Hui K, Sorgjerd K, Tanaka M, Saito T, Iwata N, Saido TC (2015) Autophagy-related protein 7 deficiency in amyloid beta (Abeta) precursor protein transgenic mice decreases Abeta in the multivesicular bodies and induces Abeta accumulation in the Golgi. Am J Pathol 185:305–13
Nixon RA (2013) The role of autophagy in neurodegenerative disease. Nat Med 19:983–997
Ohno M, Sametsky EA, Younkin LH, Oakley H, Younkin SG, Citron M, Vassar R, Disterhoft JF (2004) BACE1 deficiency rescues memory deficits and cholinergic dysfunction in a mouse model of Alzheimer's disease. Neuron 41:27–33
Qiu WQ, Walsh DM, Ye Z, Vekrellis K, Zhang J, Podlisny MB, Rosner MR, Safavi A, Hersh LB, Selkoe DJ (1998) Insulin-degrading enzyme regulates extracellular levels of amyloid beta-protein by degradation. J Biol Chem 273:32730–32738
Rajput PS, Kharmate G, Norman M, Liu SH, Sastry BR, Brunicardi CF, Kumar U (2011) Somatostatin receptor 1 and 5 double knockout mice mimic neurochemical changes of Huntington's disease transgenic mice. PLoS One 6:e24467
Saido TC (2003) Overview-Abeta metabolism: from Alzheimer research to brain aging control. In: Saido TC (ed) Abeta metabolism and Alzheimer's disease. Landes Bioscience, Georgetown, USA
Saito T, Iwata N, Tsubuki S, Takaki Y, Takano J, Huang SM, Suemoto T, Higuchi M, Saido TC (2005) Somatostatin regulates brain amyloid beta peptide Abeta42 through modulation of proteolytic degradation. Nat Med 11:434–439
Saito T, Suemoto T, Brouwers N, Sleegers K, Funamoto S, Mihira N, Matsuba Y, Yamada K, Nilsson P, Takano J, Nishimura M, Iwata N, Van Broeckhoven C, Ihara Y, Saido TC (2011) Potent amyloidogenicity and pathogenicity of Abeta43. Nat Neurosci 14:1023–1032
Saito T, Matsuba Y, Mihira N, Takano J, Nilsson P, Itohara S, Iwata N, Saido TC (2014) Single app knock-in mouse models of Alzheimer's disease. Nat Neurosci 17:661–663
Sandebring A, Welander H, Winblad B, Graff C, Tjernberg LO (2013) The pathogenic abeta43 is enriched in familial and sporadic Alzheimer disease. PLoS One 8:e55847
Sasaki H, Saito Y, Hayashi M, Otsuka K, Niwa M (1988) Nucleotide sequence of the tissue-type plasminogen activator cDNA from human fetal lung cells. Nucleic Acids Res 16:5695
Savonenko AV, Melnikova T, Laird FM, Stewart KA, Price DL, Wong PC (2008) Alteration of BACE1-dependent NRG1/ErbB4 signaling and schizophrenia-like phenotypes in BACE1-null mice. Proc Natl Acad Sci U S A 105:5585–5590
Schenk D, Barbour R, Dunn W, Gordon G, Grajeda H, Guido T, Hu K, Huang J, Johnson-Wood K, Khan K, Kholodenko D, Lee M, Liao Z, Lieberburg I, Motter R, Mutter L, Soriano F, Shopp G,

Vasquez N, Vandevert C, Walker S, Wogulis M, Yednock T, Games D, Seubert P (1999) Immunization with amyloid-beta attenuates Alzheimer-disease-like pathology in the PDAPP mouse. Nature 400:173–177

Selkoe DJ (1991) The molecular pathology of Alzheimer's disease. Neuron 6:487–498

Selkoe DJ (2001) Clearing the brain's amyloid cobwebs. Neuron 32:177–180

Shibata M, Yamada S, Kumar SR, Calero M, Bading J, Frangione B, Holtzman DM, Miller CA, Strickland DK, Ghiso J, Zlokovic BV (2000) Clearance of Alzheimer's amyloid-ss(1-40) peptide from brain by LDL receptor-related protein-1 at the blood-brain barrier. J Clin Invest 106:1489–1499

Solomon B, Koppel R, Hanan E, Katzav T (1996) Monoclonal antibodies inhibit in vitro fibrillar aggregation of the Alzheimer beta-amyloid peptide. Proc Natl Acad Sci U S A 93:452–455

Tsubuki S, Takaki Y, Saido TC (2003) Dutch, Flemish, Italian, and Arctic mutations of APP and resistance of Abeta to physiologically relevant proteolytic degradation. Lancet 361:1957–1958

Turner AJ, Isaac RE, Coates D (2001) The neprilysin (NEP) family of zinc metalloendopeptidases: genomics and function. Bioessays 23:261–269

Vogelgesang S, Cascorbi I, Schroeder E, Pahnke J, Kroemer HK, Siegmund W, Kunert-Keil C, Walker LC, Warzok RW (2002) Deposition of Alzheimer's beta-amyloid is inversely correlated with P-glycoprotein expression in the brains of elderly non-demented humans. Pharmacogenetics 12:535–541

Vogelgesang S, Warzok RW, Cascorbi I, Kunert-Keil C, Schroeder E, Kroemer HK, Siegmund W, Walker LC, Pahnke J (2004) The role of P-glycoprotein in cerebral amyloid angiopathy; implications for the early pathogenesis of Alzheimer's disease. Curr Alzheimer Res 1:121–125

Waksman G, Hamel E, Fournie-Zaluski MC, Roques BP (1986) Autoradiographic comparison of the distribution of the neutral endopeptidase "enkephalinase" and of mu and delta opioid receptors in rat brain. Proc Natl Acad Sci U S A 83:1523–1527

Wang TL, Chang H, Hung CR, Tseng YZ (1997) Attenuation of neutrophil and endothelial activation by intravenous morphine in patients with acute myocardial infarction. Am J Cardiol 80:1532–1535

Wang HY, Lee DH, D'andrea MR, Peterson PA, Shank RP, Reitz AB (2000) beta-Amyloid(1-42) binds to alpha7 nicotinic acetylcholine receptor with high affinity. Implications for Alzheimer's disease pathology. J Biol Chem 275:5626–5632

Wang DS, Iwata N, Hama E, Saido TC, Dickson DW (2003) Oxidized neprilysin in aging and Alzheimer's disease brains. Biochem Biophys Res Commun 310:236–241

Welander H, Franberg J, Graff C, Sundstrom E, Winblad B, Tjernberg LO (2009) Abeta43 is more frequent than Abeta40 in amyloid plaque cores from Alzheimer disease brains. J Neurochem 110:697–706

Wirths O, Bayer TA (2012) Intraneuronal Abeta accumulation and neurodegeneration: lessons from transgenic models. Life Sci 91:1148–1152

Wolfe MS, Xia W, Ostaszewski BL, Diehl TS, Kimberly WT, Selkoe DJ (1999) Two transmembrane aspartates in presenilin-1 required for presenilin endoproteolysis and gamma-secretase activity. Nature 398:513–517

Wong-Leung YL, Kenny AJ (1968) Some properties of a microsomal peptidase in rat kidney. Biochem J 110:5P

Xiong H, Callaghan D, Jones A, Bai J, Rasquinha I, Smith C, Pei K, Walker D, Lue LF, Stanimirovic D, Zhang W (2009) ABCG2 is upregulated in Alzheimer's brain with cerebral amyloid angiopathy and may act as a gatekeeper at the blood-brain barrier for Abeta(1-40) peptides. J Neurosci 29:5463–5475

Yamada T, Kluve-Beckerman B, Liepnieks JJ, Benson MD (1995) In vitro degradation of serum amyloid A by cathepsin D and other acid proteases: possible protection against amyloid fibril formation. Scand J Immunol 41:570–574

Yan SD, Chen X, Fu J, Chen M, Zhu H, Roher A, Slattery T, Zhao L, Nagashima M, Morser J, Migheli A, Nawroth P, Stern D, Schmidt AM (1996) RAGE and amyloid-beta peptide neurotoxicity in Alzheimer's disease. Nature 382:685–691

Yan P, Hu X, Song H, Yin K, Bateman RJ, Cirrito JR, Xiao Q, Hsu FF, Turk JW, Xu J, Hsu CY, Holtzman DM, Lee JM (2006) Matrix metalloproteinase-9 degrades amyloid-beta fibrils in vitro and compact plaques in situ. J Biol Chem 281:24566–24574

Yasojima K, Akiyama H, Mcgeer EG, Mcgeer PL (2001a) Reduced neprilysin in high plaque areas of Alzheimer brain: a possible relationship to deficient degradation of beta-amyloid peptide. Neurosci Lett 297:97–100

Yasojima K, Mcgeer EG, Mcgeer PL (2001b) Relationship between beta amyloid peptide generating molecules and neprilysin in Alzheimer disease and normal brain. Brain Res 919:115–121

Zou K, Yamaguchi H, Akatsu H, Sakamoto T, Ko M, Mizoguchi K, Gong JS, Yu W, Yamamoto T, Kosaka K, Yanagisawa K, Michikawa M (2007) Angiotensin-converting enzyme converts amyloid beta-protein 1-42 (Abeta(1-42)) to Abeta(1-40), and its inhibition enhances brain Abeta deposition. J Neurosci 27:8628–8635

Chapter 20
Biomarkers in Alzheimer's Disease: From Pathogenic Initiation to Downstream Outcomes

Sun-Ho Han, Jong-Chan Park, and Inhee Mook-Jung

Abstract Alzheimer's disease (AD) is a progressive neurodegenerative disorder and is currently the most common form of dementia in the elderly population. Even though AD-related molecular alterations begin decades before the appearance of clinical symptoms, early diagnosis is not possible due to the absence of early diagnostic biomarkers. Early therapeutic treatment interventions are also lacking. Numerous clinical trial failures of promising therapeutic candidates have forced clinicians and researchers to identify diagnostic biomarkers for early detection. Here, we summarize promising biomarker candidates for AD that have been discovered to date, based on information from functional studies of neuropathological mechanisms and clinical research in AD. Potential biomarkers include cerebrospinal fluid (CSF)-based, blood-based, and immunological biomarkers as well as biomarkers detected by genetic profiling and neuroimaging. We discuss the use of these proposed biomarkers in practical applications to diagnose and initiate therapeutic treatment, along with their advantages and limitations.

Keywords Alzheimer's disease • Biomarker • Neurodegenerative disease • CSF • Amyloid β • Tau • Inflammation • Blood-brain barrier

20.1 Introduction

Among all the neurodegenerative disorders affecting the current elderly population, Alzheimer's disease (AD) has one of the most severe pathologies leading to memory loss and cognitive dysfunction. Sporadic AD occurs most frequently, and thus AD risk increases with age. The number of deaths due to AD has been drastically rising, and the number of patients with AD is expected to reach approximately 80 million in 2050 (Humpel 2011). The absence of diagnostic tools, particularly in the early course of the disease, and the lack of effective therapeutic

---

S.-H. Han • J.-C. Park • I. Mook-Jung (✉)
Department of Biochemistry and Biomedical Sciences, College of Medicine, Seoul National University, Seoul 110-799, South Korea
e-mail: inhee@snu.ac.kr

© Springer Japan 2015

N. Mori, I. Mook-Jung (eds.), *Aging Mechanisms*,
DOI 10.1007/978-4-431-55763-0_20

treatments are the biggest hurdles that must be overcome to prevent this predicted exponential increase. In contrast to the huge impact of AD on modern society, the history of AD research is relatively short. Research has revealed specific hallmarks of AD: extracellular plaques and intracellular tangles in the brain. Amyloid β (Aβ) and tau comprise these plaques and tangles, respectively, and were discovered in the mid-1980s. These two proteins are thought to be the main causative factor of AD. Subsequently, the amyloid cascade hypothesis supported the critical role of Aβ in the pathogenesis of AD and abnormal tau phosphorylation was demonstrated to be tightly related to AD (Alonso et al. 1994; Hampel et al. 2010; Hardy and Selkoe 2002).

A biomarker is an indicator of the status of a biological condition, including health, pathogenic processes, and medical intervention. Biomarker studies are valuable, not only as a diagnostic indicator, but also for determining disease risk factors or therapeutic targets. Discovering biomarkers for AD is critical because there is neither an accurate diagnostic tool nor a therapeutic treatment currently available. Reliable biomarkers could enable clinicians to distinguish patients from healthy individuals, as well as distinguish AD from other forms of neurodegenerative disease. Differentiating AD from other neurodegenerative diseases is fairly challenging because these diseases share many common biochemical features and symptoms, such as dementia. Biomarkers that could help clinicians to identify and distinguish different stages of AD progression are critical because drug treatment or disease-modifying strategies might vary during different stages. Furthermore, biomarkers have great potential as surrogate markers, which are essential to monitor the effectiveness of new drug development as well as disease initiation and progression.

20.2 Biomarkers for Alzheimer's disease

20.2.1 Cerebrospinal Fluid (CSF)-Based Biomarkers

Early diagnosis of AD in the course of disease progression is important for devising a therapeutic strategy. Further, a clinical method to differentiate AD from other types of neurodegenerative diseases that share similar symptoms is necessary to treat and modify disease progression. Thus, three AD biochemical hallmarks (Aβ, total tau, and phospho-tau-181) have attracted much attention and have emerged as promising biomarkers for early diagnosis. Specially, biomarker levels in the CSF reflect changes in the brain during AD pathogenesis (Fig. 20.1). CSF analysis has revealed a significant decrease in Aβ42 and an increase in both total tau and phospho-tau-181 in patients with AD compared to controls (cut-off values: Aβ42 $<$ 500 pg/mL, total tau $>$600 pg/mL, and phosphor-tau-181 $>$ 60 pg/mL) (Humpel 2011) (Table 20.1). Reduced clearance of brain Aβ42 into the blood/CSF and increased accumulation as plaque deposition in the brain leads to decreased CSF Aβ42. The Aβ42/Aβ40 ratio may be a better indicator for AD diagnosis, as one study demonstrated that CSF Aβ40 was increased or unchanged, while CSF Aβ42 was decreased in AD (Shoji 2002).

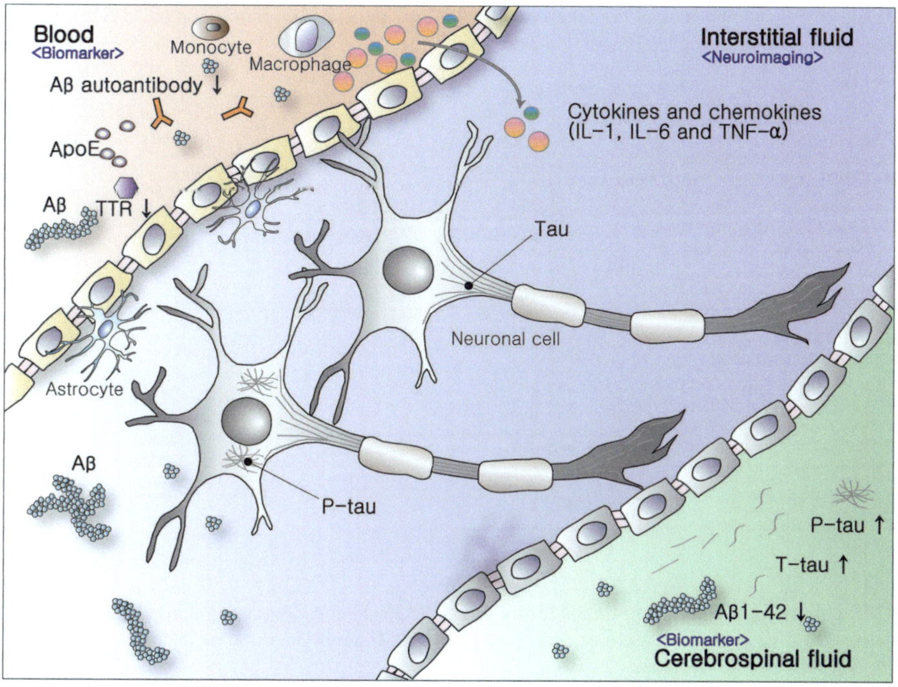

Fig. 20.1 Biomarker candidates during AD pathogenesis

However, other studies failed to demonstrate enhanced accuracy of diagnosis using the Aβ42/Aβ40 ratio (Schoonenboom et al. 2005). The total CSF tau level increases with age; however, a significantly high CSF level (cut-off value: 600 pg/mL) was observed in patients with AD compared to age-matched controls (Humpel 2011). The CSF tau level differentiated whether mild cognitive impairment (MCI) proceeded to AD or not: higher CSF tau level was detected in 90 % of patients with MCI who developed AD, but not in patients who did not (Blennow 2004; Riemenschneider et al. 2002). However, an increased total tau level in the CSF is also related to other types of neuronal damage, such as infarct size in acute stroke and Creutzfeldt-Jakob disease (Blennow 2005) (Table 20.1).

Tau has 39 possible phosphorylation sites and is hyperphosphorylated in AD. Specifically, tau hyperphosphorylation at position 181 is correlated with AD pathogenesis (Hampel et al. 2010), and the P-tau181/Aβ42 ratio discriminates patients with AD from both control subjects and individuals that have other types of dementia or neuropsychiatric diseases, such as frontotemporal lobe dementia, alcohol dementia, and major depression (Blasko et al. 2006). However, CSF growth factors and cytokines/chemokines were ineffective to discriminate patients with AD in this study. In CSF samples from patients with early AD and MCI-AD, decreased Aβ42, increased total tau, and increased tau phosphorylated at threonine 181 were detected; the CSF Aβ42 cutoff concentration for optimal determination of

Table 20.1 Changes in the CSF levels of Aβ1-42, total tau and phosphorylated tau in Alzheimer's disease and other neurodegenerative diseases

Diagnosis	CSF biomarkers Aβ1-42 (pg/ml)	Total tau (pg/ml, age)	Phosphorylated tau (pg/ml)
Control	794 ± 20	136 ± 89, (21–50 years)	23 ± 2
		243 ± 127, (51–70 years)	
		341 ± 171, (>71 years)	
AD	↓ (<500)	↑ (>450, 51–70 years; >600, >71 years)	↑ (>60)
Depression	–	–	–
Neuroinflammation	↓	–	–
Alcohol dementia	–	–	–
PD	–	–	–
CJD	↓↓	↑↑↑	–
FTD	↓↓	↑	–
LBD	↓↓	↑	↑
VaD	↓↓	↑	–
Acute stroke	–	↑↑	–

This table was modified from the tables by Humpel (2011) and Zetterberg et al. (2010)
– Normal, ↑ or ↓ Mild increase or decrease, ↑↑ or ↓↓ Moderate increase or decrease, ↑↑↑ or ↓↓↓ Marked increase or decrease
CSF cerebrospinal fluid, *Aβ1-42*, amyloid-β, *AD* alzheimer's disease, *PD* Parkinson's diasease, *CJD* Creutzfeldt-Jakob disease, *FTD* frontotemporal dementia, *LBD* Lewy body dementia, *VaD* vascular dementia

early AD as opposed to other types of dementia was 197.7 pg/mL (Lewczuk et al. 2008). The combinatorial detection of these biomarkers along with other diagnostic tools for early detection of AD pathogenesis also reportedly distinguished incipient AD in MCI patients (Blennow 2004; Zetterberg et al. 2003). Other forms of phosphorylated tau (e.g., phosphor tau-199, 231, 235, 396, and 404) may be potential biomarkers that implicate AD. Specifically, phosphor-tau-181 and phosphor-tau-231 distinguish AD from other types of psychiatric disorders, such as frontotemporal lobe dementia, Lewy body dementia (LBD), vascular dementia, and major depression (Cedazo-Minguez and Winblad 2010; Hampel et al. 2010).

20.2.2 Blood-Based Biomarkers

Although some AD-related biomarkers in the CSF are highly specific, CSF sample collection is invasive. This is particularly a problem for elderly patients who must repeatedly undergo the procedure. Thus, clinicians and researchers are desperately seeking a noninvasive, routine way to detect biomarkers. Researchers first attempted to detect known CSF biomarkers, including Aβ, in blood. However, detection of a reliable amyloid biomarker in blood has been problematic, as

Table 20.2 Summary of biomarker candidates tested in Alzheimer's disease

Categories	Biomarkers	Phenomenon	Analytical methods	References
CSF-based biomarkers	Aβ1-42	↓	ELISA, INNO-BIA AlzBio3 Luminex-based technology	Humpel (2011), Shoji (2002), and Blennow (2005)
	Total tau	↑		
	Phosphorylated tau 181	↑		
Blood-based biomarkers	Aβ	Diverse results	ELISA	Cedazo-Minguez and Winblad (2010) and Zetterberg et al. (2010)
	ApoE	↑	Western blotting	Taddei et al. (1997)
		–	ELISA	Slooter et al. (1998)
		↓	ELISA, Immunoturbidimetry, mass spectrometry and flow sorting analysis	Vliet et al. (2009), Siest et al. (2000), and Han et al. (2014)
	Adhesion molecules (VCAM-1, ICAM-1)	↑	ELISA	Ewers et al. (2010)
	Transthyretin	↓	ELISA, Western blotting	Han et al. (2011) and Velayudhan et al. (2012)
Immunological biomarkers	Peripheral blood cells	Alteration of distribution or reactivity	Flow cytometry, IHC and two photon microscopy	Richartz-Salzburger et al. (2007), Fiala et al. (2005), and Baik et al. (2014)
		Cell adhesion molecules ratio (ICAM-3/CD14, P-selectin/CD14) ↓	Multiplex ELISAs	Hochstrasser et al. (2010)
		Apoptosis ↑, antiapoptotic proteins and antioxidant caspase ↑	Flow cytometry	Ankarcrona and Winblad (2005), Bergman et al. (2002), Leuner et al. (2007), and Schindowski et al. (2006)

(continued)

Table 20.2 (continued)

Categories	Biomarkers	Phenomenon	Analytical methods	References
	Cytokines and chemokines (IL-1, IL-6 and TNF-α)	↑	ELISA	Licastro et al. (2000) and Rubio-Perez and Morillas-Ruiz (2012)
	Aβ auto-antibodies	↓	Immunoprecipitation assay	Brettschneider et al. (2005)
Genetic variation	ApoE4	Significant correlation to AD	Polymerase chain reaction	Corder et al. (1993)
	CLU and PICALM		GWAS	Harold et al. (2009)
	TOMM40			Ferencz et al. (2012)
	TREM2			Guerreiro et al. (2013)
Neuroimaging	Brain volume	↓	MRI	Petrella et al. (2003) and Matsuda (2007)
	Cerebral blood flow	↓	SPECT	
	Glucose metabolism	↓	FDG-PET	
	Brain Aβ plaques	↑	PiB-PET	Kemppainen et al. (2007), Klunk et al. (2004), and Clark et al. (2011)

– Normal, ↑ Increase, ↓ Decrease
Aβ amyloid-β, *ELISA* Enzyme-linked immunosorbent assay, *ApoE* Apolipoprotein E, *VCAM-1* vascular cell adhesion molecule 1, *ICAM-1* intrercellular adhesion molecule 1, *IL* interleukin, *TNF-α* tumor necrosis factor-α, *CLU* clusterin, *PICALM* phosphatidylinositol-binding clathrin assembly protein, *TOMM40* translocase of outer mitochondrial membrane 40, *TREM2* triggering receptor expressed on myeloid cells 2, *GWAS* genome-wide association study, *MRI* magnetic resonance imaging, *SPECT* single-photon emission computed tomography, *FDG-PET* fludeoxyglucose positron emission tomography, *PiB-PET* Pittsburgh compound B positron emission tomography

previous studies on blood Aβ level have been inconsistent and conflicting. These studies reported AD-related increases, decreases, and no change of Aβ in blood (Cedazo-Minguez and Winblad 2010; Zetterberg et al. 2010). Blood Aβ level is fairly unstable because blood is a heterogeneous environment, which includes many factors that influence Aβ concentration, including binding/aggregation status in blood through direct and indirect binding, regulation of expression, and metabolism of Aβ. In addition, diverse quantification of Aβ detection methods may affect Aβ levels in peripheral blood. For this reason, researchers tried to expand the number of potential blood biomarker candidates and have diverted their attention to other targets indirectly related to AD pathogenesis (Table 20.2).

Aβ-interacting proteins were the next target for the detection of AD biomarker candidates in blood, especially the proteins that regulate the lifespan of Aβ, including Aβ production, aggregation, and degradation. Aβ-interacting proteins play diverse roles in AD pathogenesis since Aβ elicits its toxicity through binding or unbinding to numerous partners. Apolipoprotein E (ApoE) polymorphism has been recognized as an indicator for the risk of AD (Corder et al. 1993; Tsai et al. 1994) and ApoE is also known to bind to Aβ and to be involved in Aβ aggregation and clearance. Conflicting results were reported regarding blood ApoE level, including an AD-related increase (Taddei et al. 1997), AD-related decrease (Siest et al. 2000; van Vliet et al. 2009) and no change (Slooter et al. 1998). A recent study overcame methodological variation and used two distinct quantification methods (i.e., targeted mass spectrometry and flow sorting analysis) to demonstrate an AD-related decrease in blood ApoE level (Han et al. 2014). Transthyretin (TTR) is another well-characterized protein in AD because it binds to Aβ, subsequently affecting Aβ aggregation status and degradation (Costa et al. 2008; Schwarzman and Goldgaber 1996). AD-related decrease of TTR level has been reported in blood and the correlation has been shown between decreased TTR level in blood and pathological severity of AD determined using the Mini-Mental State Examination (MMSE) (Han et al. 2011; Velayudhan et al. 2012).

In addition to components of AD clinical hallmarks, other signaling proteins may be possible biomarkers for AD. AD pathogenesis involves diverse molecular mechanisms and different signaling pathways undergo dynamic alterations. Interestingly, 18 signaling proteins detected in plasma samples were investigated to identify patients with MCI who progressed to AD within 2–6 years, with 90 % accuracy (Ray et al. 2007). These 18 signaling proteins were related to hematopoiesis, immune response, apoptosis, and neuronal support; therefore, the relevance of these proteins to AD pathogenesis implies systemic dysregulation during disease progress. This study suggested the manifestation of neurodegeneration in the peripheral environment. Further investigation using blood mononuclear cell transcriptome expression profiling analysis in subjects with sporadic AD demonstrated systemic gene expression alterations (Maes et al. 2007). Significantly decreased gene expression related to cytoskeletal maintenance, cellular stress response, trafficking, redox homeostasis, transcription, and DNA repair was observed (Maes et al. 2007). However, subsequent studies on some of those 18 blood biomarkers failed to replicate the initial findings, and lower sensitivity and specificity were observed (Marksteiner et al. 2011). Alteration of the molecular mechanism underlying AD pathology has been utilized as possible biomarkers in blood level. Microvascular abnormalities and damage are well known pathological phenomena in AD, particularly in the cerebrovascular systems such as the blood-brain barrier (BBB). Analysis of microvascular biomarker candidates using diverse vasodilators, vasoconstrictors, and adhesion molecules showed increased blood levels of vascular adhesion molecule-1 (VCAM-1) and intercellular adhesion molecule-1 (ICAM-1) in patients with AD (Ewers et al. 2010). Endothelial vasodilatory function of endothelin (ET-1), adrenomedullin, atrial natriuretic peptide, and sphingolipids were significantly modified in mild AD and in the predementia stage of MCI, suggesting the possibility that these molecules could be early diagnostic biomarker candidates.

20.2.3 Immunological Biomarkers

Apart from traditional amyloid plaque and tangle neuropathology, immune response activation and neuroinflammation contribute to the AD pathological process. Local immune response activation has been reported in the AD brain, including glial and astrocyte activation and secretion of diverse mediators from these activated immune cells, which accelerated the disease progress (Akiyama et al. 2000). In addition, brain inflammation occurs in AD, including upregulation of cytokines, complement cascades, acute phase proteins, histamine, and arachidonic metabolites (Akiyama et al. 2000; Eddleston and Mucke 1993; Gasque et al. 2000). Hence, molecules involved in immune activation during AD pathogenesis could be a potential biomarker at the peripheral level, as a growing number of studies suggest synchronization of immune activation between central and peripheral systems in AD. Altered distribution or reactivity of peripheral blood cells (Richartz-Salzburger et al. 2007) and abnormal chemokine and cytokine production also (Rubio-Perez and Morillas-Ruiz 2012) occur in AD.

Levels of various monocyte cell adhesion molecules reportedly change during AD pathogenesis, resulting in a decreased ICAM-3/cluster of differentiation 14 (CD14) ratio in monocytes from AD and MCI and decreased ratio of P-selectin/CD14 in monocytes from AD but not in monocytes from MCI (Hochstrasser et al. 2010). Defective Aβ42 clearance in AD results from altered monocyte functions, including poor differentiation into macrophages and surface-only Aβ uptake and human leukocyte antigen DR and cyclooxygenase-2 were abnormally expressed on neutrophils and monocytes in AD (Fiala et al. 2005). The combinatorial detection of immune mediators and other molecules, such as the combination of the interleukin-6 (IL-6) receptor, protein α1 fraction, cysteine, and cholesterol, was beneficial in discriminating patients with AD from control subjects (Teunissen et al. 2003). Other immune-related findings in patients with AD include: increased sensitivity of peripheral blood cells to apoptosis and increased apoptosis of CD4+ T cells and natural killer cells with increased expression of the anti-apoptotic protein B-cell lymphoma 2 (Bcl-2), the antioxidant enzyme superoxide dismutase 1 (SOD1), and different caspase subtypes (Ankarcrona and Winblad 2005; Bergman et al. 2002; Leuner et al. 2007; Schindowski et al. 2006). Infiltration of activated peripheral monocytes into the brain was recently demonstrated via two-photon in vivo imaging in a mouse model of AD, illustrating the concept that central and peripheral immune activation is synchronized and that activated peripheral monocytes play a role in the AD pathological process (Baik et al. 2014).

In addition to immune cell activation, mediators secreted by activated immune cells, such as cytokines and chemokines, play a critical role in AD pathogenesis. IL-1, IL-6, and tumor necrosis factor (TNF)-α play critical roles in the inflammatory cascade in AD, and their levels were found to be increased in the plasma from patients with AD (Licastro et al. 2000; Rubio-Perez and Morillas-Ruiz 2012). Furthermore, certain cytokines and chemokines are involved in Aβ production and metabolism, which generates a vicious cycle during AD pathogenesis (Blasko et al. 1999). Antigen/antibody reactions occur against Aβ; most neural plaques in patients with AD were found to be surrounded by naturally occurring autoantibodies against Aβ (Kellner et al. 2009). As expected, decreased Aβ42 autoantibodies were detected in AD serum (Brettschneider et al. 2005). Thus, Aβ autoantibody detection may serve as an effective biomarker for AD (Humpel 2011; Kellner et al. 2009).

20.2.4 Genetic Profiling and Neuroimaging

Genetic variation could be used as diagnostic tool and biomarker in late-onset AD. ApoE ε4 is the most prominent polymorphism risk factor of late-onset AD (Corder et al. 1993). A recent AD genome-wide association study (GWAS) suggested an association between some genetic variations with increased risk of late-onset AD, including variation in the CLU gene and 5′ region of the PICALM gene (Harold et al. 2009). In addition, the significant association of translocase of outer mitochondrial membrane 40 (*TOMM40*) with late-onset AD suggested that mitochondrial dysfunction is relevant to the pathophysiology of this disease (Ferencz et al. 2012). The significant association of triggering receptor expressed on myeloid cells 2 (*TREM2*) variants with AD has attracted much attention; the differential expression of TREM2 in the brain of an AD mouse model suggested its potential as a diagnostic biomarker (Guerreiro et al. 2013).

Gene expression profiling is promising for the diagnosis of AD as well, because altered gene expression profiles reflect both increased genetic variation and the molecular mechanism of pathogenesis. Using quantitative real-time reverse transcription polymerase chain reaction (qRT-PCR), a study investigated the RNA expression profile of 33 genes in white blood cell samples from patients with AD and healthy controls. This study revealed that five genes were significantly correlated with the degree of dementia, as determined using the MMSE (Grunblatt et al. 2009). Among these five, H3-histone and cannabinoid-receptor-2 represented the strongest correlation with dementia.

Diverse neuroimaging techniques have been used to obtain diagnostic and prognostic markers for AD: magnetic resonance imaging (MRI), single-photon emission computed tomography (SPECT), and positron emission tomography (PET). Structural MRI and SPECT and ^{18}F-fludeoxyglucose (FDG) PET functional evidence are good indicators of AD and useful in diagnosing early stage AD, in predicting MCI conversion to AD, and in estimating the rapidness of disease progression (Matsuda 2007). MRI and CT have been utilized for routine evaluation of AD, while volumetric and subtraction MR techniques have been used to monitor dementia progression (Petrella et al. 2003). Functional imaging, including SPECT and PET, have been used to diagnose dementia; functional imaging was especially useful in distinguishing AD from other types of dementia, such as vascular dementia, frontotemporal dementia, LBD, and depression, and was suitable for early diagnosis since it detected fairly subtle changes in pathologic processes (Petrella et al. 2003).

Development of a new, sensitive Aβ ligand, Pittsburgh compound B (PiB), enabled advanced neuroimaging in AD by allowing amyloid deposit quantification in the brain of living subjects (Kemppainen et al. 2007; Klunk et al. 2004). Technical developments have resulted in production of new advanced Aβ ligands, such as florbetapir, which accurately and reliably correlates with Aβ density (Clark et al. 2011). Furthermore, neuroimaging initiatives have begun a new era in AD research and have greatly contributed to neuroimaging development for early detection and biomarker monitoring in AD. The Alzheimer's disease Neuroimaging Initiative (ADNI) is a longitudinal multicenter study that has put much effort into developing standardized and optimized methods for MRI and PET imaging and analysis for AD diagnosis, as well as clinical tests and CSF biomarkers (Weiner et al. 2012).

20.3 Application of Biomarker Candidates to Diagnosis and Therapeutic Treatments

AD biomarker candidate studies are valuable not only for diagnostic purposes, but also for drug development, with the hopes of modifying disease initiation and progression. Further, biomarkers can be used as surrogate markers to monitor drug treatment efficiency and efficacy during different stages of disease progression. Tremendous effort has been devoted to obtain an accurate diagnostic biomarker for AD; quantification of CSF Aβ42, total tau, and phosphorylated tau are the most promising candidates to date. However, CSF sample collection is a painful procedure and noninvasive biomarkers for AD diagnosis are desperately needed (e.g., biomarkers in blood, urine or saliva).

A single biomarker would likely not be sufficient to diagnose AD accurately. To detect multiple biomarkers more effectively, a combinatorial approach would be more sensitive and specific. Thus, biomarker selection for combinational detection is critical because individual biomarkers must be specific enough to differentiate AD pathology and be synergistic for maximum AD diagnosis accuracy. Different biomarker combinations could be used for various goals in diverse disease progression stages. Certain combinations may be suitable for early detection of AD, and other combinations may be more suitable for predicting the possibility or rapidness of conversion from MCI to AD (Thambisetty and Lovestone 2010). Combinatorial detection using neuroimaging and blood/CSF biomarkers may be optimal as an AD diagnostic and prognostic method, as expected in the combinational detection of CSF tau, phosphorylated tau, MRI and Aβ PET (Weiner et al. 2012).

Quantification method is another critical factor in biomarker detection. Studies on identical biomarker targets produce inconsistent results, possibly resulting from differential quantification methods. Enzyme-linked immunosorbent assay (ELISA), bead-based multiplex detection, and PCR are widely used for biomarker quantification in body fluid samples. Target verification using more than one method is necessary, and a worldwide, standardized biomarker method is prerequisite for a future diagnostic platform for any newly discovered biomarker. In the future, highly sensitive proteomic analysis, such as mass spectrometry and microarray, may be useful for AD diagnosis to improve methodological specificity and sensitivity, enable early AD diagnosis, and to determine a better treatment strategy.

References

Akiyama H, Barger S, Barnum S, Bradt B, Bauer J, Cole GM, Cooper NR, Eikelenboom P, Emmerling M, Fiebich BL, Finch CE, Frautschy S, Griffin WS, Hampel H, Hull M, Landreth G, Lue L, Mrak R, Mackenzie IR, McGeer PL, O'Banion MK, Pachter J, Pasinetti G, Plata-Salaman C, Rogers J, Rydel R, Shen Y, Streit W, Strohmeyer R, Tooyoma I, Van Muiswinkel FL, Veerhuis R, Walker D, Webster S, Wegrzyniak B, Wenk G, Wyss-Coray T (2000) Inflammation and Alzheimer's disease. Neurobiol Aging 21:383–421

Alonso AC, Zaidi T, Grundke-Iqbal I, Iqbal K (1994) Role of abnormally phosphorylated tau in the breakdown of microtubules in Alzheimer disease. Proc Natl Acad Sci U S A 91:5562–5566

Ankarcrona M, Winblad B (2005) Biomarkers for apoptosis in Alzheimer's disease. Int J Geriatr Psychiatry 20:101–105

Baik SH, Cha MY, Hyun YM, Cho H, Hamza B, Kim DK, Han SH, Choi H, Kim KH, Moon M, Lee J, Kim M, Irimia D, Mook-Jung I (2014) Migration of neutrophils targeting amyloid plaques in Alzheimer's disease mouse model. Neurobiol Aging 35:1286–1292

Bergman M, Salman H, Beloosesky Y, Djaldetti M, Bessler H (2002) Are peripheral blood cells from patients with Alzheimer disease more sensitive to apoptotic stimuli? Alzheimer Dis Assoc Disord 16:156–160

Blasko I, Lederer W, Oberbauer H, Walch T, Kemmler G, Hinterhuber H, Marksteiner J, Humpel C (2006) Measurement of thirteen biological markers in CSF of patients with Alzheimer's disease and other dementias. Dement Geriatr Cogn Disord 21:9–15

Blasko I, Marx F, Steiner E, Hartmann T, Grubeck-Loebenstein B (1999) TNFalpha plus IFNgamma induce the production of Alzheimer beta-amyloid peptides and decrease the secretion of APPs. FASEB J 13:63–68

Blennow K (2004) CSF biomarkers for mild cognitive impairment. J Intern Med 256:224–234

Blennow K (2005) CSF biomarkers for Alzheimer's disease: use in early diagnosis and evaluation of drug treatment. Expert Rev Mol Diagn 5:661–672

Brettschneider S, Morgenthaler NG, Teipel SJ, Fischer-Schulz C, Burger K, Dodel R, Du Y, Moller HJ, Bergmann A, Hampel H (2005) Decreased serum amyloid beta(1-42) autoantibody levels in Alzheimer's disease, determined by a newly developed immuno-precipitation assay with radiolabeled amyloid beta(1-42) peptide. Biol Psychiatry 57:813–816

Cedazo-Minguez A, Winblad B (2010) Biomarkers for Alzheimer's disease and other forms of dementia: clinical needs, limitations and future aspects. Exp Gerontol 45:5–14

Clark CM, Schneider JA, Bedell BJ, Beach TG, Bilker WB, Mintun MA, Pontecorvo MJ, Hefti F, Carpenter AP, Flitter ML, Krautkramer MJ, Kung HF, Coleman RE, Doraiswamy PM, Fleisher AS, Sabbagh MN, Sadowsky CH, Reiman EP, Zehntner SP, Skovronsky DM, AV45-A07

Study Group (2011) Use of florbetapir-PET for imaging beta-amyloid pathology. JAMA 305:275–283

Corder EH, Saunders AM, Strittmatter WJ, Schmechel DE, Gaskell PC, Small GW, Roses AD, Haines JL, Pericak-Vance MA (1993) Gene dose of apolipoprotein E type 4 allele and the risk of Alzheimer's disease in late onset families. Science 261:921–923

Costa R, Ferreira-da-Silva F, Saraiva MJ, Cardoso I (2008) Transthyretin protects against A-beta peptide toxicity by proteolytic cleavage of the peptide: a mechanism sensitive to the Kunitz protease inhibitor. PLoS One 3:e2899

Eddleston M, Mucke L (1993) Molecular profile of reactive astrocytes – implications for their role in neurologic disease. Neuroscience 54:15–36

Ewers M, Mielke MM, Hampel H (2010) Blood-based biomarkers of microvascular pathology in Alzheimer's disease. Exp Gerontol 45:75–79

Ferencz B, Karlsson S, Kalpouzos G (2012) Promising genetic biomarkers of preclinical Alzheimer's disease: the influence of APOE and TOMM40 on brain integrity. Int J Alzheimers Dis 2012:421452

Fiala M, Lin J, Ringman J, Kermani-Arab V, Tsao G, Patel A, Lossinsky AS, Graves MC, Gustavson A, Sayre J, Sofroni E, Suarez T, Chiappelli F, Bernard G (2005) Ineffective phagocytosis of amyloid-beta by macrophages of Alzheimer's disease patients. J Alzheimers Dis 7:221–232; discussion 255–262

Gasque P, Dean YD, McGreal EP, VanBeek J, Morgan BP (2000) Complement components of the innate immune system in health and disease in the CNS. Immunopharmacology 49:171–186

Grunblatt E, Bartl J, Zehetmayer S, Ringel TM, Bauer P, Riederer P, Jacob CP (2009) Gene expression as peripheral biomarkers for sporadic Alzheimer's disease. J Alzheimers Dis 16:627–634

Guerreiro R, Wojtas A, Bras J, Carrasquillo M, Rogaeva E, Majounie E, Cruchaga C, Sassi C, Kauwe JS, Younkin S, Hazrati L, Collinge J, Pocock J, Lashley T, Williams J, Lambert JC, Amouyel P, Goate A, Rademakers R, Morgan K, Powell J, St George-Hyslop P, Singleton A, Hardy J, Alzheimer Genetic Analysis Group (2013) TREM2 variants in Alzheimer's disease. N Engl J Med 368:117–127

Hampel H, Blennow K, Shaw LM, Hoessler YC, Zetterberg H, Trojanowski JQ (2010) Total and phosphorylated tau protein as biological markers of Alzheimer's disease. Exp Gerontol 45:30–40

Han SH, Jung ES, Sohn JH, Hong HJ, Hong HS, Kim JW, Na DL, Kim M, Kim H, Ha HJ, Kim YH, Huh N, Jung MW, Mook-Jung I (2011) Human serum transthyretin levels correlate inversely with Alzheimer's disease. J Alzheimers Dis 25:77–84

Han SH, Kim JS, Lee Y, Choi H, Kim JW, Na DL, Yang EG, Yu MH, Hwang D, Lee C, Mook-Jung I (2014) Both targeted mass spectrometry and flow sorting analysis methods detected the decreased serum apolipoprotein E level in Alzheimer's disease patients. Mol Cell Proteomics 13:407–419

Hardy J, Selkoe DJ (2002) The amyloid hypothesis of Alzheimer's disease: progress and problems on the road to therapeutics. Science 297:353–356

Harold D, Abraham R, Hollingworth P, Sims R, Gerrish A, Hamshere ML, Pahwa JS, Moskvina V, Dowzell K, Williams A, Jones N, Thomas C, Stretton A, Morgan AR, Lovestone S, Powell J, Proitsi P, Lupton MK, Brayne C, Rubinsztein DC, Gill M, Lawlor B, Lynch A, Morgan K, Brown KS, Passmore PA, Craig D, McGuinness B, Todd S, Holmes C, Mann D, Smith AD, Love S, Kehoe PG, Hardy J, Mead S, Fox N, Rossor M, Collinge J, Maier W, Jessen F, Schurmann B, Heun R, van den Bussche H, Heuser I, Kornhuber J, Wiltfang J, Dichgans M, Frolich L, Hampel H, Hull M, Rujescu D, Goate AM, Kauwe JS, Cruchaga C, Nowotny P, Morris JC, Mayo K, Sleegers K, Bettens K, Engelborghs S, De Deyn PP, Van Broeckhoven C, Livingston G, Bass NJ, Gurling H, McQuillin A, Gwilliam R, Deloukas P, Al-Chalabi A, Shaw CE, Tsolaki M, Singleton AB, Guerreiro R, Muhleisen TW, Nothen MM, Moebus S, Jockel KH, Klopp N, Wichmann HE, Carrasquillo MM, Pankratz VS, Younkin SG, Holmans PA,

O'Donovan M, Owen MJ, Williams J (2009) Genome-wide association study identifies variants at CLU and PICALM associated with Alzheimer's disease. Nat Genet 41:1088–1093
Hochstrasser T, Weiss E, Marksteiner J, Humpel C (2010) Soluble cell adhesion molecules in monocytes of Alzheimer's disease and mild cognitive impairment. Exp Gerontol 45:70–74
Humpel C (2011) Identifying and validating biomarkers for Alzheimer's disease. Trends Biotechnol 29:26–32
Kellner A, Matschke J, Bernreuther C, Moch H, Ferrer I, Glatzel M (2009) Autoantibodies against beta-amyloid are common in Alzheimer's disease and help control plaque burden. Ann Neurol 65:24–31
Kemppainen NM, Aalto S, Wilson IA, Nagren K, Helin S, Bruck A, Oikonen V, Kailajarvi M, Scheinin M, Viitanen M, Parkkola R, Rinne JO (2007) PET amyloid ligand [11C]PIB uptake is increased in mild cognitive impairment. Neurology 68:1603–1606
Klunk WE, Engler H, Nordberg A, Wang Y, Blomqvist G, Holt DP, Bergstrom M, Savitcheva I, Huang GF, Estrada S, Ausen B, Debnath ML, Barletta J, Price JC, Sandell J, Lopresti BJ, Wall A, Koivisto P, Antoni G, Mathis CA, Langstrom B (2004) Imaging brain amyloid in Alzheimer's disease with Pittsburgh Compound-B. Ann Neurol 55:306–319
Leuner K, Pantel J, Frey C, Schindowski K, Schulz K, Wegat T, Maurer K, Eckert A. and Muller WE (2007) Enhanced apoptosis, oxidative stress and mitochondrial dysfunction in lymphocytes as potential biomarkers for Alzheimer's disease. J Neural Trans Suppl 72:207–215
Lewczuk P, Kornhuber J, Vanderstichele H, Vanmechelen E, Esselmann H, Bibl M, Wolf S, Otto M, Reulbach U, Kolsch H, Jessen F, Schroder J, Schonknecht P, Hampel H, Peters O, Weimer E, Perneczky R, Jahn H, Luckhaus C, Lamla U, Supprian T, Maler JM, Wiltfang J (2008) Multiplexed quantification of dementia biomarkers in the CSF of patients with early dementias and MCI: a multicenter study. Neurobiol Aging 29:812–818
Licastro F, Pedrini S, Caputo L, Annoni G, Davis LJ, Ferri C, Casadei V, Grimaldi LM (2000) Increased plasma levels of interleukin-1, interleukin-6 and alpha-1-antichymotrypsin in patients with Alzheimer's disease: peripheral inflammation or signals from the brain? J Neuroimmunol 103:97–102
Maes OC, Xu S, Yu B, Chertkow HM, Wang E, Schipper HM (2007) Transcriptional profiling of Alzheimer blood mononuclear cells by microarray. Neurobiol Aging 28:1795–1809
Marksteiner J, Kemmler G, Weiss EM, Knaus G, Ullrich C, Mechtcheriakov S, Oberbauer H, Auffinger S, Hinterholzl J, Hinterhuber H, Humpel C (2011) Five out of 16 plasma signaling proteins are enhanced in plasma of patients with mild cognitive impairment and Alzheimer's disease. Neurobiol Aging 32:539–540
Matsuda H (2007) Role of neuroimaging in Alzheimer's disease, with emphasis on brain perfusion SPECT. J Nucl Med 48:1289–1300
Petrella JR, Coleman RE, Doraiswamy PM (2003) Neuroimaging and early diagnosis of Alzheimer disease: a look to the future. Radiology 226:315–336
Ray S, Britschgi M, Herbert C, Takeda-Uchimura Y, Boxer A, Blennow K, Friedman LF, Galasko DR, Jutel M, Karydas A, Kaye JA, Leszek J, Miller BL, Minthon L, Quinn JF, Rabinovici GD, Robinson WH, Sabbagh MN, So YT, Sparks DL, Tabaton M, Tinklenberg J, Yesavage JA, Tibshirani R, Wyss-Coray T (2007) Classification and prediction of clinical Alzheimer's diagnosis based on plasma signaling proteins. Nat Med 13:1359–1362
Richartz-Salzburger E, Batra A, Stransky E, Laske C, Kohler N, Bartels M, Buchkremer G, Schott K (2007) Altered lymphocyte distribution in Alzheimer's disease. J Psychiatr Res 41:174–178
Riemenschneider M, Lautenschlager N, Wagenpfeil S, Diehl J, Drzezga A, Kurz A (2002) Cerebrospinal fluid tau and beta-amyloid 42 proteins identify Alzheimer disease in subjects with mild cognitive impairment. Arch Neurol 59:1729–1734
Rubio-Perez JM, Morillas-Ruiz JM (2012) A review: inflammatory process in Alzheimer's disease, role of cytokines. TheScientificWorldJOURNAL 2012:756357
Schindowski K, Peters J, Gorriz C, Schramm U, Weinandi T, Leutner S, Maurer K, Frolich L, Muller WE, Eckert A (2006) Apoptosis of CD4+ T and natural killer cells in Alzheimer's disease. Pharmacopsychiatry 39:220–228

Schoonenboom NS, Mulder C, Van Kamp GJ, Mehta SP, Scheltens P, Blankenstein MA, Mehta PD (2005) Amyloid beta 38, 40, and 42 species in cerebrospinal fluid: more of the same? Ann Neurol 58:139–142

Schwarzman AL, Goldgaber D (1996) Interaction of transthyretin with amyloid beta-protein: binding and inhibition of amyloid formation. CIBA Found Symp 199:146–160; discussion 160–144

Shoji M (2002) Cerebrospinal fluid Abeta40 and Abeta42: natural course and clinical usefulness. Front Biosci Virtual Libr Med 7:d997–d1006

Siest G, Bertrand P, Qin B, Herbeth B, Serot JM, Masana L, Ribalta J, Passmore AP, Evans A, Ferrari M, Franceschi M, Shepherd J, Cuchel M, Beisiegel U, Zuchowsky K, Rukavina AS, Sertic J, Stojanov M, Kostic V, Mitrevski A, Petrova V, Sass C, Merched A, Salonen JT, Tiret L, Visvikis S (2000) Apolipoprotein E polymorphism and serum concentration in Alzheimer's disease in nine European centres: the ApoEurope study. Clin Chem Lab Med 38:721–730

Slooter AJ, de Knijff P, Hofman A, Cruts M, Breteler MM, Van Broeckhoven C, Havekes LM, van Duijn CM (1998) Serum apolipoprotein E level is not increased in Alzheimer's disease: the Rotterdam study. Neurosci Lett 248:21–24

Taddei K, Clarnette R, Gandy SE, Martins RN (1997) Increased plasma apolipoprotein E (apoE) levels in Alzheimer's disease. Neurosci Lett 223:29–32

Teunissen CE, Lutjohann D, von Bergmann K, Verhey F, Vreeling F, Wauters A, Bosmans E, Bosma H, van Boxtel MP, Maes M, Delanghe J, Blom HJ, Verbeek MM, Rieckmann P, De Bruijn C, Steinbusch HW, de Vente J (2003) Combination of serum markers related to several mechanisms in Alzheimer's disease. Neurobiol Aging 24:893–902

Thambisetty M, Lovestone S (2010) Blood-based biomarkers of Alzheimer's disease: challenging but feasible. Biomark Med 4:65–79

Tsai MS, Tangalos EG, Petersen RC, Smith GE, Schaid DJ, Kokmen E, Ivnik RJ, Thibodeau SN (1994) Apolipoprotein E: risk factor for Alzheimer disease. Am J Hum Genet 54:643–649

van Vliet P, Westendorp RG, Eikelenboom P, Comijs HC, Frolich M, Bakker E, van der Flier W, van Exel E (2009) Parental history of Alzheimer disease associated with lower plasma apolipoprotein E levels. Neurology 73:681–687

Velayudhan L, Killick R, Hye A, Kinsey A, Guntert A, Lynham S, Ward M, Leung R, Lourdusamy A, To AW, Powell J, Lovestone S (2012) Plasma transthyretin as a candidate marker for Alzheimer's disease. J Alzheimers Dis 28:369–375

Weiner MW, Veitch DP, Aisen PS, Beckett LA, Cairns NJ, Green RC, Harvey D, Jack CR, Jagust W, Liu E, Morris JC, Petersen RC, Saykin AJ, Schmidt ME, Shaw L, Siuciak JA, Soares H, Toga AW, Trojanowski JQ, Alzheimer's Disease Neuroimaging Initiative Investigators (2012) The Alzheimer's disease neuroimaging initiative: a review of papers published since its inception. Alzheimers Dement J Alzheimers Assoc 8:S1–S68

Zetterberg H, Blennow K, Hanse E (2010) Amyloid beta and APP as biomarkers for Alzheimer's disease. Exp Gerontol 45:23–29

Zetterberg H, Wahlund LO, Blennow K (2003) Cerebrospinal fluid markers for prediction of Alzheimer's disease. Neurosci Lett 352:67–69

Part VII
Anti-brain Aging: Neuroprotection and Therapeutic Approaches

Chapter 21
Potentiation of Cellular Defense Capacity by Phytochemicals Activating NF-E2-Related Factor 2 for the Prevention and/or Treatment of Alzheimer's Disease

Gyu Hwan Park and Jung-Hee Jang

Abstract Increasing number of evidence suggest that nuclear factor erythroid 2-related factor 2 (Nrf2) plays a pivotal role in cellular endogenous defense against neurotoxic insults via augmentation of numerous neuroprotective genes which wok synergistically in the central nervous system (CNS). Therefore, we reviewed the current literature investigating the roles and molecular features of Nrf2 in the neuropathogenic process of Alzheimer's disease (AD) such as oxidative stress, inflammation, and apoptosis and further discussed the regulator mechanisms of Nrf2 by covering upstream modulators and downstream target proteins. Finally, as a strategy to fortify Nrf2 signaling pathway, we have introduced representative naturally occurring phytochemicals exhibiting neuroprotective potentials by activating Nrf2 in the CNS and enhancing neurohormetic stress response as a promising therapeutic target for the prevention and/or treatment of AD. Collectively, the information will provide a comprehensive understanding for the currently identified roles of Nrf2 in AD and may facilitate the design of non-clinical and clinical research establishing Nrf2 as a therapeutic target.

Keywords Alzheimer's disease (AD) • Apoptosis • Inflammation • NF-E2-related factor 2 (Nrf2) • Oxidative stress • Phytochemicals

---

G.H. Park
Research Institute of Pharmaceutical Sciences, College of Pharmacy, Kyungpook National University, Daegu 702-701, South Korea

J.-H. Jang (✉)
Department of Pharmacology, School of Medicine, Keimyung University, Daegu 704-701, South Korea
e-mail: pamy202@kmu.ac.kr

© Springer Japan 2015

N. Mori, I. Mook-Jung (eds.), *Aging Mechanisms*,
DOI 10.1007/978-4-431-55763-0_21

21.1 Introduction

Alzheimer's disease (AD) is a representative neurodegenerative disorder with progressive loss of synapses and neurons in the brain, which has been correlated with the formation of intraneuronal neurofibrillary tangles (NFTs) and extra-neuronal senile plaques (SPs), the two neuropathological hallmarks of the disease. As a major component of SPs, β-amyloid peptide (Aβ) is considered to play a causal role in development and progression of AD (Hardy and Higgins 1992). There has been compelling evidence supporting that Aβ-induced neurotoxicity is mediated through accumulation of reactive oxygen and/or nitrogen species (ROS/RNS) and oxidative damages. It has been reported that Aβ produces H_2O_2, which gives rise to most potent oxidant hydroxyl radicals via the Fenton reaction in the presence of transition metal ions such as iron and copper. Aβ has also been reported to generate nitric oxide (NO) by induction of inducible nitric oxide synthase (iNOS), which forms extremely potent oxidant peroxynitrite ($ONOO^-$) by interacting with super-oxide anion ($O_2^{-\bullet}$). Therefore, considerable attention has been focused on identifying redox-sensitive transcription factors and their target genes that are able to counteract excess ROS/RNS, thereby protecting against oxidative cell death for the prevention and/or treatment of AD.

One of the promising candidates is NF-E2-related factor 2 (Nrf2), which is a member of the cap'n'collar (CNC) family of transcription factors including Nrf1, Nrf3 and p45 NFE2. Nrf2 molecule has six functional erythroid-derived CNC homology protein (ECH) domains, named Neh1-Neh6. The redox-sensitive transcription factor Nrf2 is known to regulate expression of numerous cytoprotective genes that encode detoxifying and antioxidant enzymes and other defensive proteins against oxidative stress, inflammation, and apoptosis (Nguyen et al. 2003). In cytosol, Nrf2 activity is mainly controlled by cytoskeleton-associated inhibitory protein called Kelch-like ECH-associated protein 1 (Keap1). When cells are exposed to electrophilic or oxidative stimuli, Nrf2 can translocate to the nucleus and bind to antioxidant response element (ARE) or electrophile response element (EpRE), which leads to up-regulation of phase II detoxification enzymes and antioxidant proteins, such as γ-glutamylcysteine ligase (GCL), glutathione *S*-transferase (GST), glutathione peroxidase (GPx), heme oxygenase-1 (HO-1), NAD(P)H: quinone oxidoreductase 1 (NQO1), thioredoxins (Trxs), superoxide dismutase (SOD), and so on. Recent studies have revealed that Nrf2 activation involved in neuronal survival under diverse neurological conditions and Nrf2 inhibition has been found to decrease the cell viability by stimulating apoptosis. Although Nrf2 appears to be an important component of the cellular response to oxidative stress, the molecular events and genetic programs of Nrf2 involved in providing cells with resistance against neurotoxic insults, especially against the Aβ-induced neuronal cell death and memory impairment in AD, have not been extensively reported.

Therefore, this review is to summarize the latest literature information on the protective role of Nf2 in the neuropathogenesis of AD and on searching for phytochemicals fortifying Nrf2-mediated cellular defense capacity as therapeutics for AD. To discuss the roles of Nrf2 in AD, we have focused on the cellular processes of oxidative stress, neuroinflammation, apoptosis, autophagy, and neurogenesis in AD. Some upstream modulators of Nrf2 such as inhibitory protein Keap1, diverse kinases, and interacting proteins and some downstream target molecules including GCL, GST, HO-1, and NQO1 were described. Finally, based on the roles and molecular features of Nrf2, we have introduced Nrf2-activating phytochemicals derived from functional foods and medicinal plants for the prevention and/or treatment of AD.

21.2 Protective Roles of Nrf2 in Diverse Neuropathological Processes of AD

AD is characterized by a progressive impairment of cognitive functions including memory loss and disturbance in behaviors. At macroscopic levels, the brains of AD patients show cortical atrophy with narrowed gyri as well as widened sulci and at microscopic levels, SPs and NFTs were predominantly observed. Aβ is the main constituents of SPs, which is derived from the transmembrane amyloid precursor protein (APP) by the sequential actions of β- and γ-secretases. As a result of the imbalance between generation and clearance of Aβ from the brain, Aβ starts to aggregate in various types of extraneuronal depositions forming SPs. Intraneuronal accumulations of paired helical filaments, which mainly consist of hyperphosphorylated microtubule-associated tau, lead to the formation of NTFs. Various reports have demonstrated the fundamental roles of Nrf2 in the modulation of oxidative and/or nitrosative stress, neuroinflammation, apoptotic cell death, autophagy, and neurogenesis in cell cultures as well as animal models for AD (Fig. 21.1).

21.2.1 Oxidative Stress in AD

There has been a growing body of evidence implicating ROS, ROS-induced oxidative damages, antioxidant enzyme impairment, and excitotoxic mechanisms in the pathogenesis of AD. The brain is more vulnerable to oxidative stress because of high oxygen consumption, glucose-driven metabolic rates, polyunsaturated fatty acids contents, and transition metal ions, and relatively poor antioxidant defense molecules compared with other organs. Aβ is able to trigger the production of ROS with consequent chemical, structural and conformational modifications of Aβ by generating oxidized and cross-linked forms of Aβ, which are prone to aggregate into SPs (Galeazzi et al. 1999). ROS produced by aggregated form of Aβ can interact with almost every critical cellular macromolecules including DNA, lipids,

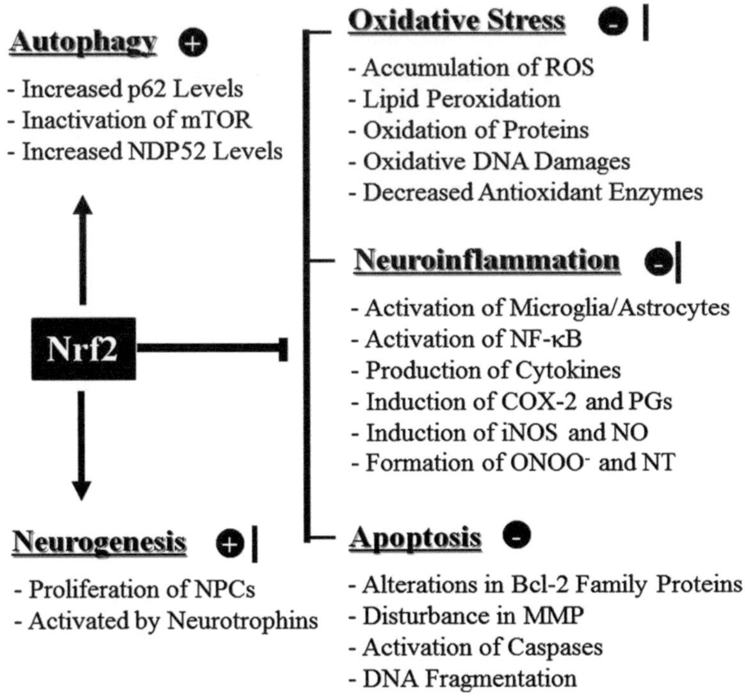

Fig. 21.1 Specific roles of Nrf2 in diverse neuropathological process of AD

and proteins, thereby causing functional as well as structural alterations in these biomolecules, which ultimately leads to neuronal cell death and tissue damages.

The products of lipid peroxidation including thiobarbituric acid reactive substances (TBARS) and/or 4-hydroxynonenal (4-HNE) were significantly increased in the brains (Sultana et al. 2006) and ventricular fluid (Lovell et al. 1997) of AD patients compared to control subjects. The oxidative damage to nuclear and/or mitochondrial DNA determined by using 8-hydroxy-2′-deoxyguanosine (8-OHdG) as a hallmark, were enhanced in the ventricular cerebrospinal fluid (Lovell et al. 1999) and parietal cortex (Mecocci et al. 1994) of AD patients. Proteins derived from the brains of AD patients are more oxidized than those derived from age-matched controls (Sultana et al. 2006). Advanced glycation endproducts (AGEs) formed by non-enzymatic reactions between glucose and long-lived protein deposits present in the SPs produce more radicals than non-glycated proteins, which augment neuronal death by increasing oxidative stress (Durany et al. 1999). Conversely, the plasma levels of several antioxidants and antioxidant enzymes such as SOD and GPx were reduced in elderly subjects with early AD or with mild cognitive impairment (MCI) (Rinaldi et al. 2003).

Nrf2 has pivotal roles as the regulator of key endogenous defenses against oxidative stress in the central nervous system (CNS). Nrf2 overexpressing neuronal cultures exhibited higher levels of antioxidant proteins, detoxifying enzymes, calcium homeostasis proteins, and growth factors (Lee et al. 2003b). Over-expression of Nrf2 in astrocytes enhances the antioxidant capacity particularly up-regulation of enzymes involved in biosynthesis, utilization, and transport of glutathione (GSH) thereby protecting neurons from oxidative stress (Shih et al. 2003). However, the lack of Nrf2 renders cultured cortical neurons and/or astrocytes susceptible to oxidative stress induced by diverse neurotoxins (Lee et al. 2003a, b).

The oxidative stress induced by neurotoxic insults in neuronal cells can be attenuated by synthetic as well as natural compounds activating Nrf2 signaling pathway. $H_2O_2/FeSO_4$- and/or 4-HNE-induced oxidative damages were inhibited by *tert*-butylhydroquinone (tBHQ), a representative inducer of Nrf2 in NT2N neurons (Eftekharzadeh et al. 2010). tBHQ treatment also decreased formation of Aβ in NT2N neuron treated with $H_2O_2/FeSO_4$- and/or 4-HNE. Oxidative stress also has been reported to increase production of Aβ. The $A\beta_{1-42}$ levels were elevated in a transgenic AD mice lacking CuZn-SOD by enhancing β-cleavage of APP (Murakami et al. 2012). Moreover, lipid peroxidation product, 4-HNE modified nicastrin, a γ-secretase substrate receptor, thereby enhancing its binding to the substrate C99, which ultimately led to Aβ accumulation (Gwon et al. 2012).

Natural products, decursin and decursinol angelate derived from *Angelica gigas* Nakai increased cellular resistance to $A\beta_{25-35}$-induced oxidative injury in PC12 cells, presumably through not only the induction of Nrf2 and downstream antioxidant enzymes (SOD, GPx, GST, and catalase (CAT)), but also the reduction of Aβ aggregation (Li et al. 2011). Allicin, an organosulfur compound present in garlic significantly ameliorated age-associated cognitive dysfunction through enhancing of Nrf2-mediated antioxidant signaling pathways such as reduced levels of ROS, lipid peroxidation, and protein carbonylation and increased levels of GSH, GPx, GCL, NQO1, and total antioxidant capacity (Li et al. 2012).

21.2.2 Neuroinflammation in AD

Oxidative stress not only causes direct oxidative damages but also triggers neuro-inflammatory responses by activating NF-κB-mediated transcription of inflammation-related genes, which sensitize vulnerable brain regions to additional stresses in AD (Akiyama et al. 2000). Elevated expression of inflammation-related molecules and their receptors has been observed in the brains of patients with AD, and recent studies have suggested that such brain-derived pro-inflammatory factors disrupt normal neurophysiology and contribute to cognitive dysfunctions.

Diverse inflammation-related components have been reported to be activated in the brains of patients with AD. Extensive activation of glia such as astrocytes and microglia has been observed in the CNS of AD patients. A series of cytokines and chemokines, including tumor necrosis factor-α (TNF-α), interleukin-1β (IL-1β), and IL-6 were released form the activated microglia and astrocytes, which are known to be associated with immunologic and inflammatory changes in the AD-affected brains (Grammas and Ovase 2001). TNF-α, IL-6, and IFN-γ further stimulated the γ-secretase activity, which led to enhanced production of Aβ and the intracellular domain of APP (Liao et al. 2004). Furthermore, prolonged exposure to pro-inflammatory cytokines can induce activation of certain kinases, which block the activation of Nrf2 thereby leading to neurological injuries. Treatment of astrocyte-rich cultures with the pro-inflammatory cytokine TNF-α transiently activated Nrf2 after 24 h, which was finally down-regulated after 72 h (Correa et al. 2012).

The principal enzymes that play a crucial role in mediating inflammatory responses are cyclooxygenase-2 (COX-2) and iNOS. COX, also known as prostaglandin H_2 synthase, is a rate-limiting enzyme in the synthesis of prostaglandins (PGs) from membrane phospholipid arachidonic acids. COX-2 protein levels were elevated in the frontal cortex region of AD patients which overlapped the cellular morphological features of apoptosis (Pasinetti and Aisen 1998). In the AD brains, neurons in the CA1–CA4 regions of the hippocampal pyramidal layer exhibited intense COX-2 signals, which were also correlated with Aβ plaques and atrophy (Ho et al. 1999). Furthermore, the neurons derived from transgenic mice overexpressing COX-2 are more vulnerable to Aβ-induced toxicity by potentiation of redox impairments (Ho et al. 1999).

NO, a potent pro-inflammatory mediator is produced by NOS-catalyzed oxidative deamination of $_{\text{L}}$-arginine to $_{\text{L}}$-citrulline. Under normal condition, NO exerts diverse physiological roles including proliferation, survival and differentiation of neurons, synaptic transmission, neural plasticity, and memory functions (Calabrese et al. 2009). However, a redox-mediated posttranslational modification to critical cysteine residues in proteins including S-nitrosylation by NO can lead to protein misfolding, mitochondrial dysfunction, synaptic loss, or apoptosis (Calabrese et al. 2009). Moreover, excess NO can further react with superoxide anion ($O_2^{-\bullet}$) to generate extremely potent oxidant peroxynitrite ($ONOO^-$). Peroxynitrite can attack diverse cellular macromolecules including DNA, proteins, and lipids in complex ways, which ultimately leads to neuronal cell death. Nitrotyrosine (NT), a combined product of tyrosine residues in protein with peroxynitrite, was detected in neurons, astrocytes and blood vessels in AD patients (Lüth et al. 2002). Moreover, NO synthesizing enzyme iNOS were highly expressed in astrocytes of AD patients and co-localized with nitrotyrosine (Lüth et al. 2002).

A growing body of evidence supports possible roles of Nrf2 during neuroinflammatory responses in addition to its strong antioxidant properties. Nrf2 can play an important role against inflammation due to its ability to antagonize the redox-sensitive transcription factor nuclear factor-κB (NF-κB) which regulates the expression of pro-inflammatory genes (Jin et al. 2008). After traumatic brain injury (TBI), Nrf2-deficient mice exhibited increase activation of NF-κB and elevated expression of pro-inflammatory cytokines such as TNF-α, IL-1β, and IL-6 and intercellular adhesion molecule-1 (ICAM-1) in the brains compared with control wild-type mice (Jin et al. 2008). In Nrf2 knockout mice administered with 1-methyl-4-phenyl-1,2,5,6-tetrahydropyridine (MPTP), a range of inflammation markers including astrogliosis, microgliosis, COX-2, iNOS, IL-6, and TNF-α were increased whereas anti-inflammatory markers such as FIZZ-1, YM-1, arginase-1, and IL-4 were decreased (Rojo et al. 2010). Nrf2 knockout mice were hypersensitive to the neuroinflammation induced by lipopolysaccharide (LPS), as determined by an increase in the mRNA and protein levels of F4/80, indicative of an increase in microglia, and upregulation of pro-inflammatory cytokines such as iNOS, IL-6, and TNF-α compared with wild-type littermates (Innamorato et al. 2008). Furthermore, APP/PS1 AD mice lacking Nrf2 exhibited increased levels of Aβ and activation of glial cells such as astrocytes and microglia (Joshi et al. 2014).

Conversely, the anti-inflammatory effects of some phytochemicals such quercetin (Kang et al. 2013), trans-isoferulic acid (Dilshara et al. 2014), and kavalactone (Terazawa et al. 2013) against LPS were partly mediated by activation of Nrf2 in BV2 cells. As a downstream anti-inflammatory target protein of Nrf2 activation by aforementioned phytochemicals, HO-1 and its product carbon monoxide (CO) have been suggested, and a HO-1 inducer as well as a CO donor decreased LPS-stimulated inflammatory responses (Terazawa et al. 2013).

21.2.3 Neuronal Cell Death in AD

Apoptosis, a programmed cell death in the neurodegenerative diseases including AD, is generally characterized by membrane blebbing, shrinkage of cells, chromatin condensation, and DNA fragmentation where dysfunction of mitochondria, alterations in the Bcl-2 family proteins and activation of caspases play pivotal roles. Bcl-2 family proteins include both proteins mediating cell survival (e.g., anti-apoptotic proteins such as Bcl-2, Bcl-X_L, Bcl-w, etc.) and others promoting cell death (e.g., pro-apoptotic proteins such as Bax, Bad, Bid, Bim, Bak, Bik, etc.). Bcl-2 family proteins preserve or disrupt mitochondrial integrity by releasing diverse mitochondrial apoptotic factors including cytochrome c, Smac/Diablo, or apoptosis-inducing factor (AIF) (Vila and Przedborski 2003). Cytochrome c provokes apoptotic protease activating factor-1 (Apaf-1)-regulated activation of caspase-9, which initiates the downstream effector caspase cascades (Obulesu and Lakshmi 2014).

Caspases can be activated by proteolytic cleavage from procaspases, which are divided into initiators (procaspases 2, 8, 9 and 10) and effectors (procaspases 3, 6, and 7) (Vila and Przedborski 2003). Recent studies have suggested that caspase activation may also play a role in facilitating neuropathological process of AD such as formation of SPs and NFTs by promoting cleavage of APP as well as tau (Rohn and Head 2009). During TNF-α-induced cell death in HeLa cells, Nrf2 was also cleaved by active caspase-3 and overexpression of the C-terminal cleavage fragment of Nrf2 containing the DNA binding and leucine-zipper domains caused apoptotic cell death (Ohtsubo et al. 1999).

It has been reported that AD risk factors-induced neuronal cell death and memory impairment are mediated by apoptotic process via activation of caspase-3 and inactivation of Nrf2 as well. Homocysteic acid (HCA), an oxidized metabolite of homocysteine, is one of the biomarkers in the brain with AD and expected to be involved in intraneuronal accumulation of $A\beta_{1-42}$. In HT22 cell, HCA increased ROS production, depleted GSH, inactivated Nrf2, and caused apoptotic cell death by deteriorating mitochondrial membrane potential (MMP), increasing Bax/Bcl-2 ratio, and activating caspase-3 (Tan et al. 2013). Epidemiological and biological data support a correlation between type 2 diabetes mellitus and AD. Therefore, patients with diabetes have a higher incidence of cognitive dysfunction and an increased risk of developing AD. In the animal model of AD induced by streptozotocin (STZ), a toxin toward insulin-producing pancreatic β cells, STZ caused neuronal cell death by increasing Bax/Bcl-2 ratio and cleaved caspase-3 and by inactivating Nrf2 (Song et al. 2014).

Recently, several groups have reported that overexpression of Nrf2 promotes cell survival by hampering the induction of apoptosis in CNS. Nrf2 overexpression protected neuronal and/or glial cells from apoptotic cell death caused by MPTP, rotenone (Lee et al. 2003b), 6-hydroxydopamine (6-OHDA), 3-morpholinosydnonimine (SIN-1) (Cao et al. 2005), ethanol (Narasimhan et al. 2011), and H_2O_2 (Li et al. 2005). In an *in vitro* cell culture system, pretreatment of astrocytes with tBHQ which induces Nrf2 nuclear translocation and coordinates up-regulation of ARE-driven genes, attenuated H_2O_2 or platelet-activating factor-induced cell death (Lee et al. 2003a; Li et al. 2005). Conversely, Nrf2 inhibition has been found to decrease the cell viability by stimulating apoptotic cell death. $Nrf2^{-/-}$ neurons were more susceptible to apoptosis induced by rotenone and ionomycin (Lee et al. 2003b).

An array of data from *in vitro* cell culture models of AD showed that Nrf2-activating phytochemicals inhibited diverse apoptotic signals involved in neuronal cell death. In PC12 cells, $A\beta_{25-35}$-induced cell death was mediated by apoptosis as revealed by significant reduction in cell viability, phosphatidylserine externalization, decreased mitochondrial transmembrane potential, cytochrome c release, activation of caspase-3, cleavage of PARP, and DNA fragmentation, which were effectively inhibited by gypenoside XVII, a novel phytoestrogen isolated from *Gynostemma pentaphyllum* by activating Nrf2-HO-1 pathway (Meng et al. 2014a). In SH-SY5Y cells, $A\beta_{25-35}$-induced cytotoxicity and apoptotic cell death such as DNA fragmentation, elevated Bax/Bcl-2 ratio, dissipation of mitochondrial membrane potential, and activation of caspase-3 were attenuated by [6]-gingerol treatment via Nrf2- GCL/HO-1 pathway (Lee et al. 2011).

21.2.4 Autophagy in AD

Autophagy is a self-degradation process of misfolded or aggregated proteins and dysfunctional cellular components. Autophagic process consists with nucleation (initiation), elongation-closure-cargo recruitment (trafficking), and maturation steps with subsequent fusion with lysosome to form autolysosome (Friedman et al. 2014). Diverse autophagy-related genes (ARGs) are involved in the autophagic process which is negatively regulated by mTOR signals (Friedman et al. 2014). Growing body of evidence indicates the defective autophagic clearance as one of the disease-causing factor in AD which is characterized by the aberrant accumulation of endogenous proteins resulting in the formation of cytotoxic aggregates and inclusions.

Activation of Nrf2 is involved in regulation of autophagic process via increasing p62, a selective substrate of autophagy recruiting target poly-ubiquitinated proteins for autophagosome-dependent degradation. When autophagy is impaired, p62 oligomers accumulates and interacts with poly-ubiquitinated proteins particularly Keap1 to form aggregates/inclusions leading to the stabilization and activation of Nrf2 (Komatsu et al. 2010; Lau et al. 2010). As the promoter region of the p62 gene contain ARE binding sites, activation of Nrf2 in turns increases p62 levels as a positive feedback loop (Jain et al. 2010).

p62-deficient MEF cells are more vulnerable to H_2O_2-induced cytotoxicity and exhibited decreased activation of Nrf2 (Du et al. 2009). In the brains of aged mice as well as human subjects and a triple transgenic AD mice, p62 levels were marked decreased whereas oxidative damage to p62 promoter was increased (Du et al. 2009). In APP/PS1 mice lacking Nrf2 (APP/PS1/Nrf2$^{-/-}$), p62 is prone to make insoluble aggregates, which leads to the accumulation of poly-ubiquitin-conjugated proteins and inhibition of autophagy (Joshi et al. 2014). Moreover, mTOR pathway, a negative regulator for autophagic process and its downstream target p70S6k were activated in APP/PS1/Nrf2$^{-/-}$ mice (Joshi et al. 2014).

Autophagy adaptor protein NDP52 also contains AREs in its promoter region which can be induced by activation of Nrf2. Therefore, overexpression of NDP52 can facilitate the clearance of hyperphosphorylated protein tau in the presence of an autophagy stimulator (Kim et al. 2014). In Nrf2-knockout mice, phosphorylated and sarkosyl-insoluble tau levels were increased in the brains where the NDP52 levels were decreased (Jo et al. 2014).

21.2.5 Neurogenesis in AD

Although neurogenesis is most active during prenatal development, it continues in some parts of the brain including the hippocampus and the subventricular zone. The stimulation of endogenous neurogenesis by trophic factors, cytokines, physical activities, environmental stimuli, and drugs is regarded as a potential therapeutic strategy for the treatment of AD (Felsenstein et al. 2014). Particularly, neurotropic factors such as nerve growth factor (NGF), glial cell-derived neurotrophic factor (GDNF), and brain-derived neurotrophic factor (BDNF) are important for the maintenance, survival, and regeneration of specific neurons in the adult brains (Allen et al. 2013).

Recent studies report a role of Nrf2 during prenatal as well as adult neurogenic process in AD. In neural stem/progenitor cells (NPCs) that can proliferate to produce new neurons in neurogenic areas throughout the lifetime, growth of neurospheres and neural differentiation were increased by overexpression of Nrf2 or by treatment with Nrf2-activating compound, pyrrolidine dithiocarbamate (PDTC) (Kärkkäinen et al. 2014). Furthermore, $A\beta_{1-42}$-induced toxicity and reduction in proliferation of neurosphere were prevented by Nrf2 overexpression, which were also aggravated by inhibition of Nrf2 (Kärkkäinen et al. 2014).

In another study, intravenously transplanted bone marrow-derived endothelial progenitor cells (BM-EPCs) migrated into the brain of rats and improved the scopolamine (SCO)-induced learning and memory deficits by suppressing deposition of Aβ and hyperphosphorylation of tau, perturbations of neurotransmitter levels, and neuroinflammation (Safar et al. 2014). Furthermore, BM-EPCs induced behavioral recovery via enhancing vascular endothelial growth factor (VEGF), NGF, BDNF levels and activating Nrf2 (Safar et al. 2014). In PC12h cells, carnosic acid (CA), a compound derived from rosemary strongly promoted neurite outgrowth by activation of Nrf2 whereas suppressed by Nrf2 knockdown (Kosaka et al. 2010). Moreover, CA increased p62 levels in an Nrf2-dependent manner while CA-induced Nrf2 activation and neural differentiation were significantly reduced by knockdown of p62 gene (Kosaka et al. 2010).

21.3 Downstream Target Molecules of Nrf2

Many basic leucine zipper (bZIP) transcription factors including Nrf, Jun, Fos, Fra and Maf bind to the ARE sequences, defined as 5′-A/^{G}TGAC/^{T}NNNGCA/^{G}-3′ thereby inducing the expression of a wide array of antioxidant proteins and phase II detoxification enzymes (Hayes and McMahon 2001). The representative proteins that are encoded by the ARE gene battery include enzymes associated with GSH biosynthesis, redox proteins with active sulfhydryl moieties, and drug-metabolizing enzymes. An array of cellular defense enzymatic and non-enzymatic systems exist to counteract ROS/RNS that lower steady-state concentrations of ROS/RNS and repair oxidative cellular damages. Nrf2-overexpressing cells have been shown to exhibit high levels of endogenous antioxidant GSH and detoxification/antioxidant enzymes such as GCL, GST, GPx, HO-1, NQO1, SOD and Trx as schematically represented in Fig. 21.2.

21.3.1 GCL

GCL is the rate limiting enzyme in the GSH biosynthesis. GSH is synthesized by sequential actions of GCL and glutathione synthetase (GS). Glutamate and cysteine are ligated by GCL to form γ-glutamylcysteine, which is then joined to glycine by GS to form GSH. The synthesis of GSH is primarily controlled at the transcriptional level of GCL. GCL is a heterodimer of a heavy catalytic subunit (GCLC, 73 kDa) and a light modulatory subunit (GCLM, 31 kDa). GSH plays crucial roles in the

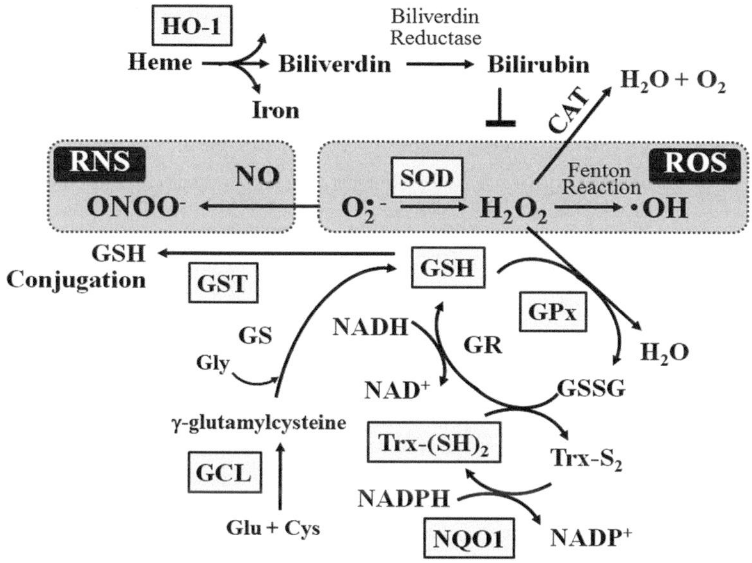

Fig. 21.2 Downstream neuroprotective proteins up-regulated by activation of Nrf2

CNS, including scavenging of ROS/RNS, redox-modulation of ionotropic receptor activity, and neurotransmission (Bains and Shaw 1997). GSH redox status is critical for transcriptional activation of specific genes involved in the redox-regulated signal transduction and modulation of cell proliferation, apoptosis and inflammation.

Diminished GSH status has been associated with normal ageing as well as with diverse neurodegenerative diseases including AD. There is compelling evidence supporting the altered GSH levels in specific brain regions of subjects affected by AD (Aksenov and Markesbery 2001). Decreased GSH content was also observed in lymphoblasts from patients carrying presenilin (PS) or APP mutations (Cecchi et al. 1999). The significantly decreased GSH content in the red blood cells from AD patients was associated with reduced GCL and GS activities (Liu et al. 2004). In in vitro cell cultures, Aβ treatment significantly decreased GSH in neurons, astrocytes, and glial cells (Abramov et al. 2003; Sharpe et al. 2002), while *N*-acetyl-L-cysteine (NAC) pretreatment has been shown to protect SH-SY5Y neuroblastoma cells from Aβ-induced oxidative cell death (Olivieri et al. 2001).

The 5′-flanking regions in both of human GCL subunits have been characterized, and putative binding sites for ARE have been identified in the promoter region (Dickinson et al. 2004). The transient overexpression of wild-type Nrf2 in HepG2 cells resulted in an increased activity of the GCL promoter reporter construct (Wild et al. 1999). $Nrf2^{-/-}$ fibroblasts transiently transfected with Nrf2 cDNA expression plasmid showed an increase in GSH levels and GCL gene transcription compared with mock-transfected controls (Chan and Kwong 2000). Overexpression of equimolar ratio of Nrf2 and c-Jun led to maximum induction of ARE-mediated GCL gene expression (Jeyapaul and Jaiswal 2000). Particularly, Nrf2 overexpression up-regulated GSH levels by modulating genes involved in biosynthesis (xCT cystine antiporter, GCL, and GS), use (GST and GR), and export (multidrug resistance protein 1) in astrocytes and increased the resistance of neuronal cells against glutamate-induced oxidative stress (Shih et al. 2003).

21.3.2 GST

GST catalyzes nucleophilic attack by reduced GSH on nonpolar xenobiotics that contain an electrophilic carbon, nitrogen, or sulfur atom. In general, conjugation of foreign compounds with GSH leads to formation of less reactive products that are readily excreted. GST displays broad substrate specificity and plays a critical role in providing protection against oxidative stress and electrophilic attack, leading to their excretion. Seven classes of cytosolic GST are characterized in mammals based on amino acid sequence similarities (alpha, mu, pi, sigma, theta, omega, and zeta). Other classes of cytosolic GST, namely beta, delta, epsilon, lambda, phi, tau, and the "U" class, have been identified in non-mammalian species (Hayes et al. 2005).

The level of GST expression is a pivotal factor in determining the susceptibility of cells to a wide array of noxious stimuli and the induction of GST families is regarded as a protective and/or adaptive response to oxidative stress and electrophilic attack. GST activity and protein expression were found to be depleted in most brain regions and ventricular cerebrospinal fluid of AD subjects (Lovell et al. 1998). The continuous intracerebroventricular (i.c.v.) infusion of Aβ resulted in a significant reduction of the GST immunoreactivity in the affected brain areas (Kim et al. 2003). GSTs are present in many organs and have been implicated in the detoxification of endogenous unsaturated aldehydes, including 4-HNE (Berhane et al. 1994). Pretreatment of primary hippocampal cells with GST was shown to enhance survival against HNE toxicity (Xie et al. 1998). Stable GST transfectants exhibited increased resistance to ferrous sulfate-H_2O_2, Aβ, and peroxynitrite-induced oxidative stress (Xie C et al. 2001). In addition, GST functions as a regulatory molecule for cellular signaling pathways and may affect cell proliferation and cell cycle control.

A substantial number of GST genes have been found to contain an ARE or related sequences (Hayes et al. 2005). Nrf2 overexpression significantly stimulated placental GST (GST-P) expression in mouse F9 embryonal carcinoma and H4IIE rat hepatoma cells (Ikeda et al. 2004). Conversely, in the livers of $Nrf2^{-/-}$ mice, the levels of mRNA encoding GSTA1, GSTA2, GSTM1, and GSM3 have been reported to be decreased (Chanas et al. 2002).

21.3.3 HO-1

HO-1, also known as heat shock protein 32 (HSP32) or inducible HO, is a representative antioxidant defense enzyme that catalyzes the rate-limiting step in heme degradation leading to the formation of carbon monoxide (CO) and biliverdin/bilirubin (Dore 2002). HO-1 expression can be transiently activated not only by its physiological substrate heme, but also by a wide variety of noxious stimuli including oxidative and nitrosative stress. Enhanced HO-1 biosynthesis exerts neuroprotective effects by converting prooxidant heme to biologically inactive by-products, such as biliverdin/bilirubin and CO.

Significant induction of HO-1 levels has been observed in AD neocortex and cerebral vessels in association with NFTs and SPs (Takeda et al. 2000; Premkumar et al. 1995). According to a recent report, there is an increase in bilirubin and its metabolites in cerebrospinal fluid of AD patients (Kimpara et al. 2000). The elevation of HO-1 in response to noxious factors suggests that the enzyme may function as a critical cytoprotective molecule. It has been recognized that induction of HO-1 activity could afford neuroprotection. The transgenic mice overexpressing HO-1 in the brain attenuated ischemic stroke-induced cellular injuries through up-regulation of cGMP and Bcl-2 levels, inhibition of p53 activation, and suppression of lipid peroxidation (Panahian et al. 1999). In complementary *in vitro* studies, cerebellar granule cells harvested from HO-1 transgenic mice were relatively resistant to glutamate- and H_2O_2-related oxidative injuries (Chen et al. 2000). Pretreatment of hybrid septal SN56 cells with HO-1 antisense oligonucleotides exacerbated the Aβ- and H_2O_2-induced cytotoxicity, which was attenuated by hemin, a HO-1 inducer (Le et al. 1999).

The ARE has been identified in the 5′-flanking regions of the mouse and human HO-1 genes, where it has been designated as a stress response element (STRE) (Alam and Cook 2003). The peritoneal macrophages from Nrf2-deficient mice have been shown to exhibit impaired expression of HO-1, A170, peroxiredoxin MSP23, and cystine membrane transporter (Ishii et al. 1999). Stable transfection of L929 cells with a dominant negative mutant form of Nrf2 were shown to decrease the HO-1 mRNA accumulation in response to $CdCl_2$, $ZnSO_4$, arsenite and tBHQ (Alam et al. 1999).

21.3.4 NQO1

NQO1, an enzyme with DT-diaphorase activity, is a ubiquitous cytosolic flavoenzyme that catalyzes obligatory two-electron reduction of quinones thus preventing their participation in redox cycling and subsequent oxidative stress. NQO1 expression is induced in response to xenobiotics, oxidants as well as antioxidants, heavy metals, ionizing radiations, and UV light. NQO1 has been found to protect cells from oxidative stress, mutagenesis, and carcinogenesis (Jaiswal 2000).

Dramatic increase in NQO1 activity has been detected in astrocytes and neurons of the hippocampus in association with AD pathology (Wang et al. 2000). NQO1 might be a part of an active antioxidant enzyme system expressed in the neuropathogenesis of AD, ameliorating cellular injury (SantaCruz et al. 2004). The role of NQO1 in H_2O_2-, platelet-activating factor- or dopamine-induced toxicity in primary astrocytes and/or neuroblastoma cell lines has been reported as well (Lee et al. 2003a; Hara et al. 2003).

Deletion mutagenesis studies with the human NQO1 gene promoter identified several *cis*-elements that are essential for the expression and induction of NQO1. One of these elements is ARE which is involved in basal expression as well as induction of NQO1 in response to β-naphthoflavone, tBHQ, and H_2O_2 (Jaiswal 2000). Nucleotide sequence analysis of human NQO1 gene ARE revealed that it contains one perfect and one imperfect AP-1 (TPA response) elements arranged as inverse repeats separated by three base pairs followed by a 'GC' box (Nioi and Hayes 2004). Nrf1 and Nrf2 positively and c-Fos and Fra1 negatively regulate the human ARE-mediated expression of NQO1 (Venugopal and Jaiswal 1996). Overexpression of Nrf1 and Nrf2 cDNA was shown to up-regulate the expression of NQO1 gene in response to antioxidants and xenobiotics (McMahon et al. 2001). Furthermore, induction of NQO1 activity by treatment with tBHQ reduced glutamate toxicity which was potentiated by the NQO1 inhibitor dicumarol in the N18-RE-105 neuronal cell line (Murphy et al. 1991). Conversely, mice lacking the Nrf2 gene showed a marked decrease in the expression and induction of NQO1 (McMahon et al. 2001).

21.4 Upstream Regulators of Nrf2

Although Nrf2 appears to be an important component of cellular defense system in AD, molecular mechanisms of its activation are complex reflecting the multiplicity of binding proteins and the existence of the large number of different gene-specific sequences recognizing Nrf2. In cytosol, Nrf2 activity is controlled by its interaction with a cytoskeleton-associated inhibitory protein called Keap1. By forming a complex with its inhibitory counterpart Keap1, Nrf2 functions as an intracellular sensor that recognizes redox signaling by detecting electrophiles or ROS/RNS. Nrf2 dissociates from its cytoplasmic inhibitory protein Keap1 and moves into the nuclei when cells are exposed to stimuli. Although the actual mechanism of dissociation of Nrf2 from Keap1 remains unresolved, it is considered to involve thiol modifications in the cysteine residues of Keap1 and Nrf2, phosphorylation of serine, threonine, or tyrosine residues in Nrf2, and interaction with other regulatory protein partners as indicated in Fig. 21.3.

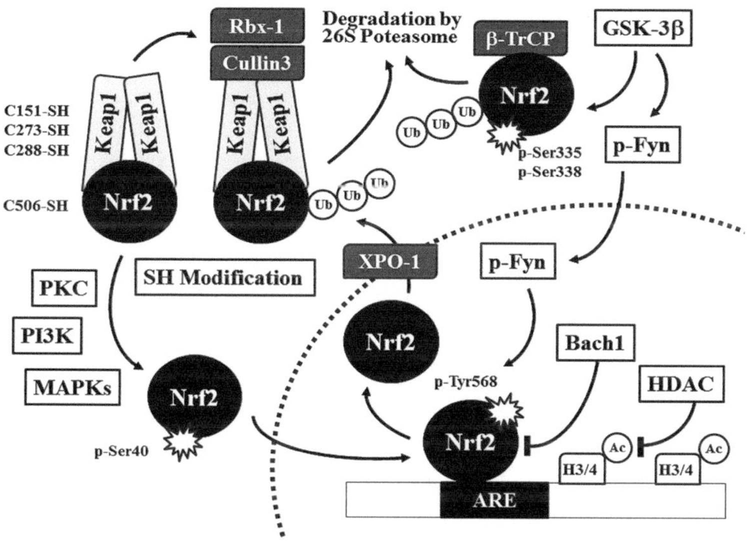

Fig. 21.3 Upstream regulatory molecules involved in the activation and/or inactivation of Nrf2

21.4.1 Thiol Modifications of Keap1 and Nrf2

Keap1 is a homodimer with five domains such as N-terminal region (NTR), Broad complex, Tramtrack, and Bric-a-Brac (BTB) domain, cysteine-rich intervening regions (IVR), double glycine or Kelch repeat (DGR), and C-terminal region (CTR) (Lee and Surh 2005). Under normal condition, Keap1 molecules dimerized through their BTB domains bind to Nrf2 (Neh2 domain) through DRG domain. The repressor protein Keap1 enhances Neh2-dependent Nrf2 ubiquitination and degradation. N-terminal BTB domain and central linker region of Keap1 bind to an E3 ubiquitin ligase complex (ring-box1, Rbx-1) via an adaptor protein Cullin3, which directs ubiquitination of Nrf2 on Neh2 domain and subsequent degradation by the 26S proteasome (Sandberg et al. 2013).

Modification of cysteine residues within Keap1 or Nrf2 could trigger the dissociation of Nrf2-Keap1 from its tethering site. The cysteine residues at C151, C273 and C288 in IVR are required for Keap1-dependent ubiquitination of Nrf2 (Zhang and Hannink 2003). Several groups subsequently have shown that mutation of either C273 and C288 disrupts the repressive activity of Keap1 against Nrf2, suggesting that modification of these two cysteine sites is critical for the suppression of Nrf2 (Wakabayashi et al. 2004). Nrf2 also contains a critical cysteine residue in its DNA-binding domain that has been shown to be redox-regulated. In human hepatoblastoma HepG2 cells, functionally inactive Nrf2 harboring mutated C506S less efficiently induced activation of ARE-mediated gene expression compared with wild-type Nrf2 (Bloom et al. 2002).

21.4.2 Phosphorylation by Protein Kinase C (PKC)

PKCs are a family of structurally related serine/threonine kinases. PKC isoforms can be divided into three groups in mammals, the classical PKC (cPKC), novel PKC (nPKC), and atypical PKC (aPKC) based on their structure in the amino-terminal regulatory domain. Structural differences of the PKC isoforms result in the requirement of distinct cofactors for each of the PKC isoforms (Gutcher et al. 2003). Calcium and diacylglycerol are required for activation of cPKCs, whereas calcium is not essential for the activation of nPKC. aPKCs do not require these cofactors when activated by certain lipid species including the products of phosphatidylinositol-3-kinase (PI3K) and ceramide. Increasing evidence indicates that cPKCs as well as nPKCs play important roles in regulating cell proliferation, growth, differentiation, apoptotic cell death, and carcinogenesis. aPKCs also have been shown to be involved in the proliferation, differentiation, cell cycle, apoptosis, cytoskeletal organization, and cell migration (Carter and Kane 2004). Therefore, natural compounds and pharmacologic interventions which can modulate PKC activity have been extensively studied to prevent or treat diseases with aberrant activation of PKC.

Phosphorylation of Nrf2 Ser40 residing in the Neh2 domain by PKC interferes with its interaction with Keap1 in HepG2 cells (Huang et al. 2000, 2002; Numazawa et al. 2003). Phosphorylation of Nrf2 by PKC promoted its dissociation from Keap1, whereas the Nrf2-S40A mutant bearing an alanine substitution at Ser40 remained associated, suggesting that PKC-catalyzed phosphorylation of Nrf2 at Ser40 is a critical signaling event in the ARE-mediated cellular antioxidant response (Huang et al. 2002). The broad-spectrum PKC inhibitors, such as staurosporine and Ro-32-0432, impaired ARE reporter gene activity stimulated by phorbol 12-myristate 13-acetate (PMA), TPA, tBHQ and β-naphthoflavone (Huang et al. 2000). aPKCs are responsible for the phosphorylation Nrf2 at Ser-40 in response to 4-HNE. The ARE reporter activity induced by 4-HNE was increased by coexpression of aPKCι, which was inhibited by an aPKC inhibitor Ro-31-8220 (Numazawa et al. 2003).

21.4.3 Phosphorylation by Phosphatidylinositol-3-Kinase (PI3K)

PI3K phosphorylates the D-3 position of inositol ring in phosphatidylinositol and has been shown to form a heterodimer consisting of a 85 kDa (adapter) and 110 kDa (catalytic) subunits. The involvement of PI3K in intracellular processes, such as cell growth, differentiation, apoptosis, calcium signaling, insulin signaling, autophagy, and neurogenesis is well known. PI3K has been implicated in neuronal as well as glial survival and differentiation by inducing inositol lipid second messengers, which bind to pleckstrin homology (PH) domains in diverse signal transduction proteins and subsequently alter their enzymatic activities and/or subcellular localization (Rodgers and Theibert 2002). PI3K/Akt pathway also can modulate mTOR activity, a major regulator for the autophagic process which eliminates abnormal and toxic protein aggregates in neurodegenerative disorders including AD (Heras-Sandoval et al. 2014).

PI3K signaling pathway modulates the rearrangement of actin microfilaments in response to tBHQ, which allows the translocation of Nrf2 bound with actin to the nucleus and stimulates rGSTA2 expression in rat hepatoma H4IIE cells (Kang et al. 2002). Pretreatment of H4IIE cells with the PI3K inhibitors, wortmannin or LY294002 abolished the induction of rGSTA2 mRNA by tBHQ, sulfur amino acid deprivation, or SIN-1 (Kang et al. 2001). In the IMR32 and SH-SY5Y neuroblastoma cell lines, the pharmacological inhibition of PI3K attenuated ARE-driven transcription in response of tBHQ and hemin, respectively (Li et al. 2002; Nakaso et al. 2003).

21.4.4 Phosphorylation by Mitogen-Activated Protein Kinases (MAPKs)

MAPKs encompass a large number of serine/threonine kinases involved in regulating a wide array of cellular responses including proliferation, differentiation, stress adaptation, and apoptosis (Owuor and Kong 2002). MAPK are divided into three multimember subfamilies based on structural differences: extracellular signal-regulated kinase (ERK), p38 MAPK, and c-Jun N-terminal kinase (JNK, also named as stress-activated protein kinase or SAPK). The roles of MAPKs during cell survival and death are controversial. However, in general, in response to growth factor stimulation ERK is activated whereas environmental stress such as ROS/RNS, ultraviolet irradiation, and endotoxin triggers activation of p38 MAPK and JNK.

Simultaneous inhibition of ERK and p38 MAPK resulted in complete inhibition of pyrrolidine dithiocarbamate (PDTC)-induced GCL transcription, suggesting that ERK and p38 MAPK cooperate in maximal GCL subunit gene induction in response to PDTC in HepG2 liver carcinoma cells (Zipper and Mulcahy 2000). In human hepatoma HepG2 and murine hepatoma Hepa1c1c7 cells, the ERK inhibition by PD98059 or dominant-negative mutant of ERK2 attenuated tBHQ- or sulforaphane-induced ARE reporter gene activity, while inhibition of p38 activation by SB203580 enhanced it (Yu et al. 1999, 2000). Gastrodin, a phenolic compound present in *Gastrodia elata* attenuated $A\beta_{1-42}$-induced neurotoxicity in primary hippocampal neurons and up-regulated Nrf2, SOD, and CAT via activation of ERK1/2 (Zhao et al. 2012).

21.4.5 Phosphorylation by Glycogen Synthase Kinase-3β (GSK-3β)

GSK-3β is a multifunctional serine/threonine kinase implicated in the diverse cellular mechanisms ranging from glycogen metabolism to neuropathological characteristics of AD particularly hyperphosphorylation of tau and formation of NFTs (Farr et al. 2014). GSK-3β expression has been reported to be elevated in the brains of AD patients prominently in neuronal cell bodies as well as their processes and co-localizes with NFTs (Pei et al. 1997). Exposure of rat hippocampal neurons to $A\beta_{25-35}$ and $A\beta_{1-42}$ enhanced phosphorylation of tau through activation of GSK-3β (Takashima et al. 1996).

GSK-3β can phosphorylate the Neh6 domain of Nrf2 and increase the expression of a scaffolding protein, β-transducin repeat-containing protein (β-TrCP) that binds Nrf2 and facilitates Nrf2 ubiquitination and degradation by Cullin1-Rbx1 complex (Rada et al. 2012). Furthermore, GSK-3β activation acts as upstream of Src kinases and phosphorylates a tyrosine kinase, Fyn leading to its nuclear localization. Activated Fyn facilitates the subsequent phosphorylation of Nrf2 at Tyr568 and eventually export Nrf2 out of the nucleus through interaction with exportin-1 (XPO-1) (Jain and Jaiswal 2007). Once excluded from the nucleus, Nrf2 is ubiquitinated and degraded.

Transgenic mice that conditionally over-express GSK-3β in hippocampal and cortical neurons showed diverse neuropathology of AD such as hyperphosphorylation of tau, neuronal death, and reactive gliosis and memory impairments (Hernández et al. 2002). Conversely, antisense oligonucleotide against GSK-3β improved age-related learning and memory deficit in SAMP8 mice by increasing nuclear location of Nrf2 and by reducing oxidative stress such as formation of protein carbonyls and protein-bound HNE (Farr et al. 2014). PDTC, a clinically tolerated inhibitor of NF-κB reduced the amount of active GSK-3β with concomitant improvement in the learning in APP/PS1 transgenic AD mice (Malm et al. 2007). Puerarin, a major isoflavone glycoside from Kudzu root (*Pueraria lobata*) protected primary hippocampal from $A\beta_{25-35}$-induced oxidative stress via inducing Nrf2-HO-1 pathway, which were reversed by treatment with a pharmacological inhibitor of GSK-3β, lithium chloride (Zou et al. 2013).

21.4.6 BTB (Broad-Complex, Tramtrack and Bric-a-Brac) and CNC Homology 1 (Bach1)

Bach1 can bind to ARE sequence in conjugation with a small Maf protein and play a role as a transcriptional repressor of Nrf2, quenching the expression of target genes. Sulfhydryl oxidizing agent, diamide-induced Nrf2 nuclear translocation was suppressed by ectopic expression of Bach1, which was abrogated by substitution of the cysteine residue (C574) in the DNA binding domain of Bach1 to serine (Ishikawa et al. 2005).

21.4.7 *Histone Deacetylases (HDACs)*

Activation of Nrf2 can also be regulated by histone acetylation. Astrocyte-rich cultures incubated with conditioned media derived from LPS-treated microglia increased HDAC activity which deacetylates histones H3 and H4 and decreased activation of Nrf2 and expression of its downstream target protein GCL (Correa et al. 2011). Conversely, HDAC inhibitors such as valproic acid and trichostatin-A elevated the acetylation levels of histone and restored activation of Nrf2.

21.5 Neuroprotective Phytochemicals via Activation of Nrf2

In general, plant antioxidants derived from daily diet or dietary supplements, have been reported to protect against ROS/RNS-related neurological disorders by counteracting oxidative/nitrosative stress, inflammation, and apoptotic cell death. However, the beneficial effects of these phytochemicals cannot be fully explained by their antioxidant activity directly scavenging free radicals. Phytochemicals trigger neurohormesis particularly by activating Nrf2 signaling pathway that makes cells more resistant to severe neurotoxic insults. Herein we have introduced neurohormetic phytochemicals for the prevention and/or treatment of AD (Fig. 21.4) and reviewed their roles and regulatory mechanisms involved in these responses.

21.5.1 *Sulforaphane*

Sulforaphane (SUL) is an isothiocyanate derived from cruciferous vegetables and has been reported to have beneficial effects against acute and chronic neuro-degenerative disorders including AD. SUL protected AD-like lesions induced by combined administration of aluminum and D-galactose in Kunming mice (Zhang et al. 2014b) and C57BL/6 mice (Zhang et al. 2014a) by attenuating cholinergic neuronal loss whereas promoting neurogenesis (Zhang et al. 2014b) and amelio-rating formation of Aβ plaques as well as protein carbonyls by restoring expression and activity of GPx (Zhang et al. 2014a). SUL attenuated SCO-induced learning and memory loss by fortifying cholinergic system as shown by the increased expression of choline acetyltransferase (ChAT), decreased activity of

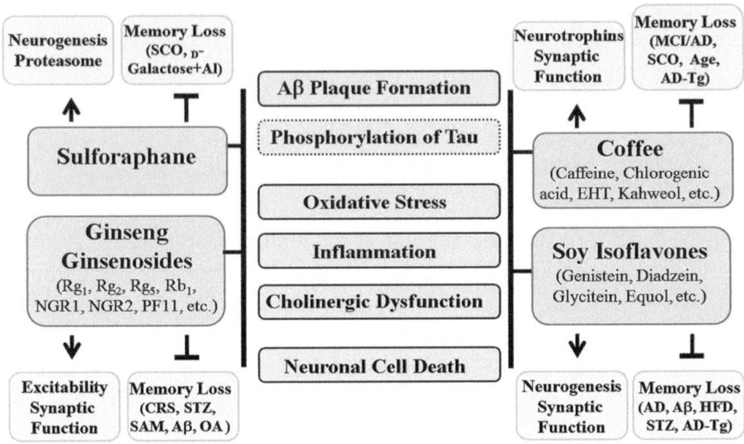

Fig. 21.4 Representative neuroprotective phytochemicals preventing and/or protecting against AD-related neurological damages

acetylcholinesterase (AChE), and elevated acetylcholine (ACh) levels, which were also confirmed in the primary cultures of cortical neurons (Lee et al. 2014).

In Neuro2A murine neuroblastoma cells, SUL exhibited neuroprotective effects against $A\beta_{1-42}$ by inducing proteasome expression (PSMB5 and PSMB6 subunits), which facilitates the clearance of the Aβ and leads to improvement of protein misfolding in AD (Park et al. 2009). Furthermore SUL protected Neuro2A and N1E 115 cells from $A\beta_{1-42}$-induced cell death, which was abolished by inhibition of proteasome activities by MG132. SUL has been also shown to enhance proteasome activities in HeLa and COS-1 cells by activation of heat shock transcription factor 1 (Hsf1)-mediated heat shock responses, particularly up-regulation of heat shock protein 27 (HSP27) (Gan et al. 2010).

SUL exerts its neuroprotective effects in neurodegenerative disorders via mechanisms related to the Nrf2/ARE-involved pathophysiological processes (Table 21.1). In the primary cultures of astrocytes, SUL showed a long-lasting elevation of endogenous antioxidant enzymes such as HO-1, NQO1, GCLC, GCLM, and TrxR1 thereby increasing resistance to H_2O_2-induced cell damages (Bergström et al. 2011). SUL effectively suppressed 6-OHDA-induced caspase-3 activation and subsequent cell death in PC12 cells by inducing the nuclear translocation of Nrf2 and subsequent expression of HO-1 via PI3K/Akt pathway (Deng et al. 2012). $A\beta_{25-35}$-induced cytotoxicity and apoptotic features including c-JNK activation, altered expression of Bcl-2 family proteins, decreased mitochondrial membrane potential, and DNA fragmentation were effectively attenuated by SUL pretreatment in SH-SY5Y cells (Lee et al. 2013). The antiapoptotic activity of SUL was mediated by inhibiting oxidative stress via activation of Nrf2 and Nrf2 target genes such as GCL, HO-1, and NQO1.

In in vitro studies, SUL protected female Fischer rats from spinal cord injury (SCI) by not only increasing levels of Nrf2 and Nrf2-target protein GCL and but

Table 21.1 Neuroprotective phytochemicals-mediated activation of Nrf2 and its downstream targets in neurological disorders

Compounds	Inducers	Experimental models	Downstream targets	Upstream regulators	References
Cruciferous vegetables					
SUL	6-OHDA	PC12 cells	HO-1	PI3K	Deng et al. (2012)
	H_2O_2	Primary astrocytes	GCL, HO-1, NQO1, TrxR1		Bergström et al. (2011)
	$A\beta_{25-35}$	SH-SY5Y cells	GCL, HO-1, NQO1		Lee et al. (2013)
	SCI	Fischer rats	GCL (by deleting HO-1, NQO1)		Wang et al. (2012)
	SCI	ICR mice	NQO1, GST		Jin et al. (2014)
	TBI	SD rats	HO-1, NQO-1		Hong et al. (2010)
	HI	SD rats	HO-1		Ping et al. (2010)
Ginseng					
Gypenoside XVII	$A\beta_{25-35}$	PC12 cells	HO-1	PI3K	Meng et al. (2014a)
Total saponins	H_2O_2	Primary astrocytes	HO-1, GST		Zhou et al. (2014)
Rg_1	Iron	SK-N-SH cells	HO-1, SOD	PI3K	Du et al. (2013)
Panaxatriol saponins	OGD	PC12 cells	HO-1	PI3K	Huang et al. (2014c)
NGR1	OGD	Primary neurons	HO-1	PI3K	Meng et al. (2014b)
Rb_1	6-OHDA	SH-SY5Y cells	HO-1	PI3K	Hwang and Jeong (2010)
NGR2	6-OHDA	SH-SY5Y cells	HO-1, GPx, GR	MEK1/2 ERK1/2	Meng et al. (2013)
AST-IV w/ Rg_1, Rb_1, and NGR1	IR	C57BL/6 mice	HO-1, SOD, GSH		Huang et al. (2014b)
Coffee					
Coffee extract	Aβ	C. elegans			
Decaffeinated coffee extract	Aβ	Drosophila	GSH		Trinh et al. (2010)
Kahweol	6-OHDA	SH-SY5Y cells	HO-1	PI3K p38 MAPK	Hwang and Jeong (2008)
Coffee		Humans (Blood lymphocytes)	NQO1, GST		Volz et al. (2012)

(continued)

Table 21.1 (continued)

Compounds	Inducers	Experimental models	Downstream targets	Upstream regulators	References
Soy					
SIF	Aβ	Wistar rats	HO-1, GCL, GSH		Xi et al. (2014)
SIF	Aβ	Wistar rats	GSH, GPx, SOD, HO-1		Ding et al. (2013)

also decreasing levels of pro-inflammatory cytokines including IL-1β and TNF-α leading to a reduction in contusion volume and improvement in coordination (Wang et al. 2012). Moreover, SUL improved SCI-induced defects in hindlimb locomotor function, histologic injury, neuronal death, and spinal cord edema in male ICR mice by activation of Nrf2 and upregulation of antioxidant enzymes such as NQO1 and GST (Jin et al. 2014). Furthermore, SUL reduced inflammatory damages such as production of pro-inflammatory cytokines such as TNF-α, IL-1β, and IL-6 by inhibition of NF-κB.

In addition, SUL significantly reduced neuronal death, contusion volume, and neurological dysfunction after TBI by activation of Nrf2, up-regulation of Nrf2-dependent antioxidant enzymes (HO-1 and NQO1), and reduction of oxidative damages to lipid (4-HNE) and DNA (8-OHdG), which were abolished in Nrf2-knockout (Hong et al. 2010). In a neonatal hypoxia-ischemia (HI) brain injury model, SUL pretreatment reduced infarct ratio, oxidative damages, activation of microglia, neuronal cell death by increasing expression of Nrf2 and HO-1 in the affected brain area (Ping et al. 2010).

21.5.2 Ginseng

Ginseng, the root of the plant Panax (C.A. Meyer Araliaceae) has been empirically and traditionally used as a general tonic in the herbal medicine of eastern countries to resist against a wide range of physical, chemical, and mental stresses. It is regarded as an adaptogen protecting from noxious stimuli by strengthening physiological functions and by restoring cellular homeostasis (Nocerino et al. 2000). Ginseng exhibits a variety of pharmacological activities including anti-stress, anticonvulsant, antipsychotic, and memory enhancing effects. Ginseng saponins (ginsenosides) are triterpene dammarane derivatives and have a basic structure with a steroid-like four-ring system and carbohydrate moieties attached. Ginsenosides are the principal active ingredients of ginseng and more than 40 different ginsenosides have been isolated and identified.

Ginenosides have reported to exhibit neuroprotective and memory enhancing effects in diverse animal models of memory defects by inhibiting oxidative stress and neuroinflammation. Ginsenoside Rg_1 significantly protected against learning and memory impairments induced by chronic restraint stress (CRS) in male Kunming mice, which was mediated by decreasing ROS generation, subsequent oxidative damages, and cell death in the frontal cortex and hippocampus (Wang et al. 2014). In STZ-injected Wistar rats, Rg_5 improved cognitive dysfunction by attenuating neuroinflammatory responses such as increased expression of TNF-α, IL-1β, iNOS, and COX-2. Moreover, Rg_5 fortified cholinergic function by reducing AChE activity whereas inducing ChAT activity and alleviated Aβ deposits in the hippocampus and cerebral cortex (Chu et al. 2014). In senescence-accelerated mouse (SAM) model, *Panax notoginseng* (Chinese ginseng) saponins inhibited the deposition of $A\beta_{1-40}$ and $A\beta_{1-42}$ and reduced the content of APP by both increasing α-secretase activity and decreasing β-secretase activity, which seemed to be mediated by enhancing up-regulation of ADAM9 and down-regulation of β-site APP cleaving enzyme (BACE, β-secretase) respectively (Huang et al. 2014a). Furthermore, in SAMP8 mice chronically treated with ginsenosides mixtures, besides decreased $A\beta_{1-42}$ levels in the hippocampus, the amount of antioxidant in the serum and the expression of plasticity-related proteins in hippocampus were markedly up-regulated (Zhao et al. 2009).

In cell culture as well as animal models of AD, ginsenosides exhibited neuroprotective effects by regulating diverse molecular mechanisms involved in the neuropathogenesis of AD. In PC12 cells, glutamate treatment decreased cell viability by increasing intracellular calcium levels, production of malondialdehyde (MDA) and NO, and expression of calpain II, caspase-3 and $A\beta_{1-40}$, which were effectively attenuated by Rg_2 (Li et al. 2007). Pretreatment of PC12 cells with Rb_1 not only inhibited $A\beta_{25-35}$-induced ROS overproduction and lipid peroxidation, but also modulated apoptotic signals by increasing the Bcl-2/Bax ratio and attenuating caspase-3 activation (Xie et al. 2010). Notoginsenoside R1 (NGR1), a major saponin isolated from *Panax notoginseng* significantly counteracted the effects of $A\beta_{25-35}$ by increasing cell survival, reducing oxidative damages and apoptosis, and suppressing stress-activated MAPK signaling pathways in PC12 cells (Ma et al. 2014).

In APP/PS1 AD animal model, NGR1 improved the learning performance by inducing the membrane excitability of CA1 pyramidal neurons in hippocampal slices via inhibition of voltage-gated K^+ currents and reversed $A\beta_{1-42}$ oligomers-induced impairments in long term potentiation (LTP) (Yan et al. 2014). Pseudoginsenoside-F11 (PF11), a component of *Panax quinquefolium* (American ginseng) protected against AD-like cognitive impairment induced by i.c.v. injection of $A\beta_{1-42}$ in male KM mice and APP/PS1 transgenic AD mice (Wang et al. 2013). PF11 significantly inhibited the expressions of APP and $A\beta_{1-40}$ in the cortex and hippocampus, restored decreased activities of antioxidant enzymes and increased oxidative damages, and suppressed apoptotic changes such as activation of JNK, p53, and caspase-3. In Sprague Dawley (SD) rats, i.c.v. injection of okadaic acid (OA), a potent phosphatase inhibitor-induced memory impairment as shown by the increased phosphorylation of tau, activation of GSK-3β, and formation of Aβ, was reversed by Rg_1 (Song et al. 2013).

As a promising neuroprotective molecular mechanism of ginseng extracts and ginsenosides, activation of Nrf2 and its downstream target genes were proposed in a diverse experimental models of neurological disorders (Table 21.1). Total saponins in leaves of *Panax notoginseng* reduced H_2O_2-induced cell death in primary rat cortical astrocytes by attenuating intracellular accumulation of ROS via activation of Nrf2 and up-regulation of downstream antioxidant enzymes such as HO-1 and GST (Zhou et al. 2014). In SK-N-SH cells, Rg_1 ameliorated iron-induced neurotoxicity by inducing nuclear translocation of Nrf2 and expression of HO-1 and SOD though PI3K/Akt signals (Du et al. 2013). Panaxatriol saponins, the main components extracted from *Panax notoginseng* and NGR1 protected PC12 cells and primary cortical neurons respectively from oxygen-glucose deprivation (OGD)-reperfusion-induced cell death via activation of PI3K/Akt-Nrf2-HO-1 pathway (Huang et al. 2014c; Meng et al. 2014b). In SH-SY5Y cells, Rb_1 and NGR2 inhibited 6-OHDA-induced apoptotic pathway by activation of Nrf2 and subsequent expression of HO-1 through PI3K/Akt and MEK1/2-ERK1/2 pathways, respectively (Meng et al. 2013; Hwang and Jeong 2010). Astragaloside-IV (AST-IV, the effective component of Astragalus) combined with ginsenoside Rg_1, ginsenoside Rb_1, NGR1 could antagonize ischemic reperfusion (IR)-induced oxidative damages, which were shown by the decreased survival rate of nerve cells, reduced antioxidant systems such as Nrf2, HO-1, SOD, and GSH, and increased contents of MDA and NO in C57BL/6 mice (Huang et al. 2014b).

21.5.3 Coffee

Epidemiologic data have indicated that coffee and caffeine could be an effective therapeutics against AD. Coffee and/or caffeine reversed or protected against AD-like cognitive dysfunction and accumulation of abnormal Aβ protein in transgenic animal models for AD (Arendash and Cao 2010). Early prospective studies in human reported significant less cognitive decline in aged men taking three cups of coffee daily (van Gelder et al. 2007) and in aged women with caffeine intake equivalent to 3+ cups of coffee per day (Ritchie et al. 2007). Recent clinical studies reported a decreased risk of AD in people drinking 3–5 cups of coffee a day during their 40s–50s suggesting an association of coffee drinking at midlife with a decreased risk of AD later in life (Gelber et al. 2011). Moreover, MCI subjects with high plasma caffeine levels greater than 1200 ng/ml did not progressed to dementia during the 2–4 year follow-up period (Cao et al. 2012).

Coffee is a complex mixture of bioactive compounds including natural alkaloid caffeine, phenolic chlorogenic acid, and diterpenes such as kahweol and cafestol. Caffeine treatment reversed pre-existing memory impairment by direct reduction of Aβ deposition in aged AD transgenic mice (AD-Tg) due to suppression of both β-secretase and PS1/γ-secretase expression (Arendash et al. 2009). In addition, administration of a transgenic AD mice with crude caffeine prevented memory impairment by reducing Aβ levels as well as the number of SPs in the hippocampus (Chu et al. 2012). In addition, crude caffeine protected primary neurons from Aβ-induced cell death by suppressing activation of caspase-3. Moreover, caffeine prevents Aβ-induced synaptotoxicity and neuronal damage by antagonizing the adenosine A2A receptor (Canas et al. 2009). A long-term study in mice demonstrated the protective effect of caffeine intake for the prevention of age-related memory decline, which seemed to be associated with changes in neurotrophic factor BDNF levels (Costa et al. 2008).

Chlorogenic acid, a main polyphenolic component in coffee, attenuated A-β-induced cell death in SH-SY5Y cells and restored SCO-induced memory impairment by inhibiting AChE activity and oxidative stress such as generation of free radicals and lipid peroxidation (Kwon et al. 2010). Instant decaffeinated coffee and chlorogenic acid protected PC12 cells from H_2O_2-mediated apoptotic changes such as cleavage of poly(ADP-ribose) polymerase (PARP), down-regulation of anti-apoptotic Bcl-X_L and activation of caspase-3 by blocking the intracellular accumulation of ROS (Butt and Sultan MT 2011). Eicosanoyl-5-hydroxytryptamide (EHT), a minor component of coffee unrelated to caffeine improved AD-related defects including cognitive impairment, elevated levels of Aβ, and hyperphosphorylation of tau in a rat model for AD induced by endogenous inhibitor of phosphatase 2A (PP2A) which involved in the dephosphorylation of tau and APP proteins (Basurto-Islas et al. 2014).

Experimental data from *in vitro*, *in vivo*, and clinical studies regarding the consumption of coffee and/or its active compounds revealed an inverse correlation with progression of AD via activation of Nrf2 (Table 21.1). Coffee and/or its active compounds ameliorate oxidative cell death in AD by inducing a series of cytoprotective genes and proteins via stimulation of Nrf2-ARE signaling pathway. In a transgenic *C. elegans* AD model, coffee extract protected against Aβ-induced toxicity by activation of skn-1 corresponding to Nrf2 in mammals and conversely inactivation of skn-1 genetically or by utilizing RNAi strongly abrogated the protective effects of coffee extract (Dostal et al. 2010). Decaffeinated coffee exerts neuroprotective effects in a Drosophila model of AD by increasing the Cnc (Nrf2 homolog in Drosophila) transcripts and subsequent up-regulation of GSH (Trinh et al. 2010). In SH-SY5Y neuroblastoma cells, kahweol, a diterpene present in coffee significantly reduced 6-OHDA-induced neuronal cell death by suppressing generation of ROS and activation of caspase-3 via up-regulation of PI3K and p38 MAPK-Nrf2-HO-1 signaling pathway (Hwang and Jeong 2008).

Male Wistar rats administered with coffee brew for 28 days exhibited greatly increased cytosolic Nrf2 levels and potentiated activities of antioxidant enzymes (SOD, CAT, and GPx) as well as total antioxidant capacity in the hepatic tissue (Vicente et al. 2014). In HT29 human colon adenocarcinoma cells, chlorogenic acid- and N-methylpyridinium-rich coffee extract (CN-CE) induced nuclear translocation of Nrf2 translocation, enhanced the transcription of ARE-dependent genes such as NQO1 and GSTA1, and increased the enzyme activity of GST. Moreover, in a pilot human intervention study, daily consumption of CN-CE for 4 weeks increased Nrf2 transcription in peripheral blood lymphocytes on average with substantial inter-individual variations (Volz et al. 2012). Coffee lipid fraction containing cafestol and kahweol act as a safeguard against some malignant cells by regulating the detoxifying enzymes, the downstream target genes of Nrf2 activation (Butt and Sultan 2011).

21.5.4 Soy Isoflavones

Soy isoflavones (SIF) such as genistein, daidzein, glycitein, and equol have been reported to exhibit beneficial effects in neurodegenerative disorders including AD. SIF is regarded as phytoestrogens which have been demonstrated in cell culture as well as animal studies to reduce AD-related neuropathology and alleviating risk of AD progressing via interaction with estrogen receptors (Soni et al. 2014). Meta-analysis of ten placebo-controlled randomized controlled trials of SIF supplementation showed that SIF supplementation seemed to have a beneficial effect on improving summary cognitive function and visual memory particularly in postmenopausal women (Cheng et al. 2014).

Soy germ phytoestrogens improved memory performance in ovariectomized rats by increasing expression of BDNF and the synaptic formation proteins such as synaptophysin, spinophilin, synapsin 1 and PSD-95 (Pan et al. 2010). Daidzein also protected against obesity-induced disruption of adult neurogenesis in the subgranular zone (SGZ) of the dentate gyrus and reversed the high fat diet (HFD)-associated apoptotic and pro-inflammatory changes in the hippocampus as well as in the dentate gyrus (Rivera et al. 2013).

SIF improved learning and memory impairments induced by i.c.v. injection of $A\beta_{1-42}$ in Wistar rats and ameliorated vascular Aβ transportation and vascular inflammation by regulating the expression of receptor for advanced glycation endproducts (RAGE), low-density lipoprotein receptor-related protein (LRP)-1, NF-κB, and inflammatory cytokines such as TNF-α and IL-1β (Xi et al. 2013). In female triple transgenic mouse model of AD, early intervention with estrogen receptor β-selective phytoestrogenic formulation (phyto-β-SERM) containing SIF prolonged survival, improved spatial recognition memory, and attenuated deposition of Aβ as well as formation of SPs (Zhao et al. 2013).

Genistein significantly attenuated Aβ-induced cytotoxicity and apoptotic cell death by decreasing intracellular levels of calcium and activation of caspase-3 in PC12 cells (Luo et al. 2012) and cultured hippocampal neurons (Zeng et al. 2004). In the hippocampal cultures, genistein up-regulated the activity of α-secretase and down-regulated the activity of β-secretase, which seemed to be mediated by enhanced expression and activation of PKC (Liao et al. 2013). Furthermore, in the primary culture of astrocytes, genistein suppressed Aβ-induced expression of pro-inflammatory mediators such as COX-2, iNOS, IL-1β, and TNF-α (Valles et al. 2010). In BV2 microglial cells, genistein significantly reversed Aβ-induced inflammatory responses such as increased expression Toll-like receptors 2 and

4, Myd88, IKK, NF-κB and IL-6 and decreased levels TGF-β and IL-10 (Yu et al. 2013).

In addition, as a molecular mechanism SIF is able to protect cells and tissues from oxidative stress related to AD. Glycitein may suppress Aβ toxicity in *C. elegans* through scavenging ROS such as H_2O_2 and hydroxyl radicals and inhibition of Aβ deposits (Gutierrez-Zepeda et al. 2005). In hippocampal neurons, genistein attenuated Aβ-induced cytotoxicity by suppressing intracellular accumulation of ROS (Zeng et al. 2004). SIF significantly attenuated D-galactose-induced oxidative damages such as lipid peroxidation in serum and brain, protein carbonylation in liver, kidney and brain, and RAGEs in serum, and expression of Aβ, PS1 and BACE1 in brain of C57BL/6 mice (Hsieh et al. 2009).

Moreover, in *in vivo* animal models of AD, SIF effectively reduced Aβ-induced oxidative stress such as excess formation of nitrotyrosine in cerebrovascular tissue of Wistar rats (Xi et al. 2014). In this study, SIF maintained cellular redox homeostasis by restoring down-regulation of Nrf2 and suppressing the increase in Keap1. Subsequent expression of Nrf2 target genes such as HO-1 and GCL were elevated by treatment with SIF, which may lead to increased GSH levels and GSH/GSSG ratio. In another study, SIF reduced oxidative stress and improved the antioxidant ability in mitochondria from rat brain damaged by injection of $A\beta_{1-42}$ in Wistar rats via activation of Nrf2 and expression of antioxidant defense enzymes (Ding et al. 2013). SIF restored the reduced glutathione/oxidized glutathione (GSH/GSSG) ratio and increased expression of GPx, MnSOD, and HO-1 in brain tissue. Studies using *in vitro* as well as *in vivo* animal models of cardiovascular disease suggest SIF-increased eNOS activity and activation of Nrf2 signaling pathway leading to up-regulation of detoxifying and antioxidant defense enzymes (Mann et al. 2009). The roles and mechanisms of SIF-induced Nrf2 activation are summarized in Table 21.1.

21.6 Summary and Future Directions

Constitutive activation of the redox-sensitive transcription factor Nrf2 is part of a self-defense program that enables neuronal cells to protect themselves against oxidative stress. However, the complete molecular mechanisms involved in the potentiation of cellular antioxidant defense capacity by Nrf2 remain to be elucidated. Continued attempts are required to identify novel molecular targets of Nrf2 function and to elucidate their cross-talk with upstream as well as downstream signaling molecules. Considering some neuronal cells can survive the accumulating oxidative damages and degenerative processes, the understanding of the molecular events that can alleviate the vulnerability of neurons and consequently can increase their resistance to oxidative stress by using Nrf2-activating phytochemicals is of great interest in the context of establishing efficient therapeutic strategies for the prevention and management of AD.

References

Abramov AY, Canevari L, Duchen MR et al (2003) Changes in intracellular calcium and glutathione in astrocytes as the primary mechanism of amyloid neurotoxicity. J Neurosci 23: 5088–5095

Akiyama H, Barger S, Barnum S et al (2000) Inflammation and Alzheimer's disease. Neurobiol Aging 21(3):383–421

Aksenov MY, Markesbery WR (2001) Changes in thiol content and expression of glutathione redox system genes in the hippocampus and cerebellum in Alzheimer's disease. Neurosci Lett 302:141–145

Alam J, Cook JL (2003) Transcriptional regulation of the heme oxygenase-1 gene via the stress response element pathway. Curr Pharm Des 9:2499–2511

Alam J, Stewart D, Touchard C et al (1999) Nrf2, a Cap'n'Collar transcription factor, regulates induction of the heme oxygenase-1 gene. J Biol Chem 274:26071–26078

Allen SJ, Watson JJ, Shoemark DK et al (2013) GDNF, NGF and BDNF as therapeutic options for neurodegeneration. Pharmacol Ther 138(2):155–175. doi:10.1016/j.pharmthera.2013.01.004

Arendash GW, Cao C (2010) Caffeine and coffee as therapeutics against Alzheimer's disease. J Alzheimers Dis 20(Suppl 1):S117–S126. doi:10.3233/JAD-2010-091249

Arendash GW, Mori T, Cao C et al (2009) Caffeine reverses cognitive impairment and decreases brain amyloid-beta levels in aged Alzheimer's disease mice. J Alzheimers Dis 17(3):661–680. doi:10.3233/JAD-2009-1087

Bains JS, Shaw CA (1997) Neurodegenerative disorders in humans: the role of glutathione in oxidative stress-mediated neuronal death. Brain Res Brain Res Rev 25:335–358

Basurto-Islas G, Blanchard J, Tung YC et al (2014) Therapeutic benefits of a component of coffee in a rat model of Alzheimer's disease. Neurobiol Aging 35(12):2701–2712. doi:10.1016/j.neurobiolaging.2014.06.012

Bergström P, Andersson HC, Gao Y et al (2011) Repeated transient sulforaphane stimulation in astrocytes leads to prolonged Nrf2-mediated gene expression and protection from superoxide-induced damage. Neuropharmacology 60(2–3):343–353. doi:10.1016/j.neuropharm.2010.09.023

Berhane K, Widersten M, Engstrom A et al (1994) Detoxication of base propenals and other alpha, beta-unsaturated aldehyde products of radical reactions and lipid peroxidation by human glutathione transferases. Proc Natl Acad Sci U S A 91:1480–1484

Bloom D, Dhakshinamoorthy S, Jaiswal AK et al (2002) Site-directed mutagenesis of cysteine to serine in the DNA binding region of Nrf2 decreases its capacity to upregulate antioxidant response element-mediated expression and antioxidant induction of NAD(P)H:quinone oxidoreductase1 gene. Oncogene 21(14):2191–2200

Butt MS, Sultan MT (2011) Coffee and its consumption: benefits and risks. Crit Rev Food Sci Nutr 51(4):363–373. doi:10.1080/10408390903586412

Calabrese V, Cornelius C, Rizzarelli E et al (2009) Nitric oxide in cell survival: a janus molecule. Antioxid Redox Signal 11(11):2717–2739. doi:10.1089/ARS.2009.2721

Canas PM, Porciúncula LO, Cunha GM et al (2009) Adenosine A2A receptor blockade prevents synaptotoxicity and memory dysfunction caused by beta-amyloid peptides via p38 mitogen-activated protein kinase pathway. J Neurosci 29(47):14741–14751. doi:10.1523/JNEUROSCI.3728-09.2009

Cao TT, Ma L, Kandpal G et al (2005) Increased nuclear factor-erythroid 2 p45-related factor 2 activity protects SH-SY5Y cells against oxidative damage. J Neurochem 95(2):406–417

Cao C, Loewenstein DA, Lin X et al (2012) High blood caffeine levels in MCI linked to lack of progression to dementia. J Alzheimers Dis 30(3):559–572. doi:10.3233/JAD-2012-111781

Carter CA, Kane CJ (2004) Therapeutic potential of natural compounds that regulate the activity of protein kinase C. Curr Med Chem 11(21):2883–2902
Cecchi C, Latorraca S, Sorbi S et al (1999) Glutathione level is altered in lymphoblasts from patients with familial Alzheimer's disease. Neurosci Lett 275:152–154
Chan JY, Kwong M (2000) Impaired expression of glutathione synthetic enzyme genes in mice with targeted deletion of the Nrf2 basic-leucine zipper protein. Biochim Biophys Acta 1517: 19–26
Chanas SA, Jiang Q, McMahon M et al (2002) Loss of the Nrf2 transcription factor causes a marked reduction in constitutive and inducible expression of the glutathione *S*-transferase Gsta1, Gsta2, Gstm1, Gstm2, Gstm3 and Gstm4 genes in the livers of male and female mice. Biochem J 365:405–416
Chen K, Gunter K, Maines MD et al (2000) Neurons overexpressing heme oxygenase-1 resist oxidative stress-mediated cell death. Neurochemistry 75:304–313
Cheng PF, Chen JJ, Zhou XY et al (2014) Do soy isoflavones improve cognitive function in postmenopausal women? A meta-analysis. Menopause, 7 Jul 2014. [Epub ahead of print]
Chu YF, Chang WH, Black RM et al (2012) Crude caffeine reduces memory impairment and amyloid β(1-42) levels in an Alzheimer's mouse model. Food Chem 135(3):2095–2102. doi:10.1016/j.foodchem.2012.04.148
Chu S, Gu J, Feng L et al (2014) Ginsenoside Rg5 improves cognitive dysfunction and beta-amyloid deposition in STZ-induced memory impaired rats via attenuating neuroinflammatory responses. Int Immunopharmacol 19(2):317–326. doi:10.1016/j.intimp.2014.01.018
Correa F, Mallard C, Nilsson M et al (2011) Activated microglia decrease histone acetylation and Nrf2-inducible anti-oxidant defence in astrocytes: restoring effects of inhibitors of HDACs, p38 MAPK and GSK3β. Neurobiol Dis 44(1):142–151. doi:10.1016/j.nbd.2011.06.016
Correa F, Mallard C, Nilsson M et al (2012) Dual TNFα-induced effects on NRF2 mediated antioxidant defence in astrocyte-rich cultures: role of protein kinase activation. Neurochem Res 37(12):2842–2855. doi:10.1007/s11064-012-0878-y
Costa MS, Botton PH, Mioranzza S et al (2008) Caffeine prevents age-associated recognition memory decline and changes brain-derived neurotrophic factor and tirosine kinase receptor (TrkB) content in mice. Neuroscience 153(4):1071–1078. doi:10.1016/j.neuroscience.2008.03.038
Deng C, Tao R, Yu SZ et al (2012) Sulforaphane protects against 6-hydroxydopamine-induced cytotoxicity by increasing expression of heme oxygenase-1 in a PI3K/Akt-dependent manner. Mol Med Rep 5(3):847–851. doi:10.3892/mmr.2011.731
Dickinson DA, Levonen AL, Moellering DR et al (2004) Human glutamate cysteine ligase gene regulation through the electrophile response element. Free Radic Biol Med 37:1152–1159
Dilshara MG, Lee KT, Jayasooriya RG et al (2014) Downregulation of NO and PGE_2 in LPS-stimulated BV2 microglial cells by trans-isoferulic acid via suppression of PI3K/Akt-dependent NF-κB and activation of Nrf2-mediated HO-1. Int Immunopharmacol 18(1): 203–211. doi:10.1016/j.intimp.2013.11.020
Ding J, Yu HL, Ma WW et al (2013) Soy isoflavone attenuates brain mitochondrial oxidative stress induced by β-amyloid peptides 1-42 injection in lateral cerebral ventricle. J Neurosci Res 91(4):562–567. doi:10.1002/jnr.23163
Dore S (2002) Decreased activity of the antioxidant heme oxygenase enzyme: implications in ischemia and in Alzheimer's disease. Free Radic Biol Med 32:1276–1282
Dostal V, Roberts CM, Link CD et al (2010) Genetic mechanisms of coffee extract protection in a Caenorhabditis elegans model of β-amyloid peptide toxicity. Genetics 186(3):857–866. doi:10.1534/genetics.110.120436

Du Y, Wooten MC, Gearing M et al (2009) Age-associated oxidative damage to the p62 promoter: implications for Alzheimer disease. Free Radic Biol Med 46(4):492–501. doi:10.1016/j.freeradbiomed.2008.11.003

Du X, Xu H, Jiang H et al (2013) Akt/Nrf2 activated upregulation of heme oxygenase-1 involves in the role of Rg1 against ferrous iron-induced neurotoxicity in SK-N-SH cells. Neurotox Res 24(1):71–79. doi:10.1007/s12640-012-9362-3

Durany N, Münch G, Michel T et al (1999) Investigations on oxidative stress and therapeutical implications in dementia. Eur Arch Psychiatry Clin Neurosci 249(Suppl 3):68–73

Eftekharzadeh B, Maghsoudi N, Khodagholi F et al (2010) Stabilization of transcription factor Nrf2 by tBHQ prevents oxidative stress-induced amyloid beta formation in NT2N neurons. Biochimie 92(3):245–253. doi:10.1016/j.biochi.2009.12.001

Farr SA, Ripley JL, Sultana R et al (2014) Antisense oligonucleotide against GSK-3β in brain of SAMP8 mice improves learning and memory and decreases oxidative stress: involvement of transcription factor Nrf2 and implications for Alzheimer disease. Free Radic Biol Med 67: 387–395. doi:10.1016/j.freeradbiomed.2013.11.014

Felsenstein KM, Candelario KM, Steindler DA et al (2014) Regenerative medicine in Alzheimer's disease. Transl Res 163(4):432–438. doi:10.1016/j.trsl.2013.11.001

Friedman LG, Qureshi YH, Yu WH et al (2014) Promoting autophagic clearance: viable therapeutic targets in Alzheimer's disease. Neurotherapeutics, 25 Nov 2014. [Epub ahead of print]

Galeazzi L, Ronchi P, Franceschi C et al (1999) In vitro peroxidase oxidation induces stable dimers of beta-amyloid (1-42) through dityrosine bridge formation. Amyloid 6(1):7–13

Gan N, Wu YC, Brunet M et al (2010) Sulforaphane activates heat shock response and enhances proteasome activity through up-regulation of Hsp27. J Biol Chem 285(46):35528–35536. doi:10.1074/jbc.M110.152686

Gelber RP, Petrovitch H, Masaki KH et al (2011) Coffee intake in midlife and risk of dementia and its neuropathologic correlates. J Alzheimers Dis 23(4):607–615. doi:10.3233/JAD-2010-101428

Grammas P, Ovase R (2001) Inflammatory factors are elevated in brain microvessels in Alzheimer's disease. Neurobiol Aging 22:837–842

Gutcher I, Webb PR, Anderson NG et al (2003) The isoform-specific regulation of apoptosis by protein kinase C. Cell Mol Life Sci 60(6):1061–1070

Gutierrez-Zepeda A, Santell R, Wu Z et al (2005) Soy isoflavone glycitein protects against beta amyloid-induced toxicity and oxidative stress in transgenic Caenorhabditis elegans. BMC Neurosci 6:54

Gwon AR, Park JS, Arumugam TV et al (2012) Oxidative lipid modification of nicastrin enhances amyloidogenic γ-secretase activity in Alzheimer's disease. Aging Cell 11(4):559–568. doi:10.1111/j.1474-9726.2012.00817.x

Hara H, Ohta M, Ohta K et al (2003) Increase of antioxidative potential by tert-butylhydroquinone protects against cell death associated with 6-hydroxydopamine-induced oxidative stress in neuroblastoma SH-SY5Y cells. Brain Res Mol Brain Res 119:125–131

Hardy JA, Higgins GA (1992) Alzheimer's disease: the amyloid cascade hypothesis. Science 256:184–185

Hayes JD, McMahon M (2001) Molecular basis for the contribution of the antioxidant responsive element to cancer chemoprevention. Cancer Lett 174:103–113

Hayes JD, Flanagan JU, Jowsey IR et al (2005) Glutathione transferases. Annu Rev Pharmacol Toxicol 45:51–88

Heras-Sandoval D, Pérez-Rojas JM, Hernández-Damián J et al (2014) The role of PI3K/AKT/mTOR pathway in the modulation of autophagy and the clearance of protein aggregates in neurodegeneration. Cell Signal 26(12):2694–2701. doi:10.1016/j.cellsig.2014.08.019

Hernández F, Borrell J, Guaza C et al (2002) Spatial learning deficit in transgenic mice that conditionally over-express GSK-3beta in the brain but do not form tau filaments. J Neurochem 83(6):1529–1533

Ho L, Pieroni C, Winger D et al (1999) Regional distribution of cyclooxygenase-2 in the hippocampal formation in Alzheimer's disease. J Neurosci Res 57(3):295–303

Hong Y, Yan W, Chen S et al (2010) The role of Nrf2 signaling in the regulation of antioxidants and detoxifying enzymes after traumatic brain injury in rats and mice. Acta Pharmacol Sin 31(11):1421–1430. doi:10.1038/aps.2010.101

Hsieh HM, Wu WM, Hu ML et al (2009) Soy isoflavones attenuate oxidative stress and improve parameters related to aging and Alzheimer's disease in C57BL/6J mice treated with D-galactose. Food Chem Toxicol 47(3):625–632. doi:10.1016/j.fct.2008.12.026

Huang HC, Nguyen T, Pickett CB et al (2000) Regulation of the antioxidant response element by protein kinase C-mediated phosphorylation of NF-E2-related factor 2. Proc Natl Acad Sci U S A 97(23):12475–12480

Huang HC, Nguyen T, Pickett CB et al (2002) Phosphorylation of Nrf2 at Ser-40 by protein kinase C regulates antioxidant response element-mediated transcription. J Biol Chem 277(45): 42769–42774

Huang J, Wu D, Wang J et al (2014a) Effects of Panax notoginseng saponin on α, β, and γ secretase involved in Aβ deposition in SAMP8 mice. Neuroreport 25(2):89–93. doi:10.1097/WNR. 0000000000000048

Huang XP, Qiu YY, Wang B et al (2014b) Effects of Astragaloside IV combined with the active components of Panax notoginseng on oxidative stress injury and nuclear factor-erythroid 2-related factor 2/heme oxygenase-1 signaling pathway after cerebral ischemia-reperfusion in mice. Pharmacogn Mag 10(40):402–409. doi:10.4103/0973-1296.141765

Huang Y, Yu J, Wan F et al (2014c) Panaxatriol saponins attenuated oxygen-glucose deprivation injury in PC12 cells via activation of PI3K/Akt and Nrf2 signaling pathway. Oxid Med Cell Longev 2014:978034. doi:10.1155/2014/978034

Hwang YP, Jeong HG (2008) The coffee diterpene kahweol induces heme oxygenase-1 via the PI3K and p38/Nrf2 pathway to protect human dopaminergic neurons from 6-hydroxydopamine-derived oxidative stress. FEBS Lett 582(17):2655–2662. doi:10.1016/j. febslet.2008.06.045

Hwang YP, Jeong HG (2010) Ginsenoside Rb_1 protects against 6-hydroxydopamine-induced oxidative stress by increasing heme oxygenase-1 expression through an estrogen receptor-related PI3K/Akt/Nrf2-dependent pathway in human dopaminergic cells. Toxicol Appl Pharmacol 242(1):18–28. doi:10.1016/j.taap.2009.09.009

Ikeda H, Nishi S, Sakai M et al (2004) Transcription factor Nrf2/MafK regulates rat placental glutathione *S*-transferase gene during hepatocarcinogenesis. Biochem J 380:515–521

Innamorato NG, Rojo AI, García-Yagüe AJ et al (2008) The transcription factor Nrf2 is a therapeutic target against brain inflammation. J Immunol 181(1):680–689

Ishii T, Itoh K, Sato H et al (1999) Oxidative stress-inducible proteins in macrophages. Free Radic Res 31:351–355

Ishikawa M, Numazawa S, Yoshida T (2005) Redox regulation of the transcriptional repressor Bach1. Free Radic Biol Med 38(10):1344–1352

Jain AK, Jaiswal AK et al (2007) GSK-3beta acts upstream of Fyn kinase in regulation of nuclear export and degradation of NF-E2 related factor 2. J Biol Chem 282(22):16502–16510

Jain A, Lamark T, Sjøttem E et al (2010) p62/SQSTM1 is a target gene for transcription factor NRF2 and creates a positive feedback loop by inducing antioxidant response element-driven gene transcription. J Biol Chem 285(29):22576–22591. doi:10.1074/jbc.M110.118976

Jaiswal AK (2000) Regulation of genes encoding NAD(P)H:quinone oxidoreductases. Free Radic Biol Med 29:254–262

Jeyapaul J, Jaiswal AK (2000) Nrf2 and c-Jun regulation of antioxidant response element (ARE)-mediated expression and induction of γ-glutamylcysteine synthetase heavy subunit gene. Biochem Pharmacol 59:1433–1439

Jin W, Wang H, Yan W et al (2008) Disruption of Nrf2 enhances upregulation of nuclear factor-κB activity, proinflammatory cytokines, and intercellular adhesion molecule-1 in the brain after traumatic brain injury. Mediators Inflamm 2008:725174. doi:10.1155/2008/725174

Jin W, Ni H, Hou X et al (2014) Tert-butylhydroquinone protects the spinal cord against inflammatory response produced by spinal cord injury. Ann Clin Lab Sci 44(2):151–157
Jo C, Gundemir S, Pritchard S et al (2014) Nrf2 reduces levels of phosphorylated tau protein by inducing autophagy adaptor protein NDP52. Nat Commun 5:3496. doi:10.1038/ncomms4496
Joshi G, Gan KA, Johnson DA et al (2014) Increased Alzheimer's disease-like pathology in the APP/ PS1ΔE9 mouse model lacking Nrf2 through modulation of autophagy. Neurobiol Aging, pii: S0197-4580(14)00599-5. doi: 10.1016/j.neurobiolaging.2014.09.004
Kang KW, Cho MK, Lee CH et al (2001) Activation of phosphatidylinositol 3-kinase and Akt by tert-butylhydroquinone is responsible for antioxidant response element-mediated rGSTA2 induction in H4IIE cells. Mol Pharmacol 59(5):1147–1156
Kang KW, Lee SJ, Park JW et al (2002) Phosphatidylinositol 3-kinase regulates nuclear translocation of NF-E2-related factor 2 through actin rearrangement in response to oxidative stress. Mol Pharmacol 62(5):1001–1010
Kang CH, Choi YH, Moon SK et al (2013) Quercetin inhibits lipopolysaccharide-induced nitric oxide production in BV2 microglial cells by suppressing the NF-κB pathway and activating the Nrf2-dependent HO-1 pathway. Int Immunopharmacol 17(3):808–813. doi:10.1016/j.intimp.2013.09.009
Kärkkäinen V, Pomeshchik Y, Savchenko E et al (2014) Nrf2 regulates neurogenesis and protects neural progenitor cells against Aβ toxicity. Stem Cells 32(7):1904–1916. doi:10.1002/stem.1666
Kim HC, Yamada K, Nitta A et al (2003) Immunocytochemical evidence that $A\beta_{1\text{-}42}$ impairs endogenous antioxidant systems in vivo. Neuroscience 119:399–419
Kim S, Lee D, Song JC et al (2014) NDP52 associates with phosphorylated tau in brains of an Alzheimer disease mouse model. Biochem Biophys Res Commun 454(1):196–201. doi:10.1016/j.bbrc.2014.10.066
Kimpara T, Takeda A, Yamaguchi T et al (2000) Increased bilirubins and their derivatives in cerebrospinal fluid in Alzheimer's disease. Neurobiol Aging 21:551–554
Komatsu M, Kurokawa H, Waguri S et al (2010) The selective autophagy substrate p62 activates the stress responsive transcription factor Nrf2 through inactivation of Keap1. Nat Cell Biol 12(3):213–223. doi:10.1038/ncb2021
Kosaka K, Mimura J, Itoh K et al (2010) Role of Nrf2 and p62/ZIP in the neurite outgrowth by carnosic acid in PC12h cells. J Biochem 147(1):73–81. doi:10.1093/jb/mvp149
Kwon SH, Lee HK, Kim JA et al (2010) Neuroprotective effects of chlorogenic acid on scopolamine-induced amnesia via anti-acetylcholinesterase and anti-oxidative activities in mice. Eur J Pharmacol 649(1–3):210–217. doi:10.1016/j.ejphar.2010.09.001
Lau A, Wang XJ, Zhao F et al (2010) A noncanonical mechanism of Nrf2 activation by autophagy deficiency: direct interaction between Keap1 and p62. Mol Cell Biol 30(13):3275–3285. doi:10.1128/MCB.00248-10
Le WD, Xie WJ, Appel SH et al (1999) Protective role of heme oxygenase-1 in oxidative stress-induced neuronal injury. J Neurosci Res 56:652–658
Lee JM, Calkins MJ, Chan K et al (2003a) Identification of the NF-E2-related factor-2-dependent genes conferring protection against oxidative stress in primary cortical astrocytes using oligonucleotide microarray analysis. J Biol Chem 278(14):12029–12038
Lee JM, Shih AY, Murphy TH et al (2003b) NF-E2-related factor-2 mediates neuroprotection against mitochondrial complex I inhibitors and increased concentrations of intracellular calcium in primary cortical neurons. J Biol Chem 278(39):37948–37956
Lee C, Park GH, Kim CY et al (2011) [6]-Gingerol attenuates β-amyloid-induced oxidative cell death via fortifying cellular antioxidant defense system. Food Chem Toxicol 49(6):1261–1269. doi:10.1016/j.fct.2011.03.005
Lee C, Park GH, Lee SR et al (2013) Attenuation of β-amyloid-induced oxidative cell death by sulforaphane via activation of NF-E2-related factor 2. Oxid Med Cell Longev 2013:313510. doi:10.1155/2013/313510

Lee JS, Surh YJ (2005) Nrf2 as a novel molecular target for chemoprevention. Cancer Lett 224 (2):171–184

Lee S, Kim J, Seo SG et al (2014) Sulforaphane alleviates scopolamine-induced memory impairment in mice. Pharmacol Res 85:23–32. doi:10.1016/j.phrs.2014.05.003

Li J, Lee JM, Johnson JA et al (2002) Microarray analysis reveals an antioxidant responsive element-driven gene set involved in conferring protection from an oxidative stress-induced apoptosis in IMR-32 cells. J Biol Chem 277:388–394

Li J, Johnson D, Calkins M et al (2005) Stabilization of Nrf2 by tBHQ confers protection against oxidative stress-induced cell death in human neural stem cells. Toxicol Sci 83(2):313–328

Li N, Liu B, Dluzen DE et al (2007) Protective effects of ginsenoside Rg_2 against glutamate-induced neurotoxicity in PC12 cells. J Ethnopharmacol 111(3):458–463

Li L, Li W, Jung SW et al (2011) Protective effects of decursin and decursinol angelate against amyloid β-protein-induced oxidative stress in the PC12 cell line: the role of Nrf2 and antioxidant enzymes. Biosci Biotechnol Biochem 75(3):434–442

Li XH, Li CY, Lu JM et al (2012) Allicin ameliorates cognitive deficits ageing-induced learning and memory deficits through enhancing of Nrf2 antioxidant signaling pathways. Neurosci Lett 514(1):46–50. doi:10.1016/j.neulet.2012.02.054

Liao YF, Wang BJ, Cheng HT et al (2004) Tumor necrosis factor-α, interleukin-1β, and interferon-γ stimulate γ-secretase-mediated cleavage of amyloid precursor protein through a JNK-dependent MAPK pathway. J Biol Chem 279(47):49523–49532

Liao W, Jin G, Zhao M et al (2013) The effect of genistein on the content and activity of α- and β-secretase and protein kinase C in Aβ-injured hippocampal neurons. Basic Clin Pharmacol Toxicol 112(3):182–185. doi:10.1111/bcpt.12009

Liu H, Wang H, Shenvi S et al (2004) Glutathione metabolism during aging and in Alzheimer disease. Ann N Y Acad Sci 1019:346–349

Lovell MA, Ehmann WD, Mattson MP et al (1997) Elevated 4-hydroxynonenal in ventricular fluid in Alzheimer's disease. Neurobiol Aging 18(5):457–461

Lovell MA, Xie C, Markesbery WR et al (1998) Decreased glutathione transferase activity in brain and ventricular fluid in Alzheimer's disease. Neurology 51:1562–1566

Lovell MA, Gabbita SP, Markesbery WR et al (1999) Increased DNA oxidation and decreased levels of repair products in Alzheimer's disease ventricular CSF. J Neurochem 72(2):771–776

Luo S, Lan T, Liao W et al (2012) Genistein inhibits $A\beta_{25-35}$-induced neurotoxicity in PC12 cells via PKC signaling pathway. Neurochem Res 37(12):2787–2794. doi:10.1007/s11064-012-0872-4

Lüth HJ, Münch G, Arendt T et al (2002) Aberrant expression of NOS isoforms in Alzheimer's disease is structurally related to nitrotyrosine formation. Brain Res 953(1–2):135–143

Ma B, Meng X, Wang J et al (2014) Notoginsenoside R1 attenuates amyloid-β-induced damage in neurons by inhibiting reactive oxygen species and modulating MAPK activation. Int Immunopharmacol 22(1):151–159. doi:10.1016/j.intimp.2014.06.018

Malm TM, Iivonen H, Goldsteins G et al (2007) Pyrrolidine dithiocarbamate activates Akt and improves spatial learning in APP/PS1 mice without affecting beta-amyloid burden. J Neurosci 27(14):3712–3721

Mann GE, Bonacasa B, Ishii T et al (2009) Targeting the redox sensitive Nrf2-Keap1 defense pathway in cardiovascular disease: protection afforded by dietary isoflavones. Curr Opin Pharmacol 9(2):139–145. doi:10.1016/j.coph.2008.12.012

McMahon M, Itoh K, Yamamoto M et al (2001) The Cap'n'Collar basic leucine zipper transcription factor Nrf2 (NF-E2 p45-related factor 2) controls both constitutive and inducible expression of intestinal detoxification and glutathione biosynthetic enzymes. Cancer Res 61: 3299–3307

Mecocci P, MacGarvey U, Beal MF et al (1994) Oxidative damage to mitochondrial DNA is increased in Alzheimer's disease. Ann Neurol 36(5):747–751

Meng XB, Sun GB, Wang M et al (2013) P90RSK and Nrf2 activation via MEK1/2-ERK1/2 pathways mediated by Notoginsenoside R2 to prevent 6-hydroxydopamine-induced apoptotic

death in SH-SY5Y cells. Evid Based Complement Alternat Med 2013:971712. doi:10.1155/2013/971712
Meng X, Wang M, Sun G et al (2014a) Attenuation of Aβ25-35-induced parallel autophagic and apoptotic cell death by gypenoside XVII through the estrogen receptor-dependent activation of Nrf2/ARE pathways. Toxicol Appl Pharmacol 279(1):63–75. doi:10.1016/j.taap.2014.03.026
Meng X, Wang M, Wang X et al (2014b) Suppression of NADPH oxidase- and mitochondrion-derived superoxide by Notoginsenoside R1 protects against cerebral ischemia-reperfusion injury through estrogen receptor-dependent activation of Akt/Nrf2 pathways. Free Radic Res 48(7):823–838. doi:10.3109/10715762.2014.911853
Murakami K, Murata N, Noda Y et al (2012) Stimulation of the amyloidogenic pathway by cytoplasmic superoxide radicals in an Alzheimer's disease mouse model. Biosci Biotechnol Biochem 76(6):1098–1103
Murphy TH, De Long MJ, Coyle JT et al (1991) Enhanced NAD(P)H:quinone reductase activity prevents glutamate toxicity produced by oxidative stress. Neurochemistry 56:990–995
Nakaso K, Yano H, Fukuhara Y et al (2003) PI3K is a key molecule in the Nrf2-mediated regulation of antioxidative proteins by hemin in human neuroblastoma cells. FEBS Lett 546:181–184
Narasimhan M, Mahimainathan L, Rathinam ML et al (2011) Overexpression of Nrf2 protects cerebral cortical neurons from ethanol-induced apoptotic death. Mol Pharmacol 80(6):988–999. doi:10.1124/mol.111.073262
Nguyen T, Sherratt PJ, Pickett CB et al (2003) Regulatory mechanisms controlling gene expression mediated by the antioxidant response element. Annu Rev Pharmacol Toxicol 43:233–260
Nioi P, Hayes JD (2004) Contribution of NAD(P)H:quinone oxidoreductase 1 to protection against carcinogenesis, and regulation of its gene by the Nrf2 basic-region leucine zipper and the arylhydrocarbon receptor basic helix-loop-helix transcription factors. Mutat Res 555:149–171
Nocerino E, Amato M, Izzo AA (2000) The aphrodisiac and adaptogenic properties of ginseng. Fitoterapia 71(Suppl 1):S1–S5
Numazawa S, Ishikawa M, Yoshida A et al (2003) Atypical protein kinase C mediates activation of NF-E2-related factor 2 in response to oxidative stress. Am J Physiol Cell Physiol 285(2):C334–C342
Obulesu M, Lakshmi MJ (2014) Apoptosis in Alzheimer's disease: an understanding of the physiology, pathology and therapeutic avenues. Neurochem Res 39(12):2301–2312. doi:10.1007/s11064-014-1454-4
Ohtsubo T, Kamada S, Mikami T et al (1999) Identification of NRF2, a member of the NF-E2 family of transcription factors, as a substrate for caspase-3(-like) proteases. Cell Death Differ 6(9):865–872
Olivieri G, Baysang G, Meier F et al (2001) *N*-acetyl-L-cysteine protects SHSY5Y neuroblastoma cells from oxidative stress and cell cytotoxicity: effects on β-amyloid secretion and tau phosphorylation. J Neurochem 76:224–233
Owuor ED, Kong AN (2002) Antioxidants and oxidants regulated signal transduction pathways. Biochem Pharmacol 64:765–770
Pan M, Li Z, Yeung V et al (2010) Dietary supplementation of soy germ phytoestrogens or estradiol improves spatial memory performance and increases gene expression of BDNF, TrkB receptor and synaptic factors in ovariectomized rats. Nutr Metab (Lond) 7:75. doi:10.1186/1743-7075-7-75
Panahian N, Yoshiura M, Maines MD et al (1999) Overexpression of heme oxygenase-1 is neuroprotective in a model of permanent middle cerebral artery occlusion in transgenic mice. J Neurochem 72:1187–1203
Park HM, Kim JA, Kwak MK et al (2009) Protection against amyloid beta cytotoxicity by sulforaphane: role of the proteasome. Arch Pharm Res 32(1):109–115. doi:10.1007/s12272-009-1124-2
Pasinetti GM, Aisen PS (1998) Cyclooxygenase-2 expression is increased in frontal cortex of Alzheimer's disease brain. Neuroscience 87(2):319–324

Pei JJ, Tanaka T, Tung YC et al (1997) Distribution, levels, and activity of glycogen synthase kinase-3 in the Alzheimer disease brain. J Neuropathol Exp Neurol 56(1):70–78

Ping Z, Liu W, Kang Z et al (2010) Sulforaphane protects brains against hypoxic-ischemic injury through induction of Nrf2-dependent phase 2 enzyme. Brain Res 1343:178–185. doi:10.1016/j.brainres.2010.04.036

Premkumar DR, Smith MA, Richey PL et al (1995) Induction of heme oxygenase-1 mRNA and protein in neocortex and cerebral vessels in Alzheimer's disease. J Neurochem 65:1399–1402

Rada P, Rojo AI, Evrard-Todeschi N et al (2012) Structural and functional characterization of Nrf2 degradation by the glycogen synthase kinase 3/β-TrCP axis. Mol Cell Biol 32(17):3486–3499. doi:10.1128/MCB.00180-12

Rinaldi P, Polidori MC, Metastasio A et al (2003) Plasma antioxidants are similarly depleted in mild cognitive impairment and in Alzheimer's disease. Neurobiol Aging 24(7):915–919

Ritchie K, Carrière I, de Mendonca A et al (2007) The neuroprotective effects of caffeine: a prospective population study (the Three City Study). Neurology 69(6):536–545

Rivera P, Pérez-Martín M, Pavón FJ et al (2013) Pharmacological administration of the isoflavone daidzein enhances cell proliferation and reduces high fat diet-induced apoptosis and gliosis in the rat hippocampus. PLoS One 8(5), e64750. doi:10.1371/journal.pone.0064750. Print 2013

Rodgers EE, Theibert AB (2002) Functions of PI 3-kinase in development of the nervous system. Int J Dev Neurosci 20(3–5):187–197

Rohn TT, Head E (2009) Caspases as therapeutic targets in Alzheimer's disease: is it time to "cut" to the chase? Int J Clin Exp Pathol 2(2):108–118

Rojo AI, Innamorato NG, Martín-Moreno AM et al (2010) Nrf2 regulates microglial dynamics and neuroinflammation in experimental Parkinson's disease. Glia 58(5):588–598

Safar MM, Arab HH, Rizk SM et al (2014) Bone marrow-derived endothelial progenitor cells protect against scopolamine-induced Alzheimer-like pathological aberrations. Mol Neurobiol, 21 Dec 2014. [Epub ahead of print]

Sandberg M, Patil J, D'Angelo B et al (2013) NRF2-regulation in brain health and disease: implication of cerebral inflammation. Neuropharmacology 79:298–306. doi:10.1016/j.neuropharm.2013.11.004

SantaCruz KS, Yazlovitskaya E, Collins J et al (2004) Regional NAD(P)H:quinone oxidoreductase activity in Alzheimer's disease. Neurobiol Aging 25:63–69

Sharpe MA, Clark JB, Duchen MR et al (2002) Amyloid fragment 25-35 causes mitochondrial dysfunction in primary cortical neurons. Neurobiol Dis 10:258–267

Shih AY, Johnson DA, Wong G et al (2003) Coordinate regulation of glutathione biosynthesis and release by Nrf2-expressing glia potently protects neurons from oxidative stress. J Neurosci 23(8):3394–3406

Song XY, Hu JF, Chu SF et al (2013) Ginsenoside Rg_1 attenuates okadaic acid induced spatial memory impairment by the GSK3β/tau signaling pathway and the Aβ formation prevention in rats. Eur J Pharmacol 710(1–3):29–38. doi:10.1016/j.ejphar.2013.03.051

Song J, Hur BE, Bokara KK et al (2014) Agmatine improves cognitive dysfunction and prevents cell death in a streptozotocin-induced Alzheimer rat model. Yonsei Med J 55(3):689–699. doi:10.3349/ymj.2014.55.3.689

Soni M, Rahardjo TB, Soekardi R et al (2014) Phytoestrogens and cognitive function: a review. Maturitas 77(3):209–220. doi:10.1016/j.maturitas.2013.12.010

Sultana R, Perluigi M, Butterfield DA et al (2006) Protein oxidation and lipid peroxidation in brain of subjects with Alzheimer's disease: insights into mechanism of neurodegeneration from redox proteomics. Antioxid Redox Signal 8(11–12):2021–2037

Takashima A, Noguchi K, Michel G et al (1996) Exposure of rat hippocampal neurons to amyloid beta peptide (25-35) induces the inactivation of phosphatidyl inositol-3 kinase and the activation of tau protein kinase I/glycogen synthase kinase-3 beta. Neurosci Lett 203(1):33–36

Takeda A, Perry G, Abraham NG et al (2000) Overexpression of heme oxygenase in neuronal cells, the possible interaction with Tau. J Biol Chem 275:5395–5399

Tan M, Ouyang Y, Jin M et al (2013) Downregulation of Nrf2/HO-1 pathway and activation of JNK/c-Jun pathway are involved in homocysteic acid-induced cytotoxicity in HT-22 cells. Toxicol Lett 223(1):1–8. doi:10.1016/j.toxlet.2013.08.011

Terazawa R, Akimoto N, Kato T et al (2013) A kavalactone derivative inhibits lipopolysaccharide-stimulated iNOS induction and NO production through activation of Nrf2 signaling in BV2 microglial cells. Pharmacol Res 71:34–43. doi:10.1016/j.phrs.2013.02.002

Trinh K, Andrews L, Krause J et al (2010) Decaffeinated coffee and nicotine-free tobacco provide neuroprotection in Drosophila models of Parkinson's disease through an NRF2-dependent mechanism. J Neurosci 30(16):5525–5532. doi:10.1523/JNEUROSCI.4777-09.2010

Valles SL, Dolz-Gaiton P, Gambini J et al (2010) Estradiol or genistein prevent Alzheimer's disease-associated inflammation correlating with an increase PPAR gamma expression in cultured astrocytes. Brain Res 1312:138–144. doi:10.1016/j.brainres.2009.11.044

van Gelder BM, Buijsse B, Tijhuis M et al (2007) Coffee consumption is inversely associated with cognitive decline in elderly European men: the FINE Study. Eur J Clin Nutr 61(2):226–232

Venugopal R, Jaiswal AK (1996) Nrf1 and Nrf2 positively and c-Fos and Fra1 negatively regulate the human antioxidant response element-mediated expression of NAD(P)H:quinone oxidoreductase1 gene. Proc Natl Acad Sci U S A 93:14960–14965

Vicente SJ, Ishimoto EY, Torres EA (2014) Coffee modulates transcription factor Nrf2 and highly increases the activity of antioxidant enzymes in rats. J Agric Food Chem 62(1):116–122. doi:10.1021/jf401777m

Vila M, Przedborski S (2003) Targeting programmed cell death in neurodegenerative diseases. Nat Rev Neurosci 4(5):365–375

Volz N, Boettler U, Winkler S et al (2012) Effect of coffee combining green coffee bean constituents with typical roasting products on the Nrf2/ARE pathway in vitro and in vivo. J Agric Food Chem 60(38):9631–9641

Wakabayashi N, Dinkova-Kostova AT, Holtzclaw WD et al (2004) Protection against electrophile and oxidant stress by induction of the phase 2 response: fate of cysteines of the Keap1 sensor modified by inducers. Proc Natl Acad Sci U S A A101(7):2040–2045

Wang Y, Santa-Cruz K, DeCarli C et al (2000) NAD(P)H:quinone oxidoreductase activity is increased in hippocampal pyramidal neurons of patients with Alzheimer's disease. Neurobiol Aging 21:525–531

Wang X, de Rivero Vaccari JP, Wang H et al (2012) Activation of the nuclear factor E2-related factor 2/antioxidant response element pathway is neuroprotective after spinal cord injury. J Neurotrauma 29(5):936–945. doi:10.1089/neu.2011.1922

Wang CM, Liu MY, Wang F et al (2013) Anti-amnesic effect of pseudoginsenoside-F11 in two mouse models of Alzheimer's disease. Pharmacol Biochem Behav 106:57–67. doi:10.1016/j.pbb.2013.03.010

Wang Y, Kan H, Yin Y et al (2014) Protective effects of ginsenoside Rg_1 on chronic restraint stress induced learning and memory impairments in male mice. Pharmacol Biochem Behav 120:73–81. doi:10.1016/j.pbb.2014.02.012

Wild AC, Moinova HR, Mulcahy RT et al (1999) Regulation of γ-glutamylcysteine synthetase subunit gene expression by the transcription factor Nrf2. J Biol Chem 274:33627–33636

Xi YD, Li XY, Ding J et al (2013) Soy isoflavone alleviates $A\beta_{1-42}$-induced impairment of learning and memory ability through the regulation of RAGE/LRP-1 in neuronal and vascular tissue. Curr Neurovasc Res 10(2):144–156

Xi YD, Li XY, Yu HL et al (2014) Soy isoflavone antagonizes the oxidative cerebrovascular injury induced by β-amyloid peptides 1-42 in rats. Neurochem Res 39(7):1374–1381. doi:10.1007/s11064-014-1319-x

Xie C, Lovell MA, Markesbery WR et al (1998) Glutathione transferase protects neuronal cultures against four hydroxynonenal toxicity. Free Radic Biol Med 25:979–988

Xie C, Lovell MA, Xiong S et al (2001) Expression of glutathione-*S*-transferase isozyme in the SY5Y neuroblastoma cell line increases resistance to oxidative stress. Free Radic Biol Med 31:73–81

Xie X, Wang HT, Li CL et al (2010) Ginsenoside Rb_1 protects PC12 cells against β-amyloid-induced cell injury. Mol Med Rep 3(4):635–639. doi:10.3892/mmr_00000308

Yan S, Li Z, Li H et al (2014) Notoginsenoside R1 increases neuronal excitability and ameliorates synaptic and memory dysfunction following amyloid elevation. Sci Rep 4:6352. doi:10.1038/srep06352

Yu R, Lei W, Mandlekar S et al (1999) Role of a mitogen-activated protein kinase pathway in the induction of phase II detoxifying enzymes by chemicals. J Biol Chem 274:27545–27552

Yu R, Mandlekar S, Lei W et al (2000) p38 mitogen-activated protein kinase negatively regulates the induction of phase II drug-metabolizing enzymes that detoxify carcinogens. J Biol Chem 275:2322–2327

Yu HL, Li XY, Zhou X et al (2013) Beta amyloid peptide (25-35) leading to inflammation through Toll-like receptors and the anti-inflammatory effect of genistein in BV-2 cells. J Mol Neurosci 51(3):771–778. doi:10.1007/s12031-013-0063-z

Zeng H, Chen Q, Zhao B et al (2004) Genistein ameliorates beta-amyloid peptide (25-35)-induced hippocampal neuronal apoptosis. Free Radic Biol Med 36(2):180–188

Zhang DD, Hannink M et al (2003) Distinct cysteine residues in Keap1 are required for Keap1-dependent ubiquitination of Nrf2 and for stabilization of Nrf2 by chemopreventive agents and oxidative stress. Mol Cell Biol 23(22):8137–8151

Zhang R, Miao QW, Zhu CX et al (2014a) Sulforaphane ameliorates neurobehavioral deficits and protects the brain from amyloid β deposits and peroxidation in mice with Alzheimer-like lesions. Am J Alzheimers Dis Other Demen. pii: 1533317514542645

Zhang R, Zhang J, Fang L et al (2014b) Neuroprotective effects of sulforaphane on cholinergic neurons in mice with Alzheimer's disease-like lesions. Int J Mol Sci 15(8):14396–14410. doi:10.3390/ijms150814396

Zhao H, Li Q, Zhang Z et al (2009) Long-term ginsenoside consumption prevents memory loss in aged SAMP8 mice by decreasing oxidative stress and up-regulating the plasticity-related proteins in hippocampus. Brain Res 1256:111–122. doi:10.1016/j.brainres.2008.12.031

Zhao X, Zou Y, Xu H et al (2012) Gastrodin protect primary cultured rat hippocampal neurons against amyloid-beta peptide-induced neurotoxicity via ERK1/2-Nrf2 pathway. Brain Res 1482:13–21. doi:10.1016/j.brainres.2012.09.010

Zhao L, Mao Z, Chen S et al (2013) Early intervention with an estrogen receptor β-selective phytoestrogenic formulation prolongs survival, improves spatial recognition memory, and slows progression of amyloid pathology in a female mouse model of Alzheimer's disease. J Alzheimers Dis 37(2):403–419. doi:10.3233/JAD-122341

Zhou N, Tang Y, Keep RF et al (2014) Antioxidative effects of Panax notoginseng saponins in brain cells. Phytomedicine 21(10):1189–1195. doi:10.1016/j.phymed.2014.05.004

Zipper LM, Mulcahy RT (2000) Inhibition of ERK and p38 MAP kinases inhibits binding of Nrf2 and induction of GCS genes. Biochem Biophys Res Commun 278:484–492

Zou Y, Hong B, Fan L et al (2013) Protective effect of puerarin against beta-amyloid-induced oxidative stress in neuronal cultures from rat hippocampus: involvement of the GSK-3β/Nrf2 signaling pathway. Free Radic Res 47(1):55–63. doi:10.3109/10715762.2012.742518

Chapter 22
Immunotherapeutic Approaches Against Amyloid-β in Drug Discovery for Alzheimer's Disease

Seung-Hoon Yang, Jiyoon Kim, and YoungSoo Kim

Abstract Alzheimer's disease (AD) is a progressive neurodegenerative disorder and the most common type of dementia. The major pathological hallmark and culprit of AD is aggregation of the amyloid-β (Aβ) peptide. Since the Aβ abnormality begins in the asymptomatic stage of AD, immunotherapeutic approaches clearing Aβ aggregates are investigated as the most promising treatment in clinical trials. Both active and passive immunization against Aβ showed significant reduction of Aβ levels in the brain and enhancement of learning and memory. Albeit pathologically effective, these immunotherapeutic vaccines need to overcome side effects such as vasogenic edema and microhemorrhages. In this chapter, we introduce the basic concept of immunotherapy for clearance of Aβ, compare putative immunotherapeutic vaccine candidates, and discuss their benefits, disadvantages, and challenges.

Keywords Alzheimer's disease • Amyloid-β • Active immunotherapy • Passive immunotherapy • Vaccination

22.1 Introduction

Alzheimer's disease (AD) is the most common phenomenon of dementia, characterized by the extensive loss of neurons and synapses and the progressive decline of memories (Alzheimer's 2012; Brookmeyer et al. 2007). AD is a polysynthetic disease involving aggregation and deposition of amyloid-β (Aβ) and

S.-H. Yang
Center for Neuro-Medicine, Brain Science Institute, Korea Institute of Science and Technology, Hwarangno 14-gil 5, Seongbuk-gu, Seoul, Republic of Korea

J. Kim • Y. Kim (✉)
Center for Neuro-Medicine, Brain Science Institute, Korea Institute of Science and Technology, Hwarangno 14-gil 5, Seongbuk-gu, Seoul, Republic of Korea

Biological Chemistry Program, Korea University of Science and Technology, 217 Gajungro Yuseong-gu, Daejeon, Republic of Korea
e-mail: yskim@bio.kist.re.kr

© Springer Japan 2015

N. Mori, I. Mook-Jung (eds.), *Aging Mechanisms*,
DOI 10.1007/978-4-431-55763-0_22

hyperphosphorylated tau, accompanied by oxidative stress, glial activation and neuronal cell death (Wyss-Coray 2006). Aβ is a short peptide of 39–43 amino acids and generated throughout the serial proteolysis of amyloid precursor protein (APP) (De Strooper et al. 2010; Selkoe 2001; Wolfe 2006). In normal neurogenesis, called the non-amyloidogenic pathway, the extracellular domain of APP is cleaved by α-secretase, leading to release the soluble extracellular fragment known as sAPP-α (Edwards et al. 2008; Pietri et al. 2013). Then, γ-secretase cleaves the truncated APP in the plasma membrane into the APP intracellular C-terminal domain (Shoji et al. 1992; Golde et al. 2013; Chang and Suh 2010). In the amyloidogenic pathway, however, the sequential cleavage by β-secretase and γ-secretase generates the Aβ peptide (Zhang et al. 2012; O'Brien and Wong 2011). In the monomeric state, Aβ is a soluble and non-toxic α-helical peptide (Takano et al. 2006; Lansbury 1997; Kirkitadze et al. 2001). However, at high concentration, the peptide undergoes a conformational change to form amyloid oligomers and fibrils. Then, these fibrils aggregate into the insoluble cluster called "plaques" in the brains of AD patients (Fig. 22.1) (Jellinger 2006; Walsh et al. 2002; Shankar et al. 2007).

Aggregation of Aβ in the brain plays a pivotal role in AD as a pathological culprit (Duran-Aniotz et al. 2013; Jin et al. 2011). Deposition of Aβ aggregates is observed in the early stage during the development of AD (Leuner et al. 2012; Gowing et al. 1994; Pigino et al. 2009). Thus, overproduction and aggregation of Aβ have been the major target of AD drug candidates (Barten et al. 2006; Pohanka 2011; Doraiswamy and Xiong 2006; Lleo et al. 2006; Michaelis 2003). However, disappointing clinical trials of amyloid inhibitors, targeting APP proteolysis or Aβ aggregation, have raised concerns for alternative therapeutic approaches. As abnormal Aβ deposition precedes cognitive decline, the newly suggested mode of action is the immunotherapy to remove toxic Aβ oligomers and plaques from the brain of AD patients. While numerous clinical trials have been investigated to reduce cerebral Aβ deposits and facilitate Aβ clearance, the strongest approach to date is immunotherapy, which can be mainly divided into active or passive (Lobello et al. 2012). Active immunization utilizes administration of synthetic Aβ peptide fragments conjugated with carrier proteins, and passive immunization uses humanized monoclonal antibodies against Aβ peptides.

22.2 Active Immunotherapy

The active immunotherapy aims specific activation of cellular and humoral immune systems such as inducing antigen producing cells, T cells, and B cells. Once APCs are initially activated by stimulation of compromised antigens, Aβ peptides, combined with an immune adjuvant to get the high immune response, transfer their immune signals to T cells. Activated T cells progressively stimulate B cells to produce specific antibodies against Aβ. These antibodies bind to the Aβ peptides, then target for clearance (Fig. 22.2) (Lemere and Masliah 2010).

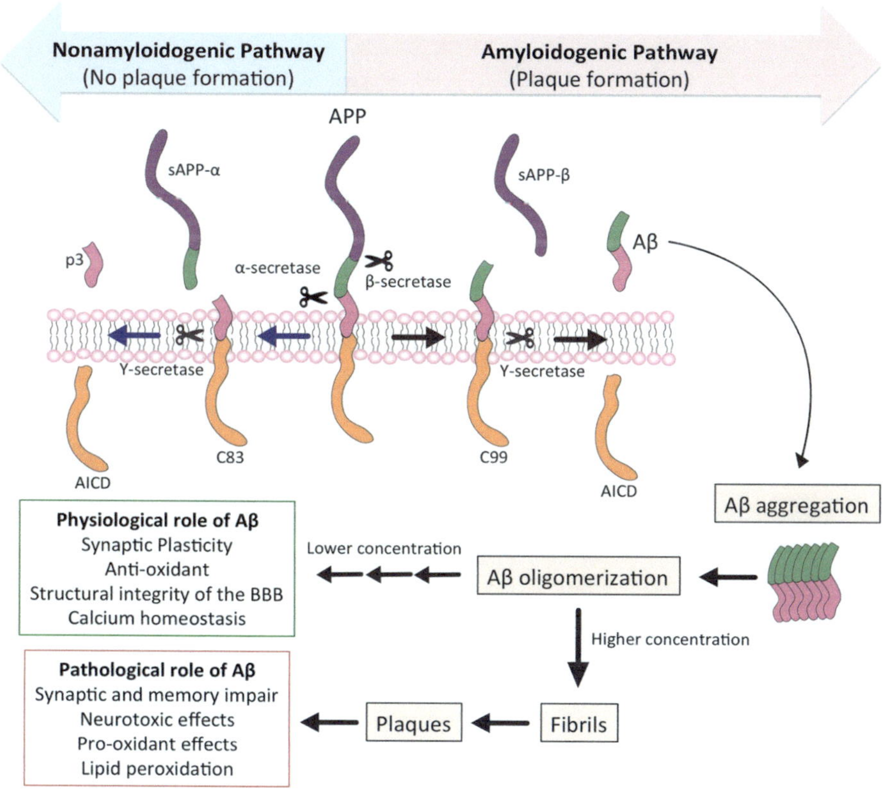

Fig. 22.1 A diagram of amyloid precursor protein (*APP*) processing. The transmembrane protein APP can be cleaved by two pathways. In the non-amyloidogenic pathway, α-secretase cleaves the extracellular domain of APP to release soluble extracellular fragments (sAPP-α). This truncated APP is then cleaved by γ-secretase to release the APP intracellular C-terminal domain (*AICD*) and p3 fragment. In the amyloidogenic pathway, β-secretase cleaves the extracellular domain of APP to release soluble extracellular fragments (*sAPP-β*). Then, γ-secretase cleaves the truncated APP of transmembrane part to generate Aβ monomers. At low concentration, Aβ, in monomer state, is less toxic and plays several physiological roles. At higher concentration level, the peptide undergoes the aggregation to form amyloid plaques and found in AD brains

In 1999, Schenk and his colleagues first reported that the active immunotherapy using synthetic Aβ peptides, with complete Freund adjuvant and incomplete Freund adjuvant, could prevent the development of Aβ deposition in the brain of PDAPP transgenic mice model with Aβ plaque pathology (Schenk et al. 1999). The therapeutic approaches were, then, extended to diverse animal models and demonstrated that active Aβ immunotherapeutic treatment can prevent the accumulation of Aβ in the brain and rescue the abnormal cognitive behaviors (Lemere et al. 2000; Weiner et al. 2000; Das et al. 2001; Sigurdsson et al. 2001; Maier et al. 2006).

Although the active immunotherapy is a powerful method due to its ability to induce long-term antibody production, low-cost efficiency, and easy handling, it has

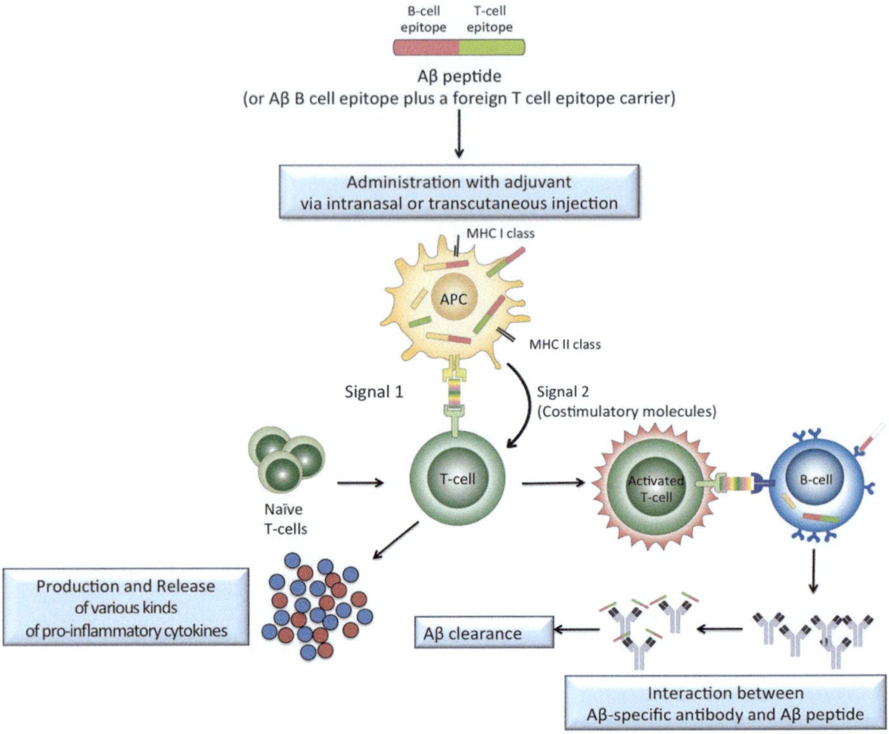

Fig. 22.2 Active immunotherapeutic vaccination approaches. Active immunization induces the humoral immune system to generate Aβ-specific antibodies. Aβ peptides conjugated with foreign T cell epitope carriers can be administrated as antigens and activate the antigen presenting cells (*APC*), which engulf and process the antigen. Then, a signal can be transmitted via activating naïve T lymphocytes to produce several kinds of pro-inflammatory mediators. Another signal with co-stimulatory molecules induces the enhancement of T lymphocytes, which leads to generate the antibodies against Aβ from B lymphocytes

the risk of detrimental immune response. For example, if T cells recognize the antigen as a self-protein, they do not induce the proper immune system. Also, activated T cells induce a release diverse in pro-inflammatory cytokines to affect the whole body defense mechanism. Moreover, since an active immunization leads to polyclonal antibodies production that recognize multiple epitopes of Aβ peptides, the antibodies may have low specificity or avidity against Aβ peptides, eventually lead to less effective immune responses (Delrieu et al. 2012a; Lannfelt et al. 2014b).

22.2.1 AN-1792

In the late 1990s, Elan Pharmaceuticals and Wyeth Corporation introduced an active immunotherapy (AN-1792) with synthetic pre-aggregated human Aβ 1–42 in animal studies. Administration of AN-1792 blocked the formation of Aβ plaques in the brain of AD transgenic mice and dramatically reduced preformed plaques in aged mice (Schenk et al. 1999). AN-1792, in addition, induced improvement of mice performance in behavior tests related to learning and memory (Bayer et al. 2005; Ferrer et al. 2004; Masliah et al. 2005; Nicoll et al. 2003). Following the promising animal studies, AN-1792 was tested in the clinical trial Phase I to assess its therapeutic effects, safety, and tolerability in AD patients and was found with no adverse side effect (Bayer et al. 2005). However, in Phase II-A trials, the clinical investigation was suspended, when several participants developed severe inflammation in the brain and the spinal cord. AN-1792 was eventually withdrawn from the clinical trials in 2002, after 18 recipients with vaccination (about 6 % of recipients) developed the brain inflammation such as meningoencephalitis (Gilman et al. 2005; Orgogozo et al. 2003; Robinson et al. 2004).

22.2.2 ACC-001

Despite the suspension of AN-1792 in the clinical trial Phase II-A for safety reasons, active Aβ vaccine is still an attractive therapeutic mode of action to treat AD and several second-generation vaccines are currently tested in clinical trials. Janssen Alzheimer Immunotherapy, a subsidiary of Johnson & Johnson, launched Vanutide cridificar (ACC-001) vaccine, as a modified version of AN-1792. ACC-001 was developed as a N-terminal short fragment of Aβ (Aβ 1–7) conjugated with a carrier protein, a non-toxic variant of diphtheria toxin, using the saponin adjuvant QS-21. ACC-001 induced a humoral immune response including antibody generation with no sign of intolerable side effects in the clinical trial Phase I (Ryan and Grundman 2009). However, this vaccine was briefly suspended in 2008, because one of the patients, in Phase II, developed skin vasculitis, indicating malfunction of immune or hypersensitive allergic responses. Although the patient recovered and the clinical trials resumed within 6 weeks, no results have been published in journals (Lemere and Masliah 2010; Okura and Matsumoto 2009). In August 2013, this immunotherapy was been discontinued from clinical development.

22.2.3 CAD106

Novartis Pharmaceuticals and Cytos Biotechnology developed an active Aβ vaccine, CAD106, composed of multiple copies of the Aβ 1–6 fragment coupled with a virus-like carrier particle (Qβ). This vaccine is designed to block activation of the autoimmune Th1-cell response and to induce the Th2-cell mediated humoral response (Winblad et al. 2012). CAD106 was confirmed in animals to inhibit the formation of Aβ plaques in the brain (Wiessner et al. 2011) and advanced to clinical trials with mild-to-moderate AD patients. In the Phase I, CAD106 induced a significant humoral antibody response when high dose of antibody was administrated. The Phase II clinical investigations of CAD106 have been reported the favorable safety, tolerability, and humoral antibody response (Kingwell 2012). Besides, chill and fever, under the permissible level, were observed in the Phase II (Winblad et al. 2012, 2014). However, numbers of concerns were raised regarding reactivity and safety during the clinical trials. First of all, the six-amino-acid synthetic Aβ fragment might not be long enough to specifically activate Th2-cells and induce humoral immune responses. Furthermore, the design of the clinical trial was re-evaluated by concerning the size and selection of patients; (1) the study was tested in the small group of subjects, (2) the duration of the vaccine administration was short to record clinical effects including safety and tolerability. In addition, intracerebral hemorrhage was found in one patient from the CAD106 administration group, who had cerebral amyloid angiopathy (Winblad et al. 2014).

22.2.4 Affitope AD02

Affitope AD02, by AFFiRiS AG, is a KLH vaccine with the six N-terminal amino acids of Aβ. By introducing the non-endogenous Aβ mimic, this vaccine was designed to exhibit a favorable safety profile and to prevent development of tolerance. The composition of Affitope AD02 enabled to prevent the autoimmune T cells activation with cross-reactivity with APP by specific recognition of Aβ (Schneeberger et al. 2009). In AD animal models, Affitope AD02 reduced levels of Aβ plaques. In the clinical trial Phase I, a favorable safety profile was observed in 24 AD patients after four-time vaccination (Brody and Holtzman 2008; Madeo and Frieri 2013; Winblad et al. 2014; Mangialasche et al. 2010). No meningoencephalitis was found during the investigation. 332 AD patients were subjected to the Phase 2 trial and limited data has been reported so far. The clinical investigation is still on-going by enrolling patients.

22.2.5 ACI-24

AC Immune SA's ACI-24 is an active tetra-palmitoylated Aβ 1–15 peptide vaccine, embedded within a liposome to eventually induce the generation of β-sheet conformation-specific antibody against Aβ (Muhs et al. 2007). In cynomolgus monkeys and APP/PS1 transgenic mice, the antibodies generated by ACI-24 had high titer level to induce the humoral immune response. In addition, ACI-24 significantly reduced concentration of soluble and insoluble Aβ and restored behavioral performances of learning and memory (Muhs et al. 2007; Winblad et al. 2014). ACI-24 is currently in the clinical trial Phase I/II for AD (Lemere 2013), so far little is known for more detail data in this stage.

Table 22.1 Active amyloid-β immunotherapeutic vaccines in clinical trials

Vaccine (company)	Epitope / Key behaviors observed	Phase status	Completion
AN-1792 (Elan/Wyeth)	Aβ 1–42	Phase II-A	March, 2002
	Blockage of plaque formation		
	Inflammation during Phase II-A (meningoencephalitis)		
ACC-001 (Janssen)	Aβ 1–7 with non-toxic diphtheria toxin	Phase II	August, 2013
	High titers of antibody without intolerance in Phase I		
	Skin vasculitis in Phase II		
CAD106 (Novartis/Cytos)	Aβ 1–6 with Qβ carrier	Phase II	December, 2012
	Prevention of the autoimmune Th1-cell activation		
	Blockage of plaque formation		
	Side effects such as chill or fever (permissible event in Phase II)		
AFFITOPE AD02 (AFFiRiS AG)	Six-amino acid peptide that mimics N-terminus of Aβ	Phase II	Ongoing
	Prevention of the autoimmune T cell activation		
	Reduction of Aβ plaques		
	Favorable safety in Phase I trials		
ACI-24 (AC Immune SA)	Tetra-palmitoylated Aβ 1–15	Phase I/II	Ongoing
	Generation of β-sheet conformation specific antibodies		
	Reduction of Aβ plaque deposition		
	Recovery of learning and memory in animal studies		
V950 (Merck)	N-terminal fragments of Aβ	Phase I	Ongoing
	Production of antibodies against N-terminal of Aβ in the serum and CSF		

22.2.6 *V950*

Merck's V950 is a multivalent vaccine that links N-terminal fragments of Aβ to an adjuvant ISCO-MATRIX. V950 was reported to induce production of antibodies against N-terminal of Aβ in the serum and CSF (Savage et al. 2010). The clinical trial Phase I was performed with 86 AD patients in 51 sites for safety and tolerability. The investigation was completed recently (October 2014) (Lemere and Masliah 2010; Winblad et al. 2014) (Table 22.1).

22.3 Passive Immunotherapy

Passive immunotherapy refers to direct injection of monoclonal antibodies without sensitizing the humoral immune system for generation of antibody responses (Brody and Holtzman 2008; Bacskai et al. 2001). Mechanisms of the anti-Aβ

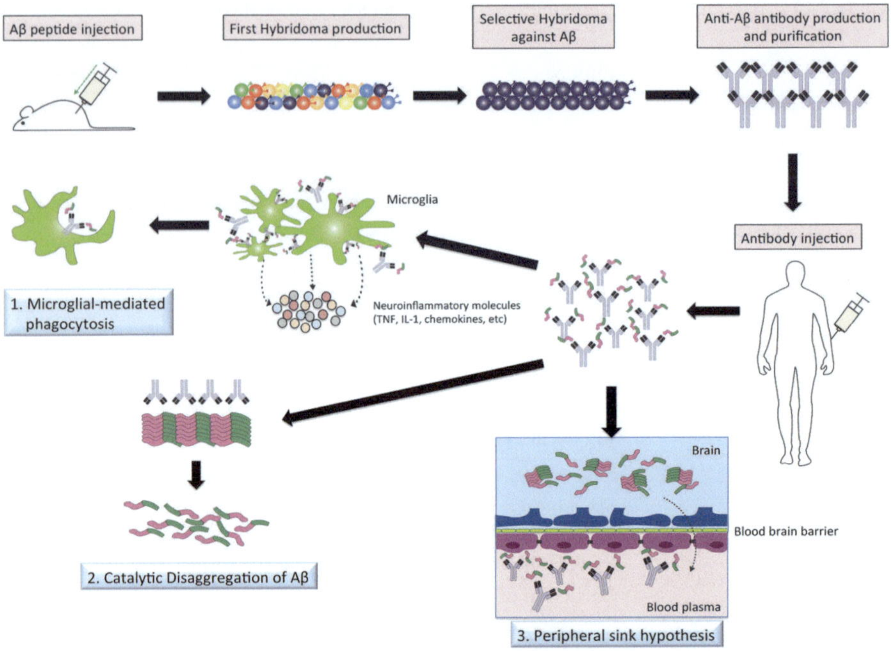

Fig. 22.3 Passive immunotherapeutic approaches and proposed mechanisms. The mice immunized with Aβ peptide to produce hybridoma cells. Then hybridoma cells are selected for proper antibodies against Aβ. The antibodies are then purified and administrated to patients with AD. The antibodies may clear Aβ through three kinds of proposed mechanisms: (1) microglial-mediated phagocytosis, (2) catalytic disaggregation of Aβ deposition, and (3) peripheral sink hypothesis in the bloodstream

passive immunotherapy can be categorized into microglial-mediated phagocytosis, catalytic disaggregation of Aβ deposition, and peripheral sink (Fig. 22.3) (Alves et al. 2014; Menendez-Gonzalez et al. 2011). In the microglial-mediated phagocytosis, antibodies directly bind to amyloid plaques and trigger microglial activation via their Fc receptors. Then, activated microglial cells rapidly facilitate the elimination of Aβ through phagocytosis. Meanwhile, they may induce neuroinflammatory events including secretion of various inflammatory mediators such as IL-1, IL-6, TNF, free radical, and chemokines (Wilcock et al. 2004; Cai et al. 2014; Kakimura et al. 2002). In catalytic disaggregation of Aβ deposition, administered antibodies bind to Aβ aggregates and catalyze the conformational change of Aβ peptides. Such actions eventually lead to disaggregation of Aβ aggregates and reduction of amyloid-induced neurotoxicity (Solomon et al. 1996, 1997; Legleiter et al. 2004; Frenkel et al. 2000; Bacskai et al. 2001). The peripheral sink hypothesis was first reported when the m266 anti-Aβ monoclonal antibody directly targeted and completely sequestered Aβ in the plasma (DeMattos et al. 2001). Peripheral administration of m266 to PDAPP transgenic mice induced a rapid elevation of plasma Aβ levels due to the change in Aβ distribution between central nervous and peripheral circulatory systems. The altered equilibrium of Aβ leaded to facilitate the peripheral clearance of Aβ in the plasma instead of Aβ deposition in the brain (Deane et al. 2003, 2005; Dodart et al. 2002).

Compared to the active immunotherapy, antibody drugs shall be beneficial to patients as the dosage of antibodies in each subject is known before administration. The amount and concentration of antibodies can be easily controlled. Moreover, the administration can be immediately stopped and the antibody will be rapidly removed if there are any signs for side effects. Besides, unnecessary cellular responses can be avoided in passive immunotherapy (Mangialasche et al. 2010; Guan et al. 2012; Lemere 2013). However, high-cost humanized monoclonal antibodies and repeated drug injection for long-term treatment is a considerable disadvantage of the passive immunotherapy (McElhaney and Effros 2009). In addition, antibody drugs may act as antigens and induce immune responses, which may lead to side effects such as glomerulonephritis and vasculitis (Lemere 2013).

22.3.1 Bapineuzumab (AAB-001) and PF-05236812 (AAB-003)

Bapineuzumab (AAB-001) is a humanized therapeutic monoclonal antibody against N-terminus of Aβ (3D6 clone, IgG1 isotype) developed by Elan, Wyeth, Johnson & Johnson (Janssen), and Pfizer (Brody and Holtzman 2008; Blennow et al. 2012; Panza et al. 2010). AAB-001 was reported to pass blood-brain barriers, to bind fibrillar and soluble Aβ, and to induce microglial-mediated phagocytosis the plaques in AD transgenic mice (Bard et al. 2000, 2003; Racke et al. 2005). However, in two large Phase II/III trials, no clinical benefit but serious side effects were reported including cerebral vasogenic edema, retinal vascular disorder, and microhemorrhages (Okura and Matsumoto 2009; Pfeifer et al. 2002; Racke et al. 2005). MRI scans revealed that vascogenic edema was found in AD patients with the high dose group (Khorassani and Hilas 2013; Sperling et al. 2012). These results led the clinical investigation of AAB-001 to the termination in 2012. One of the possibilities raised for the lack of clinical efficacy was that the administration of this vaccine was too late in the disease process to reverse the neurodegenerative changes.

PF-05236812 (AAB-003) was then developed as a derivative of bapineuzumab with a modified Fc domain to reduce effector functions on microglial activation. It was specifically designed to avoid amyloid-related imaging abnormalities (ARIA), a complication of bapineuzumab administration. The clinical trial Phase I was performed with 88 AD patients to evaluate the safety and tolerability of PF-05236812 and trial was completed in August 2014 (Moreth et al. 2013).

22.3.2 Solanezumab (LY2062430)

Eli Lilly & Co.'s Solanezumab (LY2062430) is a humanized IgG1 version of the aforementioned m266 monoclonal antibody. Unlike Bapineuzumab, Solanezumab targets the mid-domain of the Aβ peptide (Aβ 13–28) and binds selectively to soluble Aβ species (Mangialasche et al. 2010; Moreth et al. 2013; Spencer and Masliah 2014). Cognitive recovery of AD transgenic mice by m266 supports the view that soluble oligomeric Aβ is highly related to neuronal and synaptic dysfunction in AD brains. During the clinical trials, the significant increase of Aβ levels were observed in both the blood and CSF by the peripheral sink mechanism (Farlow et al. 2012; Siemers et al. 2010). Currently, solanezumab is investigated in two large clinical trial Phase III studies with a total of 2,052 subjects from 16 countries (Doody et al. 2014a, b). According to interim reports by Eli Lilly & Co., cardiac disorders and even 24 deaths were observed in Solanezumab-treated patients (Doody et al. 2014b). However, no clear relation was found between the death and Solanezumab. Although Solanezumab is considered as the first clinical evidence that anti-amyloid approach helps AD patients, it needs to consider for further development of this vaccine and the skepticism still exists on the ability of this drug to slow the rate of deterioration in patients with later-stage of diseases.

22.3.3 Gantenerumab (RO4909832, RG1450)

Gantenerumab (RO4909832, RG1450), by Roche, is a fully human IgG1 monoclonal antibody against Aβ that has a high affinity to specifically bind to cerebral amyloid plaques (Delrieu et al. 2012b). Gantenerumab appears to preferentially bind the fibrillar form of Aβ by recognizing both N-terminus (Aβ 3–12) and mid-domain (Aβ 18–27). Gantenerumab induces microglial-mediated phagocytosis by binding to small Aβ plaques (Bohrmann et al. 2012). Thus, unlike Solanezumab, Gantenerumab decreased Aβ deposition in the brain without increasing plasma Aβ levels. In 360 mild-to-moderate AD patients administrated with Gantenerumab of Phase II, it reduced brain amyloid load around 30 % by PET imaging analysis. However, 2 patients with ARIA were observed in the high dose group (Ostrowitzki et al. 2012). Recently, the Phase III was started with 1000 mild-AD patients via subcutaneous injection (Novakovic et al. 2013). A separate clinical trial is also under investigation in Phase III with prodromal AD patients through Dominantly Inherited Alzheimer Network (DIAN).

22.3.4 Gammagard (Intravenous Immunoglobulin, IVIg)

Baxter Healthcare's passive immunotherapeutic approach, Gammagard, is distinct from aforementioned monoclonal antibodies. Gammagard is an intravenous immunoglobulin (IVIg), a pooled mixture of natural human polyclonal immunoglobulin that extracted from the plasma of over one thousand blood donors. As a result, Gammagard recognizes Aβ monomers, oligomers, and fibrils (Dodel et al. 2002, 2004). IVIg is widely used for the treatment of various pathological disorders as a replacement therapy for various immunodeficiency syndromes. Since IVIg is the product from non-selective antibody collection from various normal patients, it was doubtful for the potential clinical effect on AD. In 2002, Dodel et al reported the effects of commercially available IVIg significantly reduced the level of Aβ in the CSF and blood of AD patients after 6-month administration (Dodel et al. 2002). Notably, administered anti- Aβ antibodies detected in the CSF of patients as a indication that IVIg might transfer the blood-brain barrier and directly decreased the Aβ level in the brain (Fillit et al. 2009; Relkin et al. 2009). Currently, Baxter Healthcare and Alzheimer's Disease Consortium Study (ADCS) are investigating this vaccine in Phase III. A derivative of IVIg (Octagam) is currently investigated by Octapharma in Phase II (Lobello et al. 2012; Moreth et al. 2013). However, IVIg has potential side effects for AD patients; (1) IVIg can lead to thromboemboli because it increases serum viscosity, (2) renal dysfunction or failure can be induced because IVIg products use sucrose as a stabilizing agent (Loeffler 2013), and (3) IVIg can also lead severe allergic difficulties such as breathing or skin rashes, severe headache or fever, and dark colored urine (Levy and Pusey 2000).

22.3.5 *Ponezumab*

Ponezumab, by Pfizer, is a humanized IgG2a monoclonal antibody, which recognizes the C-terminus of the Aβ 40 peptide (Aβ 33–40). Ponezumab was reported to reduce autoimmune T cell responses (Madeo and Frieri 2013). The clinical trial Phase I for safety and tolerability was completed without microhemorrhage, ARIA, or encephalitis. Ponezumab is currently in the Phase II with 234 AD patients (Freeman et al. 2012; Landen et al. 2013).

22.3.6 *Crenezumab*

Crenezumab, by Genentech, is a fully humanized IgG4 monoclonal antibody targeting both Aβ monomers and oligomers. The antibody was designed to reduce the Fc receptor-mediated microglial activation and the risk of the immune cell stimulation (Poduslo et al. 2010; van der Zee et al. 1986; Bruhns et al. 2009). Crenezumab is currently in the clinical trial Phase II with 361 AD patients (Adolfsson et al. 2012; Lemere 2013).

22.3.7 BAN2401 (mAb158)

Conformation-dependent antibodies to selectively recognize pathogenic structures have been attractive drug candidates and BioArctic developed the monoclonal antibody 158 (mAb158) against Aβ protofibrils (Englund et al. 2007; Sehlin et al. 2012). mAb158 reduced the level of Aβ protofibrils in the brain of both young and old AD transgenic mice and eventually led the reduction of Aβ plaque formation (Lord et al. 2009). Eisai acquired the antibody and developed BAN2401, an immunotherapeutic IgG1 monoclonal antibody, by further optimization. BAN2401 is currently in clinical trial Phase II with 800 AD patients (Tucker et al. 2015; Lannfelt et al. 2014a; Araki 2010).

22.3.8 Aducanumab (BIIB037)

Biogen Ided's Aducanumab is a fully human IgG1 monoclonal antibody that strongly binds to aggregated forms of Aβ. Aducanumab was reported to reduce the size of plaques in the brain of APP transgenic mice models (Lemere 2013; Moreth et al. 2013; Prins and Scheltens 2013). However, in the high dose, the antibody induced microhemorrhages. The clinical trial Phase I is currently under investigation with 160 mild-AD patients (Table 22.2).

22.4 Conclusion and Further Discussion

In this review, we investigated current active and passive anti-Aβ antibody drugs in AD drug discovery. Albeit promising, results from clinical trials suggest further optimization of these immunotherapeutics for better efficacy and lower side effects. The First issue is the selection of target epitope with high efficiency and safety (Aisen and Vellas 2013). Newer immunotherapeutic vaccines need to avoid the autoimmune response upon the anti-Aβ antibody treatment. Several strategies to overcome this issue aim to develop a combination therapy of present adjuvants or to use foreign T cell epitopes. Another issue is the need to monitor therapeutic progression, as the clearance of Aβ cannot completely reverse clinical symptoms such as neuroinflammation, which lead to neuronal cell death and cognitive impairment. Therefore, selection for proper biomarkers is important to detect pre-clinical disease with mild cognitive impairment and predict which patients may benefit from immunotherapy. Several biomarkers are currently under investigation, but more researches are required before they can clinically be useful (Mayeux and Schupf 2011). Lastly, these antibodies have to cross the blood-brain barrier (BBB) efficiently and safely. The BBB controls the passage of most proteins and small molecules from the blood into the central nervous system. Thus, the transport of

Table 22.2 Passive amyloid-β immunotherapeutic vaccines in clinical trials

Vaccine (company)	Epitope/target	Isotype	Phase status	Completion
	Key behaviors observed			
Bapineuzumab (AAB-001) PF-05236812 (AAB-003) (Elan/Wyeth/ Janssen/Pfizer)	Aβ 1–5/soluble and aggregated Aβ	IgG1	Phase III (AAB-001)	August, 2012
			Phase I (AAB-003)	August, 2014
	Aβ subjected to microglial-mediated phagocytosis			
	Patients found with vasogenic edema and microhemorrhages in Phase II/III			
Solanezumab (LY2062430) (Eli Lilly)	Aβ 13–28/soluble Aβ	IgG1	Phase III	Ongoing
	Aβ subjected to peripheral sink hypothesis			
	Significant increase of Aβ levels in the plasma			
	Patient found with cardiac arrhythmia and cardiac ischemia in Phase III			
Gantenerumab (RO4909832, RG1450) (Roche)	Aβ 3–12/aggregated Aβ	IgG1	Phase III	Ongoing
	Preferentially binding to fibril form of Aβ			
	Leading to microglial-mediated phagocytosis			
	ARIA observed at the high-dose treatment during Phase II			
IVIg (Gammagard/ Octagam) (Baxer healthcare/Octapharma)	Central and C-terminus of Aβ/ Aβ monomer, oligomer, fibrils	Pooled mixture of human polyclonal antibody	Phase III (Gammagard) Phase II (Octagam)	Ongoing
	Alternative therapy for various immunodeficiency syndromes			
	Aβ level reduced in the CSF			
	Patients found with thromboemboli, renal dysfunction, and allergic reactions			
Ponezumab (Pfizer)	Aβ 33–40/soluble and aggregated Aβ	IgG2a	Phase II	Ongoing
	Targeting C-terminus of Aβ			
	Autoimmune T-cell response reduced			
Crenezumab (Genentech)	Aβ 12–23/soluble oligomeric	IgG4	Phase II	Ongoing
	Fc receptor-mediated microglial activation reduced			
BAN2401 (Eisai)	N-terminus of Aβ/ soluble Aβ protofibrils	IgG1	Phase II	Ongoing
	Targeting the oligomeric form of Aβ			
	Aβ plaque formation reduced			
Aducanumab (BIIB037) (Biogen Idec)	Conformational Aβ/fibrillar Aβ	IgG1	Phase I	Ongoing
	Strong binding to aggregated form of Aβ			
	Plaques reduced			
	Animal models found with microhemorrhages			

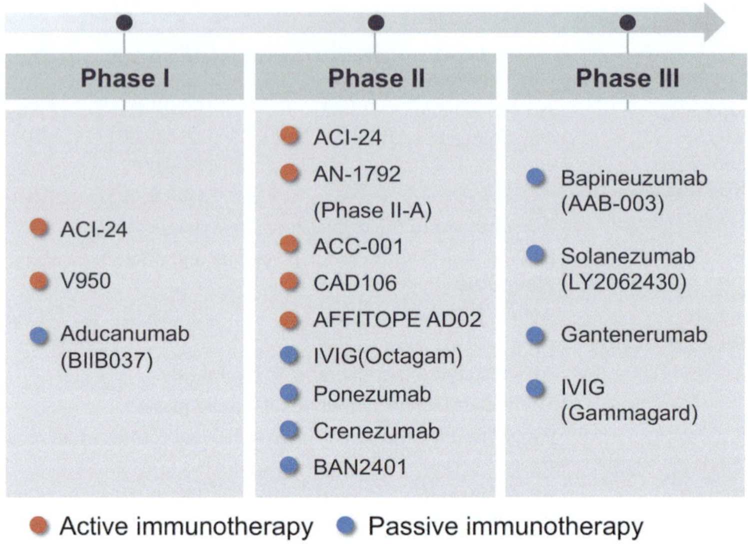

Fig. 22.4 List of current immunotherapeutic vaccines in clinical trials

monoclonal antibodies between BBB has been believed extremely difficult (Spencer and Masliah 2014). Previous studies reported that only the small portion of administered antibody crossed the BBB while the majority was metabolized in the liver or excreted through the kidney (Banks et al. 2002). As the biological drugs commonly cost higher than chemicals, increasing the BBB penetration rate will not only contribute to the therapeutic efficacy but also the medical costs for patients. Receptor-mediated BBB penetration of monoclonal antibodies into central nervous system is currently under investigation (Boado et al. 2013).

More than 100 years has been passed since the initial observation of AD. Aβ was identified as a critical pathogen of AD (Backman et al. 2004; Hardy and Higgins 1992; Okura and Matsumoto 2009; Jia et al. 2014). Among the numerous drug mechanisms regulating amyloidogenesis, the immunotherapy using the Aβ peptides or antibody against Aβ is the leading therapeutic strategy due to the clearance action (Fig. 22.4).

References

Adolfsson O, Pihlgren M, Toni N, Varisco Y, Buccarello AL, Antoniello K, Lohmann S, Piorkowska K, Gafner V, Atwal JK, Maloney J, Chen M, Gogineni A, Weimer RM, Mortensen DL, Friesenhahn M, Ho C, Paul R, Pfeifer A, Muhs A, Watts RJ (2012) An effector-reduced anti-beta-amyloid (Abeta) antibody with unique abeta binding properties promotes neuroprotection and glial engulfment of Abeta. J Neurosci 32(28):9677–9689

Aisen PS, Vellas B (2013) Passive immunotherapy for Alzheimer's disease: what have we learned, and where are we headed? J Nutr Health Aging 17(1):49–50

Alves RP, Yang MJ, Batista MT, Ferreira LC (2014) Alzheimer's disease: is a vaccine possible? Braz J Med Biol Res 47(6):438–444

Alzheimer's A (2012) Alzheimer's disease facts and figures. Alzheimers Dement 8(2):131–168

Araki S (2010) Current status of Alzheimer's disease immunotherapy and pharmacologic effect of BAN2401. Nihon Yakurigaku Zasshi 136(1):21–25

Backman L, Jones S, Berger AK, Laukka EJ, Small BJ (2004) Multiple cognitive deficits during the transition to Alzheimer's disease. J Intern Med 256(3):195–204

Bacskai BJ, Kajdasz ST, Christie RH, Carter C, Games D, Seubert P, Schenk D, Hyman BT (2001) Imaging of amyloid-beta deposits in brains of living mice permits direct observation of clearance of plaques with immunotherapy. Nat Med 7(3):369–372

Banks WA, Terrell B, Farr SA, Robinson SM, Nonaka N, Morley JE (2002) Passage of amyloid beta protein antibody across the blood-brain barrier in a mouse model of Alzheimer's disease. Peptides 23(12):2223–2226

Bard F, Cannon C, Barbour R, Burke RL, Games D, Grajeda H, Guido T, Hu K, Huang J, Johnson-Wood K, Khan K, Kholodenko D, Lee M, Lieberburg I, Motter R, Nguyen M, Soriano F, Vasquez N, Weiss K, Welch B, Seubert P, Schenk D, Yednock T (2000) Peripherally administered antibodies against amyloid beta-peptide enter the central nervous system and reduce pathology in a mouse model of Alzheimer disease. Nat Med 6(8):916–919

Bard F, Barbour R, Cannon C, Carretto R, Fox M, Games D, Guido T, Hoenow K, Hu K, Johnson-Wood K, Khan K, Kholodenko D, Lee C, Lee M, Motter R, Nguyen M, Reed A, Schenk D, Tang P, Vasquez N, Seubert P, Yednock T (2003) Epitope and isotype specificities of antibodies to beta -amyloid peptide for protection against Alzheimer's disease-like neuropathology. Proc Natl Acad Sci U S A 100(4):2023–2028

Barten DM, Meredith JE Jr, Zaczek R, Houston JG, Albright CF (2006) Gamma-secretase inhibitors for Alzheimer's disease: balancing efficacy and toxicity. Drugs R D 7(2):87–97

Bayer AJ, Bullock R, Jones RW, Wilkinson D, Paterson KR, Jenkins L, Millais SB, Donoghue S (2005) Evaluation of the safety and immunogenicity of synthetic Abeta42 (AN1792) in patients with AD. Neurology 64(1):94–101

Blennow K, Zetterberg H, Rinne JO, Salloway S, Wei J, Black R, Grundman M, Liu E, A. A. B. Investigators (2012) Effect of immunotherapy with bapineuzumab on cerebrospinal fluid biomarker levels in patients with mild to moderate Alzheimer disease. Arch Neurol 69 (8):1002–1010

Boado RJ, Lu JZ, Hui EK, Sumbria RK, Pardridge WM (2013) Pharmacokinetics and brain uptake in the rhesus monkey of a fusion protein of arylsulfatase a and a monoclonal antibody against the human insulin receptor. Biotechnol Bioeng 110(5):1456–1465

Bohrmann B, Baumann K, Benz J, Gerber F, Huber W, Knoflach F, Messer J, Oroszlan K, Rauchenberger R, Richter WF, Rothe C, Urban M, Bardroff M, Winter M, Nordstedt C, Loetscher H (2012) Gantenerumab: a novel human anti-Abeta antibody demonstrates sustained cerebral amyloid-beta binding and elicits cell-mediated removal of human amyloid-beta. J Alzheimers Dis 28(1):49–69

Brody DL, Holtzman DM (2008) Active and passive immunotherapy for neurodegenerative disorders. Annu Rev Neurosci 31:175–193

Brookmeyer R, Johnson E, Ziegler-Graham K, Arrighi HM (2007) Forecasting the global burden of Alzheimer's disease. Alzheimers Dement 3(3):186–191

Bruhns P, Iannascoli B, England P, Mancardi DA, Fernandez N, Jorieux S, Daeron M (2009) Specificity and affinity of human Fcgamma receptors and their polymorphic variants for human IgG subclasses. Blood 113(16):3716–3725

Cai Z, Hussain MD, Yan LJ (2014) Microglia, neuroinflammation, and beta-amyloid protein in Alzheimer's disease. Int J Neurosci 124(5):307–321

Chang KA, Suh YH (2010) Possible roles of amyloid intracellular domain of amyloid precursor protein. BMB Rep 43(10):656–663

Das P, Murphy MP, Younkin LH, Younkin SG, Golde TE (2001) Reduced effectiveness of Abeta1-42 immunization in APP transgenic mice with significant amyloid deposition. Neurobiol Aging 22(5):721–727

De Strooper B, Vassar R, Golde T (2010) The secretases: enzymes with therapeutic potential in Alzheimer disease. Nat Rev Neurol 6(2):99–107

Deane R, Yan SD, Submamaryan RK, LaRue B, Jovanovic S, Hogg E, Welch D, Manness L, Lin C, Yu J, Zhu H, Ghiso J, Frangione B, Stern A, Schmidt AM, Armstrong DL, Arnold B, Liliensiek B, Nawroth P, Hofman F, Kindy M, Stern D, Zlokovic B (2003) RAGE mediates amyloid-beta peptide transport across the blood-brain barrier and accumulation in brain. Nat Med 9(7):907–913

Deane R, Sagare A, Hamm K, Parisi M, LaRue B, Guo H, Wu Z, Holtzman DM, Zlokovic BV (2005) IgG-assisted age-dependent clearance of Alzheimer's amyloid beta peptide by the blood-brain barrier neonatal Fc receptor. J Neurosci 25(50):11495–11503

Delrieu J, Ousset PJ, Caillaud C, Vellas B (2012a) 'Clinical trials in Alzheimer's disease': immunotherapy approaches. J Neurochem 120(Suppl 1):186–193

Delrieu J, Ousset PJ, Vellas B (2012b) Gantenerumab for the treatment of Alzheimer's disease. Expert Opin Biol Ther 12(8):1077–1086

DeMattos RB, Bales KR, Cummins DJ, Dodart JC, Paul SM, Holtzman DM (2001) Peripheral anti-A beta antibody alters CNS and plasma A beta clearance and decreases brain A beta burden in a mouse model of Alzheimer's disease. Proc Natl Acad Sci U S A 98(15):8850–8855

Dodart JC, Bales KR, Gannon KS, Greene SJ, DeMattos RB, Mathis C, DeLong CA, Wu S, Wu X, Holtzman DM, Paul SM (2002) Immunization reverses memory deficits without reducing brain Abeta burden in Alzheimer's disease model. Nat Neurosci 5(5):452–457

Dodel R, Hampel H, Depboylu C, Lin S, Gao F, Schock S, Jackel S, Wei X, Buerger K, Hoft C, Hemmer B, Moller HJ, Farlow M, Oertel WH, Sommer N, Du Y (2002) Human antibodies against amyloid beta peptide: a potential treatment for Alzheimer's disease. Ann Neurol 52 (2):253–256

Dodel RC, Du Y, Depboylu C, Hampel H, Frolich L, Haag A, Hemmeter U, Paulsen S, Teipel SJ, Brettschneider S, Spottke A, Nolker C, Moller HJ, Wei X, Farlow M, Sommer N, Oertel WH (2004) Intravenous immunoglobulins containing antibodies against beta-amyloid for the treatment of Alzheimer's disease. J Neurol Neurosurg Psychiatry 75(10):1472–1474

Doody RS, Farlow M, Aisen PS, A. Alzheimer's Disease Cooperative Study Data, C. Publication (2014a) Phase 3 trials of solanezumab and bapineuzumab for Alzheimer's disease. N Engl J Med 370(15):1460

Doody RS, Thomas RG, Farlow M, Iwatsubo T, Vellas B, Joffe S, Kieburtz K, Raman R, Sun X, Aisen PS, Siemers E, Liu-Seifert H, Mohs R, C. Alzheimer's Disease Cooperative Study Steering, G. Solanezumab Study (2014b) Phase 3 trials of solanezumab for mild-to-moderate Alzheimer's disease. N Engl J Med 370(4):311–321

Doraiswamy PM, Xiong GL (2006) Pharmacological strategies for the prevention of Alzheimer's disease. Expert Opin Pharmacother 7(1):1–10

Duran-Aniotz C, Morales R, Moreno-Gonzalez I, Hu PP, Soto C (2013) Brains from non-Alzheimer's individuals containing amyloid deposits accelerate Abeta deposition in vivo. Acta Neuropathol Commun 1(1):76

Edwards DR, Handsley MM, Pennington CJ (2008) The ADAM metalloproteinases. Mol Aspects Med 29(5):258–289

Englund H, Sehlin D, Johansson AS, Nilsson LN, Gellerfors P, Paulie S, Lannfelt L, Pettersson FE (2007) Sensitive ELISA detection of amyloid-beta protofibrils in biological samples. J Neurochem 103(1):334–345

Farlow M, Arnold SE, van Dyck CH, Aisen PS, Snider BJ, Porsteinsson AP, Friedrich S, Dean RA, Gonzales C, Sethuraman G, DeMattos RB, Mohs R, Paul SM, Siemers ER (2012) Safety and biomarker effects of solanezumab in patients with Alzheimer's disease. Alzheimers Dement 8(4):261–271

Ferrer I, Boada Rovira M, Sanchez Guerra ML, Rey MJ, Costa-Jussa F (2004) Neuropathology and pathogenesis of encephalitis following amyloid-beta immunization in Alzheimer's disease. Brain Pathol 14(1):11–20

Fillit H, Hess G, Hill J, Bonnet P, Toso C (2009) IV immunoglobulin is associated with a reduced risk of Alzheimer disease and related disorders. Neurology 73(3):180–185

Freeman GB, Lin JC, Pons J, Raha NM (2012) 39-week toxicity and toxicokinetic study of ponezumab (PF-04360365) in cynomolgus monkeys with 12-week recovery period. J Alzheimers Dis 28(3):531–541

Frenkel D, Katz O, Solomon B (2000) Immunization against Alzheimer's beta -amyloid plaques via EFRH phage administration. Proc Natl Acad Sci U S A 97(21):11455–11459

Gilman S, Koller M, Black RS, Jenkins L, Griffith SG, Fox NC, Eisner L, Kirby L, Rovira MB, Forette F, Orgogozo JM, Team ANS (2005) Clinical effects of Abeta immunization (AN1792) in patients with AD in an interrupted trial. Neurology 64(9):1553–1562

Golde TE, Koo EH, Felsenstein KM, Osborne BA, Miele L (2013) Gamma-secretase inhibitors and modulators. Biochim Biophys Acta 1828(12):2898–2907

Gowing E, Roher AE, Woods AS, Cotter RJ, Chaney M, Little SP, Ball MJ (1994) Chemical characterization of A beta 17-42 peptide, a component of diffuse amyloid deposits of Alzheimer disease. J Biol Chem 269(15):10987–10990

Guan X, Zou J, Gu H, Yao Z (2012) Short amyloid-beta immunogens with spacer-enhanced immunogenicity without junctional epitopes for Alzheimer's disease immunotherapy. Neuroreport 23(15):879–884

Hardy JA, Higgins GA (1992) Alzheimer's disease: the amyloid cascade hypothesis. Science 256 (5054):184–185

Jellinger KA (2006) Challenges in neuronal apoptosis. Curr Alzheimer Res 3(4):377–391

Jia Q, Deng Y, Qing H (2014) Potential therapeutic strategies for Alzheimer's disease targeting or beyond beta-amyloid: insights from clinical trials. Biomed Res Int 2014:837157

Jin M, Shepardson N, Yang T, Chen G, Walsh D, Selkoe DJ (2011) Soluble amyloid beta-protein dimers isolated from Alzheimer cortex directly induce Tau hyperphosphorylation and neuritic degeneration. Proc Natl Acad Sci U S A 108(14):5819–5824

Kakimura J, Kitamura Y, Takata K, Umeki M, Suzuki S, Shibagaki K, Taniguchi T, Nomura Y, Gebicke-Haerter PJ, Smith MA, Perry G, Shimohama S (2002) Microglial activation and amyloid-beta clearance induced by exogenous heat-shock proteins. FASEB J 16(6):601–603

Khorassani F, Hilas O (2013) Bapineuzumab, an investigational agent for Alzheimer's disease. P T 38(2):89–91

Kingwell K (2012) Alzheimer disease: amyloid-beta immunotherapy CAD106 passes first safety test in patients with Alzheimer disease. Nat Rev Neurol 8(8):414

Kirkitadze MD, Condron MM, Teplow DB (2001) Identification and characterization of key kinetic intermediates in amyloid beta-protein fibrillogenesis. J Mol Biol 312(5):1103–1119

Landen JW, Zhao Q, Cohen S, Borrie M, Woodward M, Billing CB Jr, Bales K, Alvey C, McCush F, Yang J, Kupiec JW, Bednar MM (2013) Safety and pharmacology of a single intravenous dose of ponezumab in subjects with mild-to-moderate Alzheimer disease: a phase I, randomized, placebo-controlled, double-blind, dose-escalation study. Clin Neuropharmacol 36(1):14–23

Lannfelt L, Moller C, Basun H, Osswald G, Sehlin D, Satlin A, Logovinsky V, Gellerfors P (2014a) Perspectives on future Alzheimer therapies: amyloid-beta protofibrils – a new target for immunotherapy with BAN2401 in Alzheimer's disease. Alzheimers Res Ther 6(2):16

Lannfelt L, Relkin NR, Siemers ER (2014b) Amyloid-ss-directed immunotherapy for Alzheimer's disease. J Intern Med 275(3):284–295

Lansbury PT Jr (1997) Structural neurology: are seeds at the root of neuronal degeneration? Neuron 19(6):1151–1154

Legleiter J, Czilli DL, Gitter B, DeMattos RB, Holtzman DM, Kowalewski T (2004) Effect of different anti-Abeta antibodies on Abeta fibrillogenesis as assessed by atomic force microscopy. J Mol Biol 335(4):997–1006

Lemere CA (2013) Immunotherapy for Alzheimer's disease: hoops and hurdles. Mol Neurodegener 8:36
Lemere CA, Masliah E (2010) Can Alzheimer disease be prevented by amyloid-beta immunotherapy? Nat Rev Neurol 6(2):108–119
Lemere CA, Maron R, Spooner ET, Grenfell TJ, Mori C, Desai R, Hancock WW, Weiner HL, Selkoe DJ (2000) Nasal A beta treatment induces anti-A beta antibody production and decreases cerebral amyloid burden in PD-APP mice. Ann N Y Acad Sci 920:328–331
Leuner K, Muller WE, Reichert AS (2012) From mitochondrial dysfunction to amyloid beta formation: novel insights into the pathogenesis of Alzheimer's disease. Mol Neurobiol 46 (1):186–193
Levy JB, Pusey CD (2000) Nephrotoxicity of intravenous immunoglobulin. QJM 93(11):751–755
Lleo A, Greenberg SM, Growdon JH (2006) Current pharmacotherapy for Alzheimer's disease. Annu Rev Med 57:513–533
Lobello K, Ryan JM, Liu E, Rippon G, Black R (2012) Targeting Beta amyloid: a clinical review of immunotherapeutic approaches in Alzheimer's disease. Int J Alzheimers Dis 2012:628070
Loeffler DA (2013) Intravenous immunoglobulin and Alzheimer's disease: what now? J Neuroinflammation 10(1):70
Lord A, Gumucio A, Englund H, Sehlin D, Sundquist VS, Soderberg L, Moller C, Gellerfors P, Lannfelt L, Pettersson FE, Nilsson LN (2009) An amyloid-beta protofibril-selective antibody prevents amyloid formation in a mouse model of Alzheimer's disease. Neurobiol Dis 36 (3):425–434
Madeo J, Frieri M (2013) Alzheimer's disease and immunotherapy. Aging Dis 4(4):210–220
Maier M, Seabrook TJ, Lazo ND, Jiang L, Das P, Janus C, Lemere CA (2006) Short amyloid-beta (Abeta) immunogens reduce cerebral Abeta load and learning deficits in an Alzheimer's disease mouse model in the absence of an Abeta-specific cellular immune response. J Neurosci 26(18):4717–4728
Mangialasche F, Solomon A, Winblad B, Mecocci P, Kivipelto M (2010) Alzheimer's disease: clinical trials and drug development. Lancet Neurol 9(7):702–716
Masliah E, Hansen L, Adame A, Crews L, Bard F, Lee C, Seubert P, Games D, Kirby L, Schenk D (2005) Abeta vaccination effects on plaque pathology in the absence of encephalitis in Alzheimer disease. Neurology 64(1):129–131
Mayeux R, Schupf N (2011) Blood-based biomarkers for Alzheimer's disease: plasma Abeta40 and Abeta42, and genetic variants. Neurobiol Aging 32(Suppl 1):S10–S19
McElhaney JE, Effros RB (2009) Immunosenescence: what does it mean to health outcomes in older adults? Curr Opin Immunol 21(4):418–424
Menendez-Gonzalez M, Perez-Pinera P, Martinez-Rivera M, Muniz AL, Vega JA (2011) Immunotherapy for Alzheimer's disease: rational basis in ongoing clinical trials. Curr Pharm Des 17 (5):508–520
Michaelis ML (2003) Drugs targeting Alzheimer's disease: some things old and some things new. J Pharmacol Exp Ther 304(3):897–904
Moreth J, Mavoungou C, Schindowski K (2013) Passive anti-amyloid immunotherapy in Alzheimer's disease: what are the most promising targets? Immun Ageing 10(1):18
Muhs A, Hickman DT, Pihlgren M, Chuard N, Giriens V, Meerschman C, van der Auwera I, van Leuven F, Sugawara M, Weingertner MC, Bechinger B, Greferath R, Kolonko N, Nagel-Steger L, Riesner D, Brady RO, Pfeifer A, Nicolau C (2007) Liposomal vaccines with conformation-specific amyloid peptide antigens define immune response and efficacy in APP transgenic mice. Proc Natl Acad Sci U S A 104(23):9810–9815
Nicoll JA, Wilkinson D, Holmes C, Steart P, Markham H, Weller RO (2003) Neuropathology of human Alzheimer disease after immunization with amyloid-beta peptide: a case report. Nat Med 9(4):448–452
Novakovic D, Feligioni M, Scaccianoce S, Caruso A, Piccinin S, Schepisi C, Errico F, Mercuri NB, Nicoletti F, Nistico R (2013) Profile of gantenerumab and its potential in the treatment of Alzheimer's disease. Drug Des Devel Ther 7:1359–1364

O'Brien RJ, Wong PC (2011) Amyloid precursor protein processing and Alzheimer's disease. Annu Rev Neurosci 34:185–204

Okura Y, Matsumoto Y (2009) Recent advance in immunotherapies for Alzheimer disease: with special reference to DNA vaccination. Hum Vaccin 5(6):373–380

Orgogozo JM, Gilman S, Dartigues JF, Laurent B, Puel M, Kirby LC, Jouanny P, Dubois B, Eisner L, Flitman S, Michel BF, Boada M, Frank A, Hock C (2003) Subacute meningoencephalitis in a subset of patients with AD after Abeta42 immunization. Neurology 61(1):46–54

Ostrowitzki S, Deptula D, Thurfjell L, Barkhof F, Bohrmann B, Brooks DJ, Klunk WE, Ashford E, Yoo K, Xu ZX, Loetscher H, Santarelli L (2012) Mechanism of amyloid removal in patients with Alzheimer disease treated with gantenerumab. Arch Neurol 69(2):198–207

Panza F, Frisardi V, Imbimbo BP, D'Onofrio G, Pietrarossa G, Seripa D, Pilotto A, Solfrizzi V (2010) Bapineuzumab: anti-beta-amyloid monoclonal antibodies for the treatment of Alzheimer's disease. Immunotherapy 2(6):767–782

Pfeifer M, Boncristiano S, Bondolfi L, Stalder A, Deller T, Staufenbiel M, Mathews PM, Jucker M (2002) Cerebral hemorrhage after passive anti-Abeta immunotherapy. Science 298(5597):1379

Pietri M, Dakowski C, Hannaoui S, Alleaume-Butaux A, Hernandez-Rapp J, Ragagnin A, Mouillet-Richard S, Haik S, Bailly Y, Peyrin JM, Launay JM, Kellermann O, Schneider B (2013) PDK1 decreases TACE-mediated alpha-secretase activity and promotes disease progression in prion and Alzheimer's diseases. Nat Med 19(9):1124–1131

Pigino G, Morfini G, Atagi Y, Deshpande A, Yu C, Jungbauer L, LaDu M, Busciglio J, Brady S (2009) Disruption of fast axonal transport is a pathogenic mechanism for intraneuronal amyloid beta. Proc Natl Acad Sci U S A 106(14):5907–5912

Poduslo JF, Gilles EJ, Ramakrishnan M, Howell KG, Wengenack TM, Curran GL, Kandimalla KK (2010) HH domain of Alzheimer's disease Abeta provides structural basis for neuronal binding in PC12 and mouse cortical/hippocampal neurons. PLoS One 5(1):e8813

Pohanka M (2011) Cholinesterases, a target of pharmacology and toxicology. Biomed Pap Med Fac Univ Palacky Olomouc Czech Repub 155(3):219–229

Prins ND, Scheltens P (2013) Treating Alzheimer's disease with monoclonal antibodies: current status and outlook for the future. Alzheimers Res Ther 5(6):56

Racke MM, Boone LI, Hepburn DL, Parsadainian M, Bryan MT, Ness DK, Piroozi KS, Jordan WH, Brown DD, Hoffman WP, Holtzman DM, Bales KR, Gitter BD, May PC, Paul SM, DeMattos RB (2005) Exacerbation of cerebral amyloid angiopathy-associated microhemorrhage in amyloid precursor protein transgenic mice by immunotherapy is dependent on antibody recognition of deposited forms of amyloid beta. J Neurosci 25(3):629–636

Relkin NR, Szabo P, Adamiak B, Burgut T, Monthe C, Lent RW, Younkin S, Younkin L, Schiff R, Weksler ME (2009) 18-month study of intravenous immunoglobulin for treatment of mild Alzheimer disease. Neurobiol Aging 30(11):1728–1736

Robinson SR, Bishop GM, Lee HG, Munch G (2004) Lessons from the AN 1792 Alzheimer vaccine: lest we forget. Neurobiol Aging 25(5):609–615

Ryan JM, Grundman M (2009) Anti-amyloid-beta immunotherapy in Alzheimer's disease: ACC-001 clinical trials are ongoing. J Alzheimers Dis 17(2):243

Savage JM, Wu G, McCampbell A, Wessner RK, Citron M, Liang X, Hsieh S, Kinney G, Wolfe AL, Rosen BL, Renger JJ (2010) A novel multivalent Aβ peptide vaccine with preclinical evidence of a central immune response that generates antisera recognizing a wide range of abeta peptide species [abstract]. Alzheimers Dement 6:S142

Schenk D, Barbour R, Dunn W, Gordon G, Grajeda H, Guido T, Hu K, Huang J, Johnson-Wood K, Khan K, Kholodenko D, Lee M, Liao Z, Lieberburg I, Motter R, Mutter L, Soriano F, Shopp G, Vasquez N, Vandevert C, Walker S, Wogulis M, Yednock T, Games D, Seubert P (1999) Immunization with amyloid-beta attenuates Alzheimer-disease-like pathology in the PDAPP mouse. Nature 400(6740):173–177

Schneeberger A, Mandler M, Otawa O, Zauner W, Mattner F, Schmidt W (2009) Development of AFFITOPE vaccines for Alzheimer's disease (AD) – from concept to clinical testing. J Nutr Health Aging 13(3):264–267

Sehlin D, Englund H, Simu B, Karlsson M, Ingelsson M, Nikolajeff F, Lannfelt L, Pettersson FE (2012) Large aggregates are the major soluble Abeta species in AD brain fractionated with density gradient ultracentrifugation. PLoS One 7(2), e32014

Selkoe DJ (2001) Alzheimer's disease: genes, proteins, and therapy. Physiol Rev 81(2):741–766

Shankar GM, Bloodgood BL, Townsend M, Walsh DM, Selkoe DJ, Sabatini BL (2007) Natural oligomers of the Alzheimer amyloid-beta protein induce reversible synapse loss by modulating an NMDA-type glutamate receptor-dependent signaling pathway. J Neurosci 27 (11):2866–2875

Shoji M, Golde TE, Ghiso J, Cheung TT, Estus S, Shaffer LM, Cai XD, McKay DM, Tintner R, Frangione B et al (1992) Production of the Alzheimer amyloid beta protein by normal proteolytic processing. Science 258(5079):126–129

Siemers ER, Friedrich S, Dean RA, Gonzales CR, Farlow MR, Paul SM, Demattos RB (2010) Safety and changes in plasma and cerebrospinal fluid amyloid beta after a single administration of an amyloid beta monoclonal antibody in subjects with Alzheimer disease. Clin Neuropharmacol 33(2):67–73

Sigurdsson EM, Scholtzova H, Mehta PD, Frangione B, Wisniewski T (2001) Immunization with a nontoxic/nonfibrillar amyloid-beta homologous peptide reduces Alzheimer's disease-associated pathology in transgenic mice. Am J Pathol 159(2):439–447

Solomon B, Koppel R, Hanan E, Katzav T (1996) Monoclonal antibodies inhibit in vitro fibrillar aggregation of the Alzheimer beta-amyloid peptide. Proc Natl Acad Sci U S A 93(1):452–455

Solomon B, Koppel R, Frankel D, Hanan-Aharon E (1997) Disaggregation of Alzheimer beta-amyloid by site-directed mAb. Proc Natl Acad Sci U S A 94(8):4109–4112

Spencer B, Masliah E (2014) Immunotherapy for Alzheimer's disease: past, present and future. Front Aging Neurosci 6:114

Sperling R, Salloway S, Brooks DJ, Tampieri D, Barakos J, Fox NC, Raskind M, Sabbagh M, Honig LS, Porsteinsson AP, Lieberburg I, Arrighi HM, Morris KA, Lu Y, Liu E, Gregg KM, Brashear HR, Kinney GG, Black R, Grundman M (2012) Amyloid-related imaging abnormalities in patients with Alzheimer's disease treated with bapineuzumab: a retrospective analysis. Lancet Neurol 11(3):241–249

Takano K, Endo S, Mukaiyama A, Chon H, Matsumura H, Koga Y, Kanaya S (2006) Structure of amyloid beta fragments in aqueous environments. FEBS J 273(1):150–158

Tucker S, Moller C, Tegerstedt K, Lord A, Laudon H, Sjodahl J, Soderberg L, Spens E, Sahlin C, Waara ER, Satlin A, Gellerfors P, Osswald G, Lannfelt L (2015) The murine version of BAN2401 (mAb158) selectively reduces amyloid-beta protofibrils in brain and cerebrospinal fluid of tg-ArcSwe mice. J Alzheimers Dis 43(2):575–588

van der Zee JS, van Swieten P, Aalberse RC (1986) Inhibition of complement activation by IgG4 antibodies. Clin Exp Immunol 64(2):415–422

Walsh DM, Klyubin I, Fadeeva JV, Cullen WK, Anwyl R, Wolfe MS, Rowan MJ, Selkoe DJ (2002) Naturally secreted oligomers of amyloid beta protein potently inhibit hippocampal long-term potentiation in vivo. Nature 416(6880):535–539

Weiner HL, Lemere CA, Maron R, Spooner ET, Grenfell TJ, Mori C, Issazadeh S, Hancock WW, Selkoe DJ (2000) Nasal administration of amyloid-beta peptide decreases cerebral amyloid burden in a mouse model of Alzheimer's disease. Ann Neurol 48(4):567–579

Wiessner C, Wiederhold KH, Tissot AC, Frey P, Danner S, Jacobson LH, Jennings GT, Luond R, Ortmann R, Reichwald J, Zurini M, Mir A, Bachmann MF, Staufenbiel M (2011) The second-generation active Abeta immunotherapy CAD106 reduces amyloid accumulation in APP transgenic mice while minimizing potential side effects. J Neurosci 31(25):9323–9331

Wilcock DM, Munireddy SK, Rosenthal A, Ugen KE, Gordon MN, Morgan D (2004) Microglial activation facilitates Abeta plaque removal following intracranial anti-Abeta antibody administration. Neurobiol Dis 15(1):11–20

Winblad B, Andreasen N, Minthon L, Floesser A, Imbert G, Dumortier T, Maguire RP, Blennow K, Lundmark J, Staufenbiel M, Orgogozo JM, Graf A (2012) Safety, tolerability, and antibody response of active Abeta immunotherapy with CAD106 in patients with

Alzheimer's disease: randomised, double-blind, placebo-controlled, first-in-human study. Lancet Neurol 11(7):597–604
Winblad B, Graf A, Riviere ME, Andreasen N, Ryan JM (2014) Active immunotherapy options for Alzheimer's disease. Alzheimers Res Ther 6(1):7
Wolfe MS (2006) Shutting down Alzheimer's. Sci Am 294(5):72–79
Wyss-Coray T (2006) Inflammation in Alzheimer disease: driving force, bystander or beneficial response? Nat Med 12(9):1005–1015
Zhang H, Ma Q, Zhang YW, Xu H (2012) Proteolytic processing of Alzheimer's beta-amyloid precursor protein. J Neurochem 120(Suppl 1):9–21

Chapter 23
Therapeutics for Polyglutamine Diseases Through Protein Degradation Pathway: Targeting the Nucleus

Atsushi Iwata

Abstract Polyglutamine diseases are caused by cytosine-adenine-guanine (CAG) trinucleotide expansions that are translated to a polyglutamine (pQ) chain in specific genes. This pQ chains tend to destabilize the entire proteins making them aggregate. The aggregates are especially toxic in the nucleus for a variety of reasons. Thus, for treating pQ diseases, targeting the nuclear pQ aggregates for degradation seems to be a promising approach. The nuclear ubiquitin proteasome system is the only major protein degradation machinery in the nucleus since the autophagy lysosome system, one of the major cytoplasmic protein degradation machinery, cannot function in the nucleus. There are ubiquitin ligases that recognize and promote nuclear pQ degradation, thus activation of these ubiquitin ligases could be one of the therapeutic approaches. We found that one of the histone deacetylases (HDAC), HDAC3, regulates the nuclear ubiquitin proteasome system. Although we may be closer to finding a molecular therapeutic approach for pQ diseases, rigorous effort to study the basic pathomechanisms of pQ diseases are necessary in order to expand our knowledge and find better therapeutics for these devastating diseases.

Keywords Polyglutamine • Aggregate • Histone deacetylases • Proteasome • Huntington's disease • Cellular compartment

23.1 Introduction

Polyglutamine disease is a class of neurodegenerative diseases caused by the expanded cytosine-adenine-guanine (CAG) trinucleotide sequence repeats (Fig. 23.1) (Kovtun and McMurray 2008; Mirkin 2007). There are currently nine

A. Iwata, M.D., Ph.D. (✉)
Department of Neurology, Graduate School of Medicine, The University of Tokyo, 7-3-1 Hongo Bunkyo-ku, Tokyo 113-8655, Japan

Japan Science and Technology Agency, PRESTO, 4-1-8 Honcho Kawaguchi, Saitama 332-0012, Japan
e-mail: iwata-tky@umin.ac.jp

© Springer Japan 2015

N. Mori, I. Mook-Jung (eds.), *Aging Mechanisms*,
DOI 10.1007/978-4-431-55763-0_23

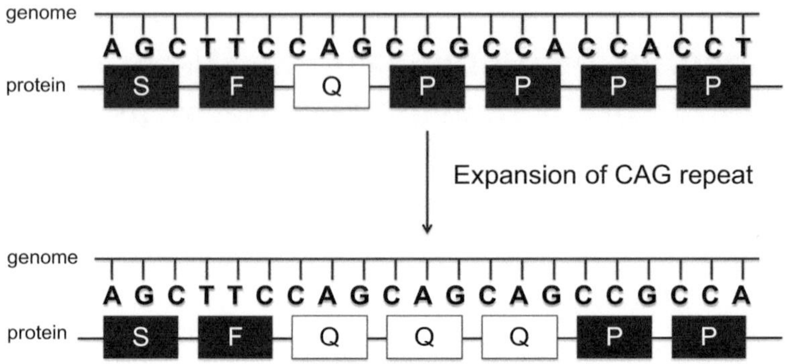

Fig. 23.1 Polyglutamine diseases are caused by expansion of CAG trinucleotide repeat

Table 23.1 Polyglutamine diseases and their specific genes, showing normal, premutational, and pathogenic CAG repeat lengths

Polyglutamine disease	Gene	Inheritance	Normal Q length	Premutation Q length	Pathogenic Q length
Huntington's disease (HD)	*HD*	AD	6–29	29–37	38–
Dentatorubral-pallidoluysian atrophy (DRPLA)	*ATN1*	AD	6–35	35–48	49–
Spinobulbar muscular atrophy (SBMA)	*AR*	XR	13–31	32–39	40–
Spinocerebellar ataxia type 1 (SCA1)	*ATXN1*	AD	6–39	40	41–
Spinocerebellar ataxia type 2 (SCA2)	*ATXN2*	AD	–30	31–32	32–
Machado-Joseph disease (MJD)	*ATXN3*	AD	12–40	41–85	52–
Spinocerebellar ataxia type 6 (SCA6)	*CACNA1A*	AD	–17	19	20–
Spinocerebellar ataxia type 7 (SCA7)	*ATXN7*	AD	4–17	29–33	36–
Spinocerebellar ataxia type 17 (SCA17)	*TBP*	AD	25–42	43–48	45–

AD autosomal dominantly inherited, *XR* X-linked recessive

neurologic disorders discovered that are classified in this category (Table 23.1). The CAG repeats are located in the coding region of the gene, which are translated into a series of glutamine (Q) residues, thus forming a chain of polyglutamine (pQ) tract. These disorders are autosomal dominant with the exception of spinobulbar muscular atrophy, which is X-linked; suggesting that the gain of function of pQ by CAG expansion is somehow causing the neurodegeneration. These disorders usually have a midlife onset and progress slowly with age. The number of CAG repeats correlate

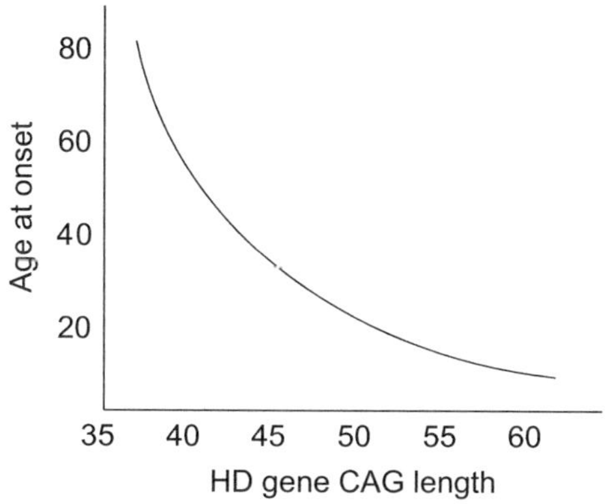

Fig. 23.2 Repeat length and age at onset have strong negative correlations in Huntington's disease (HD) (Modified from Lee et al. 2012)

with the severity and the age of onset (Fig. 23.2). Successive generations tend to have an earlier onset or express more severe symptoms, which is called anticipation. There is a certain threshold for the numbers of pQs that are pathogenic, depending on the target genes. For example, Huntington's disease (HD) occurs when the subject has 36 or more CAG repeats in the *HTT* gene and does not occur when the number of repeats are less than or equal to 35 (Table 23.1). However, the number of CAG repeats tends to increase when it is close to threshold, which affects offspring of those individuals harboring repeats nearing threshold (pre-mutation length). Individuals with a pre-mutation allele will exhibit a normal phenotype, but may have offspring who have a higher number of CAG repeats, and thus, be affected by the disease. The increase in the number of repeats occurs almost exclusively through paternal transmission. On the other hand, offspring of affected mothers are more likely to show no change or even show reduced number of CAG repeats (Kremer et al. 1995; Telenius et al. 1993; Wheeler et al. 2007). Studies in mice models suggested that this paternal expansion occurs during the late germ cell development stage (Telenius et al. 1994). The repeat number is also unstable in post-mitotic neurons, and the degree of the expansion differs among cell types or brain regions, which substantially affect the severity and age of onset among individuals (Gonitel et al. 2008; Kovtun et al. 2007).

There are various theories as to how expanded CAG repeats cause the disease. As mentioned in the earlier section, CAG repeats are translated to pQ tracts, which tend to form a beta sheet structure when long enough, and affect the structure of the entire protein containing the pQ chain. This structural alteration is thought to impair the ability of the protein to fold properly, thus making it aggregate and exhibit neurotoxicity (Perutz et al. 2002; Bevivino and Loll 2001; Tanaka et al. 2001). Various studies on neurodegeneration support that oligomeric form of aggregated proteins are more toxic to the neurons than the microscopically visible, highly aggregated inclusion bodies implicated as disease specific structures (Fig. 23.3, Table 23.2) (Arrasate et al. 2004; Iuchi et al. 2003; Takahashi et al. 2007; Legleiter et al. 2010). Moreover, these inclusion bodies themselves are thought to be a neuroprotective response by the cells bearing the aggregated proteins (Takahashi et al. 2008; Kopito 2000; Tanaka et al. 2004a).

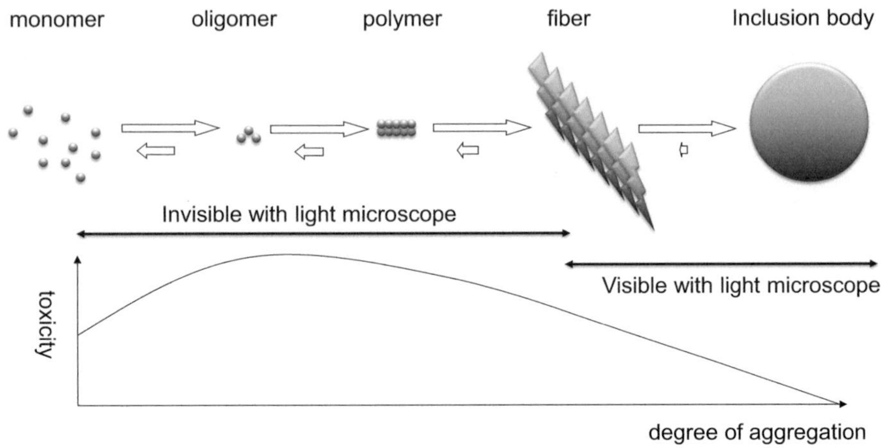

Fig. 23.3 Aggregate is formed from monomeric proteins. The more the aggregation, the less reversible it gets (note the length of *left sided arrows*). Cellular toxicity is a characteristic of oligomeric species, and the microscopically visible inclusion bodies are thought to be a protective response of the cells

Table 23.2 Neurodegenerative diseases and ubiquitin positive structures with aggregated proteins

Neurodegenerative disease	Ubiquitin positive structure	Aggregated protein
Polyglutamine diseases	Neuronal intra-nuclear inclusions	Polyglutamine containing proteins
Alzheimer's disease	Neurofibrillary tangles	Tau
Frontotemporal lobar degeneration	Ubiquitin positive inclusions	TDP-43
Amyotrophic lateral sclerosis	round bodies, skein like inclusions	TDP-43
Parkinson's disease	Lewy bodies	α-synuclein
Multiple system atrophy	Glial cytoplasmic inclusions	

23.2 Truncated pQ Proteins as a Tool for Studying Aggregated Proteins

In order to study any kind of disease, it is ideal to have simple models in the cell culture system or in laboratory animals. In this aspect, neurodegenerative diseases are of particular interest since there are no naturally occurring animal models, except for a few exceptions (ref). It has not been easy to recapitulate human pathology in mice partly because the life span of mice is approximately 2 years, which is too short to reproduce human diseases that manifest with advanced age. Thus, generation of Alzheimer's disease or Parkinson's disease animal models have been accomplished by overexpressing multiple genes related to those diseases. On the contrary, generation of polyglutamine disease mouse model was rather easy

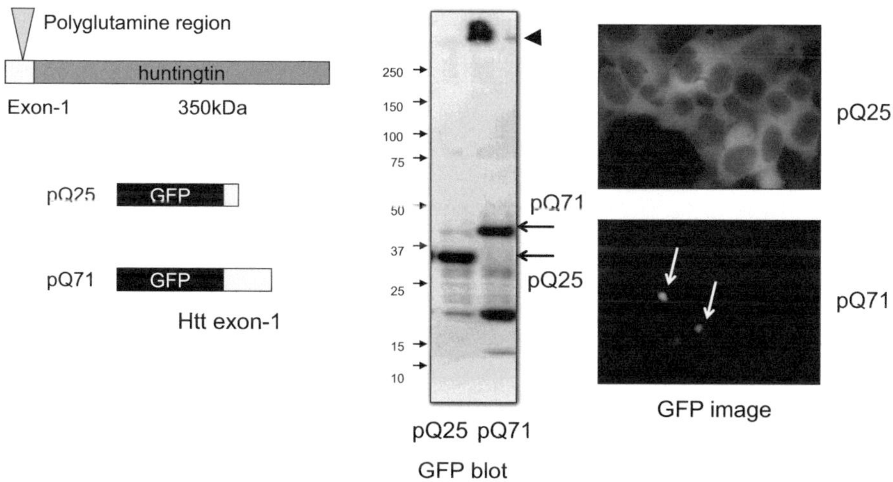

Fig. 23.4 Huntingtin exon-1 overexpression provides a valuable tool to study the aggregates and their degradation. The exon-1 contains CAG repeat region, which in this case is fused to a green fluorescent protein (GFP) and overexpressed in the cell culture system. The western blot shows monomeric bands (*arrows*) as well as high molecular weight aggregates (*arrowhead*). Microscopic images are even more dramatic with practically no cytoplasmic diffuse signals in pQ71. *Arrows* are aggregated inclusion bodies

since huntingtin (Htt), a gene product in HD, has a CAG repeat in its 1st exon, and a simple overexpression of the truncated 1st exon fragment reproduces severe phenotype and pathology resembling the human disease (Mangiarini et al. 1996). Thus, although there have been criticisms that claim this model as artificial and non-physiological, which is indeed true, this truncated over-expression model has been a gold standard in studying the pathophysiology of pQ aggregates and to screen for therapeutic measures (Fig. 23.4).

23.3 The Role of Aggregates and the Effect of Its Clearance in Polyglutamine Diseases

It has been clearly shown that the presence of aggregated truncated huntingtin affects cellular function in a Q length-dependent manner (Morley et al. 2002). Therefore, removal of those aggregates will be a treatment strategy for the disease. The proof of concept for this hypothesis was clearly shown by shutting off toxic gene expression (Yamamoto et al. 2000) or introducing chemical chaperones to reduce the amount of aggregated proteins (Tanaka et al. 2004b) resulting in a dramatic recovery of the phenotype.

The pathological hallmark of neurodegenerative disorders is the presence of ubiquitylated inclusion bodies in the neurons or in the glial cells. In the case of pQ

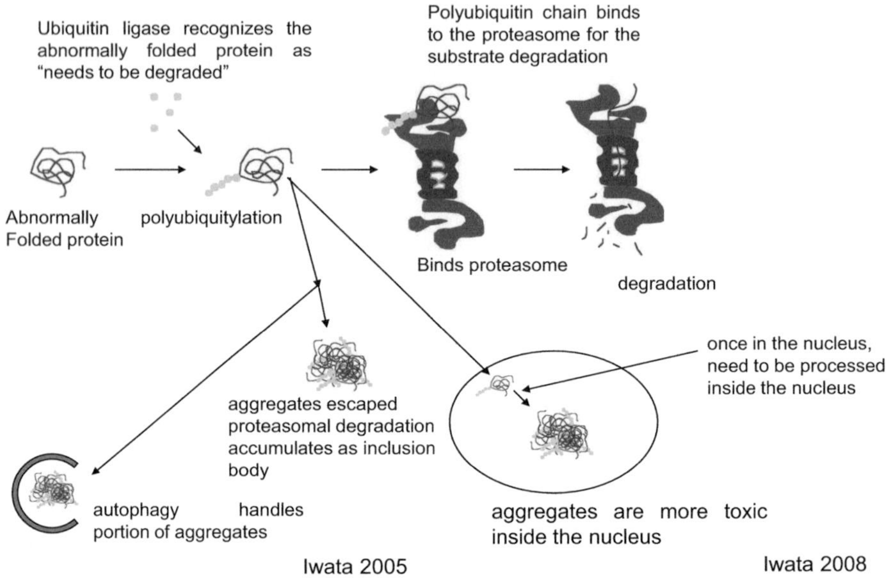

Fig. 23.5 Different degradation pathways for aggregated huntingtin. Abnormally folded huntingtin are recognized by ubiquitin ligase and ubiquitylated. The ubiquitins are conjugated to form a chain, which binds to the proteasome and facilitates the degradation of aggregated huntingtin. Some portion of the ubiquitylated aggregates escapes this degradation pathway, but are captured by the autophagosome to be degraded in the lysosome. Some of the aggregated huntingtin are transported to the nucleus. Since post-mitotic cells have no disruption of the nuclear envelope, aggregated huntingtin accumulates in the nucleus to exhibit neurotoxicity and form inclusion bodies (Iwata et al. 2005a, b; Iwata 2008)

diseases, these ubiquitylated inclusion bodies are often observed as intra-nuclear inclusion bodies. The presence of ubiquitin-positive inclusion bodies suggests that the aggregation-prone proteins are ubiquitylated by a certain ubiquitin ligase but still escape the efficient proteasome degradation (Fig. 23.5), or that the aggregates are sequestering the ubiquitylated proteins that would be degraded in the absence of these aggregates, and therefore, interfering in the cellular protein quality control system. The latter hypothesis is of particular interest since it has been shown that the presence of aggregated huntingtin inhibits the ubiquitin proteasome system (Bence et al. 2001) through a mechanism other than clogging the proteasome pore (Hipp et al. 2012), since the inhibition occurs cross-compartmentally across the cytoplasm and the nucleus (Bennett et al. 2005).

What happens when the ubiquitin proteasome system is impaired? The cellular protein degradation machinery can be classified into two large categories, the ubiquitin proteasome system, and the autophagy lysosome system (ref). It has been reported that the presence of aggregated huntingtin stimulates autophagy lysosomal pathway (Kegel et al. 2000), and autophagy is shown to be a major backup system when the ubiquitin proteasome system is compromised (Iwata et al. 2005a). Thus, these two degradation systems work harmoniously to maintain protein quality control.

23.4 The Role of Nuclear pQ Aggregates and Their Degradation

Aggregated huntingtin can be found as ubiquitylated inclusion bodies both in the nucleus, as intra-nuclear inclusions, and in the cytoplasm, as dystrophic neurites. Will the removal of aggregated huntingtin from the nucleus or from the cytoplasm be more effective? There are multiple sets of evidence suggesting that the nuclear aggregates are more toxic to the neurons than the cytoplasmic aggregates (Peters et al. 1999; Schilling et al. 2004; Katsuno et al. 2002; Klement et al. 1998). The nuclear toxicity of pQ aggregates are not completely understood, but could be due to the sequestration of transcription factors into the inclusion bodies (Shimohata et al. 2000; Nucifora et al. 2001), by disruption of chromatin organization (Seong et al. 2010; Sathasivam et al. 2001), or by inhibition of DNA replication or repair (Enokido et al. 2010).

Thus, nuclear aggregates are targets for therapeutics. However, autophagy does not function in the nucleus (Iwata et al. 2005b). Therefore, how does ubiquitin proteasome function in the nucleus? Nuclear proteins are synthesized and folded in the cytoplasm, and enter the nucleus either by passive diffusion through the nuclear pore (if small) or via active transport by importins. All of the pQ disease-related proteins are large enough to require active nuclear transport, which means that they are partially folded to escape cytoplasmic degradation and be recognized by importins, while still being aggregation-prone to form inclusion bodies in the nucleus. In the nucleus, the proteasome localizes in the euchromatin region as well as at the periphery of the heterochromatin and nucleoli (Rivett and Knecht 1993; De Conto et al. 2000), which supports the idea that ubiquitin proteasome system is the primary protein quality control system even in the nucleus.

The key component of the ubiquitin proteasome system is the ubiquitin ligase (E3), which recognizes and ubiquitylates its substrate. Several mammalian nuclear E3s have been discovered so far that are involved in nuclear pQ ubiquitylation (Iwata et al. 2009; Fu et al. 2005; Janer et al. 2006). Activation of these E3s could be one of the therapeutic approaches for pQ diseases.

As mentioned earlier, the presence of aggregates inhibits the function of ubiquitin proteasome system through an unknown pathway. Since this inhibitory effect crosses the nuclear membrane bi-directionally, at least in the cell culture system (Bennett et al. 2005), there must be a regulatory pathway spanning the two cellular compartments for the proteasome function. Since accelerated degradation of the aggregates is one of the desired therapeutic strategies, this regulatory pathway is an ideal target yet to be discovered.

23.5 Role of HDAC in HD Pathogenesis

Gene expression profile analysis has revealed genes commonly dysregulated in pQ models (Luthi-Carter et al. 2002). This dysregulation is attributed to the pQ proteins themselves since many of the CAG bearing pathogenic proteins are transcription factors or transcriptional co-activators. Rigorous study to find a common transcriptional pathway has revealed several proteins including TATA-box-binding protein-associated factor II 130 kDa (TAFII130) (Dunah et al. 2002), cAMP-response element-binding protein (CREB) (ref), tumor protein p53 (Bae et al. 2005), and CREB-binding protein (CBP) (Steffan et al. 2000) as interacting partners for huntingtin. However, still no single pathway can explain neuronal dysfunction and death.

One way to correct transcriptional dysregulation is to alter and fix gene expression by modifying histone acetylation. Multiple studies have shown that histone deacetylase (HDAC) inhibitors can improve disease phenotype in pQ disease animal models (Steffan et al. 2001; McCampbell et al. 2001; Hockly et al. 2003; Minamiyama et al. 2004; Ryu et al. 2003). However, the drugs used in these models were broad HDAC inhibitors, which is far from ideal due to their various potential side effects in human subjects.

There are four classes of HDACs in mammals defined by homology and nicotinamide adenine dinucleotide (NAD) requirement (Table 23.3). Among them, cytoplasmic HDACs are not involved in the transcriptional regulation, but are involved in the autophagic clearance of the cytoplasmic aggregates (Iwata et al. 2005a). There are nine HDACs expressed in the nucleus. Thus, specific HDAC inhibition was sought by utilizing specific inhibitors or by genetic ablation (Table 23.4). However, we are still unclear as to which HDAC should be specifically inhibited.

Table 23.3 Mammalian histone deacetylases

Class	Name	Subcellular localization	CNS expression	NAD dependence
I	HDAC1	Nucleus	+	–
	HDAC2	Nucleus	+	–
	HDAC3	Nucleus, cytoplasm	+	–
	HDAC8	Nucleus	+	–
IIA	HDAC4	Nucleus, cytoplasm	+	–
	HDAC5	Nucleus, cytoplasm	+	–
	HDAC7	Nucleus, cytoplasm, mitochondria	+	–
	HDAC9	Nucleus, cytoplasm	+	–
IIB	HDAC6	Cytoplasm	+	–
	HDAC10	Cytoplasm	+	–
III	Sirtuins (SIRT1-7)	Varies	varies	+
IV	HDAC11	Nucleus, cytoplasm	+	–

Table 23.4 Various studies testing the effect of HDAC inhibitors on pQ models

Target	Effective	Not effective
HDAC1	Small molecules (Jia et al. 2012)	n.a.
HDAC3		Heterozygous knock out mice (Moumne et al. 2012)
HDAC2	Small molecules (Mielcarek et al. 2013; Thomas et al. 2008)	n.a.
HDAC4		n.a.
HDAC6	n.a.	Knock out mice (Bobrowska et al. 2011)
HDAC7	n.a.	Knock out mice (Benn et al. 2009)
Sirtuins	Small molecules (Hathorn et al. 2011)	n.a.

23.6 HDAC3 and HD

Upon testing specific HDAC activity on pQ aggregation, we were particularly interested in HDAC3 since the result from the genetic ablation study and specific inhibitors were contradicting (Table 23.4). HDAC3 is a unique class 1 HDAC found in the cytoplasm as well as the nucleus (Gao et al. 2006; Yang et al. 2002). However, the precise functions of HDAC3 in the two cellular compartments are only vaguely known. HDAC3 directly binds to Htt with short pQ and this interaction is important for suppressing the neurotoxicity induced by HDAC3. HDAC3 loses its interaction with Htt with long Q, thus promoting the neuronal death caused by HDAC3 (Bardai et al. 2013). Moreover, HDAC3 knockdown mice are embryonic lethal, which shows that it is an essential protein during development (Moumne et al. 2012).

The results for HDAC3 inhibition in pQ models are controversial. While one study using a specific HDAC3 inhibitor showed improvement in a drosophila model (Jia et al. 2012), another study showed no effect in a mouse model of HD (Moumne et al. 2012). This discrepancy could be explained by the lack of specificity of the inhibitors and the heterozygotic knockout animals used in the study. Another possible cause for the discrepancy is that, unlike HDAC1 or 2 that only function in the nucleus, HDAC3 can shuttle between the cytoplasm and the nucleus, where it can have different roles. Therefore, the effect of HDAC3 inhibition in HD models may depend on the balance between nuclear vs. cytoplasmic aggregates, or HDAC3 activity.

To clarify these issues, we utilized highly specific HDAC3 inhibitors made by a click chemistry-based combinatorial fragment assembly technique (Suzuki et al. 2013). These HDAC3 inhibitors have an IC50 against HDAC3 that is at least 100-fold higher than those for other HDACs. By utilizing these newly developed reagents, we tested the effect of HDAC3 inhibition in cellular HD models (Iwata et al. 2009). In this study, we found that HDAC3 preferentially binds to nuclear Htt over cytoplasmic Htt. Specific HDAC3 inhibitors increased the total amount of Htt aggregates by increasing the amount of nuclear aggregates.

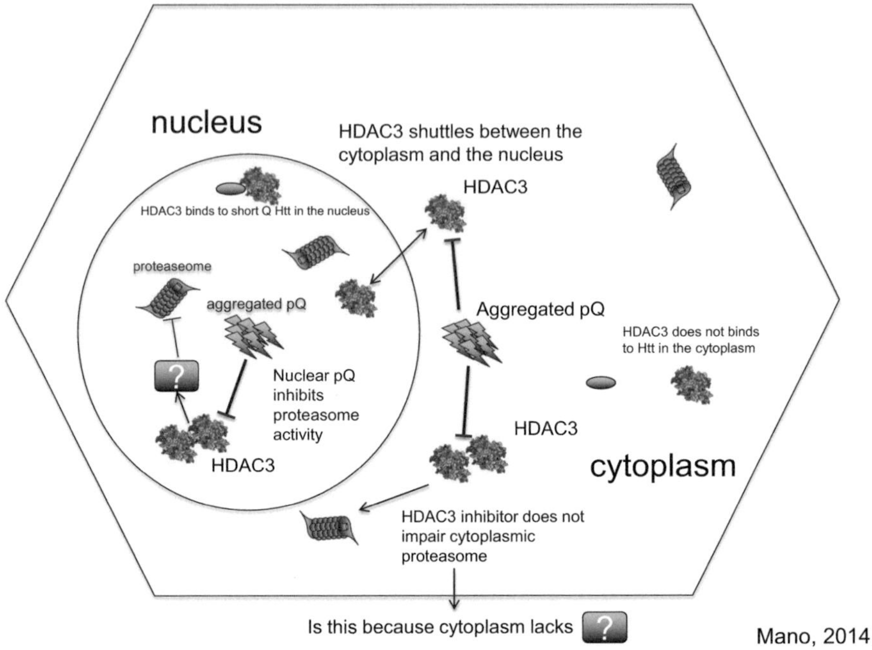

Fig. 23.6 Current model of polyglutamine degradation and its control by HDAC3 (Mano et al. 2014)

Both cytoplasmic and nuclear Htt aggregates were able to suppress endogenous HDAC3 activity, which led to a decreased nuclear proteasome activity. Therefore, we concluded that Htt aggregates impair nuclear proteasome activity through the inhibition of HDAC3 (Fig. 23.6) (Mano et al. 2014).

23.7 Conclusion

In this review, I have tried to describe the current understanding about the molecular pathogenesis of pQ diseases from a standpoint of aggregate toxicity and degradation. There are multiple schools of thought regarding this issue, and aggregates may not be the only culprit in neurodegeneration. There are aberrant transcription caused by CAG containing RNAs without any translation, or there are criticisms about aggregate being only the leftover of the damaged cells. It is needless to say that future studies are required to further elucidate the various pathomechanisms involved in pQ diseases.

Still, one thing I want to emphasize here is that even if so, pQ diseases are still a very good model studying cellular protein quality control machinery and worth looking deeper into the physiological function of the machinery especially in the nucleus, since our knowledge about that is so sparse at this very point. I believe that accumulation of small pieces of knowledge about cellular protein degradation machinery will lead us to better science and hopefully to a cure for pQ diseases.

Conflict of Interest The author has nothing to declare for COI disclosure.

References

Arrasate M, Mitra S, Schweitzer ES, Segal MR, Finkbeiner S (2004) Inclusion body formation reduces levels of mutant huntingtin and the risk of neuronal death. Nature 431(7010):805–810

Bae BI, Xu H, Igarashi S, Fujimuro M, Agrawal N, Taya Y et al (2005) p53 mediates cellular dysfunction and behavioral abnormalities in Huntington's disease. Neuron 47(1):29–41

Bardai FH, Verma P, Smith C, Rawat V, Wang L, D'Mello SR (2013) Disassociation of histone deacetylase-3 from normal huntingtin underlies mutant huntingtin neurotoxicity. J Neurosci 33 (29):11833–11838

Bence NF, Sampat RM, Kopito RR (2001) Impairment of the ubiquitin-proteasome system by protein aggregation. Science 292(5521):1552–1555

Benn CL, Butler R, Mariner L, Nixon J, Moffitt H, Mielcarek M et al (2009) Genetic knock-down of HDAC7 does not ameliorate disease pathogenesis in the R6/2 mouse model of Huntington's disease. PLoS One 4(6):e5747

Bennett EJ, Bence NF, Jayakumar R, Kopito RR (2005) Global impairment of the ubiquitin-proteasome system by nuclear or cytoplasmic protein aggregates precedes inclusion body formation. Mol Cell 17(3):351–365

Bevivino AE, Loll PJ (2001) An expanded glutamine repeat destabilizes native ataxin-3 structure and mediates formation of parallel beta -fibrils. Proc Natl Acad Sci U S A 98(21):11955–11960

Bobrowska A, Paganetti P, Matthias P, Bates GP (2011) Hdac6 knock-out increases tubulin acetylation but does not modify disease progression in the R6/2 mouse model of Huntington's disease. PLoS One 6(6):e20696

De Conto F, Pilotti E, Razin SV, Ferraglia F, Geraud G, Arcangeletti C et al (2000) In mouse myoblasts nuclear prosomes are associated with the nuclear matrix and accumulate preferentially in the perinucleolar areas. J Cell Sci 113(Pt 13):2399–2407

Dunah AW, Jeong H, Griffin A, Kim YM, Standaert DG, Hersch SM et al (2002) Sp1 and TAFII130 transcriptional activity disrupted in early Huntington's disease. Science 296 (5576):2238–2243

Enokido Y, Tamura T, Ito H, Arumughan A, Komuro A, Shiwaku H et al (2010) Mutant huntingtin impairs Ku70-mediated DNA repair. J Cell Biol 189(3):425–443

Fu L, Gao YS, Tousson A, Shah A, Chen TL, Vertel BM et al (2005) Nuclear aggresomes form by fusion of PML-associated aggregates. Mol Biol Cell 16(10):4905–4917

Gao Z, He Q, Peng B, Chiao PJ, Ye J (2006) Regulation of nuclear translocation of HDAC3 by IkappaBalpha is required for tumor necrosis factor inhibition of peroxisome proliferator-activated receptor gamma function. J Biol Chem 281(7):4540–4547

Gonitel R, Moffitt H, Sathasivam K, Woodman B, Detloff PJ, Faull RL et al (2008) DNA instability in postmitotic neurons. Proc Natl Acad Sci U S A 105(9):3467–3472

Hathorn T, Snyder-Keller A, Messer A (2011) Nicotinamide improves motor deficits and upregulates PGC-1alpha and BDNF gene expression in a mouse model of Huntington's disease. Neurobiol Dis 41(1):43–50

Hipp MS, Patel CN, Bersuker K, Riley BE, Kaiser SE, Shaler TA et al (2012) Indirect inhibition of 26S proteasome activity in a cellular model of Huntington's disease. J Cell Biol 196 (5):573–587

Hockly E, Richon VM, Woodman B, Smith DL, Zhou X, Rosa E et al (2003) Suberoylanilide hydroxamic acid, a histone deacetylase inhibitor, ameliorates motor deficits in a mouse model of Huntington's disease. Proc Natl Acad Sci U S A 100(4):2041–2046

Iuchi S, Hoffner G, Verbeke P, Djian P, Green H (2003) Oligomeric and polymeric aggregates formed by proteins containing expanded polyglutamine. Proc Natl Acad Sci U S A 100 (5):2409–2414

Iwata A, Riley BE, Johnston JA, Kopito RR (2005a) HDAC6 and microtubules are required for autophagic degradation of aggregated huntingtin. J Biol Chem 280(48):40282–40292

Iwata A, Christianson JC, Bucci M, Ellerby LM, Nukina N, Forno LS et al (2005b) Increased susceptibility of cytoplasmic over nuclear polyglutamine aggregates to autophagic degradation. Proc Natl Acad Sci U S A 102(37):13135–13140

Iwata A, Nagashima Y, Matsumoto L, Suzuki T, Yamanaka T, Date H et al (2009) Intranuclear degradation of polyglutamine aggregates by the ubiquitin-proteasome system. J Biol Chem 284 (15):9796–9803

Janer A, Martin E, Muriel MP, Latouche M, Fujigasaki H, Ruberg M et al (2006) PML clastosomes prevent nuclear accumulation of mutant ataxin-7 and other polyglutamine proteins. J Cell Biol 174(1):65–76

Jia H, Pallos J, Jacques V, Lau A, Tang B, Cooper A et al (2012) Histone deacetylase (HDAC) inhibitors targeting HDAC3 and HDAC1 ameliorate polyglutamine-elicited phenotypes in model systems of Huntington's disease. Neurobiol Dis 46(2):351–361

Katsuno M, Adachi H, Kume A, Li M, Nakagomi Y, Niwa H et al (2002) Testosterone reduction prevents phenotypic expression in a transgenic mouse model of spinal and bulbar muscular atrophy. Neuron 35(5):843–854

Kegel KB, Kim M, Sapp E, McIntyre C, Castano JG, Aronin N et al (2000) Huntingtin expression stimulates endosomal-lysosomal activity, endosome tubulation, and autophagy. J Neurosci 20 (19):7268–7278

Klement IA, Skinner PJ, Kaytor MD, Yi H, Hersch SM, Clark HB et al (1998) Ataxin-1 nuclear localization and aggregation: role in polyglutamine-induced disease in SCA1 transgenic mice. Cell 95(1):41–53

Kopito RR (2000) Aggresomes, inclusion bodies and protein aggregation. Trends Cell Biol 10 (12):524–530

Kovtun IV, McMurray CT (2008) Features of trinucleotide repeat instability in vivo. Cell Res 18 (1):198–213

Kovtun IV, Liu Y, Bjoras M, Klungland A, Wilson SH, McMurray CT (2007) OGG1 initiates age-dependent CAG trinucleotide expansion in somatic cells. Nature 447(7143):447–452

Kremer B, Almqvist E, Theilmann J, Spence N, Telenius H, Goldberg YP et al (1995) Sex-dependent mechanisms for expansions and contractions of the CAG repeat on affected Huntington disease chromosomes. Am J Hum Genet 57(2):343–350

Lee JM, Ramos EM, Lee JH, Gillis T, Mysore JS, Hayden MR et al (2012) CAG repeat expansion in Huntington disease determines age at onset in a fully dominant fashion. Neurology 78 (10):690–695

Legleiter J, Mitchell E, Lotz GP, Sapp E, Ng C, DiFiglia M et al (2010) Mutant huntingtin fragments form oligomers in a polyglutamine length-dependent manner in vitro and in vivo. J Biol Chem 285(19):14777–14790

Luthi-Carter R, Strand AD, Hanson SA, Kooperberg C, Schilling G, La Spada AR et al (2002) Polyglutamine and transcription: gene expression changes shared by DRPLA and Huntington's disease mouse models reveal context-independent effects. Hum Mol Genet 11(17):1927–1937

Mangiarini L, Sathasivam K, Seller M, Cozens B, Harper A, Hetherington C et al (1996) Exon 1 of the HD gene with an expanded CAG repeat is sufficient to cause a progressive neurological phenotype in transgenic mice. Cell 87(3):493–506

Mano T, Suzuki T, Tsuji S, Iwata A (2014) Differential effect of HDAC3 on cytoplasmic and nuclear Huntingtin aggregates. PLoS One 9(11):e111277

McCampbell A, Taye AA, Whitty L, Penney E, Steffan JS, Fischbeck KH (2001) Histone deacetylase inhibitors reduce polyglutamine toxicity. Proc Natl Acad Sci U S A 98 (26):15179–15184

Mielcarek M, Landles C, Weiss A, Bradaia A, Seredenina T, Inuabasi L et al (2013) HDAC4 reduction: a novel therapeutic strategy to target cytoplasmic huntingtin and ameliorate neurodegeneration. PLoS Biol 11(11):e1001717

Minamiyama M, Katsuno M, Adachi H, Waza M, Sang C, Kobayashi Y et al (2004) Sodium butyrate ameliorates phenotypic expression in a transgenic mouse model of spinal and bulbar muscular atrophy. Hum Mol Genet 13(11):1183–1192

Mirkin SM (2007) Expandable DNA, repeats and human disease. Nature 447(7147):932–940

Morley JF, Brignull HR, Weyers JJ, Morimoto RI (2002) The threshold for polyglutamine-expansion protein aggregation and cellular toxicity is dynamic and influenced by aging in Caenorhabditis elegans. Proc Natl Acad Sci U S A 99(16):10417–10422

Moumne L, Campbell K, Howland D, Ouyang Y, Bates GP (2012) Genetic knock-down of HDAC3 does not modify disease-related phenotypes in a mouse model of Huntington's disease. PLoS One 7(2):e31080

Nucifora FC Jr, Sasaki M, Peters MF, Huang H, Cooper JK, Yamada M et al (2001) Interference by huntingtin and atrophin-1 with cbp-mediated transcription leading to cellular toxicity. Science 291(5512):2423–2428

Perutz MF, Finch JT, Berriman J, Lesk A (2002) Amyloid fibers are water-filled nanotubes. Proc Natl Acad Sci U S A 99(8):5591–5595

Peters MF, Nucifora FC Jr, Kushi J, Seaman HC, Cooper JK, Herring WJ et al (1999) Nuclear targeting of mutant Huntingtin increases toxicity. Mol Cell Neurosci 14(2):121–128

Rivett AJ, Knecht E (1993) Protein turnover: proteasome location. Curr Biol 3(2):127–129

Ryu H, Lee J, Olofsson BA, Mwidau A, Dedeoglu A, Escudero M et al (2003) Histone deacetylase inhibitors prevent oxidative neuronal death independent of expanded polyglutamine repeats via an Sp1-dependent pathway. Proc Natl Acad Sci U S A 100(7):4281–4286

Sathasivam K, Woodman B, Mahal A, Bertaux F, Wanker EE, Shima DT et al (2001) Centrosome disorganization in fibroblast cultures derived from R6/2 Huntington's disease (HD) transgenic mice and HD patients. Hum Mol Genet 10(21):2425–2435

Schilling G, Savonenko AV, Klevytska A, Morton JL, Tucker SM, Poirier M et al (2004) Nuclear-targeting of mutant huntingtin fragments produces Huntington's disease-like phenotypes in transgenic mice. Hum Mol Genet 13(15):1599–1610

Seong IS, Woda JM, Song JJ, Lloret A, Abeyrathne PD, Woo CJ et al (2010) Huntingtin facilitates polycomb repressive complex 2. Hum Mol Genet 19(4):573–583

Shimohata T, Nakajima T, Yamada M, Uchida C, Onodera O, Naruse S et al (2000) Expanded polyglutamine stretches interact with TAFII130, interfering with CREB-dependent transcription. Nat Genet 26(1):29–36

Steffan JS, Kazantsev A, Spasic-Boskovic O, Greenwald M, Zhu YZ, Gohler H et al (2000) The Huntington's disease protein interacts with p53 and CREB-binding protein and represses transcription. Proc Natl Acad Sci U S A 97(12):6763–6768

Steffan JS, Bodai L, Pallos J, Poelman M, McCampbell A, Apostol BL et al (2001) Histone deacetylase inhibitors arrest polyglutamine-dependent neurodegeneration in Drosophila. Nature 413(6857):739–743

Suzuki T, Kasuya Y, Itoh Y, Ota Y, Zhan P, Asamitsu K et al (2013) Identification of highly selective and potent histone deacetylase 3 inhibitors using click chemistry-based combinatorial fragment assembly. PLoS One 8(7):e68669

Takahashi Y, Okamoto Y, Popiel HA, Fujikake N, Toda T, Kinjo M et al (2007) Detection of polyglutamine protein oligomers in cells by fluorescence correlation spectroscopy. J Biol Chem 282(33):24039–24048

Takahashi T, Kikuchi S, Katada S, Nagai Y, Nishizawa M, Onodera O (2008) Soluble polyglutamine oligomers formed prior to inclusion body formation are cytotoxic. Hum Mol Genet 17(3):345–356

Tanaka M, Morishima I, Akagi T, Hashikawa T, Nukina N (2001) Intra- and intermolecular beta-pleated sheet formation in glutamine-repeat inserted myoglobin as a model for polyglutamine diseases. J Biol Chem 276(48):45470–45475
Tanaka M, Kim YM, Lee G, Junn E, Iwatsubo T, Mouradian MM (2004a) Aggresomes formed by alpha-synuclein and synphilin-1 are cytoprotective. J Biol Chem 279(6):4625–4631
Tanaka M, Machida Y, Niu S, Ikeda T, Jana NR, Doi H et al (2004b) Trehalose alleviates polyglutamine-mediated pathology in a mouse model of Huntington disease. Nat Med 10 (2):148–154
Telenius H, Kremer HP, Theilmann J, Andrew SE, Almqvist E, Anvret M et al (1993) Molecular analysis of juvenile Huntington disease: the major influence on (CAG)n repeat length is the sex of the affected parent. Hum Mol Genet 2(10):1535–1540
Telenius H, Kremer B, Goldberg YP, Theilmann J, Andrew SE, Zeisler J et al (1994) Somatic and gonadal mosaicism of the Huntington disease gene CAG repeat in brain and sperm. Nat Genet 6(4):409–414
Thomas EA, Coppola G, Desplats PA, Tang B, Soragni E, Burnett R et al (2008) The HDAC inhibitor 4b ameliorates the disease phenotype and transcriptional abnormalities in Huntington's disease transgenic mice. Proc Natl Acad Sci U S A 105(40):15564–15569
Wheeler VC, Persichetti F, McNeil SM, Mysore JS, Mysore SS, MacDonald ME et al (2007) Factors associated with HD CAG repeat instability in Huntington disease. J Med Genet 44 (11):695–701
Yamamoto A, Lucas JJ, Hen R (2000) Reversal of neuropathology and motor dysfunction in a conditional model of Huntington's disease. Cell 101(1):57–66
Yang WM, Tsai SC, Wen YD, Fejer G, Seto E (2002) Functional domains of histone deacetylase-3. J Biol Chem 277(11):9447–9454

Name Index

© Springer Japan 2015
N. Mori, I. Mook-Jung (eds.), *Aging Mechanisms*,
DOI 10.1007/978-4-431-55763-0

Subject Index

© Springer Japan 2015
N. Mori, I. Mook-Jung (eds.), *Aging Mechanisms*,
DOI 10.1007/978-4-431-55763-0

MIX
Papier aus verantwortungsvollen Quellen
Paper from responsible sources
FSC® C105338

If you have any concerns about our products,
you can contact us on
ProductSafety@springernature.com

In case Publisher is established outside the EU,
the EU authorized representative is:
Springer Nature Customer Service Center GmbH
Europaplatz 3, 69115 Heidelberg, Germany

Printed by Libri Plureos GmbH
in Hamburg, Germany